T0265046

DATA FUSION MATHEMATICS
THEORY AND PRACTICE

DATA FUSION MATHEMATICS
THEORY AND PRACTICE

Jitendra R. Raol

CRC Press
Taylor & Francis Group
Boca Raton London New York

CRC Press is an imprint of the
Taylor & Francis Group, an **informa** business

CRC Press
Taylor & Francis Group
6000 Broken Sound Parkway NW, Suite 300
Boca Raton, FL 33487-2742

First issued in paperback 2017

© 2016 by Taylor & Francis Group, LLC
CRC Press is an imprint of Taylor & Francis Group, an Informa business

No claim to original U.S. Government works

ISBN-13: 978-1-4987-2097-7 (hbk)
ISBN-13: 978-1-138-74863-7 (pbk)

Library of Congress Cataloging-in-Publication Data

Raol, J. R. (Jitendra R.), 1947- author.
    Data fusion mathematics : theory and practice / Jitendra R. Raol.
        pages cm
    Includes bibliographical references and index.
    ISBN 978-1-4987-2097-7 (hardcover : alk. paper) 1. Multisensor data fusion--Mathematics. 2. Signal processing--Mathematics. 3. Statistical matching. 4. Wireless sensor nodes. 5. Wireless sensor networks. I. Title.

TK5102.9.R367 2015
621.382'2--dc23   2015026061

Visit the Taylor & Francis Web site at
http://www.taylorandfrancis.com

and the CRC Press Web site at
http://www.crcpress.com

Contents

v

Preface

Data fusion (DF), also known more formally as multi-sensory-source information fusion or multi-sensor data fusion (MSDF) or simply as sensor data fusion (SDF) is the process of combining data/information logically (and now more recently, intelligently, due to the inroads made by the theory and practice of artificial intelligence [AI] in many areas) from a number of different data sources or sensors to provide a robust and complete description of an environment, process or an object of interest (that we are monitoring), such that the overall total information after the fusion (i.e. of the fusion system) is supposed to have increased significantly as measured by some information (-based) metric. Thus, DF is of a special significance and importance in many applications where a large amount of data or information must be combined and then fused intelligently. The sensor or data aggregation (DAG), which is not quite as the same as the DF, has a special meaning and place in wireless sensor networks (WSN). In a WSN there would be a very large amount of data coming from various nodes/sensors/input channels, and one should aggregate certain data before sending these data to some other nodes/outputs/sink nodes and so on – this is DAG. Ultimately SDF is carried out to obtain enhanced information of appropriate and required quality and integrity so that the decisions that are made on the basis of this integrated/fused data/information can be highly trustworthy, and have more accuracy of prediction with reduced uncertainty about the state/status of the object/process/scenario of interest. The DF should be carried out logically and with the proper understanding of the data processing and data fusion methods. Hence, the current book presents the basic mathematics related to SDF. The DF mathematics provide credence to the data fusion process (DFP). This DFP is nothing but enhanced signal processing activity with support from other well known and proven soft technologies/disciplines: optimisation, control, estimation theory and soft computing paradigms. Image fusion is at the core level image processing, which itself is the next higher level of digital signal processing. Hence, in image fusion we deal with pixel-, and feature level fusion processes and methods. Decision making and decision fusion are at core level rooted into statistical decision theory and can use probabilistic and fuzzy logic-related arguments. All levels of DF find practical applications in (i) military systems, (ii) civilian surveillance and monitoring tasks, (iii) process control, (iv) information systems, (v) robotics, (vi) WSN, (vii) security systems and (viii) medical data (image) – fusion. Many key DF methods are particularly important towards development of autonomous systems in many of these applications. Thus, DF, also sophisticatedly known as multi-source, multi-sensor information fusion (MUSSIF), is rapidly emerging as an independent discipline to reckon with and finds further applications in (i) biomedical/bioengineering, (ii) industrial automation, (iii) aerospace engineering and (iv) environmental engineering processes and systems, in addition to the usual defence applications. The SDF offers benefits of (i) more spatial coverage of the object under observation due to (spread and distributed) geographical locations of the sensor systems, (ii) redundancy of sensors/measurements due to more number of sensors, (iii) robustness of the system's performance and fault tolerance, due to the former two aspects, (iv) increased accuracy of inferences (prediction accuracy is supposed to have increased) due to the fusion process and (v) overall assured performance of the sensors-integrated systems and fusion-system's robustness due to the all of the former aspects.

A really good study and successful application of the DF process involves a good grasp of several allied and supportive disciplines: (a) probabilistic and statistical methods, (b) signal/image processing, (c) numerical methods/algorithms, (d) sensor control and optimisation, (e) soft computing paradigms and (f) system identification, parameter estimation and filtering methods. Most important aspect of this study is to have a very good understanding of the basic mathematics related directly (or indirectly) to the DF processes and methods. There are several good books on SDF, however the treatment of mathematical aspects, that is, the DF mathematics, is highly limited or not even found explicitly in many such books. The treatment of the DF mathematics in the present volume is comprehensive and covers major DF concepts, and mathematical expressions, formulae and equations as well as their derivations (where appropriate) from the related disciplines which will help in a good understanding of SDF strategies and methods. The book follows and covers largely a mathematical approach to the extent possible. Thus, the main aim of the present volume is to fill the existing gap in the discipline of MSDF. However, the practically potential aspects and uses of these methods are also highlighted and treated, where appropriate, with a few illustrative examples via numerical simulations coded in MATLAB®. The users should have access to PC-based MATLAB software and other toolboxes such as signal processing, control systems, system identification, neural networks (NWs), fuzzy logic, genetic algorithm and image processing. Some SW for Chapters 3, 6, 8 and 10 would be made available from the data fusion mathematics (DFM)-book's website (by the publisher). Listings of some other DF-related software are provided in Appendix D. Also, the solution manual for the exercises presented in the book chapters will be available to instructors from the publisher.

In this book, we cover probability and statistical methods, fuzzy logic-based mathematics, decision theory, reliability concepts, component analyses, image algebra, tracking and filtering-related mathematics and methods, WSN/DF and soft computing paradigms (neuro-fuzzy-genetic algorithms). All these mathematical methods have a direct (sometimes indirect but closely related) bearing on the proper understanding of SDF concepts and methods. Thus, this book can be viewed as (partially) an archival and a reference volume for the DF community, as well as the primary text book for introducing the SDF subject to the undergraduate (or postgraduate) in one semester class of any discipline. The readers/users are advised to use their own discretion and judgment to apply various methods/approaches/algorithms/expressions/formulae/results discussed in the book to their practical problems, though enough care has been taken in the presentation of material.

The main users of this integrated DF mathematics will be engineers/scientists/teachers/researchers from electrical systems-, aero-, mechanical-, chemical-, civil-educational institutions, R&D laboratories and the aviation and transportation/automation/robotics industries.

MATLAB® and Simulink® are registered trademarks of The MathWorks, Inc. For product information, please contact:

The MathWorks, Inc.
3 Apple Hill Drive
Natick, MA 01760-2098 USA
Tel: 508 647 7000
Fax: 508-647-7001
E-mail: info@mathworks.com
Web: www.mathworks.com

Additional material is available from the CRC Web site: http://www.crcpress.com/product/isbn/9781498720977.

Acknowledgements

At the outset, I am grateful to Mr Herman le Roux (principal scientific officer, CSIR, SA) who in 2004–2005 realising that CSIR-NAL (India) and CSIR-SA had an MOU for scientific cooperation, then visited NAL. Herman wanted to work on SDF, an activity that I had initiated in FMCD division (Flight Mechanics and Control Division, CSIR-NAL, India) with a sponsored project from DRDO-ITR Chandipur, a few years before he visited us. After returning to his country, he initiated the process to invite me there as a visiting research scientist. I will be ever grateful to him for this gesture. However, due to some logistical reasons, I was then officially posted with another group in their CSIR unit. Herman is extremely polite and polished, someone who continues to inspire me both as a person and as a scientist and who gives me moral support even today. I am also very grateful to Dr. Ajith K. Gopal (former scientist CSIR-SA) for making my visit logistically and administratively smooth, and a memorable one, with him and his able and pleasant team of MIAS (mobile intelligent autonomous systems) scientists. I consider Herman and Dr. Gopal highly mature individuals and scientists relative to their ages and I am extremely obliged to both of them for providing me an opportunity to learn some of the emerging soft technologies in the field of robotics, which I was always curious about, but never found the time to explore. I would also like to acknowledge the role that Professor Velliyiur Subbarao played in my life. Dr. VNS was in the United States, and in 1976–1977 had temporarily returned to NAL and it was at this time that I got a chance to interact with him. The point is that even with my first class MTech I was still then, senior scientific assistant and Professor VNS felt that I should be a Scientist B. He talked to our boss, who also promptly acted on his suggestion, and advised me to apply for the then vacant position, for which the application dead line had actually gone by! I quickly applied and was subsequently selected as Scientist B. In July 2007 I retired as scientist G and the head of FMCD of CSIR-NAL. I am really grateful to Dr. VNS for such a good judgement and the presence of mind that changed the course of my technical and professional life. I am also grateful to Dr. V. P. S. Naidu (senior scientist, FMCD, CSIR-NAL) who still motivates me to continue to work in SDF. This book was initiated by him in the course of a few minutes' talk while driving home from my office. Continual moral support from Dr. I. R. N. Goudar, the former head and IST-scientist (Information Sciences & Technology) of NAL-ICAST for the last two decades is gratefully acknowledged here. I am also thankful to Dr. Mrs Girija Gopalratnam, Dr. V. P. S. Naidu and Dr. S. K. Kashyap (senior scientists, FMCD, CSIR-NAL) for providing the useful information and notes on Fusion SW and algorithms of Appendix D. Some examples of Section 3.9 have been mainly contributed by Dr. Kashyap and Section 6.10 is mainly contributed by Dr. Naidu. I am also grateful to Dr. Mrs Maya V. Karki and Dr. Mrs K. Indira (professors, Department of Electronics and Communication Engineering, MSRIT) for technical discussions and some inputs of a few examples in Chapter 7. This book is dedicated to my smart, lovely daughter Mrs Harshakumari H. Gohil and my extremely mature son Mayur for their immense lifelong moral support. Without this support from my two talented children, I would not have been able to cope with the low periods in my life, especially during the writing of this book. Thanks are also due to my wife, Virmati for efficiently managing our household matters for over four decades! My additional thanks are due to Drs. D. Amaranarayan, N. Ramakrishna, Mrs Swetha Desai, L. Prakash and Y. Nagarajan (of NAL health centre) who have efficiently monitored and maintained my good health during the writing of this book ... and always. I am also grateful to Drs. Sati Devi A.V. and Chaitra Jayadev (of Narayana Netralaya) for the excellent treatment of my eyes such that I could properly see/read and complete this book. I would have been almost blind by now, if it were not for the sincere and timely committed care of my son (and my family) and the treatment from all these doctors! I am also grateful to Jonathan Plant and his team at CRC Press, especially Amber Donley and Arlene Kopeloff for their tremendous support during this book project as well as during all the earlier ones. Syed Mohamad Shajahan and his team at Techset Composition have done commendable work of proofing of this book.

Introduction

SDF encompasses the theory, methods and tools used for creating a synergy of the information acquired from multiple sources, sensors, databases and direct information through humans – this synergy is much more than the concept of DAG. The resulting fused information of the object or process, and the decision or action is (found to be or should be), in some sense, better in terms of accuracy (in fact accuracy of prediction), and robustness, than would have been possible if any of these sources were used individually without such a synergy creation.

SDF is rapidly emerging as an independent discipline and is finding an ever-increasing role and applications in many systems/disciplines: biomedical, industrial automation, aerospace and environmental engineering processes and systems. The expected benefits of DF are: (i) more spatial coverage of the object under observation, (ii) redundancy of measurements, (iii) robustness of the system's performance, (iv) increased accuracy – in fact enhanced prediction capability – of inferences and (v) overall assured functioning and performance of the sensor-data integrated and task-oriented DF systems. The entire process of SDF involves a comprehensive study of several allied and supportive disciplines: (a) image-signal processing, (b) numerical methods and algorithms, (c) statistical/probabilistic methods, (d) sensor management, control and optimisation, (e) neural NWs, fuzzy systems and genetic algorithms, (f) system identification, parameter estimation, and Kalman filtering (and other modern filtering methods) and (g) data-structure and data base management.

The most important aspect of this synergistic study is a good understanding of the mathematics directly related to DF concepts and methods. In this book, the probability and statistical methods, fuzzy logic-based mathematics, decision theory, reliability, component analyses, image algebra, tracking and filtering-related mathematics and soft computing paradigms such as neuro-fuzzy logic-genetic algorithms (ANN-FL-GA) are presented. All these mathematical methods have a direct bearing on the proper understanding of MSDF concepts and methods, and fill the existing gap of the mathematics of DF for the fusion community at large.

There are several good books on various topics SDF [1–15], however the treatment of mathematical aspects, that is, the DF mathematics per se, is highly limited or not even found explicitly in many such books (except in References 3 and 13). The treatment of the DF mathematics in the current book is very comprehensive and covers major mathematical expressions, formulae and equations as well as their derivations (where appropriate) from the related disciplines and hence, this book can be viewed as an archival and reference volume, and in addition will also be very useful for practicing scientists and engineers. Certain other concepts of (the mathematics of) DF are briefly outlined in Appendix B. It is also emphasised here, that the treatment of mathematics in Chapters 2 to 10 is not based on theorem–proof–corollary style, because the mathematics mainly comes from several disciplines, and an attempt at unification would be a huge and a challenging task. The present approach itself is a starting point for such unification, however, there is a further scope for unification and this is left to future endeavours. A brief discussion of the chapters follows.

In Chapter 1, as a brief introduction to the topic, we discuss SDF concepts, DF models (briefly) and architectures. Several connected aspects and disciplines are briefly mentioned. More specific DF architectures in the context of decentralised systems are considered in Chapter 5.

In Chapter 2, we discuss important aspects and methods of probability and statistics. These probabilistic and information measures can be utilised in defining DF approaches, metrics and measures, and especially the weights/coefficients in the DF rule. Actually DF starts from these measures. Further, these metrics and norms are also very useful in evaluating performance of the designed and developed DF systems.

Chapter 3 discusses mainly fuzzy logic type 1 (FLT1), fuzzy implication functions and possibility theory. Fuzzy logic actually expands the scope of classical set theory and the theory of probability. Also, interval type 2 fuzzy logic (IT2FL) is discussed, which further expands the scope of type 1 fuzzy logic. We also, discuss fuzzy sets, fuzzy operators and fuzzy implication functions. The adaptive neuro-fuzzy inference system (ANFIS) is discussed. The use of fuzzy logic and ANFIS in DF is highlighted here, and in Chapters 8 and 10. Also, fusion of long wave IR and EOT images using type 1 and type 2 fuzzy (T1FL and IT2FL) logics is presented. Such applications of IT2FL for DF are perhaps novel.

In Chapter 4, we cover the mathematical treatment of many types of filtering algorithms, and target-tracking methods, and also, discuss kinematic DF methods. Single and multi-sensor tracking and fusion mathematics is presented. Gating and data association concepts and related filtering methods are described. Information filtering approaches are also presented. H-infinity filters are now being used for DF and hence these algorithms are also presented. The aspects of handling randomly missing measurements in the Kalman filter and the KF like-filtering algorithms are treated and some ramifications are derived and simulated results are presented. Two factorisation filtering algorithms for target-tracking-cum-data fusion are also briefly discussed and some practical results presented. In Chapter 5, we study decentralised DF and related filtering/estimation approaches for DF. Information filtering and Bayesian approaches are discussed. Decentralised DF architectures are very important for monitoring of large-scale structures (buildings), and health monitoring of aerospace systems, vehicles and large industrial automation plants. Various DF architectures and DF rules are given. The square root information filter for decentralised DF is considered and some numerical simulation results are presented.

Chapter 6 discusses the component analysis methods including wavelets and related methods, for example, discrete cosine transform. These concepts and methods are very useful for DF, especially for image fusion. The component analysis is also fundamentally used in many signal processing methods and the extension thereof to image fusion is a natural one. Also discussed is the curvelet method for image fusion. Also, multi-resolution singular value decomposition approach for image fusion is illustrated. A few approaches are evaluated with simulated image-data using codes written in MATLAB. Image processing and image fusion are now crucial technologies for complex modern day aerospace and other systems, for example, enhanced synthetic vision, their newer integrated versions, situation assessment requirements and field/medical robotics. Thus, understanding of image algebra and fusion mathematics is of paramount importance. These aspects are extensively dealt with in Chapter 7 by Dr. Mrs S. Sethu Selvi (professor, and head, Department of Electronics and Communications Engineering, MSRIT, Bangalore). Also, some examples of fusion in biometric systems are presented.

In Chapter 8, we present briefly decision theory and its use in the process of SDF, mainly the Bayesian approach. Decision fusion is treated as a higher level fusion method and is crucial in all applications of sensor/DF including the WSN. An approach of decision making/fusion using fuzzy logic type 1 for several aviation scenarios is presented using simulated data. Various decision fusion rules are also presented in this chapter. In Chapter 9, we discuss WSN and associated SDF and DAG approaches. These networks (NWs) have applications in structural health monitoring (including large aircraft wings, large spacecraft structures) and many surveillance systems. Distributed sensing, detection, estimation and DF are important technologies and approaches that will accrue definite benefits to the users of WSN-based utility systems, for example, security systems. Also, the signals generated by sensors of different types of modalities (principles) need to be sampled, filtered, compressed, transmitted, fused and stored. These WSNs are an important part of smart environments (in buildings, utilities, industrial places, homes, on board ships, and in the automation of transportation systems and security systems).

Finally, Chapter 10 discusses three soft computing paradigms which find increasing applications in multi-sensory DF approaches and applications: artificial neural networks (ANNs) for learning/adaptation, fuzzy logic for decision making and modelling vagueness (in data, and in rules), and genetic algorithms for global optimisation (of ANNs weights, etc.). These are very promising soft computing approaches with enormous scope for DF applications at all levels of SDF: kinematic, image and decision levels. Some hybrid methods centred on ANNs/FL/GA are also important to derive benefits from the merits and capabilities (such as learning, optimisation) of any two individual soft computing methods, and hence are briefly treated.

In Appendixes A, B, C, D, E and F we provide, respectively: (i) a few more algorithms and derivations of some filters (A), (ii) some other methods for DF and image fusion performance metrics (B), (iii) a brief note on an automatic

DF (C), (iv) some important information on commercial and other software for target tracking and sensor data/image fusion (D), (v) several definitions of sensor/DF (E) and (vi) some current research topics in the area of data/image fusion and target tracking (F).

Where appropriate, some numerical simulation examples using MATLAB are given (in Chapters 3 to 8 and 10). The end users of this integrated technology of SDF mathematics will be systems-aero-educational institutions, R&D laboratories, aerospace and other industries, the transportation/automation/robotics industries, all engineering disciplines and some branches of science, for example, the explanation of some evolutionary mechanism (of a biological species) using fuzzy logic!

Summary: Can Unified Data Fusion Mathematics Emerge?

Fundamentally the process of DF starts with combining some information (after some analysis of the original data) from individual sensors/sources using some formulae or rule, a simple rule being the average of the two 'informations', or a weighted average of these 'informations'. This information/s is either in the form of information factor or information matrix. Then we need to determine the optimal weights for this fusion rule. The Kalman filter gives the information on these weights automatically in the form of predicated covariance matrices coming from, say, two Kalman filters processing the data from two individual sensor channels. Also since an information matrix is the inverse of its covariance matrix, the information quantities are also used as weights in a DF rule. These covariance matrices are originally based on the probability definitions of uncertain phenomena/events. These events are affected by random noise processes. The covariance matrices are regarded as the statistics of the signals coming from several sensor channels. Thus, the statistics, and probability aspects of the sensor signals are connected measures, which are then used as weights for a DF rule. Since the Kalman filter can be basically derived from the Bayesian rule of probability, the latter itself then independently, becomes the basis of a DF rule. However, since vagueness (another kind of uncertainty) can be truthfully and usefully modelled by using fuzzy logic, DF rules can also be derived using the weights from fuzzy logic-based analysis. Hence, statistics, probability and fuzzy logic become naturally connected, especially because fuzzy logic generalises the concept of probability. In addition, fuzzy logic being a rule-based logic, it allows the incorporation of the human/expert's experience and intuition into this rule base. So, a designer of the SDF system can enhance the performance of the system using fuzzy logic-based analysis and control approaches. Interval type 2 fuzzy logic can further enhance the scope and performance of such a DF system. The component analysis (Fourier, discrete cosine transform, principal component analysis, wavelets, etc.) is currently very much in use for (sensor) signal processing, and system identification. Since image processing and image fusion are higher levels of signal processing aspects, the component analysis has a natural extension for image (-data) fusion. Image algebra is also an extension of signal algebra, and is useful for understanding image–data fusion approaches, and to undertake analysis of such fusion methods. Decision fusion is a higher level fusion process over kinematic DF and image fusion. Several concepts of the basic levels of DF are applicable to decision fusion also. Decision fusion involves determining statistics, obtaining the (state-) estimates from the sensor/image signal, and the use of some logic, and hence, from this point of view as well, it is a higher level of fusion activity. Also, decision fusion can be regarded as a symbol level fusion. Bayesian rule and fuzzy logic can play a very important role in decision fusion. WSN involve all the types of basic signal processing activities, and hence, many of the foregoing concepts are applicable to SDF and DAG in WSN. Many of the DF rules in decision fusion and WSN are based on probability and information measures. The neuro-fuzzy-GA paradigms are based on three distinct aspects of: (i) our own nature, (ii) our observation of natural phenomena and (iii) the evolutionary mechanism of nature, respectively. The ANNs are modelled on the basis of the biological neural networks, and ANNs have the ability to learn and adapt to the environment. Fuzzy logic models the vagueness and imperfections in the nature of our observations. Genetic algorithms (GA) are based on the evolutionary system of nature. These three soft computing paradigms are also collectively the ingredients of AI. We can then effectively build intelligent DF systems by utilising these three basic aspects of AI. ANNs use training algorithms that are rooted in some optimisation principles and techniques. GAs themselves use direct search methods for optimisation. ANFIS uses ANNs for learning the parameters of the fuzzy membership functions from the presented data. The learning algorithms can be based on classical optimisation methods or GA itself. Much of the basic mathematics of ANNs (training algorithms), fuzzy inference systems and GA is not very complicated and is rooted in basic function theory (FT), vector-space norms (VSN), classical logic chains (CLC/forward and backward) and optimisation criteria (OC). All these four aspects, FT, VSN, CLC and OC also form the basic mathematical ingredients of the DF mathematics. So, we see here that some unification and uniformity is beginning to emerge for SDF from the diverse concepts of mathematics related to statistical/probabilistic signal/image processing, filtering, component analysis, image algebra, decision making and neuro-FL–GA paradigms.

References

1. Raol, J. R. *Multi-Sensor Data Fusion with MATLAB*. CRC Press, FL, USA, 2010.
2. Edward, W. and James, L. *Multi-sensor Data Fusion*. Artech House Publishers, Boston, 1990.

3. Hall, D. L. *Mathematical Techniques in Multi-sensor Data Fusion*. Artech House, Norwood, MA, 1992.
4. Abidi, M. A. and Gonzalez, R. C. (Eds.). *Data Fusion in Robotics and Machine Intelligence*. Academic Press, USA, 1992.
5. Gregory, D. H. *Task-Directed Sensor Fusion and Planning – A Computational Approach*. Kluwer Academic Publishers, Norwell, MA, USA and The Netherlands, 1990.
6. Lawrence, A. K. *Sensor and Data Fusion: A tool for Information Assessment and Decision Making*. SPIE Press, Washington, USA, 2004.
7. Chirs, H., Xia, H. and Qiang, G. *Adaptive Modeling, Estimation and Fusion from Data: A Neuro-fuzzy Approach*. Springer, London, UK, 2002.
8. James, M. and Hugh, D-W. *Data Fusion and Sensor Management – A Decentralized Information – Theoretic Approach*. Ellis Horwood Series, 1994.
9. Dasarathy, B. V. *Decision Fusion*. Computer Society Press, USA, 1994.
10. Mitchell, H. B. *Multi-sensor Data Fusion*. Springer-Verlag, Berlin, Heidelberg, 2007.
11. Varshney, P. K. *Distributed Detection and Data Fusion*. Springer, New York, USA, 1997.
12. Clark, J. J. and Yuille, A. L. *Data Fusion for Sensory Information Processing Systems*. Kluwer, Norwell, MA, USA, 1990.
13. Goodman, I. R., Mahler, R. P. S. and Nguyen, N. T. *Mathematics of Data Fusion*. Kluwer, Norwell, MA, USA, 1997.
14. Yaakov B-S. and Li, X-R. *Multi-target Multi-sensor Tracking (Principles and Techniques)*. YBS, Storrs, CT, USA, 1995.
15. Mutambra, A. G. O. *Decentralized Estimation and Control for Multisensor Systems*. CRC Press, FL, USA, 1998.

Author

Jitendra R. Raol earned a BE and an ME in electrical engineering from MS University of Baroda, Vadodara in 1971 and 1973, respectively, and a PhD (in electrical and computer engineering) from McMaster University, Hamilton, Canada in 1986, where he was also a research and teaching assistant. He taught for 2 years at MS University of Baroda before joining the National Aeronautical Laboratory in 1975. At CSIR-NAL, he was involved in human pilot modelling in fix- and motion-based research flight simulators. He re-joined NAL in 1986 and retired on 31 July 2007 as Scientist G (and head, flight mechanics and control division at CSIR-NAL). He has visited Syria, Germany, the United Kingdom, Canada, China, the United States and South Africa on deputation/fellowships to work on research problems on system identification, neural NWs, parameter estimation, MSDF and robotics, to present several technical papers at several international conferences and has delivered guest lectures at some of these places. He had also become a Fellow of the IEE/IET (United Kingdom) and a senior member of the IEEE (United States). He is a life-fellow of the Aeronautical Society of India and a life member of the System Society of India. In 1976, he won the K. F. Antia Memorial Prize of the Institution of Engineers (India) for his research paper on nonlinear filtering. He was awarded a certificate of merit by the Institution of Engineers (India) for his paper on parameter estimation of unstable systems. He earned a best poster paper award from the National Conference on Sensor Technology (New Delhi) for a paper on SDF. He also earned a gold medal and a certificate for a paper related to target tracking (from the Institute of Electronics and Telecommunications Engineers, India). He is also one of the five recipients of the CSIR (Council of Scientific and Industrial Research, India) prestigious technology shield for the year 2003 for leadership and contributions to the development of integrated flight mechanics and control technology for aerospace vehicles in the country. The shield was associated with a plaque, a certificate and the prize equivalent to USD 67,000 for the project work. He has published more than 100 research papers and numerous technical reports. He has guest edited two special issues of *Sadhana* (an engineering journal published by the Indian Academy of Sciences, Bangalore) on (i) advances in modelling, system identification and parameter estimation, and (ii) multi-source, multi-sensor information fusion. He has also guest edited two special issues of the *Defense Science Journal* on (i) mobile intelligent autonomous systems, and (ii) aerospace avionics and allied technologies. He has guided six doctoral and eight master research scholars and he currently (technically) guides a dozen faculty members (of MS Ramaiah Institute of Technology, Bangalore) for their doctoral theses. He has co-authored an IEE/IET (London, UK) Control Series book *Modeling and Parameter Estimation of Dynamic Systems* (2004) and a CRC Press (Florida, USA) book *Flight Mechanics Modeling and Analysis* (2009). He has also authored a CRC Press book *Multisensor Data Fusion with MATLAB* (2010). He has edited (with Ajith K. Gopal) a CRC Press book *Mobile Intelligent Autonomous Systems* (2012). He has served as a member/chairman of numerous advisory, technical project review, and doctoral examination committees. He is a reviewer of more than a dozen national and international journals. His main research interests have been and are DF, system identification, state/parameter estimation, flight mechanics–flight data analysis, H-infinity filtering, ANNs, fuzzy systems, genetic algorithms and soft technologies for robotics. He has also authored a collection of 300 poems/free-verse on various topics related to science, philosophy, evolution and life itself, in search of the true meaning of human life on this planet.

1

Introduction to Data Fusion Process

The subject and the discipline of multi-source multi-sensor information (sensor/data) fusion aggregates and encompasses the theoretical concepts, methods and numerical algorithms used for exploiting the synergy in the information obtained from multiple sensor/sources, databases, knowledge bases and other information obtained by humans (often called human intelligence – HUMINT), as sensibly and intelligently as feasible. As a result of data fusion (DF), the final understanding of the object (and its predicted trajectory, if a moving object), or the process or the scene/scenario, and the decision/action is in some qualitative/quantitative sense better in terms of accuracy of prediction and robustness, compared to what would have been based on a single sensor/source (of course the quantitative measures would be preferred). This final aspect/decision/declaration is more intelligent than would have been possible if any of these sensor/sources were used individually [1,2]. DF mathematics is a sub-topic (in the sense of overall MSDF aspect), and yet a basic one, at the fundamental level, of the general subject of multi-source multi-sensor information fusion. The main process in SDF activity is to: (i) collect the observations (the measurements/data of the observables, and/or the states of a system) and samples from various similar and/or dissimilar sources and sensors, (ii) extract the required information (data analysis, filtering and estimation), (iii) draw some logical inferences (based on some comparisons and evaluation) and (iv) make some adequate and good decisions.

1.1 Data Fusion Aspects

Thus, the derived/assessed information from different sensors is combined in some sense using an appropriate fusion rule/formula with a view to obtaining an enhanced status and identity of the perceived/observed object or phenomenon. This fusion rule is often a weighted average of the individual quantitative inferences/estimates, and sometimes could be based on probabilities (Bayesian rule/theorem) and 'beliefs' (Dempster–Shafer/DS). In case of the use of fuzzy logic (FL) in decision fusion (and/ or in feature level fusion and kinematic level fusion) the fusion rule/

process could be implicit in the 'If...Then' rules of the fuzzy inference system, although explicit fusion rule can be devised and used. Even here, the weights of fusion rule/formulae can be obtained by using FL in analysis of the data samples with the intention of using the information afterwards. The MSDF primarily involves: (i) some transformations (coordinate transformations and/or use of mathematical models) between observed parameters to generate decisions regarding the location (kinematics, and even dynamics), characteristics (features/structures) and identity of an entity, and (ii) inference and interpretation (decision making and fusion) based on the detailed analysis of the observed scene/entity in the context of the object's environment.

The DF plays an ever-increasing role in many biomedical, industrial automation, aerospace and environmental engineering processes and security systems in addition to many defence applications. As we have seen earlier the DF process gives (i) more spatial coverage of the object under observation/surveillance, (ii) provides redundancy of measurements – from similar and dissimilar sensors, (iii) enhances the robustness of the system's performance – more types of measurements are available, and it is assured that at least one data set is always available, (iv) provides better prediction (with less uncertainty) of inferences – due to the use of enhanced information and (v) gives overall assured performance of the multi-sensor-integrated MSDF system. The complete process of DF involves several closely related disciplines [1,2]: (i) signal and/or image processing, (ii) computational/numerical techniques and algorithms, (iii) information-theoretic, statistical and probabilistic approaches/measures/metrics, (iv) sensors' mathematical models, sensor control-management, and sensor configuration-optimisation, (v) soft computing techniques and (vi) system identification, state and parameter estimation. Several such techniques from various fields strengthen the analytical treatment, understanding and performance evaluation of the DF systems.

The DF process is typified as: (i) low-level fusion or data (-level) fusion that combines the data from several sensor/sources (essentially of the same type) to produce a new data set that is expected to be more informative and useful than the individual ones, one can include the fusion of estimated states of a system/object by using data from several, often dissimilar, sensors (as state vector fusion), this low-level process is called the kinematic DF; (ii) intermediate mid-level fusion (or feature level fusion, in the context of image fusion) combines various features such as edges, lines, corners, textures and so on into a feature map, that is used for segmentation of images and detection of objects and so on; (iii) high-level decision fusion, combines decisions from several individual experts, methods of decision fusion being voting, FL and statistical methods [1].

We present, in next nine chapters, the mathematical treatment related to these three levels of data fusion processes (DFP).

1.2 Data Fusion Models

The process of DF is considered as a system-theoretic approach and is a synergy of sensing, signal/data/image processing, state estimation and trajectory tracking, and control and decision making. Hence, these aspects can be formulated in the form of some functional and/or task-oriented model/s called DF model [1,3–5]. Perhaps, this aspect of DF model has remained more or less a formality rather than (much of) a practice, since individual scientists and engineers who carry out the DF task/s mostly go by their own perception and the requirement of the DF task rather than following any of these functionality models to define their own approach. This is not to say that these DFP models have no place in the MSDF-soft/hard (SW/HW) technologies. Some research attempts to study the DFP have resulted in definitions and specification of several such DFP models, a few of which, for the sake of completion, are discussed very briefly in next few sub-sections.

1.2.1 Joint Directors of Laboratories Model

So far, the popular model is the JDL DF model (Joint Directors' of Laboratories Model) that originated from the department of defence, DoD (1985/1992). There are mainly three levels of DF processes: (a) level 1, (b) level 2 and (c) level 3. It has several sub-levels and auxiliary tasks of processing data, combining the information and evaluating the performance of the DF system (Figure 1.1) [1]. This DF process model is a functionality-engineered model and can be useful for many application areas: from sensor network (SNW) to human computer/machine interface (HCI). It has sub-buses to four levels of the signal and fusion processing and performance evaluation. It facilitates a higher level of interpretation of certain entities for the intended application. The DFP is conceptualised by sources of information, HCI source pre-processing, DF levels and finally DBM (data base management). This DF model is data/information centred and sort of an abstract model with some specific details. The HCI provides an interface for human input and communication of fusion results to several operators.

1.2.2 Modified Waterfall Fusion Model

The waterfall data fusion process (WFDF) model, a sort of hierarchical architecture [5], concentrates on the processing functions on the lower levels. These functions have some similarities with the JDL model: (i) sensing/signal processing relates to source pre-processing, (ii) feature extraction/pattern processing corresponds to object refinement (OR), (iii) situation assessment (SA) is connected to situation refinement (SR) and (iv) decision making relates to threat refinement (TR). The modified WFDF model as

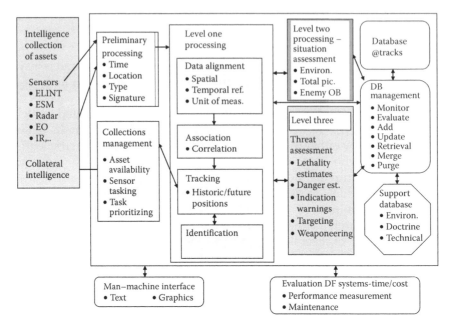

FIGURE 1.1
Basic and earlier DF model (JDL). (Adapted from Raol, J. R. *Multi-Sensor Data Fusion with MATLAB*. CRC Press, FL, 2010.) (ELINT – electronic intelligence, ESM – electronic support measures, EO – electro-optical, IR – infra red, REF – reference, MEAS – measurements, PIC – picture, OB – objective, EST – estimate, DB – database, ENVIRON – environment, DF – data fusion.)

proposed in Reference 1 with increasing sub-task complexity indicated by darker shades, is shown in Figure 1.2 [1]. This modified model (MWFDF) is more an action-oriented model as it is augmented by local feedback loops. The original WFDF model does not have such intermediate control loops, except that there is a control loop from the decision to the sensors. The idea of the overall loop has been to suggest re-calibration and re-configuration of the sensors used in robots (robotics), and WSNs to make the entire process of MSDF energy efficient. This would help in the sensor management cycle of the overall DFP.

1.2.3 The Intelligence Cycle-Based Model

Some inherent cyclic processing behaviour of the DFP is captured in the IC model [1]: (i) the planning/direction determines the intelligence aspects, (ii) the collection stage gathers appropriate information, (iii) the collation stage streamlines the gathered information, (iv) the evaluation stage utilises the information and performs the actual fusion and (v) the dissemination stage distributes the fused intelligence/inferences. The IC model can be regarded as a super set model.

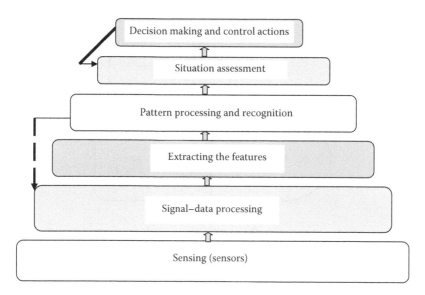

FIGURE 1.2
Modified waterfall fusion model with the modifications shown by feedback loops. (Adapted from Raol, J. R. *Multi-Sensor Data Fusion with MATLAB*. CRC Press, FL, 2010.)

1.2.4 Boyd Model

The Boyd model of Figure 1.3 [1] depicts the classical decision-support mechanism in information operations with a closed/feedback loop. As this Boyd control loop (BCL) model has a cycle of observe–orient–decide–act (OODA) features, an appropriate combination of the IC and BCL loop models would yield a better fusion process model. This combined model would be a less abstract and more action-oriented model, and one such combination results in the Omnibus (OB) model.

1.2.5 Omnibus Model

The OB model combines several beneficial features of other approaches (Figure 1.4) [1]. Interestingly this OB model seems to be a hybrid of the three models [5]: (i) BCL model, (ii) Dasarathy model and (iii) Waterfall model. The model: (a) specifies the ordering of processes, (b) makes the cyclic nature more explicit and (c) uses a general terminology. A few WFDF aspects are seen in the OB model, and it is more generalised than many other models.

1.2.6 Dasarathy Model

The Dasarathy's (functionality oriented-FODF) model is based on the fusion functions rather than the tasks (Figure 1.5a). As we already know that the three main levels of abstraction are feasible and identifiable during the DFP

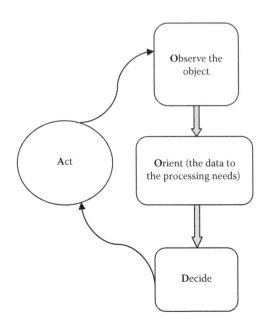

FIGURE 1.3
The Boyd-OODA cyclic loop model. (Adapted from Raol, J. R. *Multi-Sensor Data Fusion with MATLAB*. CRC Press, FL, 2010.)

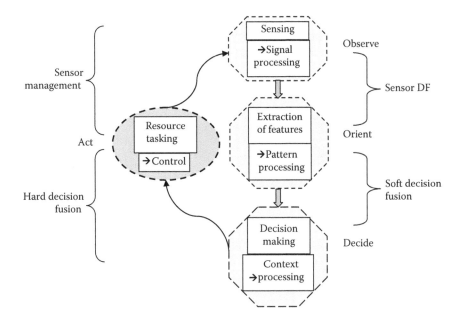

FIGURE 1.4
The expanded and more explicit Omnibus cyclic sensor-data fusion model. (Adapted from Raol, J. R. *Multi-Sensor Data Fusion with MATLAB*. CRC Press, FL, 2010.)

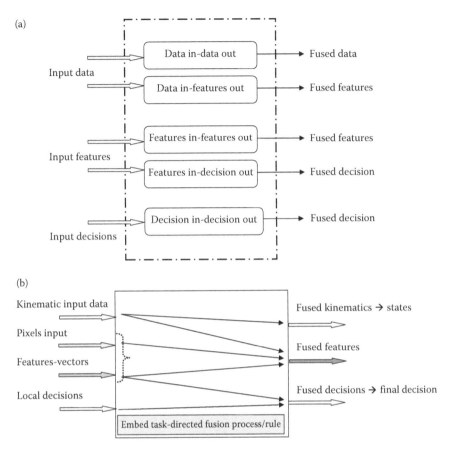

FIGURE 1.5
(a) Dasarathy's functional DF model. (b) Modified multi-sensory DF model.

as the [3]: (i) data, (ii) features and (iii) decisions, and the actual information processing (and DFP) might occur both within these three levels and also as a means of transforming between them [4]. It is a very simple, yet interesting, and also a straight-forward I/O DFP model.

A modified DFP model supported by the task-directed fusion (as an embedded) process is suggested in Figure 1.5b as a possible MSDF model (derived from Figure 1.5a).

There are a few other DFP models [5]: (i) Thomopoulos architecture (TA), (ii) multi-sensory integration fusion model (MSIF) and (iii) behaviour knowledge-based data fusion (BKDF) model. The TA model consists of three modules: (a) signal level fusion, (b) evidence level fusion and (c) dynamics level fusion. These levels can operate in a sequential manner or interchangeably. The MSIF is the process of integration of sensors in the first place. In MSIF a description of the measured phenomenon is obtained after processing the outputs of the n sensors. This is done with the aid of the information system. The DFP is then facilitated with the combination of information systems containing relevant databases and libraries. In BKDF model the process flows as [5]: (i) feature extraction, (ii) association fusion, (iii) sensor attribute fusion, (iv) analysis and aggregation and (v) representation. It is suggested that we should also try to incorporate the aspects of artificial intelligence (AI) and sensor fault detection and identification (FDI) explicitly in some DFP models in order to emphasise the increasing usage of DF in such soft technologies of AI/FDI.

1.3 Sensor Data Fusion Configurations

The sensor-DF networks (NWs) are the topologies organised with certain specified configurations/sensor arrangements (mostly physical). In many cases more than one of these configurations can be used. These are categorised [1] as follows, see Figure 1.6.

1.3.1 Complementary

The sensors are placed in order to give a more complete picture/image of the phenomenon being observed. A single object is viewed simultaneously from different regions/locations where the sensors are deployed, for example, the

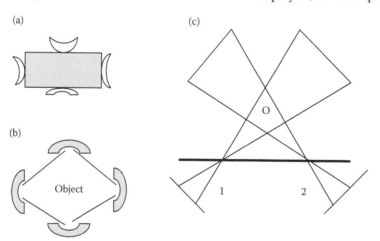

FIGURE 1.6
SNW arrangement – configurations: (a) complementary – viewing different regions outwards, (b) competitive – viewing the same enclosed area, (c) cooperative – two cameras can produce a 3D image of an object at O. (Adapted from Raol, J. R. *Multi-Sensor Data Fusion with MATLAB*. CRC Press, FL, 2010.)

four radars around a geographical area can provide a complete picture of the region where the object is located. This arrangement also provides a certain redundancy in measurements.

1.3.2 Competitive

Each sensor provides independent measurements of the same property/feature of the object. The configuration is robust and fault tolerant. The competing sensors could be identical or non-identical. One example is of three thermometers competing to measure the temperature of a body. Each sensor here gives a central numerical value with a small uncertainty band around this value. The fusion process or algorithm has to give out one value based on the three measurements and their associated uncertainties. This arrangement is very prevalent in many critical safety areas in aerospace and other industrial plants and is often termed as redundancy provision and management. This aspect is greatly studied in fault tolerant systems. This kind of redundancy is applicable to hardware, software, information coding and NWs. DF fusion concepts can even be utilised for enhancing the efficiency of fault-tolerant systems.

1.3.3 Cooperative

The data are provided by two independent sensors to derive information that might not be available from only one sensor as is done in a stereoscopic vision system: the two-dimensional (2D) images obtained from two cameras mounted at slightly different incidence angles can provide a three-dimensional (3D) image of the observed scene, after deciphering the depth information and appropriate fusion. However, some processing of the data from individual sensors is required before accurate depth information can be obtained.

1.4 Sensor Data Fusion Architectures

The sensor data fusion architectures are related to the arrangements of sensors with the emphasis on how the data at the top level (or superficial level) are processed as seen in Figure 1.7 [1]. The DF architectures for multi-sensory systems are further discussed in Chapter 5.

1.4.1 Centralised Fusion

The centralised DF architecture involves time synchronisation and bias correction of sensors data. Here, the data are transformed from the sensor-based

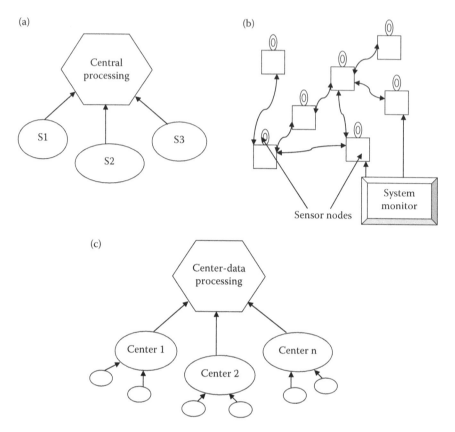

FIGURE 1.7
Sensor-data fusion processing architectures: (a) centralised processing, (b) decentralised processing, (c) hierarchical. (Adapted from Raol, J. R. *Multi-Sensor Data Fusion with MATLAB*. CRC Press, FL, 2010.)

units/coordinates to convenient, and/or the required coordinates/units system. The gating/data association are needed in case of multiple targets and multiple sensors/measurements. The decisions are based on the maximum feasible information obtainable from these sensors. Often, this is also called measurement (or data level) fusion.

1.4.2 Distributed Fusion

Here, the observations from each sensor are processed by an individual signal processing/estimation technique at each node, the node/s being the smart nodes/devices. The local tracks consisting of their estimated state vector/covariance matrices (discussed in Chapters 4 and 5) from each filter are inputted to the state vector fusion process (SVF formulae) which yields the fused state vector and fused covariance matrix. There is no need of any central fusion. The architecture is very useful for large flexible/smart structures, aircraft/spacecraft health monitoring, large automation/chemical plants, large sensor NWs and chemical industrial applications.

1.4.3 Hybrid Fusion

This architecture involves both centralised and distributed DF schemes based on the disposition of the required sensor configurations. This combination scheme is used for some applications, where certain arrangement of sensors is already pre-defined or is already in operation, and a variety of similar, dissimilar sensors/measuring systems (like radars) are used.

1.5 Data Fusion Process

Now, we can easily see the DFP at the broader level, especially by studying Figure 1.5, which in most simple terms shows the DFP, except that the fusion rule/formulae are not specified and the techniques of the data/image processing are also not specified. So, the DF is what Figure 1.5a and b illustrate at the functional level, wherein the fusion takes place in the centre boxes. The DF rule/formulae (or DFP) could be linear (non-linear), multiplicative, or could be based on some concepts of statistics, probability and/or FL. The aspects of the DF rules/formulae and certain basic methods required for processing and analysing data-signals/samples (coming from various sensors) and image analysis and corresponding fusion methods are discussed in Chapters 2 to 10 with some simple derivations and illustrative examples, wherever feasible.

Since, the field of MSDF has been evolving for last five decades, and at a rapid space recently, and has become very important and essential in many industrial systems, it is nearly an impossible task to keep track of the complete development of the field, and hence this book provides some important concepts of DF mathematics (theory and practice) which can be covered as an undergraduate seventh semester class, or in a master's second semester (perhaps of all the engineering branches). The course on sensor data fusion should be supported with a text on sensors and instrumentation, signal/image processing and computer-based algorithms (and with MATLAB-based or related SW tools, this will be an added advantage). The present volume would also be useful as a simple recipe book as well as reference book on DF mathematics, and can be considered as complementary to the many books listed as references in the introductory part of the book (References 1–15).

EXERCISES

1.1 What is the purpose of synchronisation of the data in a DFP?

1.2 Why is transformation of data/measurements needed in a DFP?

1.3 What is the meaning and purpose of the association or correlation in DF?

1.4 What is the purpose of tracking in the JDL model?

1.5 What is the purpose of a feedback loop from the decision making to the SA in the MWFDF model?

1.6 What is the use of a feedback loop from pattern processing to the signal processing in the MWFDF model?

1.7 Discuss the pros and cons of the SNW arrangements of Figure 1.7.

1.8 Why is DFP typified as low-level, mid-level and high-level processing tasks?

1.9 What exactly is done in the centralised DF architecture?

1.10 Can you stretch your imagination and give one example from real-life systems where hybrid DF architecture could be used? This can be taken as an assignment problem.

References

1. Raol, J. R. *Multi-Sensor Data Fusion with MATLAB.* CRC Press, FL, 2010.
2. Raol, J. R. and Gopal, A. K. (Eds.). *Mobile Intelligent Autonomous Systems.* CRC Press, FL, 2012.
3. Dasarathy, B. V. Sensor fusion potential exploitation – Innovative architectures and illustrative applications. *IEEE Proceedings*, 85(1), 24–38, 1997. www.ohli.de/download/papers/Dasarathy1997.pdf.
4. Farooq, M. and Gad, A. Data fusion architectures for maritime surveillance. *ISIF*, 2002. www.ohli.de/download/papers/Gad2002.pdf, accessed January 2013.
5. Esteban, J., Starr, A., Willetts, R., Hannah, P., and Bryanston-Cross, P. A review of data fusion models and architectures: Towards engineering guidelines. In *Neural Computing and Applications*, 14(4), 273–281, 2005. www.researchprofiles.herts.ac.uk/portal/en/publications, accessed April 2013.

2

Statistics, Probability Models and Reliability: Towards Probabilistic Data Fusion

2.1 Introduction

It is a fascinating fact of nature that uncertainty in one or other form abounds in many aspects of the sensing, data communications, data analysis and data fusion (DF) processes. We should provide an explicit measure of this uncertainty to enable sensory information to be fused so that the data fusion process (DFP) becomes very efficient and the state of the plant/dynamic system or of any object in question or study (for the purpose of fusion) is predictable with enhanced accuracy (of prediction). Most such methods of representing uncertainty hitherto were based on the use of probabilistic models, because these models provide a powerful and consistent means of describing uncertainty in many situations (also these probabilistic models occurred at a much earlier time in the evolutionary/developmental frame of scientific search in the human mind). This concept naturally fits into the ideas of information fusion and decision making [1]. Of course, certain practical realities of a problem often might suggest the use of some alternative method of modelling uncertainty (and/or modelling different kind of uncertainties e.g. vagueness), for example leading to fuzzy logic (FL)-based concepts and their utilisation in filtering, control and decision fusion problems. Although human perception and experience of these kind of fuzzy events date much more into the past (having had occurred many centuries earlier than gambling and other probabilistic matters), the formal mathematics and analytical approaches had not been developed until five decades ago! Although probabilistic models are good for many situations, these models cannot capture all the information required to define and describe the operations of sensing and DF processes, for example, the heuristic knowledge of a human expert in a domain area/knowledge, for which the theory of FL and related possibility theory are very suitable. Yet, probabilistic modelling techniques have played and continue to play a good and very practical role in developing DFP and methods. Hence, it is necessary to understand the probabilistic modelling methods for DF applications. Also, when uncertainty is not random, but it is deterministic and unknown (and has a bounded magnitude) some approaches based on the H-infinity norm can be considered for DF tasks, especially for basic state determination from the measurement data observed from the sensors.

We start this chapter with certain mathematical concepts (statistics) that are also useful in independent component analysis (ICA) and principal component analysis (PCA) [2] which are described in Chapter 6. With the help of statistics (i.e. certain important characteristic aspects and numerical values of the signals) and statistical methods we can look at the data from the point of distribution of measurements, that is, (i) how the data are spread out [2] in some vector space, (ii) how the data vary with time and space, (iii) how some values of these data repeat in time and space and (iv) how and why certain data are related to each other in time and space. The knowledge of these statistical concepts will be useful in understanding many broader concepts developed later for the DF process in subsequent chapters.

2.2 Statistics

As such, statistics is a (numerical) science of 'data', and it begins with data. Probability begins with trying to give a mathematical model for these data from the point of view of 'uncertainty' playing a crucial role in corrupting these data. Often we get considerable data from the observations (we make) of certain phenomena that we are interested in understanding for model building and prediction of events. These data are on the face of it as such incomprehensible, and we want to make sense out of these data, as a first-cut analysis. Then, we obtain certain statistics of these data: mean (average values), median (middle value), mode (most frequently occurring value), variance (spread around the mean), covariance (spread between various variables), autocorrelation (with respect to time), cross-correlation and standard deviation (STD) (square root of variance) [2]. Obtaining this information from the data, even if the data are manageable, helps us discover hidden and underlying patterns (e.g. correlations) in the data, and determine how much the incompleteness in the data can be tolerated, if at all there is some incompleteness. The correlations amongst the data would also indicate the correlations and relationships amongst the various events that have occurred. If the data are incomplete or some data in a chain of measurements are missing then one can use some methods to fill these gaps using some predictions/estimation methods. This is the problem of missing data and is currently an active area of research in the field of target tracking, and wireless sensor networks. The simplest method to fill this gap is to carry out interpolation, however, this cannot be done if long strings of data are missing, for which more formal and sophisticated approaches (some advanced estimation methods) are required and are already in vogue. It is interesting to note that classifying or distributing these data in certain (numerical) pockets (sequential ranges), and ascertaining that simply counting the data that fall into various categories (numerical ranges of the amplitude of the data) leads to the empirical evaluation or determination of the relative frequency which in turn is one of the interpretations of the probability. It is a simple but time consuming experimental method for determination of probability. In fact this pocketing of the data and counting how many data fell in what (-ever) ranges, gives the empirical determination of the probability density function – the frequency interpretation of the probability, whereas the basic data handling is part of the statistics. Interestingly the counting and related aspects, that can be called a part of statistics gathering are themselves the cause and the origin of the definition and interpretation of the probability, and yet, the formal definitions of statistics do involve and depend on the definition and use of the probability models! As such the statistics should be discussed before the probability is defined, however, the formal definition of the statistics of random variables/data needs to incorporate the pdf $p(.)$ of the data themselves – here, PDF is the probability distribution function and pdf is the probability density function. In fact, this pdf is a priori assumed before such statistics of the random variables are computed. For deterministic data, probabilistic descriptions are not needed; however, some statistics still can be defined and will be useful. In general, statistics are useful for analysis of the data in terms of the relationships between individual points in the data and are also useful in defining certain DF rules and processes.

2.2.1 Mathematical Expectation

This important and the fundamental quantity is defined as

$$E(x) = \sum_{i=1}^{n} x_i P(x = x_i) \qquad (2.1)$$

for discrete random variables and

$$E(x) = \int_{-\infty}^{\infty} x\, p(x)\, dx \qquad (2.2)$$

for continuous random variables with P as the probability (distribution) of variables x and p the pdf of variable x. The mathematical expectation (ME), $E(.)$, is a weighted mean, the weights coming from the PDF/pdf. In our usual computation of the mean of a signal/data, the pdf is assumed to be uniformly distributed, that is, the weights are assumed to be the same for each data point. The expected value of sum of two variables is the sum of their individual expected values $E(X + Y) = E(X) + E(Y)$, similarly $E(a \times X) = a \times E(X)$. As said earlier, if the pdf is uniform distribution, then equal weights have been assigned to all the individual components. In DF one can use the concept of the weighted average as a simplest DF rule. The weights could be derived from the covariance matrices resulting from the Kalman filter (Chapter 4). These weights can also be derived from using the concept of FL [3] (Chapters 3 and 10). The concept (of addition/averaging) is also useful for DF when the information filtering is employed for target tracking (Chapters 4 and 5). The information is a directly additive process and hence, the information matrices from the results of two information filters (related to the two sensor situation) can be added to obtain the total/fused information.

Referring to the ME, and resulting average value of a variable, there are other related definitions: (i) the median which is a middle value, the smallest number such that at least half the numbers are no greater than it, the median, that is, if the values have an odd number of entries, then the median is the middle entry after sorting the values in an increasing order; and (ii) the mode is a most common or frequently occurring value and there could be more than one mode. In estimation theory and methods based on probabilistic models, the mode defines the value of x (the state/parameter that is estimated) for which the probability, pdf, of observing the random variable is maximum.

2.2.2 Variance, Covariance and STD

The variance is a measure of the spread of the data, in fact it gives an average spread (one value of the STD, or a matrix) around the mean of the same data, in the form of the energy, represented by the (square) of the signal. Thus, the covariance between two variables is defined as

$$\mathrm{Cov}(x_i, x_j) = E\{[x_i - E(x_i)] \, [x_j - E(x_j)]^\mathrm{T}\} \tag{2.3}$$

For one variable the variance is obtained as a (average) spread of the component values of the data around its own mean. We see that here, the aspect of probability is incorporated in the definition of ME E, via Equations 2.1 and 2.2. If the x is a parameter vector then we get the parameter- (estimation-) error covariance matrix.

The square root of the diagonal elements of the covariance matrix gives the STDs of the errors in estimation of states or parameters, of course neglecting the effects of cross-covariance elements. Also, it is important to note that the inverse of the covariance matrix is the indication of the information content in the signals about the parameters or states: large covariance means higher uncertainty and low information/confidence in the estimation results; and smaller covariance means lower uncertainty and hence more information/confidence in the resultant estimates of states/parameters. Thus, the covariance matrix and the information matrix are the inverse of each other. The covariance/information matrices are naturally and automatically generated by Kalman/information filters (KF/IF) (Chapters 4 and 5). In fact, the state estimates obtained by KF/IF are generated based on these matrices. These covariance/information matrices/values are used as weights in the DF rule, as we will see in Chapters 4 and 5. Hence, we quickly gather that these fundamental definitions of average, covariance (matrices) and STDs are very useful in multi-sensory (MS) DF processes.

2.2.3 Correlations and Autocorrelation Function

The correlation defines the relation between two time series-data or sequences. The correlation coefficient (CC) is defined as [3]

$$\rho_{ij} = \frac{\text{cov}(x_i, x_j)}{\sigma_{x_i}\sigma_{x_j}}; \quad -1 \le \rho_{ij} \le 1 \tag{2.4}$$

The CC is zero for independent variables x_i and x_j, and for surely correlated sequences/variables it is equal to 1. If a variable y is dependent on several x_i, then the CC for each of x_i can be used to determine the degree of this correlation with y as

$$\rho(y, x_i) = \frac{\sum_{k=1}^{N} (y(k) - \underline{y})(x_i(k) - \underline{x}_i)}{\sqrt{\sum_{k=1}^{N} (y(k) - \underline{y})^2} \sqrt{\sum_{k=1}^{N} (x_i(k) - \underline{x}_i)^2}} \tag{2.5}$$

If $|\rho(y, x_i)|$ tends to one then y is highly and linearly related to particular x_i, (under bar indicates the mean value). For $x(t)$, a (random) signal, we can define the autocorrelation function as

$$R_{xx}(\tau) = E\{x(t)x(t + \tau)\} \tag{2.6}$$

Here, τ is the (one unit-) time lag and with increasing time, if R_{xx} shrinks then it means that the nearby values of the signal x are not (highly) correlated. The autocorrelation function of a white process (noise) is an impulse function, and this, R, for discrete-time signals/residuals is given by

$$R_{rr}(\tau) = \frac{1}{N - \tau} \sum_{k=1}^{N-\tau} r(k) \, r(k + \tau); \quad \tau = 0, \ldots, \tau_{max} \text{ (discrete-time lags)} \tag{2.7}$$

The autocorrelation function is an important statistic and is used in order to check if the residuals/innovations (from KF) are a white process or not. If the innovations process is white then it indicates that the filter has performed satisfactorily (Chapter 4).

2.3 Probability Models

Probability concepts and models are very useful in understanding KF/IF methods and fusion processes, especially the performance results, so that we can interpret these results with confidence, and have faith in the resultant DF processes, fusion algorithms (and software) and DF systems.

The probability of an event is the fraction of all the outcomes that occur in the event and hence, the probability of an event is the number of favourable outcomes divided by the total number of (all possible) outcomes. This is the frequency (i.e. counting repetitions) interpretation of the probability. There are four basic aspects for the definition of probabilistic concepts/models [1]: (i) we need to conceive and perform some experiment so that there is a primary action which would result in a series of observations/measurements, (ii) then, here will be some results of the experiment and we will call them an outcome, (iii) there will be enumeration of all possible outcomes which are permissible and (iv) an event that is a subset of the sample space, and this subset has a natural and non-technical explanation – all these aspects can be described by using probabilistic notions. For this purpose, a pdf $p_y(.)$ is defined on a random variable x (often written as $p(x)$), and this random variable may be a scalar or a vector, and may be either discrete or continuous in measure. The pdf $p(.)$ is considered as a probabilistic model of the quantity x that could be observation (measurement) or state variable (or even a parameter considered as a random variable – because this parameter is determined from the measurements that are affected by random noise). The $p(x)$ is considered valid if (i) it is positive: $p(x) > 0$ for all x; and (ii) it sums or integrates to a total probability of 1

$$\int p(x)\,dx = 1 \tag{2.8}$$

The probability measure is then defined such that it obeys the following axioms over the set E (an even set) (of super set S) [1]:

    a. $0 \le P(E) \le 1$ for all $E \subset S$

    b. $P(\phi) = 0$ and $P(S) = 1$

    c. If $E, F \subset S$ and $E \cap F = \phi$, then $P(E \cup F) = P(E) + P(F)$

d. If E_1, E_2, ... is an infinite sequence of mutually exclusive events then $P(E_1 \cup E_2 \cup ...) = P(E_1) + P(E_2) + \cdots$ The axiom (c) implies that the event E and event F do not depend on each other. We also have the following results (not necessarily the axioms):

e. If E is an event, then $P(E) + P(\overline{E}) = 1$

The 'over bar' signifies the complement of the main event.

Equivalently we have $P(\overline{E}) = 1 - P(E)$

f. If E and F are events with E as a subset of F, then obviously, $P(E) \leq P(F)$

g. The probability of the union of pair-wise disjoint events/occurrences is equal to the sum of their individual probabilities

h. If E and F have the probability measure P, then $P(E \cup F) = P(E) + P(F) - P(E \cap F)$

In (h) above, the events are not necessarily disjoint. One useful and common probability measure is the uniform model, in which case the every outcome in S is equally likely to occur. Also, the joint pdf $p(x, y)$ is defined, and integrating pdf $p(x, y)$ over the variable x gives the marginal PDF, $P_y(y)$ and integrating the joint pdf over y gives the marginal PDF, $P_x(x)$. Further, the conditional pdf $p(x|y)$ is defined by

$$p(x|y) \triangleq \frac{p(x,y)}{p(y)} \tag{2.9}$$

In Equation 2.9, the conditional pdf has properties of a pdf with x, the dependent variable, given that y takes on specific fixed values. This means that, if the variable x is dependent on y, then we are seeking to answer what is the probability of occurrence of x, if the probability of occurrence of the event y is already specified. This conditional pdf given by Equation 2.9 is based on the joint pdf of x and y. Now, if the event y has already occurred (i.e. by some experiment, and the empirical data have been gathered and are being utilised) then the $p(x|y)$ is related to the likelihood function (it may be the likelihood function). The conditional pdf $p(y|x)$ is similarly defined. The chain rule of conditional distributions is used to expand a joint pdf in terms of conditional and marginal distributions. We also have from Equation 2.9

$$p(x,y) = p(x|y)p(y) \tag{2.10}$$

The chain rule is extended to many variables

$$p(x_1, \ldots, x_n) = p(x_1|x_2, \ldots, x_n) \ldots p(x_{n-1}|x_n)p(x_n) \tag{2.11}$$

For a continuous random variable x, the $p(x)$ (pdf) exists such that for every pair of numbers $a <= b$, we have $P(a <= X <= b)$ = area under $p(x)$ between a and b. The Gaussian pdf is given as

$$p(x) = \frac{1}{\sqrt{2\pi} \, \sigma} \exp\left(-\frac{(x - m)^2}{2\sigma^2} \right)$$

with mean m and variance σ^2 of the Gaussian distribution. For the measured/recorded random variables, z, given the state x (or parameters), the Gaussian pdf is given as

$$p(z|x) = \frac{1}{(2\pi)^{n/2}|R|^{1/2}} \exp\left(-\frac{1}{2}(z - Hx)^T R^{-1}(z - Hx) \right) \qquad (2.12)$$

In Equation 2.12, R is the covariance matrix of measurement noise. The expression $(z - Hx)$ is the measurement equation flipped over to obtain the expression of the measurement noise (i.e. $z = Hx$ + measurement noise). Equation 2.12 is the basis of derivation of the Kalman (tracking) filter that is discussed in Chapter 4. Equation 2.12 can also be used for DF and in another simple way the KF itself is a data fuser [3], as explained in Chapter 4. Of course, it is not mandatory to use Equation 2.12 to derive the KF equations, especially if we have the assumption that the underlying pdf for noise processes is the Gaussian pdf model. The reason is that the Gaussian pdf is completely characterised by its mean and variance/covariance, and hence in the KF we propagate and update these means and covariance matrices only.

2.4 Probabilistic Methods for DF

In the next few sub-sections, we briefly discuss various aspects of probabilistic approaches for DF.

2.4.1 Bayesian Formula

Suppose, we have a hypothesis space S and data set X, then we can define three probabilities: (i) $P(s)$ is the probability of 's' being the correct hypothesis before seeing and using any data, $P(s)$ is the prior probability of 's' (e.g. the chance of a storm is 90% if we are close to the sea and at some latitude yet no data has been obtained); (ii) $P(X)$ is the probability of having the data X and (iii) $P(X|s)$ is the probability of the occurrence of the data X, given 's' and it is the likelihood of 's' with respect to X [4]. The Bayesian theorem (BT) relates the posterior probability of a hypothesis given the data, X, with the three probabilities mentioned earlier: (a) $P(X|s)$ is the likelihood, (b) $P(s)$ prior probability and (c) $P(X)$ is the evidence, the data are being seen. Then we have the posterior probability given by

$$p(s|X) = \frac{p(X|s) \cdot p(s)}{p(X)} \qquad (2.13)$$

A method that searches for the hypothesis with maximum $P(s|X)$ is (called) maximum a posteriori method or MAP: $H_{map} = \arg\{\max_s (P(s|X))\}$. If we assume that priori probabilities are equally likely (e.g. uniform distribution) then we get the method of maximum likelihood estimate or MLE method: $H_{ml} = \arg\{\max_s (P(X|s))\}$. The Bayesian formulation can use subjective probabilities: (i) it provides the probability of a hypothesis being true, the evidence being given; (ii) one can incorporate a priori knowledge of the likelihood of a hypothesis being true at all and (iii) it does not require the knowledge of PDFs [4]. Thus, we see that the BT is the most important result in our study of probabilistic models and their applications to estimation and DF. Let us consider two random variables x (could be a state variable) and z (could be measurement variable) on which a joint pdf, $p(x, z)$ is defined, then using the chain rule of conditional probabilities, this density function is expanded as [1]

$$p(x,z) = p(x|z)p(z) = p(z|x)p(x) \qquad (2.14)$$

Alternatively, rearranging, Equation 2.14 in terms of one of the conditional densities, BT can be expressed as

$$p(x|z) = p(z|x)p(x)/p(z) \qquad (2.15)$$

The importance of this result lies in the interpretation of the pdfs $p(x|z)$, $p(z|x)$ and $p(x)$. Assume that it is necessary to determine the various likelihoods of different values of an unknown state x (of X). There are prior notions about what values of x we can expect, and these are encoded in the relative likelihoods in the prior pdf, $p(x)$. If we want more information about the state x, we need to make certain measurements z (of Z), with the practical assumption that the z is somehow related to x (i.e. $z = Hx + $ noise, called a measurement/data equation), in that case we model these measurements as a conditional pdf $p(z|x)$. This signifies, for each fixed state x (of X), the likelihood that the measurement z will be made, that is, the probability of z given x. Then the new likelihoods associated with the state x should be computed from the original prior information pdf, $p(x)$, and the information provided/gained by the measurements we have made. This is specified by the posterior pdf, $p(x|z)$. This posterior describes the likelihoods associated with x given the observations z, that is, the state information is now enhanced because of the measurements made and utilised in the process of computation of x using some forms of Equations 2.12 and 2.13. The pdf $p(z)$ in Equation 2.15 serves to normalise the posterior Bayesian expression. The BT provides a direct means of combining observed information with prior beliefs/notions about the state of the situation (/process/plant/system/states/environs). In many DF algorithms the conditional distribution $p(z|x)$ serves the role of a sensor model, and it is interpreted as: (i) in building a sensor model, the pdf is constructed by fixing the value of $x = x$ and then asking what pdf in the variable z would result?; thus, $p(z|x)$ is considered as a distribution on z; thus, if we knew the true range to a target (x-distance/position), then $p(z|x)$ is the distribution on the actual measurement of this range and (ii) once the sensor model exists or is determined, the actual measurements are made and $z = z$ is fixed; from this, we infer the state x; thus the distribution $p(z|x)$ is now considered as a distribution in x [1]. In case (ii), the distribution is known as the likelihood function and the dependence on x is made clear by writing $L(x) = p(z|x)$.

2.4.2 DF Based on Bayesian Rule

Say, our primary interest is tracking a moving object, a target or a mobile robot, using the measurements that we have made about this object. Then, for the measurements z for this tracking problem with the target state as x, a Gaussian pdf-measurement model is given [1,3] as follows:

$$p(z|x) = \frac{1}{\sqrt{2\pi\,\sigma_z^2}}\exp\left(-\frac{(z-x)^2}{2\sigma_z^2}\right) \tag{2.16}$$

In Equation 2.16, the 'sigma' denotes the STD of the random measurement noise process. When this model is built, the state is considered a fixed one, and the distribution is then considered as a function of z. This is an act of model postulation or specification. Now, when the measurements are made and incorporated into the distribution, it becomes a function of x. The Gaussian prior is also given as [1,3]

$$p(x) = \frac{1}{\sqrt{2\pi\,\sigma_x^2}}\exp\left(-\frac{(x-x_p)^2}{2\sigma_x^2}\right) \tag{2.17}$$

Then using the Bayesian rule, the posterior, Equation 2.15, after making measurements is given as

$$p(x/z) = \text{Const.}\frac{1}{\sqrt{2\pi\sigma_z^2}}\exp\left(-\frac{(z-x)^2}{2\sigma_z^2}\right)\frac{1}{\sqrt{2\pi\sigma_x^2}}\exp\left(-\frac{(x-x_p)^2}{2\sigma_x^2}\right)$$

$$= \frac{1}{\sqrt{2\pi\sigma_f^2}}\exp\left(-\frac{(x-x_f)^2}{2\sigma_f^2}\right) \tag{2.18}$$

In Equation 2.18, we have the weighted sum/mean of the fused information given as

$$x_f = \frac{\sigma_x^2}{\sigma_x^2 + \sigma_z^2} z + \frac{\sigma_z^2}{\sigma_x^2 + \sigma_z^2} x_p \tag{2.19}$$

and

$$\sigma_f^2 = \frac{\sigma_z^2 \sigma_x^2}{\sigma_x^2 + \sigma_z^2} = \left(\frac{1}{\sigma_z^2} + \frac{1}{\sigma_x^2} \right)^{-1} \tag{2.20}$$

Equations 2.19 and 2.20 primarily show how to obtain the improved state estimate when the measurements are made and incorporated in the process using BT. In addition, we can easily see that Equation 2.19 can be regarded as the expression for the 'fused' state, if the z were the state estimate from one sensor and if x_p were the state estimate from the other sensor. So, we see that Equation 2.19 gives the 'fused' state and is regarded as the valid fusion rule/ formula, since it is derived analytically and is based on probabilistic definitions, it is not a heuristic or ad hoc rule. In a similar manner, Equation 2.20 emerges as for the 'fused' variance of the state x, and the data z [1,3]. From the previous arguments, Equation 2.20 can also be regarded as the 'fused' covariance if the estimates from two individual sensors are used for DF. This development itself is the starting point for the DF using sensor measurements. Thus, similar development is very clearly possible for two measurements z_1 and z_2 and so on. In Equation 2.19, the weights for variables x and z are automatically the variances of z and x, respectively. Thus, the Bayesian rule itself provides the fundamental formal way of DF.

Thus, extending the foregoing development from the Bayesian rule one obtains the general application rule, for the independent likelihood pool (ILP)

$$p(x|Z_n) = \{p(Z_n)\}^{-1} p(x) \prod_{i=1}^{n} p(z_i|x) \tag{2.21}$$

The conditional probabilities $p(z/x)$ are kept as functions of z and x, and for the fusion the information obtained from different sources/sensors is assumed to be independent, only the state is considered as a common state for the sensors, because one state-space model is used, considering the situation in which we have one target and more channels of measurements. Thus, for a set of measurements one obtains the standard equations as follows:

$$p(x|Z_n) = \{p(Z_n)\}^{-1} p(Z_n|x) p(x)$$

$$p(x|Z_n) = \frac{p(z_1, \ldots, z_n|x)p(x)}{p(z_1, \ldots, z_n)} \tag{2.22}$$

In Equation 2.22, the joint distribution-pdf of Z is required to be known completely. From the foregoing one can see once again that the Bayesian rule can be used for sensor DF. We observe from Equations 2.21 and 2.22 that a recursive Bayesian updating can be easily obtained for sensor DF [1] as follows:

$$p(x, Z_k) = p(x|Z_k)p(Z_k) \tag{2.23}$$
$$= p(z_k, Z_{k-1}|x)p(x)$$
$$= p(z_k|x)p(Z_{k-1}|x)p(x) \tag{2.24}$$

Further, one obtains by equating the first expression, Equation 2.23, and the expression of Equation 2.24 the following equation:

$$p(x|Z_k)P(Z_k) = p(z_k|x)p(Z_{k-1}|x)p(x) \tag{2.25}$$
$$= p(z_k|x)p(x|Z_{k-1})p(Z_{k-1}) \tag{2.26}$$

With $p(Z_k)/p(Z_{k-1}) = p(z_k|Z_{k-1})$ and rearranging one gets

$$p(x|Z_k) = p(z_k|x)p(x|Z_{k-1})/p(z_k|Z_{k-1}) \tag{2.27}$$

Equation 2.27 is a recursive formulation of the Bayesian rule and only the posterior likelihood $p(x|Z_{k-1})$, which contains a complete summary of all past information is required to be stored.

2.4.3 Distributed DF Based on Bayesian Rule

In the case of distributed DF (discussed further in Chapter 5) the sensor mathematical models (SMMs) are maintained locally at each sensor site, in the form of their respective likelihood functions. When a measurement is made, the likelihoods are used to provide a likelihood function $L_i(x)$ describing a probability distribution over the true state of the situation (/scenario/target/mobile robot) [1]. This likelihood is transmitted to the fusion centre (FC). Thus, a typical sensor communicates to the FC in terms of the underlying state (more correctly the state estimate) rather than in terms of the raw (but, yet in engineering units) measurements/data (REUD). This reduces the communications overheads on the transmission channels.

In an another approach to distributed DF using Bayesian rule each sensor computes a likelihood, and then combines this, locally with the prior from the previous time step, and produces a local posterior distribution on the state x. This information is then transmitted to the FC which then recovers the new measurement/information by dividing each posterior by the transmitted (global) prior. After using a normalised product the FC produces a new global posterior. This posterior is transmitted back to the sensors and the process is repeated in a recursive manner. In DF problems with Bayesian networks (NWs) one can use 'log-likelihoods' (LLs) for computational convenience, where only the additions and subtractions (rather than multiplications and divisions, respectively) can be used. Also, the LLs are closely related to the formal definition of information.

2.4.4 LLs-Based DF

In many DF situations and in all the Bayesian NWs, it is easier to work with the log of a probability (density function, pdf), rather than the probability (PDF, or P) itself. These LLs are computationally more convenient than the probabilities, since the additions and subtractions are employed in fusing probability-related information. The LL and the conditional LL (CLL) are defined as [1] $l(x) \equiv \log\{p(x)\}$ and $l(x|y) \equiv \log\{p(x|y)\}$. We know that the LL is always less than zero. It is only equal to zero when all the probability mass is assigned to a single value of x; $l(x) \leq 0$. The LL is useful and efficient for implementing the probability computations. For example, taking logs of both sides of Equation 2.15, we obtain BT in terms of LLs as $l(x|z) = l(z|x) + l(x) - l(z)$.

2.5 Reliability in DF

The crucial test of information fusion (IF)/sensor DF (SDF)/DF depends on how well the knowledge generated by the DF process represents reality. This certainly depends on [1,3,5,6]: (i) how adequate data are, (ii) how good and adequate is the uncertainty model used for analysis and (iii) how accurate, and appropriate prior knowledge is. Thus, it is important to consider the reliability of these underlying mathematical models (i.e. of the sensors themselves) used and reliability of the DFP itself. It is necessary to account for this fact, since the different models might have different reliability values/confidence in data received. An improper selection of metrics or poor estimation of the likelihood functions can provide an inadequate belief/notion model and this would lead to unreliable beliefs to be combined for the sensor data tasks. When combining information provided by many sources/sensors, we should take into account the range and the limitations of the (a prior) belief model/s used for each source/ sensor, and the natural way is to establish reliabilities computed within the framework of the model selected. This can be achieved by using certain reliability coefficients, that introduce the second level of uncertainty (uncertainty of evaluation of uncertainty) and represent a measure of the adequacy (or even an inadequacy) of the model used and the state of the environment considered. Also, it is feasible to use the reliability coefficients as weights in the DF rule.

Other definitions of reliability are {from http://www.businessdictionary. com/definition/reliability.html. Definitions (3); Page 1}: (i) 'The ability of an apparatus, machine, or system to consistently perform its intended or required function or mission, on demand and without degradation or failure'; (ii) for manufacturing-'The probability of failure-free performance over an item's useful life, or a specified timeframe, under specified environmental and duty-cycle conditions', often expressed as MTBF, mean time between failures or coefficient of reliability, also called quality over time; (iii) 'Consistency and validity of test results determined through statistical methods after repeated trials'.

The reliability as a higher order uncertainty is defined as per the following concepts [5]: (i) the reliability is understood as relative stability of the first-order uncertainty, that is, this reliability is often measured by the performance of each source (e.g. by the recognition or false alarm rates) and (ii) to measure accuracy of predicted beliefs/notions (reliability coefficients represent adequacy of each belief/notion model to reality). If the sources/sensors are not (highly) reliable, the fusion operators (FO) have to account for their reliability values and are represented as FO = function of degree of beliefs, reliability coefficients and so on, thus, the reliability coefficients control influence of the respective sources/sensors on fusion results. Numerically, the reliability coefficient is very close to zero if the source is unreliable, and very close to 1 if it is more reliable. It is possible to assign a numerical degree of reliability to each source, each value of reliability may be 'relative' or 'absolute' and they may or may not be linked by an equation such as sum of reliability coefficients being equal to 1. Sometimes, only an order of the reliabilities of the sources is known but no precise values are known. It may be a good idea to represent the reliability by a fuzzy variable, and an appropriate fuzzy membership function (MF).

2.5.1 Bayesian Method

The Bayesian method studied earlier can be used for incorporating reliability for DF. In this approach the degrees of belief/notion are represented by a priori belief, conditional pdf and posterior probabilities [3,5]. Then the decision is made based on the computed posterior probabilities, the fusion being performed by the Bayesian rule and total reliability of the sources is presumed.

2.5.1.1 Weighted Average Method

In this method the linear opinion pool (LOP) can be used [3,5]

$$P(x|Z_k;R_i) = \sum_i R_i P(x|z_{ki}) \tag{2.28}$$

The reliability coefficients, R_i associated with the sources, are used as the weights. Here, x is a hypothesis and z is the measurements/feature vector. Other methods are logarithmic opinion pools as given here [3,5]

$$P(x|Z_k;R_i) = \sum_i R_i \log\{P(x|z_{kj})\} \tag{2.29}$$

and

$$P(x|Z_k;R_i) = P(x)\prod_i \{P(x|z_{ki})/P(x)\}^{R_i} \tag{2.30}$$

The above formulae can be compared with the ones that occur in the context of the information pooling methods for the SDF, see Section 2.6.3.

2.5.2 Evidential Method

In this method (often called DS theory or DS method) the decision is based on the following formula [3,5]:

$$m^{1,2}(c) \propto \sum_{a\cap b=c} m^1(a)m^2(b) \tag{2.31}$$

Instead of probabilities (like in Bayesian method), the 'mass', m, is assigned, and this mass is looked upon like a probability, but it is not quite a probability. We have the following formula with R, as the reliability coefficients:

$$m(a) = \sum_i R_i m_i(a) \tag{2.32}$$

2.5.3 FL-Based Approach

The information obtained from a sensor is represented by a possibility distribution [3,5], see also Section 3.8:

$$\Pi : A- > [0,1] : \max_{a \in A} \Pi(a) = 1 \tag{2.33}$$

In general, the combination rules in FL are based on t-norms and s-norms (t-co-norms, see Chapter 3). The disjunctive rule is used when at least one source of data is reliable, but it is not known which one. In that case we have the 'OR' or 'max' rule of the fuzzy operator theory (Chapter 3/Chapter 10). If we have equally reliable sources, then a conjunctive operation 'AND' can be used. This is also 'min', or 'inf' (infimum, intersection) operator. Then the decision fusion rule based on the FL theory, including reliability R_i, is given as

$$\Pi_p(a) = \sum_i R_i \Pi_i(a) \tag{2.34}$$

The reliability coefficients required in these formulae are obtained/determined by [5]: (i) utilising domain knowledge, (ii) training of the available signals/data/image data, (iii) using a degree of consensus among various sources and/or (iv) experts' knowledge of the subjective probabilities/judgements. It is expected that the fusion system performance would be more realistic (not optimistic or pessimistic) as a result of incorporation of the reliability information into the fusion process, or at least the fusion results would be representative of the actual facts associated with the (reliable/unreliable!) information, rather than being too optimistic/pessimistic.

2.5.4 Markov Models for Reliability Evaluation

The Markov models (MM) can be used to assess and propagate the reliability aspects in time across/along the sensor NWs. The concept of a hidden Markov model (HMM) is applicable to a stochastic process that has an underlying, but hidden stochastic state transition process/model [4]. This underlying process is only inferred through a set of symbols displayed/emitted sequentially by the stochastic process. The point is, given the set of symbols exhibited, the state/sequence of states of the hidden/underlying process is sequentially determined. An algorithmic solution can exist, based on the available evidence. Several applications of the concept of HMM are: (i) signal processing/speech recognition, symbol identification in mobile communication, character recognition in images; (ii) SDF and target tracking; (iii) bioinformatics/gene finding; (iv) manufacturing/fault detection (fault detection and identification [FDI]) and (v) environment/weather prediction.

The HMM problem can be formulated with [4]: (a) the number of states, (b) the number of distinct emission symbols in each state, (c) the state transition probability matrix, (d) the emission symbol probability matrix and (e) the initial state distribution. For any given system, an MM consists of (i) a list of the possible states of that system, (ii) the possible transition paths between those states and (iii) the rate parameters of those transitions. In reliability analysis the transitions usually consist of failures and repairs, and when representing a MM graphically (i) each state is usually depicted as a 'bubble' and (ii) with arrows denoting the transition paths between states. The λ denotes the rate (-parameter) of the transition from State 0 to State 1. We also denote by $P_j(t)$, the probability of the system being in state j at time t. If the device is healthy at some initial time $t = 0$, the initial probabilities of the two states are $P_0(0) = 1$ and $P_1(0) = 0$. Thereafter, the probability of State 0 decreases at the constant λ, which means that if the system is in State 0 at any given time, the probability of making the transition to State 1 during the next increment of time dt would be $\lambda^* dt$. Hence, the overall probability that the transition from State 0 to State 1 will occur during a specific incremental interval of time 'dt' is given by multiplying: (i) the probability of being in State 0 at the beginning of that interval and (ii) the probability of the transition during an interval dt given that it was in State 0 at the beginning of that increment. This represents the incremental change dP_0 in probability of State 0 at any given time, so we have the following fundamental relations:

$$dP_0 = -(P_0)(\lambda dt) \tag{2.35}$$

$$\frac{dP_0}{dt} = -\lambda P_0 \tag{2.36}$$

This development signifies that a transition path from a given state to any other state reduces the probability of the source state at a rate equal to the transition rate parameter multiplied by the current probability of the state. Since, the total probability of both states must equal 1, it follows that the probability of State 1 must increase at the same rate that the probability of State 0 is decreasing, yielding the equations for this simple model as follows:

$$\frac{dP_0}{dt} = -\lambda P_0, \quad \frac{dP_1}{dt} = \lambda P_0, \quad P_0 + P_1 = 1 \tag{2.37}$$

The solution of these equations, with the initial conditions $P_0(0) = 1$ and $P_1(0) = 0$, is given as

$$P_0(t) = e^{-\lambda t}, \quad P_1(t) = 1 - e^{-\lambda t} \tag{2.38}$$

Obviously the form of the solution is called exponential transitions, because the transition times are exponentially distributed. It is clear that the total probability of all the states is conserved, and the probability behaves as if it simply 'flows' from one state to another. Also, the rate of occurrence of a given state equals the flow rate of probability into that state divided by the probability that the system is not already in that state. In the simple example shown above the rate of occurrence of State 1 is given by

$$(\lambda P_0)/(1 - P_1) = \lambda \qquad\qquad (2.39)$$

2.5.5 Reliability in Least-Squares Estimation

Also, the concept of reliability has been employed in the least-squares method and KF [6]. The reliability measures are normally functions of the time of proper and continuous functioning of a device/sensor as used in the study of quality control of these devices and systems. The commonly used measure in such cases is the mean time between failures (MTBF). However, interestingly the reliability measures for the estimation are quite different from the ones that are used for products of industrial processes. In case of the estimation processes, the concept of reliability is related to the controllability of measurements. This means we have the ability to detect the outliers and to estimate the effects of these outliers on the estimates of parameters or states of the dynamic systems. Thus, in the context of (state-) estimation, the concept of the reliability of estimates refers to internal reliability and external reliability. A very straightforward and fundamental statistical rule of 3σ (three STDs) has been employed to identify the measurement outliers. The real way is to study the relation between the measurement errors and their residuals. One should study the aspects in the context of reliability in estimation from the point-of-view of the effect of the measurement errors on the measurement residuals in various combinations – pairing of these individual errors on the measurement residuals – direct effects and cross effects. The internal reliability specifies to have a desired model (and its property) that facilitates the detection of systematic errors. It should also facilitate the localisation of the outliers without requiring more information. This means that the model should have a self-checking ability, and thus this definition of the internal reliability gives a measure of the capability of the system to detect the measurement outliers with a given probability. The study of the internal reliability is thus based on the specified model and can be performed as such without the measurements and their residuals. The external reliability then measures the response of the specified model to the undetected model errors, like the outliers. With this the effect of the undetectable errors on the estimates can be studied. It should be possible to exploit and use this reliability aspect in Kalman and other filters as well. Hence, the discussion of Section 2.5 is very important in sensor DF, target tracking and filtering.

2.6 Information Methods

The probabilities (and equivalently the LLs) are defined on states and/or measurements. One interesting aspect is that the amount of information contained in a given PDF can be advantageously used for estimation as well as for DF [1,3]. In fact, the information is a measure of the compactness of a probability distribution. If a probability distribution is spread evenly across many states, then its information content is fairly low, and if a probability distribution is highly peaked (and narrow) on a few states, then its information content is relatively high. As we have seen earlier, this also means that if the variance in the data is large, the uncertainty is large and the information content is low, and if the variance is low, the uncertainty is low and hence the information content is high. In fact, according to the definition of (Fisher's) information, the uncertainty measure (represented by variance/covariance and the related spread of the pdf) is inversely proportional to the information (matrix). Information is hence, a function of the probability distribution, rather than the underlying state, however, interestingly, the information can be obtained approximately from the states by computing covariance matrix (of these states errors) and then inverting this matrix to get an information matrix. These information measures/matrices play a very important role in many and perhaps all the DF systems. Two such definitions of information are of a value in the DF systems: (i) the Shannon information (defined via entropy) and (ii) the Fisher information (FI).

2.6.1 Entropy and Information

It is important for any sensor network (SNW) and/or DF system to perform efficiently in certain/particular (and yet, uncertain) environments [1,3]. The SNW/DF system should process the available information efficiently and share this information in a way to enhance the decision accuracies. One approach is to measure the value of information (VOI) obtained from various sensors and then fuse these, if the value (i.e. significant importance/ appreciation is gained) is added in terms of decision accuracy. If there is no gain or any value addition, then one need not fuse these data at all. This aspect is based on the information theoretic metric and it is an entropy measure. The SDF activity here is based on the fact that the VOI has improved the decision accuracy. Entropy sees information as a frequency of change in (the digital numbers), for example, the studied images. In fact, when a new set of data is added and used for analysis/inferences, then the total entropy (uncertainty) will be reduced compared to the previous/old entropy (uncertainty) and the difference is regarded the gain in the information. So, the use of entropy is (and should be) viewed from this context, that is, reduction of entropy is equivalent to the gain in the information – this is quantitative expression of information.

Given a finite set of the numbers of messages, any monotonic function of this number is used as a measure of the information when one message is chosen from the set. The information in fact is modelled as a probabilistic process. The occurrence of the random event x is the probability $p(x)$ of the message. The $I(x)$ is the self-information of x, and this is related to the inverse of the probability. Since if the event x always occurs, then $p(x) = 1$, and no new information can be transferred

$$I(x) = \log \frac{1}{p(x)} = -\log\{p(x)\} \tag{2.40}$$

This definition of information is intuitively appealing from the engineering point-of-view, and the average (self-) information in the set of messages with N outputs is given by [3]

$$I(x) = -Np(x_1)\log\{p(x_1)\} - Np(x_2)\log\{p(x_2)\},\ldots, Np(x_n)\log(\{p(x_n)\} \tag{2.41}$$

The average information per source output is given by H

$$H = -N\sum_{i=1}^{n} p(x_i)\log\{p(x_i)\} \tag{2.42}$$

The H is also known as Shannon's entropy/equation, in general the value of N is set to 1, so we obtain, using natural logarithm

$$H = -\sum_{i=1}^{n} p(x_i)\ln_2\{p(x_i)\} \tag{2.43}$$

From the foregoing development one can see that the entropy (directly related to covariance or uncertainty of a random variable X having a probability density function $p(x)$) is defined as

$$H(x) = -E_x\{\log p(x)\} \tag{2.44}$$

One can see that when the ME is expanded in Equation 2.44, we obtain Equation 2.43. Hence, it is (−ve) expected value of the logarithm of the pdf of the random variable X, and the entropy is roughly thought of as a measure of disorder or lack of information. Now, let $H(\beta) = -E_\beta[\log p(\beta)]$, the entropy prior to collecting data 'z' and $p(\beta)$ the prior pdf of β, that is, pdf of a parameter of interest. When the data 'z' are collected and used we have

$$H(\beta|z) = -E_{\beta/z}\{\log p(\beta|z)\} \tag{2.45}$$

Then the measure of the average amount of information provided/gained by the experiment with data 'z' on the parameter β is given by

$$I = H(\beta) - E_z\{H(\beta, z)\} \qquad (2.46)$$

Equation 2.46 gives the 'mean information' in 'z' about β. Here, we note that the entropy implies the dispersion/covariance of the pdf function and hence the uncertainty. Thus, the information is seen as the difference between the prior uncertainty, which is say generally large, and the expected posterior uncertainty which is now reduced due to the fact that the new data has brought in some new/more information about the parameter or variable of interest. This means that due to experimentation (collection), and the use of data z, the posterior, uncertainty (expected to reduce) is hence reduced and in the overall sense the information is gained. The information is a non-negative measure. It is zero, if $p(z,\beta) = p(z)p(\beta)$, that is, if the data are independent of the parameters, meaning thereby the data do not contain any information regarding that parameter. Thus, the gain in the information has happened as the reduction of entropy or reduction of uncertainty due to some more data, new sensor, or an additional sensor bringing in new information, by way of decreasing uncertainty, and hence the entropy. The entropy should be looked upon from this point of view rather than as 'direct' information.

2.6.2 Fisher Information

Another measure of information that is normally used in probabilistic modelling and (state/parameter) estimation is the FI, which may only be defined on continuous distributions, as the second derivative of the LL [1]

$$I(x) = \frac{d^2}{dx^2} \log\{p(x)\} \qquad (2.47)$$

Here, $I(x)$ is called the Fisher information matrix (FIM), if x is a vector, and describes the information content about the values of x contained in the distribution $p(x)$. The FI/FIM measures the surface of a bounding region containing probability. Thus, it measures compactness of a density function like entropy – entropy measures a volume and is a single number, whereas FIM is a series of numbers measuring the axes of the bounding surface, and hence, could be a matrix. The FI/FIM is useful in the estimation of continuous valued quantities, x. The FI also plays an important role in MS estimation problems, because it provides a direct means of accounting for the dependencies between the different estimates as it makes explicit the information available in the LL.

2.6.3 Information Pooling Methods

We suppose that there are M sources of the information (or the sensors etc.) with the associated vector x_m. We need to now compute a global posterior pdf $p(y|x_1, x_2, \ldots, x_m)$. Then we have the following three possibilities [7]. In most situations of MSDF, the ILP is the very appropriate way of combining information. This is because the prior information tends to be from the same origin. However, if there are dependencies between these information sources the LOP can be used. These methods can be appropriately used in decision fusion (Chapter 8).

2.6.3.1 Linear Opinion Pool

LOP is a measure of value in the form of sum of the weighted inputs, the inputs being allotted to each information source. The posteriors are combined linearly

$$p(y|x_1, x_2, \ldots, x_M) = \sum_{m=1}^{M} \omega_m p(y|x_m) \qquad (2.48)$$

Here, ω_m is a weight with $0 \leq \omega_m \leq 1$ and all the weights add to 1 and it attaches the importance to the individual information source. Interestingly the same formula can be used to model the reliability of the information from each source. The weights can also be used to weigh out the faulty or less reliable information sources/sensors.

2.6.3.2 Independent Opinion Pool

In independent opinion pool (IOP), the information conditioned on the measurement set is independent

$$p(y|x_1, x_2, \ldots, x_M) \propto \prod_{m=1}^{M} p(y|x_m) \qquad (2.49)$$

This pool is valid when the priors are obtained independently (or nearly so) based on subjective prior information.

2.6.3.3 Independent Likelihood Pool

In this ILP case each information source has common prior information. The information is obtained from the same origin and this is more correctly specified by

$$p(y|x_1, x_2, \ldots, x_M) \propto \frac{p(x_1, x_2, \ldots, x_M | y) p(y)}{p(x_1, x_2, \ldots, x_M)} \qquad (2.50)$$

We can assume that the likelihoods from each source of information ($m = 1$, 2, ..., M) are independent. This is because the parameter that is common is only the state of the system. Then we have the following expression:

$$p(y|x_1, x_2, \ldots, x_M | y) = p(x_1|y) p(x_2|y) \ldots p(x_M|y) \qquad (2.51)$$

Then we have as the ILP the following equation:

$$p(y|x_1, x_2, \ldots, x_M) \propto p(y) \prod_{m=1}^{M} p(x_m|y) \qquad (2.52)$$

2.7 Probability Concepts for Expert System and DF

The knowledge-based expert systems (KBSs) have certain important characteristics [8]: (i) they serve as useful aids to the decision makers, (ii) they use available facts, rules of thumb and (iii) they make inferences based on incomplete/uncertain information. The merits of such KBSs/software are: (i) they can provide more hypotheses, (ii) can ask for more information from the user and (iii) can add/delete knowledge and/or the expert-based rules. At the top level, the FL-based analysis, control and DF systems can be regarded as KBSs. In this section, we discuss the DF aspects of the expert systems in the probabilistic framework/setting.

In classical approaches to DF the quality of signals-data is characterised by root mean square error/standard deviation (RMSE/STD) and correlation. In normal cases, the sensor data from several sensor/sources are combined using weightings that are based on the STD and correlation in measurement errors (and mostly in errors in states/parameters, as measured by covariance matrices, as discussed in Chapters 4 and 5). In other cases the experts determine the estimates and indicate the quality of the estimates, based on their intuition and experience. Hence, for the purpose of DF it would be a good idea to express the signal information (SI) and the human estimates (HE i.e. the estimates made by the human experts) in similar forms. The point is that if a rule-based expert system is used to aid in the DF, it would be prudent to express the values and weighting of the SI and the HE in a form that can be interpreted by the rules. What this means is that the rules should include a process of [8]: (i) weighting conflicting or time-varying reports, (ii) weighting correlated data and (iii) propagating confidence limits via a hierarchy of the rules.

2.7.1 Probabilistic Rules and Evidence

A typical rule use is: If evidence E is True, Then hypothesis H is True. In case the probability information is used (as likelihood) the rule would be: If evidence E is True, Then hypothesis H is True with probability P_1. Another way could be: If evidence E is not True, Then hypothesis H is True with probability P_0. In case the probability information is used (as evidence) the rule would be: the evidence E is True with probability P_e. From the foregoing information, we can have the probability of the hypothesis being true P_h, as calculated by [8]

$$P_h = P_1 P_e + P_0(1 - P_e) = (P_1 - P_0)P_e + P_0 \tag{2.53}$$

From Equation 2.53, one can easily ascertain the following linear relation between two STDs:

$$\sigma_h = (P_1 - P_0)\sigma_e \tag{2.54}$$

We denote the two events/items of evidences as 'a' (A) and 'b' (B). The basic rule form is: If A is True, AND If B is True, then H ('h') is True. (See Sections 8.8.2.1 and 8.8.2.2 for such specific rules in decision fusion using FL for situation assessment.) In Tables 2.1 through 2.4 various possibilities of the (joint) events A and B are presented [8], which show how to compute the (joint) probabilities, that is, metrics when the events/evidences A and B occur independently (Table 2.2), maximally dependent (Table 2.3) or minimally dependent (Table 2.4).

2.7.2 Propagation of Confidence Limits

In all the standard and conventional estimation performance evaluations, the RMSE/STD (and hence, the variances) are used for arriving at confidence limits in the estimates. This is more so in cases where the estimation/filtering/fusion processes and decisions are based on probability models of

TABLE 2.1

Procedure for Computing Joint Probabilities: Logical AND and OR Operations

Statistical Dependence	AND Operation Prob (a and b)	OR Operation Prob (a or b)
Independence	$P_a P_b$	$P_a + P_b - P_a P_b$
Maximum dependence	Minimum (P_a, P_b)	Maximum (P_a, P_b)
Minimum dependence	Maximum ($P_a + P_b - 1, 0$)	Minimum ($P_a + P_b, 1$)

Source: Adapted/modified from Rauch, H.E. *AI Magazine*, 5(3), 55–60, 1984.

Note: P_a and P_b are the probabilities that events a and b are true. The consequences of logical AND and OR operations are computed with independence, maximum dependence and minimum dependence two events (or two items of the evidences).

TABLE 2.2

Procedure for Computing Joint Probabilities: Independent Events

	$P_b = 0.5$	$1 - P_b = 0.5$
	True	Not true
$P_a = 0.5$ True	$P_a P_b$ (e.g. $= 0.25$)	$P_a(1 - P_b)$
$1 - P_a = 0.5$ Not true	$(1 - P_a)P_b$	$(1 - P_a)(1 - P_b)$

Source: Adapted/modified from Rauch, H.E. *AI Magazine*, 5(3), 55–60, 1984.

Note: The sum of the four joint probabilities is equal to 1. The sum of the columns is equal to the probability at the top, and the sum of the rows is equal to the probability at the left. This procedure involves the products of the two events *a* and *b* since they are independent.

TABLE 2.3

Procedure for Computing Joint Probabilities: Maximum Dependence

	$P_b = 0.5$	$1 - P_b = 0.5$
	True	Not true
$P_a = 0.5$ True	$\max(P_a, P_b)$	$P_a - \max(P_a, P_b)$
$1 - P_a = 0.5$ Not true	$P_b - \max(P_a, P_b)$	$\max(1 - P_a, 1 - P_b)$

Source: Adapted/modified from Rauch, H.E. *AI Magazine*, 5(3), 55–60, 1984.

Note: This procedure involves the maximum and minimum operations when the events/items have maximum dependence.

TABLE 2.4

Procedure for Computing Joint Probabilities: Minimum Dependence

	$P_b = 0.5$	$1 - P_b = 0.5$
	True	Not true
$P_a = 0.5$ True	$P_a - \max(P_a, 1 - P_b)$	$\max(P_a, 1 - P_b)$
$1 - P_a = 0.5$ Not true	$\max(1 - P_a, P_b)$	$1 - P_a - \max(1 - P_a, P_b)$

Source: Adapted/modified from Rauch, H.E. *AI Magazine*, 5(3), 55–60, 1984.

Note: The procedure when events/items *a* and *b* have minimum dependence, and it is the same when *a* has maximum dependence with NOT *b*.

either the systems dynamics and/or the noise processes. Let us have the following specifications: P_a, P_b, P_e be the individual probabilities, and δ_a, δ_b and δ_e as the errors in the respective probability estimates. We also have the σ_a, σ_b and σ_c as the, respectively, STDs. Then we have the following expression for the probability with errors incorporated (for the logical AND operation) [8]

$$P_e + \delta P_e = (P_a + \delta P_a)(P_b + \delta P_b) \tag{2.55}$$

The error in the probability is given as

$$\delta P_e = (\delta P_a)P_b + (\delta P_b)P_a + (\delta P_a)(\delta P_b) \tag{2.56}$$

Then we get the following expression for the mean square error (MSE):

$$\delta_e^2 = (\delta_a^2)P_b^2 + (\delta_b^2)P_a^2 + (\delta_a^2)(\delta_b^2) \tag{2.57}$$

Similarly, the expression for the MSE when logical OR operation is used can be determined, see Table 2.5. Also, since, the operations of maximum/minimum could be non-linear, an ad hoc procedure is required to be developed. The normalised variables are defined as follows:

$$\delta V = \frac{P_b - P_a}{(\sigma_a^2 + \sigma_b^2)^{1/2}} ; \quad \delta U = \frac{P_b + P_a - 1}{(\sigma_a^2 + \sigma_b^2)^{1/2}} \tag{2.58}$$

respectively, for minimum and maximum dependence. Here, it is assumed that the probabilities in Equation 2.58 are exactly on the boundary of maximum, minimum. If these normalised variables are more than $L\sigma$ from the boundary, these non-linearities of the boundaries are neglected. The parameter Q is ad hoc to adjust for the non-linearity.

TABLE 2.5

Computation of Confidence Limits for Various Dependences

Statistical Dependence	AND Operation Prob (*a* and *b*)	OR Operation Prob (*a* or *b*)
Independence	$P_a^2\sigma_b^2 + P_b^2\sigma_a^2 + \sigma_a^2\sigma_b^2$	$(1 - P_a)^2\sigma_b^2 + (1 - P_b)^2\sigma_a^2 + \sigma_a^2\sigma_b^2$
Maximum or minimum dependence	$0.5(\sigma_a^2 + \sigma_b^2) + Q$	$0.5(\sigma_a^2 + \sigma_b^2) - Q$
Maximum dependence: $Q = 0.5(\sigma_a^2 - \sigma_b^2)$ $\text{minimum}\left[\frac{\delta V}{L}, 1\right]$ $\delta V = \dfrac{P_b - P_a}{(\sigma_a^2 + \sigma_b^2)^{1/2}}$	Minimum dependence: $Q = 0.5(\sigma_a^2 + \sigma_b^2)\text{minimum}\left[\frac{\delta U}{L}, 1\right];$ if $\delta U \geq 0$ $Q = 0.5(\sigma_a^2 + \sigma_b^2)\text{maximum}\left[\frac{\delta U}{L}, -1\right];$ if $\delta U < 0$ $\delta U = \dfrac{P_b + P_a - 1}{(\sigma_a^2 + \sigma_b^2)^{1/2}}$	

Source: Adapted/modified from Rauch, H.E. *AI Magazine*, 5(3), 55–60, 1984.

Note: Here, $P_b > P_a$ with corresponding (standard deviations) variances: σ_a^2; σ_b^2. L is an arbitrary constant = 2.

2.7.3 Combining-Fusion of Multiple Reports

When we receive the multiple reports/measurement data/information (RMI) related to the same object, we need to combine these reports knowing well that there is some uncertainty associated with each set of this multiple information. In this case, we have the following probability:

$$P = \frac{\sum_i w_i P_i}{w} \qquad (2.59)$$

and

$$w = \sum_i w_i \qquad (2.60)$$

Here, the ws are the weights that are functions of the STDs/uncertainties in the received RMI and are given as

$$w_i = \frac{1}{\sigma_i^2} \qquad (2.61)$$

Often all these RMIs or some of them might be correlated with covariance of the errors as R_{ij}. Equivalently, we have the information matrix Y as the inverse of R. Then we have the following expressions for the weights to be used in Equation 2.59:

$$w = \sum_i w_i = \frac{1}{\sigma^2} \qquad (2.62)$$

$$w_i = \sum_j Y_{ij} \qquad (2.63)$$

When the errors in RMIs are not correlated, the R matrix is diagonal. The recursive formulae to combine the RMIs are as follows:

$$P = \frac{(w_1 P_1 + w_2 P_2)}{w} \qquad (2.64)$$

$$w = w_1 + w_2 = \frac{1}{\sigma^2} \qquad (2.65)$$

$$w_i = \frac{(1 - (\rho \sigma_i / \sigma_j))(1 - \rho^2)^{-1}}{\sigma_i^2} \qquad (2.66)$$

In expression (2.66), the ρ brings in the correlation when the new RMI is incorporated into the previously combined RMIs. Equation 2.64 itself very much looks like a weighted DF rule.

Once again, we note that the beginning of the MSDF (combination) rules starts from here. This is evident in the KF, since in itself is a (measurement level) data fuser as very clearly explained in Reference 3 and in Chapter 4. In addition the state vector fusion concept (Chapter 4) is also developed on similar lines [3]. Further, it must be noted here that the basic smoother (equations) are developed by appropriately combining the forward pass (run of the) KF estimate and the backward pass (run of the) KF estimate [9,10]. We also emphasise here that the derivation of the basic KF measurement part itself starts with the following expression:

$$\hat{x} = K_1 \tilde{x} + K_2 z \tag{2.67}$$

Equation 2.67 suggests very profoundly that the new estimate is postulated as a weighted combination (fusion) of the previous (state-) estimate and the newly made/arrived observation/measurement [9,10]. Then, the optimal values of the gains (Ks) are determined to obtain the unbiased estimate of the state and minimise the MSEs in the measurements (i.e. the new measurement – the predicted measurement based on the previous estimate). This finally leads to the celebrated KF. So, we see that the DF concept is at the heart of the KF, BT and many estimation techniques.

2.8 Probabilistic Methods for DF: Theoretical Examples

In this section, we consider several theoretical aspects of probabilistic approaches with direct relevance to estimation theory and related SDF [11]. Based on the fundamental approaches presented in this section, one can develop yet more sophisticated estimation and DF algorithms and DF tracking-systems (some such techniques are already available in the literature). The direct data available from the sensors are called the microscopic data (micro data), and the derived data/information (e.g. expectations of the signals) are the macroscopic data (macro data).

2.8.1 Maximum Entropy Method

The MEM (maximum entropy method) is used to specify a probability to an unknown data quantity with the macro data. Let us have X as the data/quantity (DQ) under study. Assume, we have M sensors which give M values: μ_m, $m = 1, 2, \ldots, M$ which represent the mean values of M known functions: $\phi_m(X)$, $m = 1, 2, \ldots, M$. These values are related to the unknown DQ. Then we have the following expectation [11]:

$$E\{\phi_m(X)\} = \int \phi_m(x)p(x)dx = \mu_m; \quad m = 1, 2, \ldots, M \tag{2.68}$$

We now want to represent by a probability our partial knowledge of X. We have a class of possible solutions and we can use MEM to select one from this class. So we have the following entropy situation [11]:

$$H(p) = -\int p(x)\ln(p(x))dx \tag{2.69}$$

The idea is to maximise H subject to the following constraint:

$$\int \phi_m(x)p(x)dx = \mu_m; \quad m = 1, 2, \ldots, M \tag{2.70}$$

The solution is given by the following expression [11]:

$$p(x) = \frac{1}{Z(\Theta)}\exp\left[-\sum_{m=1}^{M}\theta_m\phi_m(x)\right] = \frac{1}{Z(\Theta)}\exp[-\Theta^t\Phi(x)];$$

$$Z(\Theta) = \int \exp\left[-\sum_{m=1}^{M}\theta_m\phi_m(x)\right]dx \tag{2.71}$$

We also have the following gradient to determine the parameters as

$$-\frac{\partial \ln Z(\Theta)}{\partial \theta_m} = \mu_m; \quad m = 1, 2, \ldots, M \tag{2.72}$$

2.8.2 Maximum Likelihood Method

For the maximum likelihood method (MLM), let us have $x = [x_1, x_2, \ldots, x_N]$ values of X and the parametric form of the probability given by $p(x,\Theta)$. Then, the idea is to determine the parameters theta, Θ. The solution is obtained by the method of moments by writing a set of equations relating to the theoretical and empirical moments, and solving them to obtain the following solution [11]:

$$\int G_m(\Theta) = E\{X^m\} = \int x^m p(x;\Theta)dx = \frac{1}{N}\sum_{j=1}^{N}x_j^m; \quad m = 1, 2, \ldots, M \tag{2.73}$$

In the MLM, we consider the probability as the function of the parameters and obtain the following expression for the parameter estimates:

$$\hat{\Theta} = \arg \max_{\Theta} \{L(\Theta|x)\} \quad \text{and} \quad L(\Theta|x) = p(x;\Theta) = \prod_{j=1}^{N} p(x_j;\Theta) \quad (2.74)$$

In Equation 2.74, L is the likelihood function. In the case of the generalised family of exponentials we have the following two important expressions [11]:

$$p(x;\Theta) = \frac{1}{Z(\Theta)} \exp\left[-\sum_{m=1}^{M} \theta_m \phi_m(x)\right] = \frac{1}{Z(\Theta)} \exp[-\Theta^T \phi(x)] \quad (2.75)$$

$$L(\Theta) = \prod_{j=1}^{N} p(x_j;\Theta) = \frac{1}{Z^N(\Theta)} \exp\left[-\sum_{j=1}^{N}\sum_{m=1}^{M} \theta_m \phi_m(x_j)\right] \quad (2.76)$$

Then the ML estimates are obtained from the solution of the following equation:

$$\frac{\partial \ln Z(\Theta)}{\partial \theta_m} = \frac{1}{N}\sum_{j=1}^{N} \phi_m(x_j); \quad m = 1,2,\ldots,M \quad (2.77)$$

Z is as per Equation 2.71, and we see that there is some interesting relationship between MEM and ML, if we compare Equations 2.72 and 2.77.

2.8.3 ML and Incomplete Data

We assume that the sensor gives M data/signal values y related to the samples of x via the algebraic equation $y = Ax$, and $M < N$. Then, we want to determine the parameter vector *theta*. We now write the joint pdf as follows:

$$p(x;\Theta) = p(x|y;\Theta)p(y;\Theta); \quad \forall Ax = y \quad (2.78)$$

We take expectation of Equation 2.78 to obtain the following expression:

$$\ln p(y;\Theta) = E_{x|y;\Theta'}\{\ln p(x;\Theta)\} - E_{x|y;\Theta'}\{\ln p(x|y;\Theta)\} \quad (2.79)$$

or in terms of likelihood we have the following expression:

$$L(\Theta) = Q(\Theta;\Theta') - V(\Theta;\Theta') \quad (2.80)$$

Thus, from Equation 2.80, we obtain the following expression for the differentials of likelihoods [11]:

$$L(\Theta) - L(\Theta') = [Q(\Theta;\Theta') - Q(\Theta';\Theta')] + [V(\Theta;\Theta') - V(\Theta';\Theta')] \quad (2.81)$$

We have the following inequality:

$$[V(\Theta;\Theta') \leq V(\Theta';\Theta')] \quad (2.82)$$

We then have the EM (expectation maximisation, EMM) procedure as follows [11]:

$$\left[\begin{array}{l} E(\text{expectation}) : Q(\Theta;\hat{\Theta}(k)) = E_{x|y;\Theta(k)}\{\ln p(x;\Theta)\} \\ M(\text{maximisation}) : \hat{\Theta}(k+1) = \arg\ \max_{\Theta}\ \{Q(\Theta : \hat{\Theta}(k))\} \end{array}\right] \quad (2.83)$$

The algorithm of Equation 2.84 would converge to a local ML. In the case of the generalised family of exponentials we have the following two important expressions for the EMM [11]:

$$\left[\begin{array}{l} E : Q(\Theta;\Theta') = E_{x|y;\Theta'}\{\ln p(x;\Theta)\} = -N\ln Z(\Theta) - \sum_{j=1}^{N}\Theta^T E_{x|y;\Theta'}\{\phi(x_j)\} \\ M : -\dfrac{\partial \ln Z(\Theta)}{\partial \theta_m} = \dfrac{1}{N}\sum_{j=1}^{N}E_{x_j|y;\Theta(k)}\{\phi_m(x_j)\};\quad m = 1,2,\ldots,M \end{array}\right] \quad (2.84)$$

We again see that there are some interesting relationships between MEM and MLM and EMM if we compare Equations 2.72, 2.77 and 2.84.

Next, if we want to estimate the parameters and also x, then we have the following expressions [11]:

$$\left[\begin{array}{l} E : Q(\Theta;\hat{\Theta}(k)) = E\{\ln p(x;\Theta)|y;\hat{\Theta}(k)\} \\ \hat{x}(k) = E\{x|y : \hat{\Theta}(k)\} \\ M : \hat{\Theta}(k+1) = \arg\ \max_{\Theta}\ \{Q(\Theta;\hat{\Theta}(k))\} \end{array}\right] \quad (2.85)$$

2.8.4 Bayesian Approach

Now we have the situation as in the previous examples but with the measurements corrupted with noise $y = Ax + v$. We use the Bayesian approach with the probability distribution of noise to define $p(y|x;\Theta_1) = p_v(y - Ax;\Theta_1)$ and combine this with the prior pdf $p(x;\Theta_2)$ via the Bayesian formula and obtain the following posterior pdf:

$$p(x|y;\Theta_1,\Theta_2) = \frac{p(y|x;\Theta_1)p(x;\Theta_2)}{\int p(y|x;\Theta_1)p(x;\Theta_2)dx} \tag{2.86}$$

The posterior of Equation 2.86 contains all the information on x, and we can use this expression to make an inference on x. Hence, we define the following estimators [11]:
MAP:

$$\hat{x} = \arg\max_x\{p_{x|y}(x|y;\Theta_1,\Theta_2)\} \tag{2.87}$$

Posterior mean (PM):

$$\hat{x} = E_{x|y}\{x\} = \int x p_{x|y}(x|y;\Theta_1,\Theta_2)dx \tag{2.88}$$

Marginal posterior modes (MPM):

$$\hat{x} = \arg\max_{x_i}\{p(x_i|y;\Theta_1,\Theta_2)\} \tag{2.89}$$

In Equation 2.89, we have the following expansion:

$$p(x_i|y;\Theta_1,\Theta_2) = \int p_{x|y}(x|y;\Theta)dx_1\ldots dx_{i-1}\ldots dx_{i+1}\ldots dx_n \tag{2.90}$$

In practical situations, we have two difficult aspects to be dealt with: (i) to assign the probabilities $p(y|x;\Theta_1)$ *and* $p(x;\Theta_2)$, and (ii) to determine the parameter vector $\Theta(theta) = (\Theta_1,\Theta_2)$. The assignment of the prior pdf can be done using MEM or an engineering judgement. For the estimation of the parameters *theta* we have the following schemes based on the joint posterior pdf [11]:

$$p(x,\Theta|y) \propto p(y|x;\Theta)p(x|\Theta)p(\Theta)$$
$$\propto p(x,y|\Theta)p(\Theta)$$
$$\propto p(x|y,\Theta)p(\Theta) \tag{2.91}$$

a. *The joint MAP*

$$(\hat{\Theta}, \hat{x}) = \arg\max_{(\Theta,x)}\{p(x,\Theta|y)\} \tag{2.92}$$

Given the measurements y we obtain the estimates of x and the parameters *theta* by solving Equation 2.92.

b. *The generalised ML*

$$\hat{x}(k) = \arg\max_{x}\{p(x|y;\Theta(k-1)\} \tag{2.93}$$

$$\hat{\Theta}(k) = \arg\max_{\Theta}\{p(\hat{x}(k)|y,\Theta)p(\Theta)\}$$

By solving Equations 2.93, we obtain the estimates of x and the parameters *theta* given the noisy measurements y and the initial condition of the parameters.

c. *The marginalised MLM (MML)*

$$\hat{\Theta} = \arg\max_{\Theta}\left\{\int p(y|x)p(x;\Theta)dx\right\} \tag{2.94}$$

$$\hat{x} = \arg\max_{x}\{p(x|y;\hat{\Theta})\}$$

Using the first part of Equation 2.94, we obtain estimate of the parameters *theta*, given the measurements y, then using these '*theta*' estimates and the measurements again in the second part of Equation 2.94, we obtain the estimate of x.

d. *The MML-EM method*

Since, the analytical expression is generally not possible, one can consider the data set (y, x) as the complete set, and y as the incomplete data set. Then we can use the EMM to obtain the following estimator [11]:

$$\begin{bmatrix} E : Q(\Theta;\hat{\Theta}(k)) = E_{x,y;\Theta(k)}\{\ln p(x,y;\Theta)\} \\ M : \hat{\Theta}(k+1) = \arg\max_{\Theta}\{Q(\Theta;\hat{\Theta}(k)\} \\ MAP : \hat{x} = \arg\max_{x}\{p(x|y : \hat{\Theta})\} \end{bmatrix} \tag{2.95}$$

So, from Equation 2.95, first we obtain the EM solution to the parameter estimation *theta* using the measurements y and the initial values of the *theta* (i.e. by the MLM-EM procedure), and then we obtain the estimates of x using the estimates of *theta* parameters and the measurements again by the MAP process.

2.8.5 DF Aspects/Examples

We consider here a few DF aspects and see how the approaches discussed in Sections 2.8.1 to 2.8.4 can be utilised.

2.8.5.1 Sensors with No Noise

We presume that sensor S_1 gives the data up to N samples, and the sensor S_2 gives the data samples up to M, otherwise there are gaps in the data and these are related [11] by $y = Ax + b$:

$$\text{Sensor } S_1 : x_a = (x_1, x_2, \ldots, x_N) \qquad x_b = ??$$
$$\text{Sensor } S_2 : y_a = ?? \qquad\qquad\qquad y_b = (y_1, y_2, \ldots, y_M) \qquad (2.96)$$

Thus, given the situation as in Equation 2.96, the point is how to fuse/use these data and obtain the estimation of x_b and the parameters *theta*. In fact, we have to predict the data samples

$$x_b = (x_{N+1}, x_{N+2}, \ldots, x_{N+M}) \qquad (2.97)$$

The following schemes can be used for the solutions [11]:

a. *Approach 1: MLM → (and then) MAP*   We use x_a to estimate the *theta*, the parameters of $p(x, \Theta)$ by the MLM. We use this information to then estimate x_b from y_b

$$\hat{\Theta} = \arg\max_{;\Theta} \{L_a(\Theta) = \ln p(x_a; \Theta)\} \qquad (2.98)$$

$$\hat{x}_b = \arg\max_{x_b} \{p(x_b | y_b; \hat{\Theta})\} \qquad (2.99)$$

Equation 2.98 is the MLM, and Equation 2.99 is the MAP.

b. *Approach 2: ML → JMAP, GML or ML-EM*   In this approach, we use both the available data samples to estimate x_b as follows:

$$\hat{\Theta} = \arg\max_{;\Theta} \{L_a(p(x_a; \Theta))\} \qquad (2.100)$$

$$(\hat{x}_b, \hat{\Theta}) = \arg\max_{(x_b, \Theta)} \{p(x_b, \Theta | x_a, y_b)\} \qquad (2.101)$$

In Equation 2.100, we use MLM method and then from Equation 2.101, we have the approach of either JMAP, GML or ML-EM. We see that due to additional data samples the *theta* estimate is refined by Equation 2.101.

2.8.5.2 *Fusion of Homogeneous Sensor Data*

We presume that the data samples are available via linear models and are (of) two types on the same unknown values x

$$y = H_1x + b_1 \quad \text{and} \quad z = H_2x + b_2 \tag{2.102}$$

One example is that of the x-ray tomography, where x is the mass density of the object. Then, y and z are the high/low-resolution projections. We then use the Bayesian method to handle this aspect

$$p(x|y,z) = \frac{p(y,z|x)p(x)}{p(y,z)} \tag{2.103}$$

The main problem in Equation 2.103 is to specify $p(y, z|x)$. Things become easy if we presume that the errors in these two data sets are independent. We have the following assumptions [11]:

$$
\begin{aligned}
p(y|x;\sigma_1^2) &\propto \exp\left[-\frac{1}{2\sigma_1^2}(y - H_1x)^2\right] \\
p(z|x;\sigma_2^2) &\propto \exp\left[-\frac{1}{2\sigma_2^2}(z - H_2x)^2\right] \\
p(x;m,P) &\propto \exp\left[-\frac{1}{2}(x - m)^T P^{-1}(x - m)\right]
\end{aligned}
\tag{2.104}
$$

In Equation 2.104, the parameters are assumed to be known: $\sigma_1^2, \sigma_2^2, m, P$. P is the covariance matrix of the estimation errors. Then MAP estimate can be obtained as follows:

$$\hat{x} = \arg\max_x \{p(x|y,z)\} = \arg\min_x \{J(x) = J_1(x) + J_2(x) + J_3(x)\} \tag{2.105}$$

The cost functions in Equation 2.105 are defined as follows [11]:

$$J_1(x) = \frac{1}{2\sigma_1^2}(y - H_1x)^2 \tag{2.106}$$

$$J_2(x) = \frac{1}{2\sigma_2^2}(z - H_2 x)^2 \qquad (2.107)$$

$$J_3(x) = \frac{1}{2}(x - m)^T P^{-1}(x - m) \qquad (2.108)$$

2.8.6 Some Realistic DF Problems

We now consider DF problems that are more realistic. Here, the data are of two different kinds: an x-ray data related to the mass density x, and the ultrasound probing system data related to the acoustic reflectivity r. The data equations are as follows:

$$y = H_1 x + b_1 \quad \text{and} \quad z = H_2 r + b_2 \qquad (2.109)$$

We again assume that the two sets are independent data sets. We again use the Bayesian formulae as follows [11]:

$$p(x,r|y,z) = \frac{p(y,z|x,r)p(x,r)}{p(y,z)} = \frac{p(y|x)p(z,r)p(x,r)}{p(y,z)} \qquad (2.110)$$

We also have the following relation for $p(y,z)$:

$$p(y,z) = \iint p(y|x)p(z|r)p(x,r)drdx \qquad (2.111)$$

Again the main problem in Equation 2.111 is to specify $p(y|z)$, $p(z|r)$ and $p(x,r)$. If there is some mathematical relation between x and r, then the situation is similar to the previous case. Let us have such a relation postulated as $r_j = g(x_{i+1} - x_i)$. If the g is a linear function then, we have the following relationships:

$$y = H_1 x + b_1; \quad z = H_2 r + b_2; \quad r = Gx \rightarrow y = H_1 x + b_1; \quad z = GH_2 x + b_2$$
$$\qquad (2.112)$$

We assume that the body object is constituted of the related quantities

$$b = (r,x) = (q,a,x) \qquad (2.113)$$

The variable q is a binary vector that represents the positions of the discontinuities/edges in the body and 'a' contains the reflectivity values [11]

$$q_j = 0 \rightarrow r_j = 0 \quad \text{and} \quad r_j = \{g(x_{j+1} - x_j) \quad \text{if } |x_{j+1} - x_j| > \alpha$$
$$q_j = 1 \rightarrow r_j = a_j \qquad\qquad\quad \{0 \qquad\qquad\qquad \text{otherwise}$$
(2.114)

With the above model, we can write the following expressions:

$$p(b,r) = p(x,a,q) = p(x|a,q)p(a|q)p(q)$$
(2.115)

Then using the Bayesian formula we have the following relationships [11]:

$$p(x,a,q|y,z) \propto p(y,z|x,a,q)p(x,a,q)$$
$$= p(y,z|x,a,q)p(x|a,q)p(a|q)p(q)$$
(2.116)

We can consider several assumptions as follows [11]:

a. That y and z are conditionally independent

$$p(y,z|x,a,q) = p(y|x)p(z|a)$$
(2.117)

b. Gaussian pdf for the measurement noise processes (see Equation 2.109)

$$p(y|x;\sigma_1^2) \propto \exp\left[-\frac{1}{2\sigma_1^2}(y - H_1x)^2\right]$$
$$p(z|a;\sigma_2^2) \propto \exp\left[-\frac{1}{2\sigma_2^2}(z - H_2a)^2\right]$$
(2.118)

c. Bernoulli law for q

$$p(q) \propto \sum_{i=1}^{n} q_i^\lambda (1 - q_i)^{1-\lambda}$$
(2.119)

d. Gaussian pdf for $a|q$

$$p(a|q) \propto \exp\left[-\frac{1}{2\sigma_a^2}a^T Qa\right], \quad Q = \text{diag}[q_1, q_2, \ldots, q_n]$$
(2.120)

e. Markov model for the data x

$$p(x|a,q) \propto \exp[-U(x|a,q)]$$
(2.121)

With the above expressions defined/specified, we have the following relationships:

$$p(x,a,q|y,z) \propto p(y|x)p(z|a)p(x|a,q)p(a|q)p(q) \qquad (2.122)$$

Then, we have the following eight estimation schemes [11]:

i. Simultaneous estimation of all the unknowns using JMAP

$$(\hat{x},\hat{a},\hat{q}) = \arg\max_{(x,a,q)} \{p(x,a,q|y,z)\} \qquad (2.123)$$

Given the data y and z the JMAP would give the estimates of x, a and q.

ii. First estimate the positions of the discontinuities q and then obtain the estimates of x and a

$$\hat{q} = \arg\max_{q}\{p(q|y,z)\}$$
$$(\hat{x},\hat{a}) = \arg\max_{(x,a)}\{p(x,a|y,z,\hat{q})\} \qquad (2.124)$$

Given the data y and z, we estimate q, then use these estimates and the same data (y, z) again to estimate x and a.

iii. First estimate the positions of the discontinuities q (with only data z) and then obtain the estimates of x, and a

$$\hat{q} = \arg\max_{q}\{p(q|z)\}$$
$$(\hat{x},\hat{a}) = \arg\max_{(x,a)}\{p(x,a|y,z,\hat{q})\} \qquad (2.125)$$

Given the data z, we estimate q, then use these estimates and the same data (y, z) to obtain the estimates of x and a.

iv. First estimate a and q with only data z, and then obtain the estimates of x

$$(\hat{q},\hat{a}) = \arg\max_{q,a}\{p(q,a|z)\}$$
$$\hat{x} = \arg\max_{x}\{p(x|y,\hat{a},\hat{q})\} \qquad (2.126)$$

Given the data z, we estimate a and q, then use these estimates and the same data y to obtain the estimate of x.

v. First, we estimate q using z, then we use this estimate of q and again z to estimate a. Then, we finally estimate x using the estimates of a, and q and the measurements y

$$\hat{q} = \arg\max_{q} \{p(q|z)\}$$
$$\hat{a} = \arg\max_{a} \{p(a|z,\hat{q})\}$$
$$\hat{x} = \arg\max_{x} \{p(x|y,\hat{a},\hat{q})\}$$

(2.127)

vi. First, we estimate q using z and then using this estimate and the data y we estimate x

$$\hat{q} = \arg\max_{q} \{p(q|z)\}$$
$$\hat{x} = \arg\max_{x} \{p(x|y,\hat{q})\}$$

(2.128)

vii. Estimate r using z and then estimate q and x using the estimate of r and the measurements y

$$p(r|z) \propto p(z|r)p(r)$$

(2.129)

$$p(x,q|\hat{r},y) \propto p(y|x)p(x,q|\hat{r})$$

(2.130)

For Equation 2.129, we obtain the following estimator for r:

$$\hat{r} = \arg\max_{r} \{p(r|z)\} = \arg\min_{r} \{J_1(r|z)\}$$

(2.131)

where, the cost function J is given as follows:

$$J_1(r|z) = (z - H_2r)^2 + \lambda \sum_{j} (r_{j+1} - r_j)^2$$

(2.132)

For Equation 2.130, we obtain the following estimator for x and q:

$$(\hat{x},\hat{q}) = \arg\max_{(x,q)} \{p(x,q|y,\hat{r})\} = \arg\min_{(x,q)} \{J_2(x,q|y,\hat{r})\}$$

(2.133)

Again the cost function is given as follows:

$$J_2(x,q|y,\hat{r}) = (y - H_1x)^2 + \lambda \sum_{j} (1 - q_j)(x_{j+1} - x_j)^2 + \alpha_1 \sum_{j} q_j(1 - \hat{r}_j) + \alpha_2 \sum_{j} q_j\hat{r}_j$$

(2.134)

viii. We use data z and detect the locations of some of the boundaries. We also use x-ray data to make an intensity image preserving the positions of these discontinuities

$$z \rightarrow \text{ estimate of } r \rightarrow \text{ estimate of } q \rightarrow (\text{with } y) \rightarrow \text{ estimate of } x \quad (2.135)$$

Then the estimator for r is given as

$$\hat{r} = \arg \max_{r} \{p(r|z)\} = \arg \min_{r} \{J_1(r|z)\} \quad (2.136)$$

with the cost function given as

$$J_1(r|z) = \|z - H_2 r\|^2 + \lambda_1 \|r\|^p, \quad 1 < p < 2 \quad (2.137)$$

The other approach is as follows:

$$\hat{x} = \arg \max_{x} \{p(x|y,\hat{q}) = \arg \min_{x} \{J_2(x|y;\hat{q})\} \quad (2.138)$$

$$J_2(x|y,\hat{q}) = \|y - H_1 x\|^2 + \lambda_2 \sum_{j} (1 - q_j)|x_{j+1} - x_j|^p, \quad 1 < p < 2 \quad (2.139)$$

In summary, the following inferences from the foregoing development (of Section 2.8) of the probabilistic approach to SDF [11] can be drawn: (i) the MEM is used when pdf is specified for some expected values; (ii) The MLM is used with a parametric form of pdf, $p(x, \Theta)$ and we directly use the measurement data, and are interested in estimating the parameter vector theta; (iii) MLM-EMM is used when we have incomplete measurements; (iv) Bayesian approach is highly preferable when we have noisy measurement data, and it is very useful for practical DF problems and (v) the compound MM is handy to represent the data signals and images in a Bayesian method of DF.

2.9 Bayesian Formula and Sensor/DF: Illustrative Example

We now, consider an example [1] of Bayesian formula application to estimating a discrete parameter on the basis of one measurement and some prior information. The situation is modelled by a single state x which can take on one of three values:

i. $x_1 \to x$ is a target of type 1

ii. $x_2 \to x$ is a target of type 2

iii. $x_3 \to$ target is not visible

A single sensor observes x and returns three possible values of measurements:

iv. $z_1 \to$ measurement of a target of type 1

v. $z_2 \to$ measurement of a target of type 2

vi. $z_3 \to$ no target is observed

The sensor model is described by the probability matrix $P_1(z|x)$ in Table 2.6. This model matrix is a function of both x and z, for a fixed value of the true state, and it describes the probability of a particular measurement being made (e.g. the rows of the matrix), and when a specific measurement is made, it describes a probability distribution over the values of the true state (e.g. the columns of the matrix); then it is the likelihood function $L(x)$. The posterior distribution of the true state x after making a measurement $z = z_i$ is given as follows:

$$P(x|z_i) = \alpha P_1(z_i|x)P(x) \qquad (2.140)$$

Since, the total probability sums up to 1, the constant has to be adjusted accordingly, and is called the normalising constant. We can assume that we do not have any prior information about the possible likelihood of target of types 1 and 2, and hence, we can set the prior probability vector to $P(x) = (0.333, 0.333, 0.333)$. Now, we observe $z = z_1$, then the posterior distribution will be given by $P(x|z_1) = (0.45, 0.45, 0.1)$; the first column of the matrix, Table 2.6, the likelihood function given z_1 has been observed. Next, we use this posterior distribution as the prior for a second observation $P(x) = (0.45, 0.45, 0.1)$. Again, the measurement $z = z_1$, is made, and the new posterior distribution is obtained as follows [1]:

$$\begin{aligned}
P(x|z_1) &= \alpha P_1(z_1|x)P(x) \\
&= \alpha(0.45, 0.45, 0.1) \cdot (0.45, 0.45, 0.1) \\
&= \alpha(0.2025, \ 0.2025, \ 0.01) \\
&= 2.4096 \ (0.2025, \ 0.2025, \ 0.01) \\
&= (0.488, 0.488, 0.024) \qquad (2.141)
\end{aligned}$$

TABLE 2.6

Probability-Likelihood Matrix: $P_1(z|x)$

	z_1	z_2	z_3
x_1	0.45	0.45	0.1
x_2	0.45	0.45	0.1
x_3	0.1	0.1	0.8

In the second row of Equation 2.141, '·' is the element-wise product operator and 'α' is computed as the inverse of the sum of the individual probability numbers, of the third row, within the parentheses. This is a normalisation factor in order to see that the sum of the individual probabilities (the last row in Equation 2.141) is one. There is an increase in the probability of both targets of type 1 and type 2, but at the expense of the no-target hypothesis. This sensor is good at detecting targets, and yet not good at distinguishing between targets of different types. Now, we assume that a second sensor is obtained that also makes the same three measurements as the first sensor, however, with the likelihood matrix $P_2(z_2|x)$ [1] as given in Table 2.7. As we have seen earlier, the first sensor was good at detecting targets but not in distinguishing different type of targets. Now, this second sensor has poor overall detection probabilities but has good target discrimination capabilities ($p = 0.45, 0.1$). With a uniform prior distribution, if we observe $z = z_1$ with this second sensor, the posterior distribution on possible true states is given by $P(x|z_1) = (0.45, 0.1, 0.45)$ (the first column of the first 3×3 matrix of Table 2.7). Thus, it makes sense to combine the information from these two sensors so as to get a system with good detection as well as good discrimination. From the product of the two likelihood functions, we obtain the combined likelihood function as $P_{12}(z_1, z_2|x) = P_1(z_1|x)P_2(z_2|x)$. When we observe $z_1 = z_1$ using the first sensor and $z_2 = z_1$ with the second sensor (with a uniform prior probability distribution), the posterior likelihood of x is expressed as [1]

$$
\begin{aligned}
P(x|z_1, z_1) &= \alpha P_{12}(z_1, z_1|x) \\
&= \alpha P_1(z_1|x)P_2(z_1|x) \\
&= \alpha \times (0.45, 0.45, 0.1) \cdot (0.45, 0.1, 0.45) \\
&= (0.6924, 0.1538, 0.1538)
\end{aligned}
\tag{2.142}
$$

We see that (as compared to taking two observations of z_1 with sensor 1, in which the outcome posterior was 0.488, 0.488, 0.0240, Equation 2.141), it is now seen that sensor 2 adds good target discrimination (0.6924, 0.1538) at the cost of a slight loss of detection performance for the same number of measurements. If we repeat the calculations for each z_1, z_2 measurement pair, we obtain the

TABLE 2.7

Probability-Likelihood Matrices (Combined Likelihood Matrices: CLM)

	$P_2(z_2|x)$			CLM for $z_1 = z_1$			CLM for $z_1 = z_2$			CLM for $z_1 = z_3$		
	z_1	z_2	z_3	z_1	z_2	z_3	z_1	z_2	z_3	z_1	z_2	z_3
x_1	0.45	0.10	0.45	0.6924	0.1538	0.4880	0.6924	0.1538	0.4880	0.1084	0.0241	0.2647
x_2	0.10	0.45	0.45	0.1538	0.6924	0.4880	0.1538	0.6924	0.4880	0.0241	0.1084	0.2647
x_3	0.45	0.45	0.10	0.1538	0.1538	0.0240	0.1538	0.1538	0.0241	0.8675	0.8675	0.4706

Source: Modified from Durrant-Whyte, H. Multi-sensor data fusion – Lecture Notes. Australian Centre for Field Robotics, University of Sydney, NSW 2006, Australia, January 2001.

CLM – combined likelihood matrix as in Table 2.7; the combined sensor providing substantial improvements in overall system performance. If we observe target 1 with the first sensor (the array block $z_1 = z_1$) and again observe target 1 with the second sensor (the first column of the same block), then the posterior distribution, in the three hypotheses, is $P(x|z_1,z_2) = (0.6924, 0.1538, 0.1538)$, definitely signifying that target 1 is the most probable one detected. However, if we observe a target of type 2 with the second sensor after having observed a target of type 1 with the first sensor, we obtain the posterior as $(0.1538, 0.6924, 0.1538)$, thereby signifying that type 2 target has higher probability. This happens because although sensor 1 observed a target of type 1, the likelihood function for sensor 1 implies that it is poor at distinguishing between types of targets, and hence information from sensor 2 is used. In case now we observe no target with sensor 2, after having detected type 1 target with the first sensor, the posterior with both measurements is obtained as $(0.488, 0.488, 0.024)$; we still believe that there is a target (we know that sensor 1 is better at target detection than sensor 2), but we do not have an idea whether it is target 1 or 2, as sensor 2 has not been able to make a valid detection. The analysis for sensor 1 detecting target 2 is identical to that for detection of target 1. If sensor 1 gets no detection, but sensor 2 detects type 1 target, then the posterior likelihood is obtained as $(0.108, 0.024, 0.868)$. Here, we still believe, there is no target because we know sensor 1 is better at providing this information (and sensor 2 confirms this even though it has detected target type 1). As such, the joint likelihood matrix is never constructed, because with $n = 3$ sensors, and $m = 3$ possible measurements and $k = 3$ outcomes, the joint likelihood matrix has $k \times mn = 27$ entries. Practically, the likelihood matrix is constructed for each sensor and combined when a measurement is obtained. Then, the storage reduces to n arrays of dimension $k \times m$, at the cost of a k dimensional vector multiply of the instantiated likelihood functions.

EXERCISES

2.1 What is the condition for a discrete random sequence to be white?

2.2 Which one is the special case of what between MAP and MLM?

2.3 If the two events E and F are independent, how is this formula modified: $P(E \cup F) = P(E) + P(F) - P(E \cap F)$?

2.4 Why is the weight of z proportional to the variance of x and why does the weight of x belong to the variance of z in the DF rule given by Equation 2.19? Explain the significance of this aspect.

2.5 Establish that the fusion rule in the information domain is a simple one, if we consider the information factor (we can call it inforiance) as the inverse of the variance considering Equation 2.20

$$\sigma^2 = \frac{\sigma_z^2 \sigma_x^2}{\sigma_x^2 + \sigma_z^2} = \left(\frac{1}{\sigma_z^2} + \frac{1}{\sigma_x^2} \right)^{-1}.$$

2.6 What is the probability formula of the intersection of two independent events?

2.7 For two independent events, E and F, why $P(E|F) = P(E)$ and $P(F|E) = P(F)$?

2.8 What is the probability distribution of the non-informative prior (in Bayesian formula)?

2.9 If we have the likelihood function belonging to the exponential pdf, then which is a good choice for the prior for the Bayesian rule?

2.10 Give the formula for the minimum mean square error (MMSE-estimator), using Equation 2.118

$$p(y|x;\sigma_1^2) \propto \exp\left[-\frac{1}{2\sigma_1^2}(y - H_1x)^2\right].$$

References

1. Durrant-Whyte, H. Multi sensor data fusion – Lecture Notes. Australian Centre for Field Robotics, University of Sydney NSW 2006, Australia, January 2001.
2. Lindsay, I.S. A tutorial on principal components analysis. 26 February 2002. www.ce.yildiz.edu.tr/personal/songul/file/1097/principa., accessed November 2012.
3. Raol, J.R. *Multisensor Data Fusion with MATLAB*. CRC Press, FL, 2010.
4. Wickramarachchi, N. and Halgamuge, S. Selected areas in data fusion. Ppts. saman@unimelb.edu.au, accessed November 2012.
5. Rogova, G.L. and Nimier, V. Reliability in information fusion: Literature survey. In *Proceedings of the 7th International Conference on Information Fusion* (Eds. Svensson, P. and Schubert, J.), International Society of Information Fusion, CA, USA, pp. 1158–1165, 28 June–1 July, Stockholm, Sweden, 2004.
6. Wang, J.G. Reliability analysis in Kalman filtering. *Journal of Global Positioning Systems*, 8(1), 101–111, 2009.
7. Punska, O. Bayesian approaches to multi-sensor data fusion. Signal Processing and Communications Laboratory, Department of Engineering, University of Cambridge, August 1999.
8. Rauch, H.E. Probability concepts for an expert system used for data fusion. *AI Magazine*, 5(3), 55–60, 1984.
9. Gelb, A. (Ed.) *Applied Optimal Estimation*. MIT Press, MA, 1974.
10. Raol, J.R., Girija, G. and Singh, J. *Modelling and Parameter Estimation of Dynamic Systems*. IET/IEE Control Series Books, Vol. 65, IEE/IET Professional Society, London, UK, 2004.
11. Mohammad-Djafari, A. Probabilistic methods for data fusion. *Maximum Entropy and Bayesian Methods; Fundamental Theories of Physics*. Springer, Netherlands, Vol. 98, 57–69, 1998. http://link.springer.com/chapter/10.1007%2F978-94-011-5028-6_5; http://djafari.free.fr/pdf/me97_2.pdf, accessed April 2013.

3

Fuzzy Logic and Possibility Theory-Based Fusion

3.1 Introduction

Since it is strongly felt that the concepts of fuzzy logic (FL) would find increasing use in target tracking, state estimation, sensor data fusion (DF) and decision fusion, it would be feasible to develop explicit DF processes, procedures and rules based on the mathematics of FL, interval type 2 fuzzy logic (IT2FL) and the related possibility theory.

Classically and traditionally crisp logic is used in decision and related analyses, with of course the associated limitation of crisp logic which is the inherent process of quantisation in crisp logic. The theory of probability and probability models (Chapter 2) are based on crisp logic via classical set theory. There are only two discrete states or possibilities in crisp logic: 0 or 1 (yes or no; –1 or +1; on or off) whereas in FL there are infinite possibilities between these two extremes of 0 and 1. What is meant by the crisp logic is that for an event there are only two distinct states: that a particular event will occur or not. Before the event occurs, we try to determine the probability that will tell us about the scope of the event to occur. Once that particular event has occurred, the probability vanishes. However, once that event has occurred, there still might be certain types of uncertainties (related to that event itself) present. It is for modelling such uncertainties that FL can be used. So, FL as well as possibility theory extend beyond crisp logic and probability!

In contrast to crisp logic, the FL is a multi-valued logic, and graphically/ pictorially it can be defined by, say a triangle, leading to a triangular form of the membership function (MF). There are several other types of membership functions (MFs) (Figure 3.1). This is not to say that these MFs are themselves fuzzy, they are in fact fully determinable. What the MF really does is to give a value between zero and one (and not either zero or one as in crisp logic, of course these both values are also possible for MF) to the variable (on x-axis). This value is the gradation, or the MF grad value/s, also called as MFG. Thus, the x-axis variable, say temperature, is now fuzzified and its belongingness to a given MF (low, medium or high temperature) could be any value as per

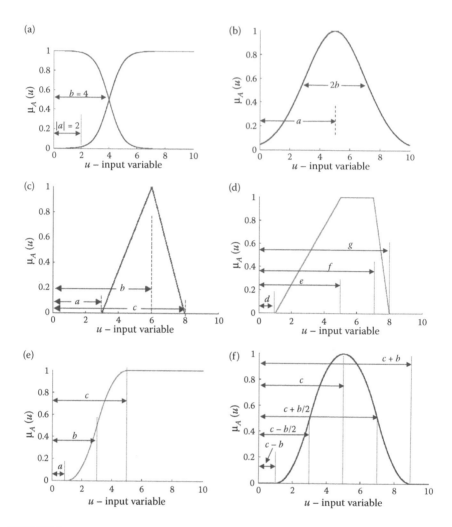

FIGURE 3.1
Different MFs for use in T1FS/FL that fuzzify the input-crisp variables into the fuzzy variables. (a) Sigmoid MF, (b) Gaussian MF, (c) triangular MF, (d) trapezoidal MF, (e) S-shaped MF, (f) Π-shaped MF.

the chosen MF. Thus, in fact the temperature variable is now given a 'range' and the concept of vagueness is introduced. This goes beyond probability whether the person has temperature or not and what the probability was of her having temperature, may be high probability. But, once the person has temperature, the vagueness of the temperature sets in: very low, low, medium, high, very high and so on; now we assign the MFG values to each one of this linguistic notion of the temperature. This is fuzzification. In fact, the theory of possibility [1] (via fuzzy sets [FSs]) is based on FL, in a similar manner as the theory of probability is based on crisp logic (via classical set theory). Crisp logic is a special case of the FL, as 0 and 1 are the extremes of the FL-MF values (the y-axis of the MF). This means that crisp logic is included in the FL as a special case.

FL-based methods/systems (FLS, including FL-based control systems) have found good use and applications in industrial control systems, some home appliances (e.g. washing machines), robotics and aerospace engineering. It is now regarded as one of the very important paradigms of soft computing (Chapter 10). FL-based controllers are used for keeping: (i) output variables between reasonable limits and (ii) control actuation (i.e. control input or related variable to the dynamic system) between the limits [2]. FL also finds its use in design and operation of intelligent control systems and knowledge-based intelligent robots. FL as such deals with vagueness rather than uncertainty, for example, if a person has a 'severe' headache, then there is a 'good chance' that she has a migraine rather than anything else. The fuzzy inference system (FIS) operates over the MFs and the (fuzzy) rule base with the help of fuzzy implication functions (FIFs). Then, via the defuzzification process FLS gives crisps outputs that are further used for the intended purpose. These If ... Then ... rules that are obtained mainly from human experts who have been dealing with such systems for several years and have gathered a lot of practical and intuitive experience of operation, analysis and design of these systems.

FL can be used and incorporated in several ways in engineering problems and systems as in: (a) Kalman filter, for doing KF tuning adaptively using FIS, (b) ANFIS (adaptive neuro-fuzzy inference system) for scheduling of gains in the control laws of a dynamic system (e.g. aircraft) and for design of control systems and (c) system identification/parameter estimation/state estimation for target tracking. FL directly or via ANFIS can be incorporated in target-image-centroid (TIC) tracking algorithms to deal with vagueness in image representation. Increasingly now, FL is being used for sensor DF, and mainly to represent uncertainty/vagueness in measured images and modelling of the noise processes that affect these real images and other data. FL-based techniques in the form of approximate reasoning aid decision making with very powerful reasoning capabilities: (i) image-analysis-cum-detection of edges, feature extraction, classification and clustering, (ii) estimation of parameter of unknown dynamic systems and (iii) decision fusion-cum-situation and threat assessment for aiding defence systems, and in C-4-I-2 (C4I2), that is, command, control, communication and computer information and intelligence. FL is also increasingly used for robot-path/motion planning, control/coordination of robots and sensor/actuator fault identification and reconfiguration [3].

Any development of a FL-based system would require to: (a) select fuzzy sets (FSs) and their appropriate MFs – this is a fuzzification process; (b) create a rule base with the help of human/design experts for mapping of input–output (I/O); to avoid contradictions in the rules some human intervention is sometimes needed at certain later stages, and to tune various adjustable

parameters in the FLS and the rule base itself; (c) select suitable fuzzy operators; (d) select FIF and aggregation methods and (e) select an appropriate
defuzzification method (to come back to crisp domain since most systems
need crisp values). In this chapter, we briefly discuss MFs, FL operators, FIFs
and rule-based systems, FIS, defuzzification method, and consider the use of
FL in sensor DF. We also, briefly discuss type 2 FS (T2FS), interval type 2 FS
(IT2FS), and give mathematical details for the latter, with a hope that these
would find increasing use in future DF systems. In Chapter 8, we discuss the
use of FL for decision fusion in aviation scenarios. Later in Chapter 10, we
discuss FL further as a soft computing paradigm, and possible usage in DF.

3.2 Fuzzy Logic Type 1

An FL is an internal extension/expansion of a crisp set within the limits
of the crisp set, making the FL much richer and truly multi-valued logic.
However, this basic FL is called fuzzy logic type 1 (T1FL). Then, crisp logic
can be called fuzzy logic type 0, for the sake of formality. An FS allows a
partial membership of a member of a set (belonging to the set). A FS A on a
universe of discourse (UOD) U with elements u is expressed as [2]

$$A = \int \{\mu_A(u)/u\} \quad \forall u \in U \tag{3.1}$$

or

$$A = \sum \{\mu_A(u)/u\} \quad \forall u \in U \tag{3.2}$$

It should be noted here that the notations $\int$ and $\sum$ are used to just represent
a fuzzy set (FS) and do not have anything to do with usual integration and
summation interpretations, although if need be so, they can be thus interpreted or these symbols can be used to represent the unions. Here, $\mu_A(u)$ is
an MF function (value) of u on the set A and provides a mapping of the UOD
U (this is a superset of collection of fuzzy variables) on the closed interval
[0, 1]. The 0 and 1 values (of crisp logic) are the extremes of the FL MF, and
hence the crisp logic is a special case of the FL. The μ_A (see Figure 3.1) is a
measure of the degree (MOD) to which (or by which) u belongs to set A, that
is, $\mu_A(u):U \rightarrow [0,1]$. It tells how much is the 'belongingness' of the u to the FS.
The fuzzy variables take on different labels defined by linguistic values such
as very low, low, medium, normal, high and very high with each represented
by different MFs. Different linguistic variables/values can be defined based
on the given problem.

3.2.1 MFs for Fuzzification

There are several types of MFs shown in Table 3.1 and Figure 3.1a–f that are being used in FL. Since for T1FSs the MFs are not fuzzy at all, we do not call them fuzzy MFs. For T2FSs and IT2FSs, although the MFs are themselves fuzzy, we still continue to use the term MF, since then from the context it will be clear that the MFs for such T2FSs and IT2FSs are themselves fuzzy

TABLE 3.1

Membership Functions for Fuzzy Logic Type 1 (T1FL)

Type of the MF	Formula	Shape
Sigmoid shaped	$\mu_A(u) = 1/(1 + e^{-a(u-b)})$	Figure 3.1a
Gaussian	$\mu_A(u) = e^{-(u-a)^2/2b^2}$	Figure 3.1b
Triangular	$\mu_A(u) = \begin{cases} 0 & \text{for } u \leq a \\ \dfrac{u-a}{b-a} & \text{for } a \leq u \leq b \\ \dfrac{c-u}{c-b} & \text{for } b \leq u \leq c \\ 0 & \text{for } u \geq c \end{cases}$	Figure 3.1c
Trapezoidal	$\mu_A(u) = \begin{cases} 0 & \text{for } u \leq d \\ \dfrac{u-d}{e-d} & \text{for } d \leq u \leq e \\ 1 & \text{for } e \leq u \leq f \\ \dfrac{g-u}{g-f} & \text{for } f \leq u \leq g \\ 0 & \text{for } u \geq g \end{cases}$	Figure 3.1d
S shaped	$\mu_A(u) = S(u; a, b, c) = \begin{cases} 0 & \text{for } u \leq a \\ \dfrac{2(u-a)^2}{(c-a)^2} & \text{for } a \leq u \leq b \\ 1 - \dfrac{2(u-c)^2}{(c-a)^2} & \text{for } b \leq u \leq c \\ 1 & \text{for } u \geq c \end{cases}$	Figure 3.1e
Π shaped	$\mu_A(u) = \begin{cases} S(u; c-b, c-b/2, c) & \text{for } u \leq c \\ 1 - S(u; c, c+b/2, c+b) & \text{for } u > c \end{cases}$	Figure 3.1f
Z shaped	$\mu_A(u) = \begin{cases} 1 & \text{for } u \leq a \\ 1 - 2\dfrac{(u-a)^2}{(a-b)^2} & \text{for } u > a \text{ and } u \leq \dfrac{a+b}{2} \\ 2\dfrac{(b-u)^2}{(a-b)^2} & \text{for } u > \dfrac{a+b}{2} \text{ and } u \leq b \\ 0 & \text{for } u > b \end{cases}$	The z-curve is defined as a reflection of the s-curve

MFs. In the sigmoid MF, u is the fuzzy variable in the UOD U and a, b are constants that give a shape to the sigmoid(-al) function. This function is also used in artificial neural networks (ANNs) as a non-linear activation function (Chapter 10). In the Gaussian MF, a, b signify the mean and standard deviation of the function, respectively – it is distributed about parameter a, and parameter b decides the width of the function. In the triangular MF, parameters a and c are the bases and b signifies where the peak of the MF occurs. In the trapezoidal MF, the parameters d and g define the bases of the trapezoid and the parameters e and f define the shoulders at the top end. In the S-shaped MF the parameters a and c define the extremes of the sloped portion of the function and b signifies the point at which $\mu_A(u) = 0.5$. This function looks like the s-curve defined in the literature using a composition of a linear function and a cosine function. In that case the z-curve is defined as a reflection of this s-curve. However, the s-curve and the s-curve look like a sigmoid function. In the Π shaped MF the parameter c locates the 'peak' and parameters $c - b$ and $c + b$ locate the extremes of the slopes (left and right) of the curve. At $u = c - b/2$ and $u = c + b/2$ the MFG of the function is equal to 0.5. This curve/function can also be implemented as a combination of a z-curve and an s-curve. Even though the shape looks like the Bell shape/Gaussian function, normally the top part of the curve is slightly flattened out. In the Z shaped MF the parameters a and b define the extremes of the sloped portion of the function (given only in Table 3.1 and not shown in Figure 3.1).

3.2.2 FS Operations

As we have studied in classical logic and Boolean algebra, for the most elementary crisp set the basic operations/operators are the intersection AND, the union OR and the complement NOT (e.g. in computer logic/Boolean logic). If A and B are the two subsets of U, the intersection of these subsets denoted by $A \cap B$ (this itself is a new resultant set that) contains all the elements that are common in/between A and B, that is, $\mu_{A \cap B}(u) = 1$ if $u \in A$ AND $u \in B$, otherwise the result is null, that is, zero. The union of A and B, denoted $A \cup B$, contains all elements in either in A or B; that is, $\mu_{A \cup B}(u) = 1$ if $u \in A$ or $u \in B$. The complement of A denoted by $\bar{A}$ contains all the elements which are not in A, that is, $\mu_{\bar{A}}(u) = 1$ if $u \notin A$ and $\mu_{\bar{A}}(u) = 0$ if $u \in A$.

Since, in the FL the variable u is fuzzified and the MFGs will vary with a grade between 0 and 1, the (classical-logic) operators/operations AND, OR and NOT now have expanded meanings, and there could be more than one definitions of AND, OR and NOT. For FL the corresponding (to AND, OR, NOT), the operators specified are *min*, *max* and *complement* [2] and are defined as follows:

$$\mu_{A \cap B}(u) = \min\left[\mu_A(u), \mu_B(u)\right] \text{(intersection)} \qquad (3.3)$$

$$\mu_{A \cup B}(u) = \max\left[\mu_A(u), \mu_B(u)\right] \text{(union)} \tag{3.4}$$

$$\mu_{\bar{A}}(u) = 1 - \mu_A(u) \text{(complement)} \tag{3.5}$$

Another way to define AND and OR operators in FL [2], respectively, is

$$\mu_{A \cap B}(u) = \mu_A(u)\,\mu_B(u) \tag{3.6}$$

$$\mu_{A \cup B}(u) = \mu_A(u) + \mu_B(u) - \mu_A(u)\,\mu_B(u) \tag{3.7}$$

3.2.3 Fuzzy Inference System

For FL, first the rules are defined or constructed. These rules can be learned from domain/human experts (depending on the situation) or can be devised from the data of the system that is under study. A fuzzy rule is 'If u is A, Then v is B'. In fact the rule itself is not fuzzy, especially for the fuzzy logic type 1 (T1FL). However, in T2FLS, these rules themselves can also be fuzzy, if required so. The 'If' part of the rule 'If u is A' is called the antecedent or premise. The 'Then' part of the rule, 'v is B' is called the consequent or conclusion part. The core aspect in the use of FL is FIS that via FIFs defines mapping from input FSs into output FSs. FIS determines the degree to which the antecedent is satisfied for each rule. If the antecedent of a given rule has more than one clause, for example, 'If u_1 is A_1 AND/OR u_2 is A_2, Then v is B', fuzzy operators (t-norm/s-norm, respectively) are applied to obtain one value that represents the result of the antecedent for that rule. The FIS can take different forms depending on the manner in which inference rule is specified. It is also possible that one or more rules may fire at the same time, in such a case, outputs of all rules are then aggregated, that is, FSs that represent the output of each rule are combined into a single FS. In FIS, the rules are fired concurrently and the order of firing does not affect the output. Figure 3.2 shows a schematic of an FIS for multi-input/single-output (MISO) system [2]. The fuzzifier/fuzzification maps input values into corresponding memberships via MFs, and is very much essential to activate rules that are in terms of linguistic variables. The fuzzifier MF takes input values and determines the degree to which these numbers belong to each of the FSs.

3.2.3.1 Steps of Fuzzy Inference Process

Next, we consider the ith fuzzy rule, with more than one part in antecedent, for a MISO system specified as

$$R^i : \text{IF } u \text{ is } T_u^i \quad \text{AND } v \text{ is } T_v^i \quad \text{THEN } w \text{ is } T_w^i \tag{3.8}$$

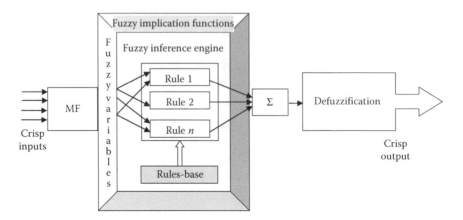

FIGURE 3.2
FIS – fuzzy inference system/process; $\Sigma \rightarrow$ aggregation.

with u, v and w as the fuzzy or linguistic variables whereas T_u, T_v and T_w are their linguistic values (low, high, large etc.) [2].

Step 1: Fuzzify the inputs u and v using MF ($\mu^i(u)$, and $\mu^i(v)$) for ith rule, it means that appropriate MFs are specified and used.

Step 2: Since antecedent part of every rule has more than one clause, FL operator is used to resolve the antecedent to a single number between 0 and 1. This will give degree of support (or firing strength) for ith rule. The firing strength then is expressed as

$$\alpha^i = \mu^i(u) * \mu^i(v) \tag{3.9}$$

Here, * represents triangular norm. The most popular triangular-norms (t-norms) used are standard intersection (min) and algebraic product (AP) (.):

$$\alpha^i = \min(\mu^i(u), \mu^i(v)) \quad \text{or} \quad \alpha^i = \mu^i(u) \cdot \mu^i(v) \tag{3.10}$$

Step 3: Use FIF to shape the consequent part, the output FS, based on the antecedent. The input to the implication process is a single number (α) given by the antecedent. Then the output is an FS. Popular methods are: (a) mini-operation rule of fuzzy implication (MORFI-Mamdani) and (b) product-operation rule of fuzzy implication (PORFI):

$$\mu^i(w)' = \min(\alpha^i, \mu^i(w)) \tag{3.11}$$

$$\mu^i(w)' = \alpha^i \cdot \mu^i(w) \tag{3.12}$$

Step 4: Since more than one rule, that is, more than one output FS, can be fired at a time, it is essential to combine the corresponding output FSs into a single composite FS. This is known as aggregation. The inputs to aggregation are outputs of implication and output of aggregation is a single FS that represents the output variable. The order in which rules are fired is not important in aggregation step. The common aggregation rule is the max, (standard union [SU]) method. Suppose rule 3 and rule 4 are fired at a time then the composite output FS is expressed as

$$\mu(w) = \max(\mu^3(w)', \mu^4(w)') \tag{3.13}$$

It should be kept in mind that Equation 3.13 represents the final output MF.

Step 5: In order to get crisp value of output variable w, a defuzzification step is carried out. The input to this process is the output from/of the aggregation and the output is a single crisp number.

This final step 5 converts output FSs, T1FSs values into crisp numbers (TOFS). If any rule has more than one clause in the antecedent part then these clauses are combined using any one of the definitions from either t-norm (used if the rules are connected by connectivity AND) or s-norm (triangular-conorm) (used if the rules are connected by the connectivity OR), and the FIF process gives the fuzzified output for each fired rule. These outputs are combined using the aggregation process.

3.2.4 Triangular-Norm

The corresponding operator for AND is *min*, Equation 3.3 and another possibility is given by Equation 3.6, these being the special cases of the t-norm in case of FL. The intersection of two FSs A and B is specified by binary mapping T (t-norm) on the unit interval as a function of the form $T:[0,1]X[0,1] \rightarrow [0,1]$ or more specifically

$$\mu_{A \cap B}(u) = T(\mu_A(u), \mu_B(u)) \tag{3.14}$$

t-Norm operator is used: (a) to combine the clauses in the *antecedent* part of a given rule, 'If u_1 is A_1 AND u_2 is A_2'; and (b) to map the input FSs into output FSs. These t-norms for fuzzy intersections are given as [2]

1. Standard intersection (SI):

$$T_{SI}(x,y) = \min(x,y) \tag{3.15}$$

2. AP:

$$T_{AP}(x,y) = x \cdot y \text{ (by Zadeh)} \tag{3.16}$$

3. Bounded difference/product (BD):

$$T_{BD|BP}(x,y) = \max(0, x + y - 1) \tag{3.17}$$

4. Drastic intersection/product (DI):

$$T_{DI|DP}(x,y) = \begin{cases} x & \text{when } y = 1 \\ y & \text{when } x = 1 \\ 0 & \text{otherwise} \end{cases} \tag{3.18}$$

We have $x = \mu_A(u)$, $y = \mu_B(u)$ and $u \in U$. It is assumed that the FSs (A and B) are normalised such that their membership grades are between 0 and 1. The definition of t-norms should satisfy certain axioms for the entire MFG such as x, y and z should be in the range [0, 1] [2].

3.2.5 s-Norm

The corresponding operator for OR is *max*, Equation 3.4, and another possible OR is given by Equation 3.7 for the FL. These are the special cases of the s-norm. The triangular-conorm or s-norm is defined as the union of two FSs A and B, specified by a binary operation on the unit interval as a function of the form [2]:

$$S:[0,1]X[0,1] \to [0,1] \quad \text{or more specifically} \quad \mu_{A \cup B}(u) = S(\mu_A(u), \mu_B(u)) \tag{3.19}$$

This s-norm fuzzy operator is used: (a) to combine the clauses in the antecedent part of a given rule, If u_1 is A_1 OR u_2 is A_2; and (b) in fuzzy implication process. The s-norms used as fuzzy unions [2] are:

1. SU:

$$S_{SU}(x,y) = \max(x,y) \tag{3.20}$$

2. Algebraic sum (AS):

$$S_{AS}(x,y) = x + y - x \cdot y \text{ (Zadeh)} \tag{3.21}$$

3. Bounded sum (BS):

$$S_{BS}(x,y) = \min(1, x + y) \tag{3.22}$$

4. Drastic union (DU):

$$S_{DU}(x,y) = \begin{cases} x & \text{when } y = 0 \\ y & \text{when } x = 0 \\ 1 & \text{otherwise} \end{cases} \tag{3.23}$$

5. Disjoint sum (DS):

$$S_{DS}(x,y) = \max\left\{\min(x, 1 - y), \min(1 - x, y)\right\} \tag{3.24}$$

with $x = \mu_A(u)$, $y = \mu_B(u)$ and $u \in U$.

3.2.6 Defuzzification

After the defuzzification process, we obtain a crisp value from/using the fuzzy outputs from FIS. A commonly used method is centroid or centre of gravity or centre of area (COA) method. The defuzzifier determines the COG v' of FS B and uses that value as the output of the FL. For continuous aggregated FS, it is given by

$$v' = \frac{\displaystyle\int_S v\mu_B(v)dv}{\displaystyle\int_S \mu_B(v)dv} \tag{3.25}$$

where S denotes the support of $\mu_B(v)$.

For the discrete FSs it is given as

$$v' = \frac{\displaystyle\sum_{i=1}^{n} v(i)\mu_B(v(i))}{\displaystyle\sum_{i=1}^{n} \mu_B(v(i))} \tag{3.26}$$

The other methods for defuzzification are: (i) maximum-decomposition method, (ii) centre of maxima (COM) or mean of maximum (MOM), (iii) smallest of maximum (SOM), (iv) largest of maximum (LOM) and (v) height defuzzification.

3.2.7 Fuzzy Implication Functions

The core of the FL-based system is FIS in which the 'If ... Then ...' rules are processed using FIF. These FIF finally obtains output as FS. A FIF plays a key role in successful design of the FLS. It becomes necessary to select an appropriate FIF from the existing methods. These FIFs should satisfy some of the intuitive criteria of GMP/GMT (generalised modus ponens/generalised modus tolens) [2] so that this FIF can be fitted into the logical development of any system using FL (Chapter 10). In any FIS, FIF provides mapping from input to output FSs so that fuzzified inputs can be mapped to desirable output FSs. Basically, a If ... Then ... rule is interpreted as a fuzzy implication. Thus, FIF essentially translates these rules into the meaningful output sets. Let us consider a simple rule

$$\text{If } u \text{ is } A, \quad \text{Then} v \text{ is } B \tag{3.27}$$

'IF u is A' is known as antecedent or premise and 'THEN v is B' is consequent part of fuzzy rule. The crisp variable u, fuzzified by set A (in UOD U), is an input to FIS whereas crisp variable v represented by set B (in UOD V) is an output from FIS. The fuzzified output of FIS, computed using sup-star composition is given by

$$B = RoA \tag{3.28}$$

With 'o' as a compositional operator and 'R' as a fuzzy relation in the product space $U \times V$. From Equation 3.28, the MF-functional relationship is given by

$$\mu_B(v) = \mu_R(u,v)o \ \mu_A(u) \tag{3.29}$$

A fuzzy implication, denoted by $\mu_{A \to B}(u,v)$, is also a kind of relation that provides mapping from the input to output variables. Equation 3.29 can be re-written as

$$\mu_B(v) = \mu_{A \to B}(u,v) \ o \ \mu_A(u) \tag{3.30}$$

There are seven standard interpretations of the If ... Then ... rules to define the FIF process [2], and by employing different combinations of t-norms and s-norms one can get a variety of ways to interpret the If ... Then ... rules. This means that a number of different fuzzy implications can be derived, if required. However, all the existing and the new FIFs might not satisfy completely (i.e. all) the intuitive criteria of GMP/GMT [2].

3.3 Adaptive Neuro-Fuzzy Inference System

In this section, we briefly study the ANFIS system that can be easily incorporated into any filtering/estimation algorithm so that FL can be used in augmentation for improving the performance of the combined system [2]. Also, the ANFIS can be used to incorporate FL in control systems and for parameter estimation. This facility of ANFIS can be used for sensor DF also. ANFIS utilises the rule-based procedure to represent the system (data-) behaviour in the absence of a precise model of the system.

Assume, we have a collection of I/O data sets and we want to build a FIS that would approximate these data as well as use the rule base and output the final result. This type of system then should consist of some MFs and If ... Then ... rules with adjustable parameters (that define the MF); these parameters (Table 3.1) can be chosen so as to adapt the MFs to the input data. This will account for the variations and the vagueness in the data values. This means that the MFs (in fact the parameters that define the structure and shape of the MFs) are now adaptive to the variations in the I/O data sets. For this purpose, the neuro-adaptive learning mechanisms can provide for the (fuzzy) modelling procedure to learn information about/from these data sets. Basically, this will facilitate computation of suitable MFs and their defining parameters/constants (Table 3.1) that would allow the associated FIS to track the given I/O data. This leads to the ANFIS which is a class of the adaptive networks that are functionally equivalent to FIS. It uses a hybrid learning method to determine the parameters. It also uses the I/O data to determine the MFs' parameters, and consists of FIS – these parameters are tuned using either a BP (back-propagation) algorithm or a combination BP and least squares (LS) method [2,3]. These parameters are updated via the learning and iterative cycles that are facilitated by a gradient vector (of the chosen cost function). This vector also provides a measure of how well the FIS is modelling the data set for a given set of parameters which are adjusted so as to reduce the error-measure/cost function (given by the sum of the squared difference of the actual and desired outputs). Thus, the MFs are adaptively tuned/determined by using ANN and the I/O data of the given system in an iterative fashion. The FIS is shown in Figure 3.2. The process of system parameter estimation using ANFIS is depicted in Figure 3.3, with the structural steps (network [NW]) of ANFIS shown in Figure 3.4.

We consider a rule base: (i) if u_1 is A_1 and u_2 is B_1, then $y_1 = c_{11}u_1 + c_{12}u_2 + c_{10}$; and (ii) if u_1 is A_2 and u_2 is B_2, then $y_2 = c_{21}u_1 + c_{22}u_2 + c_{20}$; with u_1, u_2 as non-fuzzy inputs and y as the desired output. The MFs of FSs A_i, B_i, $i = 1, 2$ be μ_A, μ_B. In Figure 3.4, Π (pi) is the product operator to combine the AND process of sets A and B with N as the normalisation and C_s as the output MF parameters. The ANFIS NW steps are: (i) each neuron 'i' in L_1 (Layer 1)

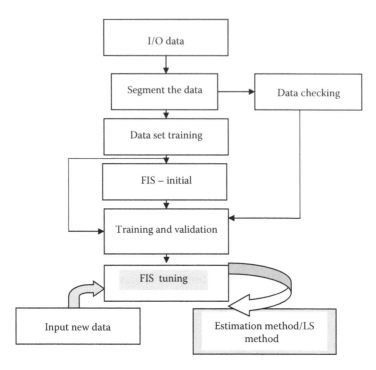

FIGURE 3.3
Scheme of using ANFIS system for parameter estimation of a dynamic system.

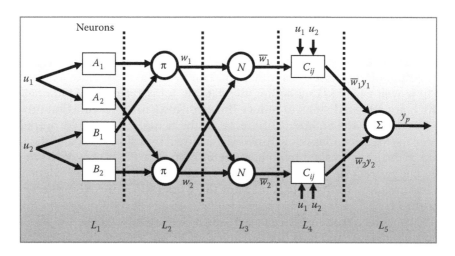

FIGURE 3.4
The stages of the ANFIS for (i) parameter estimation, (ii) incorporation of FL in any state estimation/filtering algorithm and (iii) use in sensor data fusion. (Based on the MATLAB Fuzzy Tool Box/ANFIS Demo and adapted from Raol, J. R. and Singh, J. *Flight Mechanics Modeling and Analysis*. CRC Press, FL, 2009.)

is adaptive with a parametric activation function, its output being the grade of MF to which the given input satisfies the MF → μ_A, μ_B A generalised MF

$$\mu(u) = \frac{1}{1 + \left| (u - c)/a \right|^{2b}} \qquad (3.31)$$

is used with the parameters (a, b, c) as the premise parameters; (ii) every node in L_2 (Layer 2) is a fixed node with the output (w_1) as the product Π of all incoming signals: $w_1 = \mu_{A_i}(u_1)\mu_{B_i}(u_2)$, $i = 1,2$; (iii) output of L_3 for each node is the ratio of the ith rule's firing strength relative to the sum of all rules' firing strengths: $\bar{w}_i = w_i/w_1 + w_2$; (iv) every node in L_4 is an adaptive node with a node output: Here, 'c' are consequent parameters; and in the last step (v) every node in L_5 is a fixed node which sums all incoming signals $y_p = \bar{w}_1 y_1 + \bar{w}_2 y_2$, where y_p is the predicted output. When these premise parameters get fixed, the overall output would be a linear combination of the consequent parameters. The output that is linear in the consequent parameters is written as

$$\begin{aligned} y_p &= \bar{w}_1 y_1 + \bar{w}_2 y_2 = \bar{w}_1(c_{11}u_1 + c_{12}u_2 + c_{10}) + \bar{w}_2(c_{21}u_1 + c_{22}u_2 + c_{20}) \\ &= (\bar{w}_1 u_1)c_{11} + (\bar{w}_1 u_2)c_{12} + \bar{w}_1 c_{10} + (\bar{w}_2 u_1)c_{21} + (\bar{w}_2 u_2)c_{22} + \bar{w}_2 c_{20} \qquad (3.32) \end{aligned}$$

Then a hybrid training/estimation method/algorithm adjusts the consequent parameters in a forward pass and the premise parameters are updated in the backward pass: (a) in the forward pass the NW inputs propagate forward until L_4, where the consequent parameters are identified by the LS method, and (b) in the backward pass, the errors propagate backward (while the time-computation cycle is always a forward phenomenon) and the premise parameters are updated by a gradient descent method. In fact the errors do not propagate, just that they are computed for the NW in ahead and the NW in back of the current NW. The MATLAB-ANFIS programming steps are: (a) generation of initial FIS by using INITFIS = genfis1(TRNDATA). The 'TRNDATA' is a matrix with $N + 1$ columns where the first N columns contain data for each FIS and the last column contains the output data – INITFIS is a single output FIS and (b) training of FIS: [FIS,ERROR,STEPSIZE,CHKFIS,CHKERROR] = anfis(TRNDATA, INITFIS,TRNOPT,DISPOPT,CHKDATA). Vector TRNOPT specifies training options, vector DISPOPT specifies display options during training, CHKDATA prevents overfitting of the training data set, and CHKFIS is the final tuned FIS. A few illustrative examples of use of ANFIS are given in this chapter and Chapter 10.

3.4 Fuzzy Logic Type 2

Conventionally what is well known to us are the T1FL and T1FS. In that case the crisp logic can be regarded as FL type '0' (T0FL) for uniform notation. The type '0' can handle only the conventional uncertainty, that is, random noise in the classical sense as described by probability models, whereas the T1FL can handle the vagueness. T1FL can be also be used to incorporate the heuristic knowledge of the human expert via fuzzy If … Then … rules. Then the output of T1FL – that is, the defuzzified output – is like a deterministic variable, that is, the T0FL. In probability density function (pdf) (Chapter 2) the variance provides a measure of dispersion/uncertainty about the mean and hence captures more information in the probabilistic models. In that case a T1FL is analogous to a probabilistic system only through the first moment, the mean. Then the so-called type 2 fuzzy logic (T2FL) is analogous to a probabilistic system via first and second moments [2]. Thus, the T2FL provides an additional design degree of freedom (DOF) in Mamdani and TSK FL systems. The provision of this extra DOF is useful when such FL-based systems are used in cases when lots of uncertainties (and that too of different types) are present [4,5]. In that case the resulting T2FLS can potentially provide better performance than a T1FL. This additional design DOF would be very useful for modelling sensors/communications channels that would be affected by vagueness and impreciseness, and then provide efficient and realistic DF and decision fusion (DeF) systems.

The T2FL can handle uncertainties by modelling them and minimising their effects on the control system's performance. When these uncertainties disappear T2FL reduces to T1FL. The T2FL consists of fuzzifier (with T2FL MFs), inference mechanism/engine (FIS, If … Then … rules, and FIF for operations on FSs that are characterised by MFs), the type reducer (TR) and finally the defuzzification operation (DFO). The O/P (output processor) of a T2FLS has two components: (a) T2FSs are transformed into T1FSs by means of TR and then (b) the T1FS is transformed into a crisp number by means of DFO, that is, into T0FS.

The T1FL does not directly handle rule uncertainties as it uses T1FSs (that are certain to some extent). T2FL is very useful where it is difficult to determine an exact MF for a FS. Thus, the T2FL can be used to handle the uncertainties in the rules themselves (if so desired) and even measurement uncertainties. As such a T2FL provides more DOF for design than a T1FL, and this is because its T2FSs are described by more parameters than are T1FS (in fact the MFs for the T2FS have one extra DOF i.e. a third dimension than the T1FL MFs). This aspect itself opens up new possibilities of representing and modelling more or several types of uncertainties in measurements as well as in fuzzy rule base thereby enlarging the scope of the T2FL in applications to MSDF including situation assessment (SA) and wireless sensor networks (WSN).

In Section 3.9, we present example of fusion of long-wave infrared (LWIR) and electro-optical (EO) images using type 1 and type 2 FLs in sufficient detail.

3.4.1 Type 2 and Interval Type 2 Fuzzy Sets

As we have seen that in T1FLS the MFG for each element is a crisp value in the range [0, 1], whereas the T2FL is characterised by a three-dimensional (3D) MF and a footprint of uncertainty (FOU) [4,5] (Figure 3.5); the MFG for each element of this set is an FS in [0,1] itself. This T2FS is bounded from below by the lower MF (LMF) and from above by the upper MF (UMF). The area between the LMF and UMF is the FOU. Thus, we see that the new and third dimension of T2FS and the FOU provide additional degrees of freedom that make it possible to directly model and treat the uncertainties. This is the IT2FL that use the T2FS in either their inputs or outputs and thus provides the potential to have suitable framework to handle the uncertainties in real-world problems. The T2FSs encompass a large number of T1FSs. We can quickly realise that in T2FLS the DOM (degree of membership) is in itself fuzzy (it has a range of values rather than one value), and is represented by a secondary membership function (SMF). If the SMF is at its maximum of 1 at every point, we call it as interval type 2 set (IT2FS, most people use this IT2FL and hence we continue to denote is as IT2FLS). The SMF in T1FS has only one value in its domain corresponding to the primary membership function/value (PMF/V) at which the SMF grade equals 1. Thus, in T1FS for each value of x, there is no uncertainty associated with the PMV.

In IT2FLSs (we may still use the notation T2FS) as the primary membership takes values within the interval [a, b], where each point in this interval has an associated SMF of 1 [4,5]. Hence, there is maximum uncertainty associated

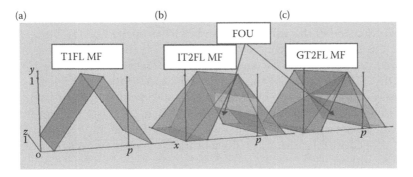

FIGURE 3.5
Types of FL MFs: (a) type-1 FL/FS (T1FL), (b) interval type-2 FL/FS (IT2FL) and FOU, (c) general type-2 FL/FS (T2FL) and FOU. (Adapted/modified from Hagras, H. and Wagner, C. Introduction to interval type-2 fuzzy logic controllers – Towards better uncertainty handling in real world applications. eNewsletter, IEEE SMC Society, No. 27, June 2009. www. http://www.my-smc.org/news/back/2009_06/SMC-Hagras.html, accessed November 2012. With permission.)

with the SMF. However, in general, type 2 fuzzy sets (T2FS) per se are not discussed further in the present book, the uncertainty represented by SMF can be modelled with any degree between T1FLSs and IT2FSs, for example, by the triangular SMF (Figure 3.5). Since, the IT2FS MFs are fuzzy and have a FOU, they can model and handle the linguistic and numerical uncertainties associated with the I/O of the FL control/systems (FLS). The merit is that the use of IT2FS will result in the reduction of the rule base when compared to using T1FS. This is because the uncertainty represented in the FOU in IT2FL allows us cover the same range as T1FS with a smaller number of labels. The additional DOFs given by the FOU enables IT2FS to produce outputs that cannot be achieved by T1FSs (with the same number of MFs). In fact, a IT2FS gives rise to an equivalent T1 MFG that is negative or larger than unity. Each I/O will be represented by a large number of T1FSs that are embedded in the T2FSs, and hence the use of large number of T1FSs to describe the I/O variables permits a detailed description of the analytical control. This addition of the extra levels of classification made possible by using T2FS gives a much smoother control response. In addition, the IT2FL can be thought of as a collection of many different embedded T1FSs.

The T2FL is computationally intensive, however, this computation can be simplified a lot by using IT2FL that enables us to design an FLS in less time and with less effort. Hence, the majority of FLSs are based on IT2FL. Traditionally, the T1FL is not visualised as a 3D set. However, its conception and visualisation in 3D is straightforward when we consider each of its points to be associated with the value 1 in the third dimension [4,5]. This means that a complete confidence is attached to each achieved MFG associated with a given crisp input (see Figures 3.5 and 3.6) [4,5]. In Figure 3.5, the same input p is applied to the three types of FSs: T1FS, IT2FS (interval type) and T2FS, resulting in a DOM that is specific to the type of FL. The extent of uncertainty/distribution linked with the degree is a (whitish) patch in Figure 3.5. This is explicitly depicted in Figure 3.6. This figure shows the SMFs – the third dimension of the T1FS, IT2FS and T2FS as elicited by the same input p.

The FIS of the IT2FL control/system depicted in Figure 3.7 consists of a fuzzifier, FIE, rule base, TR and defuzzifier. The IT2FS rules are the same as in the T1FS system, but the antecedents and/or the consequents will be represented by IT2FL. The FIS combines the fired rules and gives a mapping from input IT2FS to the output IT2FS sets. Then TR needs to be performed. This can be done in any of the two ways: (i) use the iterative Karnik–Mendel (KM) procedure/algorithm, or (ii) use the Wu–Mendel (WM) uncertainty bounds method.

3.4.2 IT2FL Mathematics

If we blur the T1FS MF, for example, say the triangular one by shifting the points on the triangle either to the left or to the right, then we obtain the type 2 MF (T2MF) as seen in Figure 3.5 [6]. This blurring need not necessarily be

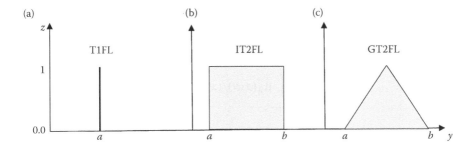

FIGURE 3.6
SMFs (secondary membership functions, see here the z-dimension) depicting the third dimension for: (a) type-1 (T1FL), (b) interval type-2 (IT2FL), (c) general type-2 (T2FL).

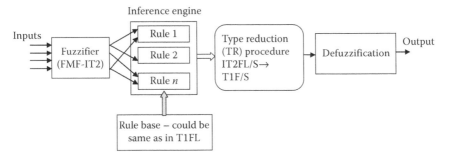

FIGURE 3.7
IT2FL fuzzy inference system (IT2FL/S-FIS).

of the same amounts, or it could be regular, irregular or uniform, the latter being preferred. Then, at a specific value of x, say x^i, there is no single value for the MF; instead it takes on values wherever the vertical line intersects the blur. All these values need not be weighted the same, so we assign an amplitude distribution to all of these points. As a result of this process, we obtain 3D MF-a T2MF, this characterises a T2FS. We emphasise here that much of the mathematics of T2FSs and IT2FSs is based on Reference 6. (In this section, we generally maintain the format and style of representing the various equations, formulae, and expressions from Reference 6; this is done in order to maintain the uniformity of such representations in the literature on FL, wherein sometimes the mathematics becomes too messy. Any attempt to make arbitrary changes in such formulations, just for the sake of making changes, would affect the flow, regularity and standardisation of the symbols used in Reference 6, since these formulations/equations/expressions have multiple levels of subscripts and superscripts.)

So, we have the following definition of the T2FS [6]:

$$\tilde{A} = ((x,u), \mu_{\tilde{A}}(x,u)); \quad \forall x \in X, \forall u \in J_x \subseteq [0,1] \tag{3.33}$$

In Equation 3.33, we have $\tilde{A}$ as the T2FS, and it is characterised by a T2MF, $\mu_{\tilde{A}}(x,u)$ where $x \in X$ and $u \in J_x \subseteq [0,1]$. Also, we have $0 \leq \mu_{\tilde{A}}(x,u) \leq 1$, then T2FS can also be expressed as follows:

$$\tilde{A} = \int_{x \in X} \int_{u \in J_x} \mu_{\tilde{A}}(x,u)/(x,u) \; J_x \subseteq [0,1] \qquad (3.34)$$

In Equation 3.34, the integral signs denote the union operator over all the permissible x and u, and for the discrete UOD (DUOD), the integral sign is replaced by the summation (sigma) sign. In Equation 3.33, we see two conditions: (i) $\forall u \in J_x \subseteq [0,1]$ is in line with the T1FL constraint, that is, $0 \leq \mu_{\tilde{A}}(x,u) \leq 1$. This is because when the uncertainty disappears a T2FL should degenerate to T1FS, that is, T2MF should reduce to T1MF; (ii) the condition $0 \leq \mu_{\tilde{A}}(x,u) \leq 1$ is also in line with the fact that the amplitudes of MF should be within 0 and 1, both inclusive. We have now the following definition for the IT2FS [6]:

$$\tilde{A} = \int_{x \in X} \int_{u \in J_x} 1/(x,u), \quad J_x \subseteq [0,1]\} \qquad (3.35)$$

We emphasise here that IT2FS/IT2FL are more prevalent and usable in FLS/FL control and are easier to handle than T2FL. Next, for each value of x, say x', the two-dimensional (2D) plane, whose axes are u and the $\mu_{\tilde{A}}(x',u)$, is called a vertical slice of the $\mu_{\tilde{A}}(x,u)$. Then, a SMF is considered as a vertical slice of $\mu_{\tilde{A}}(x,u)$, it is $\mu_{\tilde{A}}(x = x',u)$ for $x' \in X$, $\forall u \in J_{x'} \subseteq [0,1]$ and is given as follows:

$$\mu_{\tilde{A}}(x = x',u) \equiv \mu_{\tilde{A}}(x') = \int_{u \in J_{x'}} 1/u, \; J_{x'} \subseteq [0,1] \qquad (3.36)$$

Invoking the concept of secondary sets, one can interpret IT2FS as the union of all the secondary sets. Thus, one can express $\tilde{A}$ in a vertical-slice manner as follows [6]:

$$\tilde{A} = (x, \mu_{\tilde{A}}(x)); \quad \forall x \in X \qquad (3.37)$$

or

$$\tilde{A} = \int_{x \in X} \mu_{\tilde{A}}(x)/x = \int_{x \in X} \left[\int_{u \in J_x} 1/u \right] \Big/ x; \; J_x \subseteq [0,1] \qquad (3.38)$$

Next, we define the domain of a SMF. It is called the PMF of x. Thus, J_x is the PM of x and $J_x \subseteq [0,1]$; $\forall x \in X$. The amplitude of a SMF is called the SMFG (-grade), and these for an IT2FS are all equal to 1. In case X and J_x are discrete then we can express the right most part of Equation 3.38 as follows [6]:

$$\tilde{A} = \sum_{x \in X} \left[\sum_{u \in J_x} 1/u \right] / x,$$

$$= \sum_{i=1}^{N} \left[\sum_{u \in J_{xi}} 1/u \right] / x_i, = \left[\sum_{k=1}^{M_1} 1/u_{1k} \right] / x_1 + \cdots + \left[\sum_{k=1}^{M_N} 1/u_{Nk} \right] x_N \quad (3.39)$$

In Equation 3.39, '$+$' sign denotes union operation, and we see from this equation that x is discretised into N values and at each of x value, u is also discretised into M_i values. The discretisation along each u_{ik} is not necessarily the same, if it is same then $M_1 = M_2 = \cdots = M_N = M$. The uncertainty in the PMF of an IT2FS, $\tilde{A}$, has the bounded region and is called the FOU, and this FOU is the union of all the PMFs expressed as follows:

$$FOU(\tilde{A}) = \bigcup_{x \in X} J_x \quad (3.40)$$

This FOU is a vertical-slice representation of the FOU (see Figure 3.5). This is so because each of the PMs is a vertical slice. The FOU is a complete description of an IT2FS, since the SMFGs of an IT2FS do not convey any new information.

The UMF and LMF are next defined as follows:

$$\bar{\mu}_{\tilde{A}}(x) \equiv \overline{FOU(\tilde{A})}; \quad \forall x \in X$$
$$\underline{\mu}_{\tilde{A}}(x) \equiv \underline{FOU(\tilde{A})}; \quad \forall x \in X \quad (3.41)$$

Thus, we have $J_x = [\underline{\mu}_{\tilde{A}}(x), \bar{\mu}_{\tilde{A}}(x)]$; $\forall x \in X$ for IT2FS. For DUODs X and U, an embedded IT2FS has N elements. The set embedded such functions/FSs is given as follows:

$$\tilde{A}_e = \sum_{i=1}^{N} [1/u_i]/x_i, \quad u_i \in J_{x_i} \subseteq U = [0,1] \quad (3.42)$$

In Equation 3.42, we have $\prod_{i=1}^{N} M_i \tilde{A}_e$ embedded functions. Also, for DUODs X and U an embedded T1FS has N elements. The set is given as follows:

$$A_e = \sum_{i=1}^{N} u_i/x_i, \quad u_i \in J_{x_i} \subseteq U = [0,1] \quad (3.43)$$

In Equation 3.43, also, we have $\prod_{i=1}^{N} M_i A_e$ embedded functions. We have now the representation theorem [6]: for an IT2FS with discrete X and U, we have the following expressions:

$$\tilde{A} = \sum_{j=1}^{n_A} \tilde{A}_e^j \tag{3.44}$$

with $j = 1, 2, \ldots, n_A$ and

$$\tilde{A}_e^j = \sum_{i=1}^{N} [1/u_i^j]/x_i, \quad u_i^j \in J_{x_i} \subseteq U = [0,1] \tag{3.45}$$

with

$$n_A = \prod_{i=1}^{N} M_i \tag{3.46}$$

In Equation 3.46, M_i denotes the discretisation levels of SMF variables. Equation 3.44 gives a wavy slice of the set $\tilde{A}$. We make the following observations regarding Equation 3.44: (i) the MF of a IT2FS is 3D wavy, (ii) one can take all the possible wavy slices and take their union to reconstruct the original 3D MF, (iii) then, the same points that occur in different wavy slices, only occur once in the set-theoretic union, (iv) then it means collecting all of the embedded IT2FSs into a bunch of such T2FS and (v) equivalently one can collect all of the embedded T1FS into a bunch of such T1FSs. We have also the following relationship (from Equation 3.44):

$$\tilde{A} = 1/FOU(\tilde{A}) \tag{3.47}$$

with the following expressions for the set:

$$FOU(\tilde{A}) = \sum_{j=1}^{n_A} A_e^j = \{\underline{\mu}_{\tilde{A}}(x), \ldots, \bar{\mu}_{\tilde{A}}(x)\}; \quad \forall x \in X_d$$
$$[\underline{\mu}_{\tilde{A}}(x), \bar{\mu}_{\tilde{A}}(x)]; \quad \forall x \in X \tag{3.48}$$

with

$$A_e^j = \sum_{i}^{N} u_i^j/x_i, \quad u_i^j \in J_{x_i} \subseteq U = [0,1] \tag{3.49}$$

In Equation 3.48, X_d is the DUOD and X is the continuous UOD, and the equation gives an interval set of functions implying that it contains an uncountable number of functions that would completely fill the space between $\bar{\mu}_{\tilde{A}}(x) - \underline{\mu}_{\tilde{A}}(x); \quad \forall x \in X$.

3.4.3 The Set Theoretic Operations for IT2FS

We here consider the operators of intersection, union and complement for the IT2FS/IT2FL:

1. The union of two IT2FLS sets is given as follows:

$$\tilde{A} \cup \tilde{B} = 1/[\underline{\mu}_{\tilde{A}}(x) \vee \underline{\mu}_{\tilde{B}}(x), \bar{\mu}_{\tilde{A}}(x) \vee \bar{\mu}_{\tilde{B}}(x)]; \quad \forall x \in X \qquad (3.50)$$

2. The intersection operation is given as

$$\tilde{A} \cap \tilde{B} = 1/[\underline{\mu}_{\tilde{A}}(x) \wedge \underline{\mu}_{\tilde{B}}(x), \bar{\mu}_{\tilde{A}}(x) \wedge \bar{\mu}_{\tilde{B}}(x)]; \quad \forall x \in X \qquad (3.51)$$

3. The complement operation is given as

$$\overline{\tilde{A}} = 1/[1 - \underline{\mu}_{\tilde{A}}(x), 1 - \bar{\mu}_{\tilde{A}}(x)] \quad \forall x \in X \qquad (3.52)$$

The proof of the operations of Equations 3.50 through 3.52 is Reference 6.

3.4.4 Further Operations on IT2FS

The IT2FS-FIS is shown in Figure 3.7. The additional block is the TR block and it reduces the T2FSs to T1FSs. In IT2FS all the antecedents and consequent FSs are T2. The structures of the rules remain exactly the same as in the case of T1FL. The rules have the following form:

$$R^l : \text{If } x_1 \text{ is } \tilde{F}_1^l \quad \text{and} \quad \dots \, x_p \text{ is } \tilde{F}_p^l, \quad \text{Then } y \text{ is } \tilde{G}^l; \quad l = 1, 2, \dots, M \qquad (3.53)$$

We consider initially $l = 1$ and the rule is activated by a crisp number (i.e. singleton fuzzification, SF) [6].

1. *SF and one antecedent* Let us have the single rule

$$\text{If } x_1 \text{ is } \tilde{F}_1, \quad \text{Then } y \text{ is } \tilde{G} \qquad (3.54)$$

in the DUOD. Now, decompose $\tilde{F}_1$ into n_{F1} embedded IT2FSs $\tilde{F}_{1e}^{j_1}(j_1 = 1, 2, ..., n_{F1})$. Similarly for G. According to the representation theorem Equation 3.44, we have the following expressions for F and G:

$$\tilde{F}_1 = \sum_{j1=1}^{n_{F1}} \tilde{F}_{1e}^{j1} = 1/FOU(\tilde{F}_1) \tag{3.55}$$

where we have

$$FOU(\tilde{F}_1) = \sum_{j1=1}^{n_{F1}} F_{1e}^{j1} = \sum_{j1=1}^{n_{F1}} \sum_{i=1}^{N_{x1}} u_{1i}^{j1}/x_{1i}; \quad u_{1i}^{j1} \in J_{x1i} \subseteq U = [0,1] \tag{3.56}$$

Similarly, for the output (consequent) part we have

$$\tilde{G} = \sum_{j=1}^{n_G} \tilde{G}_e^{j} = 1/FOU(\tilde{G}) \tag{3.57}$$

$$FOU(\tilde{G}) = \sum_{j=1}^{n_G} G_e^{j} = \sum_{j=1}^{n_G} \sum_{k=1}^{N_y} w_k^j/y_k; \quad w_k^j \in J_{yk} \subseteq U = [0,1] \tag{3.58}$$

As a result of Equations 3.57 and 3.58, we have $n_{F_1} \times n_G$ number of possible combinations of embedded T1 antecedent and consequent FSs. With this the totality of the fired output sets for all the possible combinations of these embedded T1 antecedent and consequent FSs will be a bunch of functions $B(y)$ with

$$B(y) \underset{\equiv}{\Delta} \sum_{j_1=1}^{n_{F1}} \sum_{j=1}^{n_G} \mu_B(j_1, j)(y); \quad \forall y \in Y_d \tag{3.59}$$

In Equation 3.59, the summation operation signifies the union. The relation between the bunch of the functions $B(y)$ and the FOU of the T2 fired output FS can be summarised as: the bunch of the functions $B(y)$ obtained using T1FS operations is the same as the FOU of the T2FS fired output obtained T2FS operations.

2. *SF and multiple antecedents* Let us have $\tilde{F}_1, \tilde{F}_2, ..., \tilde{F}_p$ as the IT2FSs in DUODs $X_{1d}, X_{2d}, ..., X_{pd}$ and $\tilde{G}$ as the IT2FS in the UOD Y_d. We decompose Fs to obtain the IT2FSs as follows:

$$\tilde{F}_i = \sum_{j_i=1}^{n_{F_i}} \tilde{F}_{ie}^{j_i} = 1/FOU(\tilde{F}_i), \quad i = 1, 2, ..., p \tag{3.60}$$

The Cartesian product $\tilde{F}_1 \times \tilde{F}_2 \times ... \times \tilde{F}_p$ has $\Pi_{i=1}^p n_{F_i}$ combinations of the embedded T1FSs $F_{ie}^{j_i}$. Let us have F_e^n as the nth combination of these embedded T1FSs as follows:

$$F_e^n = F_{1e}^{j_1} \times \cdots \times F_{pe}^{j_p}; \quad 1 \le n \le \prod_{i=1}^p n_{F_i} \quad \text{and} \quad 1 \le j_i \le n_{F_i} \tag{3.61}$$

Equation 3.61 requires a combinatorial mapping from $(j_1, j_2, ..., j_p) \to n$. We express Equation 3.61 as follows:

$$F_e^n = F_{1e}^{j_1(n)} \times \cdots \times F_{pe}^{j_p(n)}, \quad 1 \le n \le \prod_{i=1}^p n_{F_i} \quad \text{and} \quad 1 \le j_i(n) \le n_{F_i} \tag{3.62}$$

Then, we have

$$\mu_{F_e^n}(x) = T_{m=1}^p \mu_{F_{me}^{j_m(n)}}(x_m), \quad 1 \le n \le \prod_{i=1}^p n_{F_i} \quad \text{and} \quad 1 \le j_m(n) \le n_{F_m} \tag{3.63}$$

with

$$n_F \equiv \prod_{m=1}^p n_{F_m} \tag{3.64}$$

In Equation 3.63, the usual t-norm of FL is used. We obtain $n_F \times n_G$ combinations of antecedent and consequent embedded T1FSs with n_G embedded T1FSs for the consequent. This generates a bunch of $n_F \times n_G$ fired output consequent T1FS functions as follows:

$$B(y) = \sum_{n=1}^{n_F} \sum_{j=1}^{n_G} \mu_B(n, j)(y); \quad \forall y \in Y_d \tag{3.65}$$

The results of type-1 non-SF and multiple antecedents are based on the results of case presented in (b) above [6].

3. *Type-2 non-SF and multiple antecedents* In this case, the p-dimensional input rule is given by the IT2FS $\tilde{A}_x$ with the assumption that MF is separable and $\tilde{X}_i$ denotes the labels of the IT2 FSs describing each of the p inputs. Specifically, $\tilde{X}_1, \tilde{X}_2, ..., \tilde{X}_p$ are IT2FSs in the DUOD of X_d. Then, further we obtain the following decomposition [6]:

$$\tilde{X}_i = \sum_{\gamma_i}^{n_{X_i}} \tilde{X}_{ie}^{\gamma_i}; \quad i = 1, 2, ..., p \qquad (3.66)$$

The domain of the each of the components of Equation 3.66, $\bar{X}_{ie}^{\gamma_i}$ is the embedded T1FS, $X_{ie}^{\gamma_i}$. The Cartesian product $\tilde{X}_1 \times \tilde{X}_2 \times \cdots \times \tilde{X}_p$ has $\prod_{\delta=1}^{p} n_{X_\delta}$ combinations of the embedded T1FSs, that is, $X_{ie}^{\gamma_i}$. So, if we have X_e^k as the kth combination of these embedded T1FS, then we have the following expression:

$$X_e^k = X_{1e}^{\gamma_1} \times \cdots \times X_{pe}^{\gamma_p}; \quad 1 \leq k \leq \prod_{\delta=1}^{p} n_{X_\delta} \quad \text{and} \quad 1 \leq \lambda_\delta \leq n_{X_\delta} \qquad (3.67)$$

Equation 3.67 requires a combinatorial mapping from $\{\lambda_1, \gamma_2, ..., \gamma_p\} \rightarrow k$; only we need to understand that it is possible to obtain such a mapping. So, we have the following expression for the mapping:

$$X_e^k = X_{1e}^{\gamma_1(k)} \times \cdots \times X_{pe}^{\gamma_p(k)}; \quad 1 \leq k \leq \prod_{\delta=1}^{p} n_{X_\delta} \quad \text{and} \quad 1 \leq \lambda_\delta \leq n_{X_\delta} \qquad (3.68)$$

Then we have the following additional requirements:

$$\mu_{X_e^k}(x) = T_{m=1}^p \, \mu_{X_{me}}^{\gamma_m(k)}(x_m); \quad 1 \leq k \leq \prod_{\delta=1}^{p} n_{X_\delta} \quad \text{and} \quad 1 \leq \lambda_m(k) \leq n_{X_m} \qquad (3.69)$$

and

$$n_X \equiv \prod_{\delta=1}^{p} n_{X_\delta} \qquad (3.70)$$

There are n_G embedded T1FSs for the consequent, $n_F = \prod_{m=1}^p n_{F_m}$ embedded T1FS for the antecedents. We also have n_X embedded T1FSs for the inputs. Hence, we have $n_X \times n_F \times n_G$ combinations of the input, the antecedent and consequent embedded T1FSs that obtain a bunch $B(y)$ of these combinations as the fired output consequent T1FSs functions and so we have the following expression for $B(y)$:

$$B(y) = \sum_{k=1}^{n_X} \sum_{n=1}^{n_F} \sum_{j=1}^{n_G} \mu_{B(k,n,j)}(y) \tag{3.71}$$

4. *Multiple rules*   Usually, we have M rules and often more than one rule fire when some input is applied to the FLS. In this case, we should incorporate l in all the expressions of IT2FSs. Then, in this case we have the following expressions [6]:

$$\tilde{B}^l = 1/FOU(\tilde{B}^l) \tag{3.72}$$

$$FOU(\tilde{B}^l) = [\underline{\mu}_{\tilde{B}^l}(y), \overline{\mu}_{\tilde{B}^l}(y)]; \quad \forall y \in Y \tag{3.73}$$

$$\underline{\mu}_{\tilde{B}^l}(y) = \inf_{\forall k,n,j} (\underline{\mu}_{B^l(k,n,j)}(y)) = [T_{m=1}^p (\sup_{x_m \in X_m} \underline{\mu}_{X_m}(x_m) * \underline{\mu}_{F_m^l}(x_m))] * \underline{\mu}_{G^l}(y); \quad \forall y \in Y \tag{3.74}$$

$$\overline{\mu}_{\tilde{B}^l}(y) = \sup_{\forall k,n,j} (\underline{\mu}_{B^l(k,n,j)}(y)) = [T_{m=1}^p (\sup_{x_m \in X_m} \overline{\mu}_{X_m}(x_m) * \overline{\mu}_{F_m^l}(x_m))] * \overline{\mu}_{G^l}(y); \quad \forall y \in Y \tag{3.75}$$

Next, we assume that the l rules/sets are combined using the union operation, then we have the following relations:

$$\tilde{B} = 1/FOU(\tilde{B}) \tag{3.76}$$

$$FOU(\tilde{B}) = [\underline{\mu}_{\tilde{B}}(y), \overline{\mu}_{\tilde{B}}(y)]; \quad \forall y \in Y$$

$$\underline{\mu}_{\tilde{B}}(y) = \underline{\mu}_{\tilde{B}_1}(y) \vee \underline{\mu}_{\tilde{B}^2}(y) \vee \cdots \vee \underline{\mu}_{\tilde{B}^M}(y) \tag{3.77}$$

Similarly, we can obtain the expression for the UMF values by replacing in Equation 3.77, the LMF values/grades by the UMF values/grades.

5. *Output processing*   The TR (see Figure 3.7) is the first step of output processing and it computes the centroid of an IT2FS. Next, we require that the IT2FLS must reduce to a T1FLS. We know that several T1-defuzzification methods are based on computing the centroid of a T1FS. We define the centroid of an IT2FS as the collection

of the centroids of all of its embedded IT2FSs. We have the following formulae to compute the centroid [6]:

$$C_{\hat{B}} = 1/\{c_s, \ldots, c_l\} \tag{3.78}$$

with

$$c_s = \min_{\forall \theta_i \in [\underline{\mu}_{\hat{B}}(y_i), \overline{\mu}_{\hat{B}}(y_i)]} \frac{\sum_{i=1}^{N} y_i \theta_i}{\sum_{i=1}^{N} \theta_i} \tag{3.79}$$

$$c_l = \max_{\forall \theta_i \in [\underline{\mu}_{\hat{B}}(y_i), \overline{\mu}_{\hat{B}}(y_i)]} \frac{\sum_{i=1}^{N} y_i \theta_i}{\sum_{i=1}^{N} \theta_i} \tag{3.80}$$

In Equation 3.78, we have the smallest and the largest elements denoted by s and l subscripts. The general formulae for the smallest and the largest values are given as [6]

$$c_s = \frac{\sum_{i=1}^{S} y_i \overline{\mu}_{\hat{B}}(y_i) + \sum_{i=S+1}^{N} y_i \underline{\mu}_{\hat{B}}(y_i)}{\sum_{i=1}^{S} \overline{\mu}_{\hat{B}}(y_i) + \sum_{i=S+1}^{N} \underline{\mu}_{\hat{B}}(y_i)} \tag{3.81}$$

$$c_l = \frac{\sum_{i=1}^{L} y_i \underline{\mu}_{\hat{B}}(y_i) + \sum_{i=L+1}^{N} y_i \overline{\mu}_{\hat{B}}(y_i)}{\sum_{i=1}^{L} \underline{\mu}_{\hat{B}}(y_i) + \sum_{i=L+1}^{N} \overline{\mu}_{\hat{B}}(y_i)} \tag{3.82}$$

In Equations 3.81 and 3.82, S and L are the switching points that can be determined by KM algorithm [6]. There are as many TR methods as there are T1FL-defuzzification techniques, because each one of the TR methods is associated with one of the latter. After the TR is performed, one uses the defuzzification as follows:

$$y(x) = 0.5\{c_s(x) + c_l(x)\} \tag{3.83}$$

Also, one can use an ad hoc TR formula as follows [6]:

$$y(x) \cong 0.5\{\underline{\mu}_{\hat{B}}(y) + \overline{\mu}_{\hat{B}}(y)\} \tag{3.84}$$

3.5 Fuzzy Intelligent Sensor Fusion

A fuzzy intelligent method, combining fuzzy aggregator and fuzzy predictor (FA/FP) for multi-sensory (MS) fusion based on information about the sensors' characteristics is presented in Reference 1, where the concentration is on accuracy and bandwidth of the associated sensors because these parameters would have considerable influence on the performance of such a sensory system. Often, many of the applications require sensors with high accuracy and good sensitivity in response to high-frequency changes in the measured variables. One approach is based on aggregation of information from certain sensors with complementary characteristics in accuracy and bandwidth. One can deal with only S_1 and S_2, the two sensors with different bandwidths (and accuracies). In the usual case, a weighted average of (the measurement data of/from) S_1 and S_2 can provide an appropriate estimation of the measured variable as per the following formula:

$$f(S_1, S_2) = \frac{w_1 S_1 + w_2 S_2}{w_1 + w_2} \tag{3.85}$$

Then, the FA determines weight of each sensor through a rule-based fuzzy system. When the weights are normalised, the FLS computes only the weight of S_1 as the output. The differential of S_2 measurement and the difference between numerical values of S_1 and S_2 are selected as the inputs of the FLS. Differential of S_2, with boost of the high-frequency components, recovers some of the signal's information lost by the low-bandwidth sensor.

This provides an appropriate value for judging the rate of changes in the measured variable (because of its low noise). The sensor S_1, being of low accuracy, would give misleading information about the signal's changes, but its difference with the numerical value measured by S_2, can give some information about the reliability of S_1 (in cases with low changes in value of S_2) [1]. To determine the weights for each sensor, one should use the system's inputs appropriately. The sensor S_1 should have a greater weight, when the changes in the measured variable are rapid. This is because these changes cannot appear in output of sensor S_2 due to its slow response. Also, when the variable changes in a smooth way, the weight of sensor S_2 should be greater to prevent the uncertainty of S_1 (from affecting the estimation). If the slope of changes in the measured variable is slow, and the difference between the values of S_1 and S_2 is large, it is because of high inaccuracy of S_1. Therefore, its weight (of S_1) should be considerably reduced. In the case of large changes in the measured variable and also large difference between two sensors' numerical values, it can be concluded that the difference has occurred because of the slow response of S_1. Hence, the weight of S_1 should be quite large. The foregoing inferences in the form of fuzzy rules can be expressed as below [1]:

If abs (S_1-S_2) is small and abs (dS_2/dt) is small, Then w_1 should be small.

If abs (S_1-S_2) is small and abs (dS_2/dt) large, Then w_1 should be large.

If abs (S_1-S_2) is large and abs (dS_2/dt) is small, Then w_1 should be very small.

If abs (S_1-S_2) is large and abs (dS_2/dt) is large, Then w_1 should be very large.

Next, the predicted input is used to improve the performance of the fuzzy fusion system (FFS). In a system that needs a good amount of accuracy the FP can be beneficial. One can compare the prediction with the output of FS prediction (FP). The FP is an intelligent system with n inputs that are n previous consecutive samples of the FA's outputs [1]. The system's structure is expressed as $f_1 = (x_1, x_2, ..., x_{n-1}) = x_n$. At each sampling instant the system's parameters are modified to achieve closer estimation of the function f to x_n. Then using this online training algorithm one gets an extrapolation of the function f to x_{n-1}, that is, predicting the next sample of the measured variable. Thus, after sufficient training steps the prediction is made more reliable. The gradient descent-based method is used for this process of estimation [1]. The method of FA/FP for sensor DF taking into account the sensor (model) accuracies was found to be working very well [1].

3.6 FL-Based Procedure for Generating the Weights for a DF Rule

We presume, here, that the original data received from each sensor (say, two sensors) are processed by a respective KF (Chapter 4) to estimate the states (position, velocity and acceleration) of a target. The residual error signal for each channel is generated by taking the difference of the measured and estimated positions of the target for that particular channel. The average estimation error is computed by [2]

$$\bar{e}_{sidn}(k) = \frac{e_{x_{sidn}}(k) + e_{y_{sidn}}(k) + e_{z_{sidn}}(k)}{3} \tag{3.86}$$

Here *sidn* is the sensor identity number and these error signals are generated by

$$\left. \begin{array}{l} e_{x_{sidn}}(k) = x_{m_{sidn}}(k) - \hat{x}_{sidn}(k) \\ e_{y_{sidn}}(k) = y_{m_{sidn}}(k) - \hat{y}_{sidn}(k) \\ e_{z_{sidn}}(k) = z_{m_{sidn}}(k) - \hat{z}_{sidn}(k) \end{array} \right\} \tag{3.87}$$

We have in Equation 3.87, $x_{m_{sidn}}$, $y_{m_{sidn}}$, $z_{m_{sidn}}$ as the target position measurements in x-, y-, z-axes, and $\hat{x}_{sidn}$, $\hat{y}_{sidn}$, $\hat{z}_{sidn}$ as the corresponding estimated positions obtainable from the KF. Then the weights to be used in DF rule are generated by the following procedure [2].

1. *Fuzzification*   The normalised error signals are fuzzified [between 0 and 1] using corresponding MFs labelled by linguistic variables. The MFs for both the error signals (from each sensor) are kept same and the variables have the attributes: ZE – zero error, SP – small positive, MP – medium positive, LP – large positive and VLP – very large positive. Figure 3.8 shows these MFs for error signals $\bar{e}_{sidn}$ and weights w_{sidn}.

2. *Rule generation using FIS*   The rules are created based on the magnitude of error signals reflecting the uncertainty in sensor measurements. Some rules for sensors S_1 and S_2 are given [2] as

   S_1:

$$\text{If } \bar{e}_1 \text{ is LP AND } \bar{e}_2 \text{ is VLP } \quad \text{then } w_1 \text{ is MP}$$

$$\text{If } \bar{e}_1 \text{ is ZE AND } \bar{e}_2 \text{ is MP } \quad \text{then } w_1 \text{ is LP} \qquad (3.88)$$

   S_2:

$$\text{If } \bar{e}_1 \text{ is ZE AND } \bar{e}_2 \text{ is VLP } \quad \text{then } w_2 \text{ is ZE}$$

$$\text{If } \bar{e}_1 \text{ is ZE AND } \bar{e}_2 \text{ is ZE } \quad \text{then } w_2 \text{ is MP} \qquad (3.89)$$

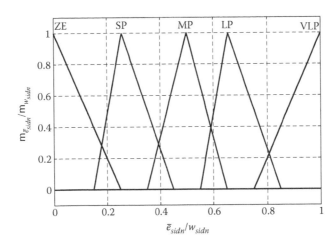

FIGURE 3.8
MFs for the error-weights w_{sidn} for DF rule.

TABLE 3.2

FL Rule Base for Generating Weights in a DF Formula

$\bar{e}_1$	$\bar{e}_2$ (Sensor 1)					$\bar{e}_2$ (Sensor 2)				
	ZE	SP	MP	LP	VLP	ZE	SP	MP	LP	VLP
ZE	MP	MP	LP	LP	VLP	MP	MP	SP	SP	ZE
SP	MP	MP	MP	LP	LP	MP	MP	MP	SP	SP
MP	SP	MP	MP	MP	LP	LP	MP	MP	MP	SP
LP	ZE	SP	SP	MP	MP	VLP	LP	LP	MP	MP
VLP	ZE	ZE	SP	MP	MP	VLP	VLP	LP	MP	MP

Table 3.2 gives the fuzzy rule base for the outputs weights w_1 and w_2 for S_1 and S_2, respectively.

3. *Defuzzification* The crisp values of w_1 and w_2 are obtained by defuzzifying the aggregated output FSs using the method of COA.

Thus, the weights to be used in a DF formula are generated by the FIS for S_1 and S_2, with the normalised values of the error signals $\bar{e}_1$ and $\bar{e}_2$ as the inputs to the FIS (associated with each sensor). Then the fused states are given by

$$\hat{X}_f(k) = w_1(k)\hat{X}_1(k) + w_2(k)\hat{X}_2(k) \tag{3.90}$$

The DF rule based on FL is validated for target tracking in Reference 2.

3.7 FL-ANFIS for Parameter Estimation and Generation of DF Weights: Illustrative Examples

In this section, we briefly discuss how the FL can be used for estimating parameters of a dynamic system from data signals, as well as for generating the weights required in a DF rule for some image data.

3.7.1 ANIFS-Based Parameter Estimation

In this section, some aspects of parameter estimation [7] of a system using FL are briefly discussed. The procedure used is: (i) generation of initial FIS and (ii) training of FIS. In the step, (i) the set of training data is generated using 'genfis1' of the MATLAB ANFIS toolbox (Section 3.3). The function used is INITFIS = genfis1(TRNDATA). Here, the TRNDATA is a matrix of $N + 1$ columns where the first N columns contain the data for each FIS input. The last column contains the output data. The INITFIS is a single-output FIS. Initially, the simulated data of a dynamic system are partitioned into the training set and the check data set. Then, the training data are used to get the tuned FIS which is validated with the help of the check data, that is, in the step (ii), the training of the FIS is carried out. The function used is: [FIS, ERROR, STEPSIZE, CHKFIT, CHKERROR] = anfis (TRNDATA, INITFIS, TRNOPT, DISPOPT, CHKDATA). The TRNOPT specifies the training options. DISPOPT specifies the display options during the training. CHKDATA is to prevent the overfitting of the training data set. Then, the tuned FIS is used to predict the system output for new input data of the same class and the parameters are estimated using the delta method [7]: (i) after the ANFIS is trained the perturbed data (with say, *delta* = 0.01) are presented to the ANFIS again (while input 1 perturbed data are presented, the input 2 data are kept at zero and vice versa for the input 2 data), (ii) with this we obtain the perturbed output data and (iii) the ratios of the averages of these respective output differences (with respect to the unperturbed data) to the perturbation size gives the parameters *a* and *b* (assuming a simple algebraic mathematical model) as is seen from the following example.

3.7.1.1 Parameter Estimation from an Algebraic Model

Generate the simulated data using the equation: $y = a + bx_1 + cx_2$ with the known parameters as: $a = 1$, $b = 2$ and $c = 1$. The output time history and the output error are shown in Figure 3.9, and the estimated parameters for

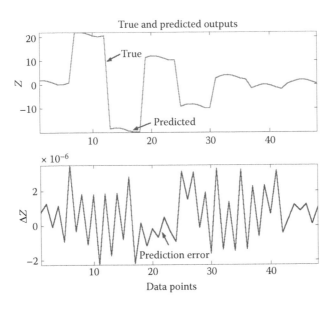

FIGURE 3.9
Output time history and the output (prediction) error for parameter estimation using FL-ANFIS (Example 3.1).

TABLE 3.3

Parameter Estimation Results Using FL-ANFIS (Example 3.1)

Parameters	True Values	Estimated Values Using Delta Method for Different Noise Levels		
		SNR = Inf	SNR = 100	SNR = 10
a	1	0.9999	0.9999	0.9999
b	2	2.0000	2.0000	2.0000
c	1	0.9999	0.9999	0.9999

various SNRs (signal to noise ratio; using the foregoing procedure of ANFIS) are given in Table 3.3, from which it is clear that the ANFIS works very well for parameter estimation of the algebraic system.

3.7.2 ANIFS for Deciphering a Linear DF Rule for Images

For image-case, we consider images generated by random numbers as well as true real-life images. We propose some linear fusion rule and fuse the two input images to get an output image. Then, we use these I/O images in the ANFIS and then estimate the coefficients/weights of the proposed rule. The procedure is an extension of that described in Section 3.7.1 to the case of image data.

3.7.2.1 Determination of DF Rule Using ANFIS and Random Image Data

CASE STUDY 3.1: Known Rule and Determined Rule

In this section, we consider an example of ANFIS system described in Section 3.3 to decipher a DF rule in the context of two images. The two input images given to the ANFIS system are the random images generated using random intensity matrices and are shown in Figure 3.10. The ANFIS requires the output for training for which we use the AS of the two input images as $y = a * input image1 + b * input image2$. The parameters

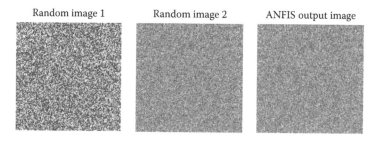

Random image 1   Random image 2   ANFIS output image

FIGURE 3.10

Random input images and the output image used in ANFIS for determination of DF rule (Example 3.2, Case Study 3.1).

$a = 1$ and $b = 0.5$. The output image is also shown in Figure 3.10. The code **ANFISJRRImfusion.m** is used for training. These image matrices were converted to column (or row) vectors by concatenating the successive columns and the 2D images were stored as one-dimensional (1D) images. These input 1D data (of the two random images) were given to the ANFIS system, likewise the training output image was also converted to an 1D string of the intensity values. For training we used 3 Gaussian/Bell MFs and 7 epochs. We obtained the following ANFIS information: (i) number of nodes: 35, (ii) number of linear parameters: 27, (iii) number of non-linear parameters: 18, (iv) total number of parameters: 45, (v) number of training data pairs: 16,384, (vi) number of checking data pairs: 0 and (vii) number of fuzzy rules: 9. After the ANFIS was trained the perturbed data (with *delta* = 0.01) were presented to the ANFIS again (while input 1 perturbed data were presented, the input 2 data were kept at zero and vice versa for the input 2 data). With this we obtain the perturbed output data, and the ratios of the averages of these respective differences (with respect to the unperturbed data) to the perturbation size gives the parameters a and b for respective input image 1 and the input image 2. The entire exercise is done with 1D image strings. We see that the estimation results are very good and we obtained the parameters: $a = 1$ and $b = 0.5$. We see that ANFIS is effectively utilised to determine the image fusion rule.

CASE STUDY 3.2: Unknown Rule and Determined Rule for Verification

Now, we consider an example of ANFIS system to decipher an unknown DF rule in the context of two random-data images. The two input images given to the ANFIS system are the random images generated using random intensity matrices: *rand('seed',1234)*; *im1 = rand(128,128)*; *rand('seed',4321)*; *im2 = rand(128,128)*; for the output image we used *randn('seed',2468)*; *imo = randn(128,128)*. This means that the input random images are generated by uniformly distributed random numbers (intensities) while the output image is generated by using the normally distributed random numbers (intensities). At this stage, we do not know the DF rule between the output and the two input images. The code **ANFISJRRImfusDFrule** is used for the training. We tried several options of choosing number of MFs and the number of epochs. The results of this study are given in Table 3.4, and the stabilisation quickly occurs. This means that the ANFIS has determined the coefficients a and b consistently. So, we choose $a = 0.7$ and $b = 0.36$ as the DF rule coefficients and determined the output image sequence using the formula: $y = 0.7 * x_1 + 0.36 * x_2$. We ran the ANFIS once again to estimate these coefficients, and we got the exact results (in bold numerals) as is seen from Table 3.4. For the last row of Table 3.4 we obtained the following ANFIS information: (i) number of nodes: 75, (ii) number of linear

TABLE 3.4

Determination of Coefficients of Unknown DF Rule Using FL-ANFIS (Example 3.2/Case Study 3.2)

No. of MFs	No. of Epochs	Parameter a	Parameter b	Remarks	
3	7	0.0311	0.1282	Estimated by ANFIS	–
5	5	0.6797	0.3579	-do-	The results start stabilising from now
5	7	0.7012	0.3654	-do-	
5	9	0.7076	0.3748	-do-	
–	–	**0.7000**	**0.3600**	True values	Known DF rule
5	9	**0.7000**	**0.3600**	Estimated by ANFIS	DF rule verified

parameters: 75, (iii) number of non-linear parameters: 30, (iv) total number of parameters: 105, (v) number of training data pairs: 16,384, (vi) number of checking data pairs 0 and (vii) number of fuzzy rules: 25. The plot of input/output random-intensities images is shown in Figure 3.11. We have thus established the efficacy of the ANFIS system to really determine the coefficients of an unknown DF rule, which was subsequently verified by using the ANFIS. At this stage it is not claimed that the procedure will work for the complicated rules for which one needs to make further studies.

3.7.2.2 Determination of DF Rule Using ANFIS and Real-Blurred Images

The two input images given to the ANFIS system are the images generated using one real-true image and then partially blurring the same true image with 'imfilter' (image processing-MATLAB tool box). The ANFIS requires the output for training for which we use the AS of the two input images as $y = a * input\ image1 + b * input\ image2$. The parameters $a = 0.4$ and $b = 0.6$. The code **ANFISJRRFruitsDFRule.m** is used for training. The I/O image matrices were converted to column vectors by concatenating the successive

Random image 1   Random image 2   ANFIS output image

FIGURE 3.11

Random-data images used for determining the unknown DF rule using ANFIS (Example 3.2, Case Study 2).

columns and the 2D images were stored as 1D images. These input 1D data (of the two random images) were given to the ANFIS system, and likewise the training output image was also converted to an 1D string of the intensity values. For training we used 3 Gaussian/Bell MFs and 4 epochs. We obtained the following ANFIS information: (i) number of no\des: 35, (ii) number of linear parameters: 27, (iii) number of non-linear parameters: 18, (iv) total number of parameters: 45, (v) number of training data pairs: 65,536, (vi) number of checking data pairs: 0 and (vii) number of fuzzy rules: 9. After the ANFIS was trained the perturbed data (with *delta* = 5 *and* 7) were presented to the ANFIS again (while input1 perturbed data were presented, the input2 data were kept at zero and vice versa for the input2 data). With this, we obtain the perturbed output data. We also obtained the ratios of the averages of the respective differences (with respect to the unperturbed data) to the perturbation delta. This gives the parameters '*a*' and '*b*' for respective input image 1 and the input image 2. The entire exercise is done with 1D image strings as discussed earlier. We see that the estimation results are good and we obtained the estimated parameters as: $a = 0.4033$ and $b = 0.5850$. We see that ANFIS is effectively utilised to determine the image fusion rule for the real images and the input images were partially blurred. Various images as input/output of this ANFIS exercise are shown in Figure 3.12.

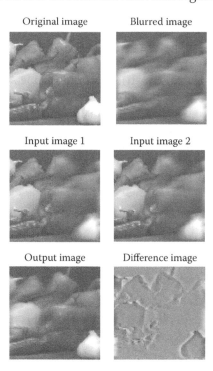

FIGURE 3.12
Real/blurred images as input to ANFIS and the output image (Example 3.3).

3.8 Possibility Theory

Possibility theory is a theoretical framework for modelling uncertainty [8]. The dual measures, the possibility and the necessity are used in order to model available information. It contrast to the probabilistic setting, the possibility is maxitive measure-framework. As of now it is very clear that the management of imprecise information cannot be separated from the management of uncertainty. An imprecise item of information can also be affected with uncertainty. The possibility theory (of Zadeh) provides a bridge between the representation of imperfect information and the evaluation of the uncertainty of the results of interest to us. The imperfect information is represented by the FSs.

3.8.1 Possibility Distribution

A possibility distribution (PODI) on a UOD, U is a mapping Π (*pi*) from U to the unit interval $[0, 1]$. We can also have the measurement scales as totally ordered scales such as $a_0 = 0 < a_1 <, \ldots , < a_n = 1$. By this ordered scale, a finite number of levels of possibility are distinguished between 'impossible' (0) and 'completely possible' (1) [8]. This scale can be numerically encoded, that is, $\{0, 0.1, 0.2, \ldots , 0.8, 0.9, 1\}$, where only the order relation between numbers is considered meaningful, and not the absolute values. For any u in U, the Π represents to what extent it is possible that u is the actual value of x. This function represents a flexible restriction of the values of x with the conventions: (i) $\Pi(u) = 0 \rightarrow$ that $x = u$ is impossible and (ii) $\Pi(u) \neq 0 \rightarrow$ that $x = u$ is fully allowed, that is, it is preferred, plausible and so on. The flexibility of the possible values of x is modelled by allowing $\Pi(u)$ between 0 and 1 for some values u. The quantity $\Pi(u)$ represents the degree of possibility of the assignment $x = u$ meaning some values of u are more possible than some other values. The closer to 1 $\Pi(u)$ is, the more possible u is the actual value of the variable. If U is the complete range of x, at least one element of U should be fully possible as a value of x, so that there exists u such that $\Pi(u) = 1$ (Then Π is said to be normal/normalised). Precise information is modelled by PODIs for which u_0 (of UOD, U) is such that $\Pi(u_0) = 1$ and $\Pi(u) = 0$ for $u \neq u_0$.

3.8.2 Possibility Set Functions

A PODI evaluates to what extent each element, u, of U can be the actual value of x, and it is possible to determine to what extent it is possible that the actual value of x belongs to any subset A of U. This is done by a measure of possibility Π which is given by the following expression [8]:

$$\Pi(A) = \sup_{u \in A} \pi(u), \; \forall A \subseteq U \qquad (3.91)$$

A possibility measure is a mapping Π from $\Pi(U) = 2^U$ to $[0, 1]$ that satisfies the characteristic conditions [8]: (i) $\Pi(\Phi) = 0$, (ii) $\Pi(U) = 1$ (due to the normalisation of π, $\sup_{u \in U} \pi(u) = 1$) and (iii) the property of maxitivity. The latter property is defined as

$$\Pi(A \cup B) = \max\{\Pi(A), \Pi(B)\}, \forall A, \forall B \tag{3.92}$$

The maxitivity condition, Equation 3.92, should hold for infinite families of sets so as to recover the condition of Equation 3.91. We can easily see that $\Pi(A) = 1$ suggests that it is completely possible that the actual value u of x belongs to A. However, it does not give any certainty about this aspect. In order to gain more information about the location of u, we need to use complementary information about x, also based on the PODI. For this we need to use a necessity measure (NM) that is a mapping, dual from the possibility measure in the sense of the following [8]:

$$N(A) = 1 - \Pi(A^c) \tag{3.93}$$

In Equation 3.93, A^c is a complement of A in U. The NM is obtained from the PODI as follows:

$$N(A) = \inf_{u \notin A}\{1 - \pi(u)\}, \quad \forall A \subseteq U \tag{3.94}$$

The NM tells the extent to which we are certain that the actual value of u of x is in the subset A of U. This implies that if we are more certain that u is in A, then it is less possible that u is in A^c. Then we obtain the following expressions:

$$N(\phi) = 0, N(U) = 1; \quad N(A \cap B) = \min(N(A), N(B)), \forall A \quad \text{and} \quad B \tag{3.95}$$

We need to define some other measures, for example, a measure of 'guaranteed possibility' (GP):

$$\Delta(A) = \inf_{u \in A} \pi(u) \tag{3.96}$$

The GP determines to what extend all the values in A are actually possible for x. This signifies that any value in A is at least possible for x at the degree $\Delta(A)$. Here, we see that the *delta* is a stronger measure than the measure *pi*, since $\Delta \leq \Pi$. This is so because the *pi* only estimates the existence of at least one value in A which is compatible with the available knowledge. The evaluation made by *delta* concerns all the values in A. We should also note that N and *delta* are not related. There is a dual measure of potential certainty (PC)

$$\Lambda(A) = 1 - \Delta(A^c) = \sup_{u \notin A}(1 - \pi(u)) \tag{3.97}$$

Equation 3.97 determines to what extent there exists at least one value in the complement of A which has a low degree of possibility. This is a necessary condition to have $x \in A$ somewhat certain. It is not a sufficient condition except if A^c has only one element. We have $N \leq \Lambda$, and the set functions Δ and Λ are monotonically decreasing, whereas N and Π are monotonically increasing.

3.8.3 Joint Possibility Distribution, Specificity and Non-Interactivity

A PODI may be assigned to the variables $\{x_1, x_2, ..., x_n\}$ with the spanning on $\{U_1, U_2, ..., U_n\}$. Then, the projection would compute the restriction induced by the joint PODI (JPODI), $\pi_{(x_1, x_2, ..., x_n)}$ on the collection variables: $\{x_1, x_2, ..., x_k\}$ as the possibility of the event $x_1 = u_1, ..., x_k = u_k$ and is given by the following expression [8]:

$$
\begin{aligned}
\pi_{(x_1, x_2, ..., x_k)}(u_1, ..., u_k) &= \Pi(\{u_1\} \times \cdots \times \{u_k\} \times U_{k+1} \times \cdots \times U_n) \\
&= \sup_{u_{k+1}, ..., u_n} \{\pi_{(x_1, ..., x_n)}(u_1, ..., u_k, u_{k+1}, ..., u_n)\}
\end{aligned}
\tag{3.98}
$$

Another important case of a PODI is the conditional PODI, and it should satisfy the following restriction:

$$
\forall u, \exists v, \pi_{yx}(v \mid u) = 1
\tag{3.99}
$$

Equation 3.99 signifies that whatever the value of x, there exists at least one completely possible value for y. Thus, *pi* defines a normal possibility measure on V. If we have π and π' such that $\pi < \pi'$ (i.e. $\pi \leq \pi'$ and $\pi(u) < \pi'(u)$ for some u), then π is said to be more *specific* than π'. This is in the sense that no value u is considered as less possible for x according to π' than to π. The concept underlies the idea that a PODI reflects the current available information about x. Hence, any PODI, π is provisional in nature, and it is likely to improve with further information. If $\pi \leq \pi'$, the information π' is redundant and can be dropped. Such features are typical of possibility theory and do not have counterpart in probability theory [8]. When the available information stems from several sensors/sources that can be considered as reliable, the PODI that accounts for it is the least-specific PODI that satisfies the set of restrictions induced by the pieces of information given by the different sources. This is the *principle of minimal specificity* (PMS). If several PODIs π^1, π^2, ... ,π^n for x become available (from n sources), the PMS leads to assume the following:

$$
\pi = \min_{i=1, n}(\pi^i)
\tag{3.100}
$$

The condition of Equation 3.100 is a straightforward consequence of the inequalities $\pi \leq \pi^i$, $i = 1, \ldots ,n$. The complete reliability of the sources entails that $\sup_u \min_{i=1,n} \pi^i(u) = 1$; otherwise the sources are conflicting as soon as $\pi(u) < 1$, $\forall u$, leading to sub-normalisation corresponding to partial inconsistency. The PMS allows PODI to be defined for a pair of variables $\rightarrow$ two variables x and y are associated with PODI π_x and π_y, respectively. The PMS leads to JPODI π as follows [8]:

$$\pi(u,v) = \min(\pi x(u), \quad \pi y(v)) \tag{3.101}$$

Also, the following inequalities are valid:

$$\pi(u,v) \leq \pi x(u), \quad \forall v \tag{3.102}$$

$$\pi(u,v) \leq \pi y(v), \quad \forall u \tag{3.103}$$

These inequalities do not account for the relations between the variables x and y. The π is separable and the variable are non-interactive. From the JPODI of two variables $\pi_{(x,\,y)}$, we can compute marginal possibility distributions (MPD) by the following projections [8]:

$$\pi_x(u) = \sup_v \pi_{(x,y)}(u,v), \quad \forall u \tag{3.104}$$

$$\pi_y(v) = \sup_u \pi_{(x,y)}(u,v), \quad \forall v \tag{3.105}$$

We can also verify the following inequality:

$$\pi_{(x,y)}(u,v) \leq \min\{\pi_x(u), \pi_y(v)\}, \quad \forall u, \forall v \tag{3.106}$$

The genuine degree of possibility of $(x = u, y = v)$ can be smaller than the value $\min(\pi_x(u), \pi_y(v))$, it may happen that $(x = u, y = v)$ is impossible because of a link between x and y, while $\pi(u, v)$ defined by Equation 3.101 is different from zero. Also, note that if there is an unknown link between x and y, then the PODI $\min(\pi_x, \pi_y)$ provides upper bounds on the degrees of possibility and gives results that are always correct but might not be informative.

The PMS plays a role similar to the maximum entropy (ME) principle in the framework of probability theory [8]. Both represent a means of determining the degree of confidence (possibility or probability) that is attached to the value of a variable (given incomplete information regarding it). Both involve some idea of least commitment. Also, independence between variables plays the same role in probability theory as non-interactivity in possibility theory.

However, the stochastic independence does not lead to bounding properties as non-interactivity does, because the stochastic independence assumes
an actual absence of correlation while non-interactivity expresses a lack of
knowledge about the link between the variables. In most of the rule-based
systems (FL/approximate reasoning), non-interactivity between variables is
assumed and the definition of JPODI is used.

3.8.4 Possibility and Necessity of Fuzzy Events

A fuzzy event (FE) is specified by a FS. We denote the MF of a FS, A as μ_A.
Possibility and necessity measures (P/NM) can be extended to fuzzy events
A as follows [8]:

$$\Pi(A) = \sup_{u \in U}\{\min[\mu_A, \pi(u)]\}$$
$$N(A) = \inf_{u \in U}\{\max[\mu_A, 1 - \pi(u)]\} \tag{3.107}$$

We need to maintain the duality $N(A) = 1 - \Pi(A^c)$. It is easily seen that the
definitions for non-FS A are particular cases of Equation 3.107. The conjunc-
tions other than *min* can be used in the definition of the possibility of a FE.
The use of *min* is in agreement with the idea of an ordinal treatment of the
possibility degrees. Let us have the normalised FS A and B defined on U and
V, respectively (i.e. $\exists u$, $\mu_A(u) = 1$ and $\exists v$, $\mu_B(v) = 1$). Also, given is a separable
PODI $\pi_{(x,\ y)} = \min(\pi_x(u),\ \pi_x(v))$ defined on $U \times V$. Then, we have the following
decomposability properties [8]:

$$\Pi(A \times B) = \min\left(\Pi_x(A), \Pi_y(B)\right) \tag{3.108}$$

$$N(A \times B) = \min\left(N_x(A),\ N_y(B)\right) \tag{3.109}$$

$$\Pi(A + B) = \max\left(\Pi_x(A), \Pi_y(B)\right) \tag{3.110}$$

$$N(A + B) = \max\left(N_x(A),\ N_y(B)\right) \tag{3.111}$$

In the equations above Π_x and N_x (and Π_y and N_y) are the PM and NM.
These are associated with π_x on U (and π_y on V). The Cartesian product $A \times B$
is defined by $(A \times B)(u, v) = \min(A(u), B(v))$, and the co-product is of the form
$A + B = (A^c \times A^c)^c$. The latter is obtained from Equations 3.108 and 3.109 by
changing *min* to *max*. These two quantities are used for comparing a FS A to
a reference one B letting $\pi = \mu_B(\cdot)$ and we make this explicit by the notation

$\Pi_B(A)$ and $N_B(A)$ instead of $\Pi(A)$ and $N(A)$. If we have the numerical values, then they are combined into a scalar measure of compatibility of B with respect to A. Then, the following is obtained: $comp(A;B) = [\Pi_B(A) + N_B(A)]/2$, so that that $comp(A;B) = 1$. Even though it summarises the uncertainty in only one number, $comp(A;B)$ would lose a part of information when A is FS, and hence the two numbers cannot be recovered from the knowledge of $comp(A;B)$ (except when A is not FS). Comp$(A;B)$ does not enjoy nice decomposability properties. $\Pi_B(A)$ is symmetrical in both arguments since it expresses to what extent $A \cap B$ (defined point wise as $\min\{\mu_A(\cdot), \mu_B(u)\}$) is not empty. However, $N_B(A)$ is not symmetrical, since it expresses a degree of inclusion of B into A. The set functions Π_B and N_B are still *max* and *min*-decomposable, respectively, on FEs. Also, $\max\{\Pi_B(A), 1-N_B(A)\}$ is not equal to 1 when A is FS, since we may have $\Pi_B(A) < 1$ and $\Pi_B(A^c) < 1$. However, $\Pi_A(A) \geq N_A(A)$ is always satisfied for A and B normalised. Also, $\Pi_A(A) = 1$, $N_A(A) \geq 1/2$ and $comp(A;A) \geq 3/4$ for any normalised FS A of U, using the unit interval as a possibility scale. The inequality $N_A(A) \geq 1/2$ is not surprising for A as FS. $N_B(A) = 1$ iff support$(B) = \{u, \mu_B(u) > 0\} \subseteq$ core$(A) = \{u, \mu_A(u) = 1\}$. Here, we are fully certain that a value restricted by $\pi = \mu_B(\cdot)$ satisfies A completely (at degree 1). $\Pi_B(A)$ and $N_B(A)$ are the upper and lower expectations in the sense of Sugeno integral of a fuzzy-valued compatibility index $COMP(A;B)$ given as follows [8]:

$$COMP(\mu_A, \mu_B)(t) = \{ \sup_{u:\mu_A(u)=t} \{\mu_B(u)\}$$
$$= \{0 \text{ if}(\mu_A(t))^{-1}\phi\} \qquad \forall t \in [0,1] \tag{3.112}$$

The COMP is the FS of possible values of the DOM in A of an ill-located element of U with possible values restricted by $\pi = \mu_B(.)$. It is an imprecise membership grade; $\Pi_B(A)$ and $N_B(A)$ are recovered from $COMP(A;B)$ as follows:

$$\Pi_B(A) = \sup_{t \in [0,1]}\{\min(t, COMP(\mu_A, \mu_B)(t))\} \tag{3.113}$$

$$N_B(A) = \inf_{t \in [0,1]}\{\max(t, 1 - COMP(\mu_A, \mu_B)(t))\} \tag{3.114}$$

3.8.5 Conditional Possibility

A conditional PM $\Pi(.|B)$ is defined as satisfying the Bayesian-like relationship: $\Pi(A \cap B) = \min(\Pi(A|B), \Pi(B))$ [8]. Compare with Equations 2.14 and 2.15. The PMS suggests the definition of $\Pi(.|B)$ as the greatest solution of this equation: if $B \neq \varnothing \rightarrow \Pi(A|B) = 1$, if $\Pi(A \cap B) = \Pi(B)$ and $\Pi(A|B) = \Pi(A \cap B)$

otherwise. The dual conditional NM is simply $N(A|B) = 1 - \Pi(A^c|B)$, and is ordinal conditioning, and is adapted to finite settings, with a qualitative (ordinal) possibility scale. The corresponding conditional PODI $\pi(\cdot|B)$ is defined by

$$\pi(u|B) = 1, \quad \text{if } u \in B \quad \text{and} \quad \pi(u) = \Pi(B) \tag{3.115}$$

$$= \pi(u), \quad \text{if } u \in B \quad \text{and} \quad \pi(u) < \Pi(B) \tag{3.116}$$

$$= 0, \quad \text{if } u \notin B \tag{3.117}$$

With the continuous numerical universes, this form of conditioning induces undesirable discontinuities, and conditioning is better defined using the product rule $\Pi(A \cap B) = \Pi(A|B)\Pi(B)$ (again compare with Equations 2.14 and 2.15). It agrees better with probabilistic concept (Chapter 2) and fits with DS rule of conditioning in belief function theory. The corresponding conditional distribution is then just $\pi(\cdot|B) = \pi(\cdot)/\Pi(B)$. Both views of conditioning enable joint JPODI to be expressed in terms of minimum (or product, respectively) of conditional PODIs given as follows [8]:

$$\pi(x_1, \ldots, x_n) = \min_{i=1,\ldots,n} \{\pi(x_i|x_{i+1} \ldots x_n)\} \text{ (ordinal case)} \tag{3.118}$$

$$\pi(x_1, \ldots, x_n) = \prod_{i=1,\ldots,n} \{\pi(x_i|x_{i+1} \ldots x_n)\} \text{ (numerical case)} \tag{3.119}$$

Here, for $i = n$, $\pi(x_i|x_{i+1} \ldots x_n)$ is replaced by the marginal PODI on the span of x_n.

It might be feasible to develop explicit DF processes, procedures and rules based on the mathematics of IT2FSs and the possibility theory presented in this chapter. As we have seen in Chapters 2, 4 and 5, the Bayesian theory has been very useful in developing DF rules and estimation techniques that are being successfully used in decision making and target tracking. The Bayesian nets (BNWs) are also being used in DF and related SA (situation awareness) work extending the early probabilistic expert systems. In the similar manner, possibilistic graphical models may become the rigorous approach to build fuzzy expert systems [8]. Since, the main problem is to properly combine local knowledge, a natural way is to start with a rigorous representation of a knowledge base as a JPODI on a space of attributes. Then conditions can be defined under which this JPODI can be split into smaller parts without losing information. In Reference 9, POSSINFER-SW exploits possibilistic NWs and it has been used for DF.

3.9 Fusion of Long-Wave IR and EOT Images Using Type 1 and Type 2 Fuzzy Logics: Illustrative Examples

In this example, a method based on T1FL and IT2FL [6,10–15] is used to fuse images obtained from EVS (enhanced synthetic vision) prototype system consisting of two video channels [16]: (i) auto-mobile grade LWIR sensor and (ii) electro-optical (EOT) colour camera. The output of EVS is in RS170 format which is digitised into image frames using a 4-channel 2255 frame grabber. Image registration, a pre-requisite before image fusion, is required to bring the images, captured by LWIR and EOT cameras having different view points (or angles), to a common view point. The mage from the EOT is registered on the reference image from the LWIR. Image fusion combines two images from different sources, from LWIR and EOT cameras, by giving appropriate weights to each source level: like Alpha blending or a principal component analysis (PCA) (Chapter 6) based method or to a pixel level of each source: Laplacian pyramid, wavelet (Chapter 6), Fl and so on.

3.9.1 FL Systems: Takagi-Sugeno-Kang Inference Method

Takagi-Sugeno-Kang (TSK) or in short the Sugeno or TSK method, introduced in 1985, is similar to Mamdani-based FLS in many aspects. The fuzzification and combining the consequents of rule using FL operator t-norm (intersection) are the same for Mamdani and Sugeno-based systems. There is no aggregation operation in the Sugeno-based system (SBS) as the fuzzfied output is singleton (i.e. 1 or 0) and crisp output is of either constant or linear (output is a linear function of inputs) type for each rule. Unlike in Mamdani-based system (MBS) where fuzzified output is obtained by applying fuzzy operation t-norm (min, prod, etc.) between firing strength and output FS (singleton type) for a fired rule, SBS supports only 'prod' fuzzy operation for inference. The defuzzification method for SBS is either weighted average (wtaver) or weighted sum (wtsum) only. Suppose, there are M rules fired and the fuzzified output is represented by $w_1, w_2, \ldots, w_M$ and crisp output is represented by $z_1, z_2, \ldots, z_M$, then final crisp output wtaver is given by [16]

$$z = \frac{\sum_{i=1}^{M} w_i z_i}{\sum_{i=1}^{M} w_i} \qquad (3.120)$$

and final crisp output wtsum is given by

$$z = \sum_{i=1}^{M} w_i z_i \qquad (3.121)$$

where $z_i = a_i u_1 + b_i u_2 + c_i$ is a crisp output for ith rule fired, u_1 and u_2 are the crisp input signals and $[a_i, b_i, c_i]$ are coefficients. For zero-order Sugeno model z_i is a constant $(a_i = b_i = 0)$. In case of SBS IT2FL, the type-reducer block would not be present, the implication method will only be based AP fuzzy operation, and defuzzification sub-block will only consist of 'wtaver' and 'wtsum'. Uncertainty in input can be represented and handled by blurring T1FL MF as shown in Figure 3.13. For a given input 'u' there are finite, as such infinite values representing its possibilities in the membership set values. For the entire value of $u \in U$a 3D, an MF is created that is known as T2MF which basically characterises T2FLS. The IT2MF can be obtained by setting a secondary MF, that is, $\mu_{\bar{A}}(x,u)$ to the one shown in Figure 3.14 which for simplicity can be represented by Figure 3.15 characterised by lower and upper MFs.

In IT2FL, there are three different types of fuzzifiers: singleton, non-singleton type 1 (NST1) and non-singleton type 2 (NST2) [16].

Singleton: In singleton fuzzifier, Figure 3.16, the fuzzified inputs $(\bar{f}, \underline{f})$ can be computed by finding the intersection point of singleton with LMF and UMF, respectively

$$\begin{aligned} \bar{f} &= \bar{\mu}_A(u) \\ \underline{f} &= \underline{\mu}_A(u); \quad x = u \in U, \ U \subseteq [0,1] \end{aligned} \qquad (3.122)$$

where $u = x = 0.6$ is the current input signal.

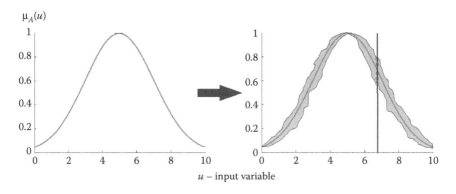

FIGURE 3.13
T1FL MF to blurred T1FL MF → T2FLMF. (From Kashyap, S. K. Fusion of LWIR and EO images using fuzzy logic type 1 and type 2. MSDF (internal report/unclassified) Rep. 1305/ESVS01 (FMCD, CSIR-NAL, Bangalore), November 27, 2013. With permission.)

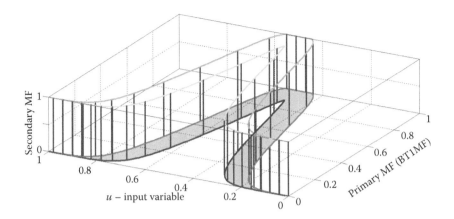

FIGURE 3.14
IT2FL – 3D MF. (From Kashyap, S. K. Fusion of LWIR and EO images using fuzzy logic type 1 and type 2. MSDF (internal report/unclassified) Rep. 1305/ESVS01 (FMCD, CSIR-NAL, Bangalore), November 27, 2013. With permission.)

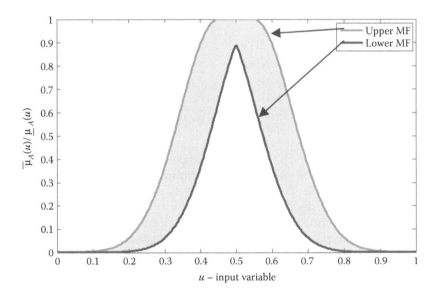

FIGURE 3.15
Gaussian MF for IT2FL/S. (From Kashyap, S. K. Fusion of LWIR and EO images using fuzzy logic type 1 and type 2. MSDF (internal report/unclassified) Rep. 1305/ESVS01 (FMCD, CSIR-NAL, Bangalore), November 27, 2013. With permission.)

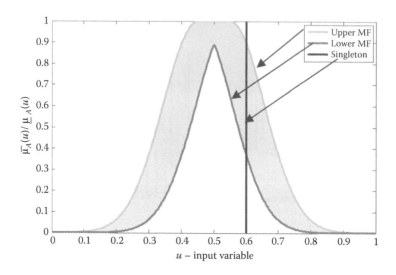

FIGURE 3.16
Singleton fuzzifier. (From Kashyap, S. K. Fusion of LWIR and EO images using fuzzy logic type 1 and type 2. MSDF (internal report/unclassified) Rep. 1305/ESVS01 (FMCD, CSIR-NAL, Bangalore), November 27, 2013. With permission.)

NST1: In NST1 fuzzifier, Figure 3.17, the fuzzified inputs $(\bar{f}, \underline{f})$ can be computed by

$$\bar{f} = \sup\left\{\int_{UXV} \bar{\mu}_A(u) * \mu'_F(v)/(u,v)\right\}$$

$$\underline{f} = \sup\left\{\int_{UXV} \underline{\mu}_A(u) * \mu'_F(v)/(u,v)\right\}; \quad u \in U, v \in V, U \subseteq [0,1], V \subseteq [0,1]$$

(3.123)

where $\mu'_F(v)$ represents NST1 fuzzifer, * indicates t-norm (intersection) fuzzy operator, 'sup' indicates supremum, maximum of output obtained after * operation (Figure 3.18). The NST1 fuzzifier is nothing but Gaussian function with its mean defined by input signal $u = x = 0.6$ and spread defined by sigma. Here, the standard intersection (min) t-norm operator is used.

NST2: In case of the NST2 fuzzifier (Figure 3.19), the fuzzified inputs $(\bar{f}, \underline{f})$ can be computed by

$$\bar{f} = \sup\left\{\int_{UXV} \bar{\mu}_A(u) * \bar{\mu}'_F(v)/(u,v)\right\}$$

$$\underline{f} = \sup\left\{\int_{UXV} \underline{\mu}_A(u) * \underline{\mu}'_F(v)/(u,v)\right\}; \quad u \in U, v \in V, U \subseteq [0,1], V \subseteq [0,1]$$

(3.124)

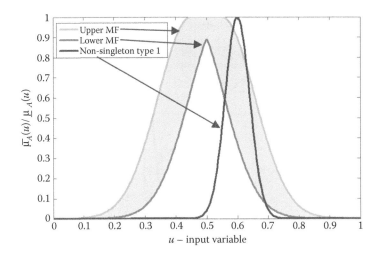

FIGURE 3.17
Non-singleton T1FL fuzzifier. (From Kashyap, S. K. Fusion of LWIR and EO images using fuzzy logic type 1 and type 2. MSDF (internal report/unclassified) Rep. 1305/ESVS01 (FMCD, CSIR-NAL, Bangalore), November 27, 2013. With permission.)

where $\bar{\mu}'_F(v)$ and $\underline{\mu}'_F(v)$ represent upper and lower NST2 (UNST2, LNST2) fuzzifier, * indicates the t-norm (intersection) fuzzy operator (Figure 3.20). The NST2 fuzzifier is nothing but two Gaussian functions with common mean defined by input signal $u = x = 0.6$ and individual spread defined by corresponding sigma. Here, again the standard intersection (min) t-norm operator is used.

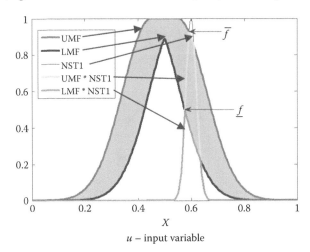

FIGURE 3.18
Non-singleton T1FL fuzzifier operation. (From Kashyap, S. K. Fusion of LWIR and EO images using fuzzy logic type 1 and type 2. MSDF (internal report/unclassified) Rep. 1305/ESVS01 (FMCD, CSIR-NAL, Bangalore), November 27, 2013. With permission.)

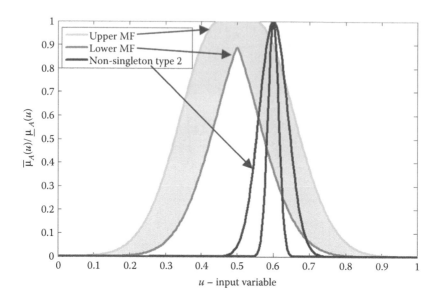

FIGURE 3.19
Non-singleton T2FL fuzzifier. (From Kashyap, S. K. Fusion of LWIR and EO images using fuzzy logic type 1 and type 2. MSDF (internal report/unclassified) Rep. 1305/ESVS01 (FMCD, CSIR-NAL, Bangalore), November 27, 2013. With permission.)

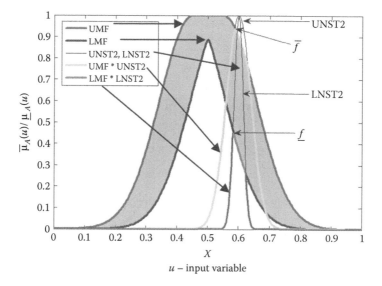

FIGURE 3.20
Non-singleton T2FL fuzzifier operation. (From Kashyap, S. K. Fusion of LWIR and EO images using fuzzy logic type 1 and type 2. MSDF (internal report/unclassified) Rep. 1305/ESVS01 (FMCD, CSIR-NAL, Bangalore), November 27, 2013. With permission.)

3.9.2 IT2FS Operations and Inference

A rule is typically defined in IT2FLS with two inputs and one output:

$$|————\text{-antecedent}—————|——\text{consequent}——|$$

$$\text{IF } x_1 \text{ is } \tilde{A}_1 \text{ AND } x_2 \text{ is } \tilde{A}_2, \text{ THEN } y \text{ is } \tilde{B} \tag{3.125}$$

where $\tilde{A}_1$ and $\tilde{A}_2$ are the inputs and $\tilde{B}$ is the output of T2FSs. These FLSs are defined by UMF and LMF. If the *antecedent* of a given rule has more than one clause (IF x_1 is $\tilde{A}_1$ AND x_2 is $\tilde{A}_2$, THEN y is $\tilde{B}$), fuzzy operators (t-norm/s-norm) are applied to obtain one number that represents the result of the *antecedent* for that rule. Let us say, the inputs are $x_1 = 0.7$ and $x_2 = 0.3$. In case of T2FLS, we have the following three combinations for antecedent combiner [16]:

Case 1: Inputs x_1 and x_2 are fuzzified using singleton fuzzifier and the antecedent combiner is realised using the following equations:

$$\begin{aligned}
\bar{f}_1 &= \bar{\mu}_{A_1}(u_1) \\
\underline{f}_1 &= \underline{\mu}_{A_1}(u_1); \quad u_1 = x_1 \in U, \quad U \subseteq [0,1] \\
\bar{f}_2 &= \bar{\mu}_{A_2}(u_2) \\
\underline{f}_2 &= \underline{\mu}_{A_2}(u_2); \quad u_2 = x_2 \in U, \quad U \subseteq [0,1] \\
\bar{f} &= \bar{f}_1 * \bar{f}_2 \\
\underline{f} &= \underline{f}_1 * \underline{f}_2
\end{aligned} \tag{3.126}$$

where '*' represents the t-norm fuzzy operator which could be 'min', 'prod' and so on and Figure 3.21 illustrates the combiner's operation.

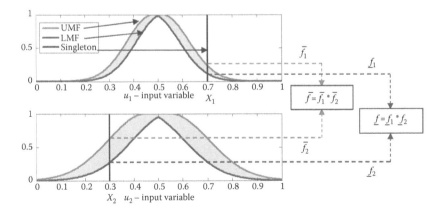

FIGURE 3.21
Antecedent combiner – singleton fuzzifier. (From Kashyap, S. K. Fusion of LWIR and EO images using fuzzy logic type 1 and type 2. MSDF (internal report/unclassified) Rep. 1305/ESVS01 (FMCD, CSIR-NAL, Bangalore), November 27, 2013. With permission.)

Case 2: Inputs u_1 and u_1 are fuzzified through the non-singleton T1FL fuzzifier. The antecedent combiner is realised using the following equations:

$$\bar{f}_1 = \sup\left\{\int_{UXV} \bar{\mu}_{A_1}(u_1) * \mu'_{F_1}(v_1)/(u_1, v_1)\right\}$$

$$\underline{f}_1 = \sup\left\{\int_{UXV} \underline{\mu}_{A_1}(u_1) * \mu'_{F_1}(v_1)/(u_1, v_1)\right\}; \quad u_1 \in U, v_1 \in V, U \subseteq [0,1], V \subseteq [0,1]$$

$$\bar{f}_2 = \sup\left\{\int_{UXV} \bar{\mu}_{A_2}(u_2) * \mu'_{F_2}(v_2)/(u_2, v_2)\right\}$$

$$\underline{f}_2 = \sup\left\{\int_{UXV} \underline{\mu}_{A_2}(u_1) * \mu'_{F_2}(v_1)/(u_2, v_2)\right\}; \quad u_2 \in U, v_2 \in V, U \subseteq [0,1], V \subseteq [0,1]$$

$$\bar{f} = \bar{f}_1 * \bar{f}_2$$

$$\underline{f} = \underline{f}_1 * \underline{f}_2$$

$$(3.127)$$

where '*' represents the t-norm fuzzy operator which could be 'min', 'prod' and so on. Here, the standard intersection (min) t-norm operator is used and Figure 3.22 illustrates the combiner's operation.

Case 3: Inputs u_1 and u_2 are fuzzified through the non-singleton T2FL fuzzifier. The antecedent combiner is realised using the following equations:

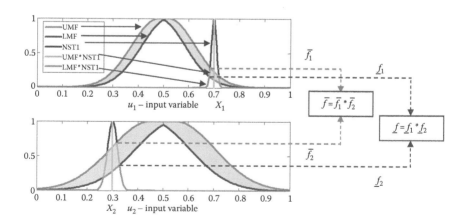

FIGURE 3.22

Illustration of antecedent combiner – non-singleton T1FL fuzzifier. (From Kashyap, S. K. Fusion of LWIR and EO images using fuzzy logic type 1 and type 2. MSDF (internal report/unclassified) Rep. 1305/ESVS01 (FMCD, CSIR-NAL, Bangalore), November 27, 2013. With permission.)

$$\bar{f}_1 = \sup\left\{\int_{UXV} \bar{\mu}_{A_1}(u_1) * \bar{\mu}'_{F_1}(v_1)/(u_1,v_1)\right\}$$

$$\underline{f}_1 = \sup\left\{\int_{UXV} \underline{\mu}_{A_1}(u_1) * \underline{\mu}'_{F_1}(v_1)/(u_1,v_1)\right\}; \quad u_1 \in U, v_1 \in V, U \subseteq [0,1], V \subseteq [0,1]$$

$$\bar{f}_2 = \sup\left\{\int_{UXV} \bar{\mu}_{A_2}(u_2) * \bar{\mu}'_{F_2}(v_2)/(u_2,v_2)\right\}$$

$$\underline{f}_2 = \sup\left\{\int_{UXV} \underline{\mu}_{A_2}(u_1) * \underline{\mu}'_{F_2}(v_1)/(u_2,v_2)\right\}; \quad u_2 \in U, v_2 \in V, U \subseteq [0,1], V \subseteq [0,1]$$

$$\bar{f} = \bar{f}_1 * \bar{f}_2$$

$$\underline{f} = \underline{f}_1 * \underline{f}_2 \tag{3.128}$$

where '*' represents the t-norm fuzzy operator which could be 'min', 'prod' and so on, here, the standard intersection (min) t-norm operator is used, and Figure 3.23 illustrates the combiner's operation.

Fuzzy inference provides fuzzified output set (shaped one) based on some operation (such as SI or AP etc.) between single value (known as firing strength) given by antecedent and output FS. In case of IT2FLS, there are two such values $(\bar{f}, \underline{f})$ obtained from antecedent operation; the fuzzified output sets for given rule can be obtained by the following equations:

$$\bar{\mu}_B^s(v) = \bar{f} * \bar{\mu}_B(v)$$

$$\underline{\mu}_B^s(v) = \underline{f} * \underline{\mu}_B(v); \quad v \in V \subseteq [0,1] \tag{3.129}$$

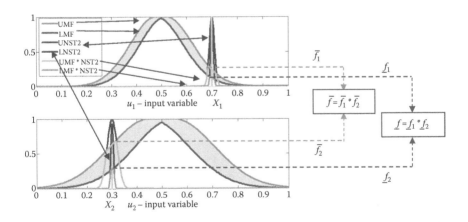

FIGURE 3.23

Illustration of antecedent combiner – non-singleton T2FL fuzzifier. (From Kashyap, S. K. Fusion of LWIR and EO images using fuzzy logic type 1 and type 2. MSDF (internal report/unclassi-fied) Rep. 1305/ESVS01 (FMCD, CSIR-NAL, Bangalore), November 27, 2013. With permission.)

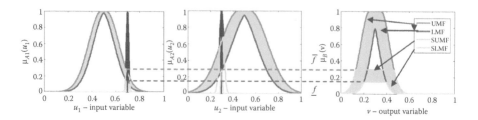

FIGURE 3.24
Illustration of fuzzy inference – non-singleton T2FL fuzzifier. (From Kashyap, S. K. Fusion of LWIR and EO images using fuzzy logic type 1 and type 2. MSDF (internal report/unclassified) Rep. 1305/ESVS01 (FMCD, CSIR-NAL, Bangalore), November 27, 2013. With permission.)

Here, '*' represents t-norm operators (SI, min, or AP etc.) If SI (min) is selected then, the fuzzified output sets (shaded UMF and shaded LMF) will be given by, Figure 3.24:

$$\bar{\mu}_B^s(v) = \min(\bar{f}, \bar{\mu}_B(v)); (\text{SUMF})$$
$$\underline{\mu}_B^s(v) = \min(\underline{f}, \underline{\mu}_B(v)); (\text{SUMF})$$

$$(3.130)$$

3.9.3 FSs-Type Reduction

Suppose lth rule of IT2FLS for DISO (dual input single output) system is defined by

$$R^l : \text{ IF } x_1 \text{ is } \tilde{A}_1^l \text{ AND } x_2 \text{ is } \tilde{A}_2^l , \text{ THEN } y \text{ is } \tilde{B}^l \qquad (3.131)$$

where total rules provided for a system are 'M', that is, $l = 1, \ldots ,M$; then, the outputs of FIS would be 'M' fuzzified $\tilde{B}^{S^l}$ T2 FLSs, and these FSs are converted into single T1 FLSs by FSs type reduction (TR) techniques Then, crisp output is obtained by applying defuzzification techniques on type reduced FS. The most popular TR method is the centre of sets (CoS) by Karnik and Mendel (KM). There are many versions of KM algorithms: Karnik Mendel Algorithm (KMA), enhanced KMA (EKMA) and enhanced KMA with new initialisation (EKMANI).

3.9.3.1 Centre of Sets (CoS)

Let the firing strengths of lth rule of IT2FS be given by $\bar{f}^l, \underline{f}^l$, then the centroid (using COA) of lth output fuzzified set is given by

$$\bar{y}^l = \frac{\sum_{k=1}^{N} \bar{\mu}_B^{S^l}(v(k)) * v(k)}{\sum_{k=1}^{N} \bar{\mu}_B^{S^l}(v(k))}$$

$$\underline{y}^l = \frac{\sum_{k=1}^{N} \underline{\mu}_B^{S^l}(v(k)) * v(k)}{\sum_{k=1}^{N} \underline{\mu}_B^{S^l}(v(k))}$$

(3.132)

where k is the number of samples (discrete points) of output FS and $v(k)$ is the crisp value (in ascending order) in output space $V \subseteq [0,1]$.

3.9.3.2 Karnik Mendel Algorithm (KMA)

The KMA is divided into two parts: (i) left and right one for computing, $\bar{y}$ and (ii) for computing $\underline{y}$ based on sets $\bar{f}^l, \underline{f}^l$ and $\bar{y}^l, \underline{y}^l$ with $l = 1, \dots, M$. Let us first compute $\bar{y}$ by using the following steps:

Step 1: Sort the $\bar{y}^l$s in ascending order, that is, $\bar{y}_a^l = [\bar{y}^1 \le \bar{y}^2 \le \bar{y}^3, \dots, \le \bar{y}^M]$ and rearrange corresponding values of $\bar{f}^l, \underline{f}^l$ and with naming as $\bar{f}_a^l, \underline{f}_a^l$, respectively.

Step 2: Compute initial value of vector $\bar{f}_s^l = \bar{f}_a^l + \underline{f}_a^l / 2$ with $l = 1, \dots, M$ and scalar variable $\bar{y}_s$ using following equation:

$$\bar{y}_s = \frac{\sum_{l=1}^{M} \bar{f}_s^l * \bar{y}_a^l}{\sum_{l=1}^{M} \bar{f}_s^l}$$

(3.133)

Assign the computed value of scalar variable $\bar{y}_s$ to new scalar variable $\bar{y}'_s$, that is, $\bar{y}'_s = \bar{y}_s$.

Step 3: Find out the index R such that $\bar{y}_a^R \le \bar{y}'_s \le \bar{y}_a^{R+1}$.

Step 4: Compute $\bar{y}_s$ using following equation:

$$\bar{y}_s = \frac{\sum_{l=1}^{R} \bar{f}_a^l * \bar{y}_a^l + \sum_{l=R+1}^{M} \underline{f}_a^l * \bar{y}_a^l}{\sum_{l=1}^{R} \bar{f}_a^l + \sum_{l=R+1}^{M} \underline{f}_a^l}$$

(3.134)

Assign the computed value of scalar variable $\bar{y}_s$ to new scalar variable $\bar{y}''_s$, that is, $\bar{y}''_s = \bar{y}_s$.

Step 5: If $\bar{y}''_s \ne \bar{y}'_s$ then go to Step 6 else stop and assign $\bar{y} = \bar{y}''_s$.

Step 6: Set $\bar{y}'_s = \bar{y}''_s$ and then go step 3.

The procedure to compute the second part (i.e. y) of the algorithm is similar to that of $\bar{y}$. In the case of SBS with IT2FLS, $\bar{y}$ and y are computed using sets $\bar{f}^l$, f^l and $\bar{y}^l$, y^l and methods 'wtaver' or 'wtsum', where w is nothing but $\bar{f}^l$ or f^l and z is $\bar{y}^l$ or y^l.

3.9.3.3 Defuzzification

The final crisp output of IT2FLS is computed by taking the average of $\bar{y}$ and y as follows:

$$y = \frac{\bar{y} + y}{2} \tag{3.135}$$

3.9.4 Implementation of Image Fusion Using MATLAB FLS Toolbox

In this section, T1FL and IT2FL are applied to obtain fused image based on input images from EOT camera (registered) and LWIR sensor of EVS prototype, and the pixel level fusion is carried out using grey images from these sensors [16]. These grey images are first normalised to crisp value between 0 and 1 and then fed as inputs to FLS to get fused output. The T1FL is designed using inbuilt MATLAB based toolbox named 'fuzzy'. Similarly, IT2FL is designed using third party MATLAB toolbox named 'fuzzy2' (copyright of MathWorks, Inc); see Appendixes 3A to 3C. Figure 3.25 shows the GUIs of these FL systems. Using these GUIs, a user can design FLS by selecting: (i) FLS architecture (Mamdani or Sugeno), (ii) fuzzifier (singleton, NST1, NST2)

(a)   (b)

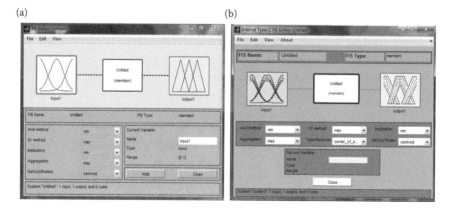

FIGURE 3.25
GUI of T1FLS (a) and IT2FLS (b). (From Kashyap, S. K. Fusion of LWIR and EO images using fuzzy logic type 1 and type 2. MSDF (internal report/unclassified) Rep. 1305/ESVS01 (FMCD, CSIR-NAL, Bangalore), November 27, 2013. With permission.)

in case of IT2FLS, (iii) number of inputs, (iv) number of linguistic labels for each input, (v) MF type and its parameters (which decide shape and spread) for each label, (vi) rule base (given by a domain expert), (vii) FL operators for antecedent combiners, (viii) inference, (ix) aggregation, (x) TR, type-reduction and (xi) defuzzification. Figure 3.26 shows the MFs of T1FLS.

The linguistic labels for input FS of both the sensors (EOT and LWIR) and output FS are designated as 'VL', 'L', 'M', 'H' and 'VH': where, 'V', 'L', 'M', 'H' stand for very, low, medium and high, respectively, representing range of grey level of input image. The MF selected for each label is of type 'trim'. Figure 3.27 shows the FMFs for IT2FLS. In this case, MF selected for each label is of Gaussian type. Since, there are two inputs (EOT and LWIR) and each input has got five linguistic labels, hence, maximum number of rules created will be $5 \times 5 = 25$. Table 3.5 shows the rule base (presently kept the same for T1FLS and IT2FLS) designed for image fusion application. The interpretation of one rule (shaded) from the rule base is as follows:

IF EOT is VL AND LWIR is VL THEN fused is VL

Figure 3.28 shows the surface plots of T1FLS and T2FLS.

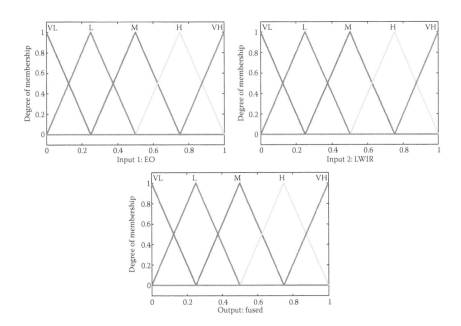

FIGURE 3.26
MFs for T1FLS. (From Kashyap, S. K. Fusion of LWIR and EO images using fuzzy logic type 1 and type 2. MSDF (internal report/unclassified) Rep. 1305/ESVS01 (FMCD, CSIR-NAL, Bangalore), November 27, 2013. With permission.)

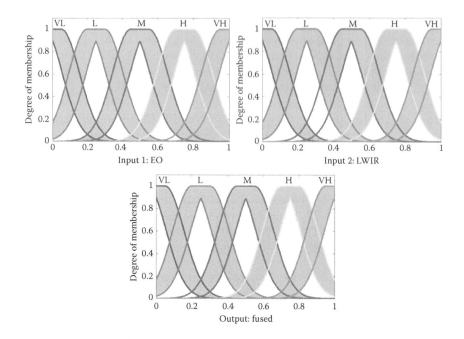

FIGURE 3.27

FMFs for IT2FLS. (From Kashyap, S. K. Fusion of LWIR and EO images using fuzzy logic type 1 and type 2. MSDF (internal report/unclassified) Rep. 1305/ESVS01 (FMCD, CSIR-NAL, Bangalore), November 27, 2013. With permission.)

TABLE 3.5

Rule Base with Total 25 Rules

		LWIR				
		VL	L	M	H	VH
EO	VL	VL	L	L	M	M
	L	VL	L	L	M	M
	M	L	L	M	H	H
	H	M	M	H	H	VH
	VH	M	M	H	H	VH

Source: From Kashyap, S. K. Fusion of LWIR and EO images using fuzzy logic type 1 and type 2. MSDF (internal report/unclassified) Rep. 1305/ESVS01 (FMCD, CSIR-NAL, Bangalore), November 27, 2013. With permission.

3.9.5 Results and Discussion

The main objective of the present example is to (i) demonstrate image fusion using T1FLS and IT2FLS and (ii) performance evaluation of these FLSs with respect to their design parameters such as: (a) number of rules, (b) FL operators for antecedent combiner, (c) inference and aggregation, (d) TR and

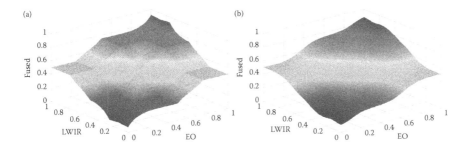

FIGURE 3.28
Surface plot of T1FLS (a) and IT2FLS (b) with 25 rules. (From Kashyap, S. K. Fusion of LWIR and EO images using fuzzy logic type 1 and type 2. MSDF (internal report/unclassified) Rep. 1305/ ESVS01 (FMCD, CSIR-NAL, Bangalore), November 27, 2013. With permission.)

(e) defuzzification methods. Tables 3.6 through 3.9 list design cases for MBS/ SBS T1FLS and IT2FLS. For IT2FLS, the function 'evalfis2.m' (Appendix 3C) is written by the contributor of this example [16]. This function is used for obtaining the fused image from the given input images.

The evaluation of the fused image for these design cases is carried out ana- lytically (through performance metrics) and qualitatively (by visual inspec- tion). The performance metrics used in evaluation are: (i) average entropy, (ii) average contrast, (iii) average luminance, (iv) image score, (v) signal to noise ratio, (vi) image contrast, (vii) correlation, (viii) energy, (ix) homogeneity, (x) spatial frequency and (xi) spectral activity measure (see Appendix B.7). The qualitative evaluation is carried out by finding out four best fused images from each SBS/MBS: (i) T1FLS, (ii) IT2FLS with singleton fuzzifier, (iii) IT2FLS with non-singleton fuzzifier T1FL and (iv) IT2FLS with non-singleton fuzzifier T2. At the second level, the top four images are sorted with one best from items (i), (ii), (iii) and (iv) of evaluation metrics. At the final level, the best image is obtained from four images obtained from the second level. Figure 3.29 shows the input images of EOT and LWIR taken from EVS proto- type using a ground experiment on an airport runway. Figures 3.30 through 3.37 show the fused images for; (i) T1FLS MBS, (ii) T1FLS SBS, (iii) IT2FLS MBS singleton, (iv) IT2FLS MBS NST1, (v) IT2FLS MBS NST2, (vi) IT2FLS SBS singleton, (vii) IT2FLS SBS NST1 and (viii) IT2FLS SBS NST2 design cases (as per Tables 3.6 through 3.9) respectively. Within a figure a number circled indicates the respective case number as mentioned in Tables 3.6 through 3.9.

3.9.5.1 Qualitative Analysis

The fused images depicted in Figures 3.30 through 3.37 were shown to a group of people for evaluation (while not disclosing the cases considered for generating the images). The criteria of evaluation were: (a) identification of runway, (b) runway markers, (c) runway number and (d) runway lights. The

TABLE 3.6

Design Cases for MBS T1FLS

Cases	Antecedent Combiner	Inference	Aggregation	Defuzzification
1	min	min	max	centroid
2	min	min	max	bisector
3	min	min	max	mom
4	min	min	max	lom
5	min	min	max	som
6	prod	min	max	centroid
7	prod	min	max	bisector
8	prod	min	max	mom
9	prod	min	max	lom
10	prod	min	max	som
11	min	prod	max	centroid
12	min	prod	max	bisector
13	min	prod	max	mom
14	min	prod	max	lom
15	min	prod	max	som
16	prod	prod	max	centroid
17	prod	prod	max	bisector
18	prod	prod	max	mom
19	prod	prod	max	lom
20	prod	prod	max	som
21	min	min	sum	centroid
22	min	min	sum	bisector
23	min	min	sum	mom
24	min	min	sum	lom
25	min	min	sum	som
26	prod	min	sum	centroid
27	prod	min	sum	bisector
28	prod	min	sum	mom
29	prod	min	sum	lom
30	prod	min	sum	som
31	min	prod	sum	centroid
32	min	prod	sum	bisector
33	min	prod	sum	mom
34	min	prod	sum	lom
35	min	prod	sum	som
36	prod	prod	sum	centroid
37	prod	prod	sum	bisector
38	prod	prod	sum	mom
39	prod	prod	sum	lom
40	prod	prod	sum	som

Source: From Kashyap, S. K. Fusion of LWIR and EO images using fuzzy logic type 1 and type 2. MSDF (internal report/unclassified) Rep. 1305/ESVS01 (FMCD, CSIR-NAL, Bangalore), November 27, 2013. With permission.

TABLE 3.7

Design Cases for SBS T1FLS

Cases	Antecedent Combiner	Inference	Aggregation	Defuzzification
1	min	prod	max	wtaver
2	min	prod	max	wtsum
3	prod	prod	max	wtaver
4	prod	prod	max	wtsum

Source: From Kashyap, S. K. Fusion of LWIR and EO images using fuzzy logic type 1 and type 2. MSDF (internal report/unclassified) Rep. 1305/ESVS01 (FMCD, CSIR-NAL, Bangalore), November 27, 2013. With permission.

TABLE 3.8

Design Cases for MBS IT2FLS

Cases	Antecedent Combiner	Inference	Type Reduction	Defuzzification
1	min	min	CoS	(3.135)
2	min	min	height	(3.135)
3	min	prod	CoS	(3.135)
4	min	prod	height	(3.135)
5	prod	min	CoS	(3.135)
6	prod	min	height	(3.135)
7	prod	prod	CoS	(3.135)
8	prod	prod	height	(3.135)

Source: From Kashyap, S. K. Fusion of LWIR and EO images using fuzzy logic type 1 and type 2. MSDF (internal report/unclassified) Rep. 1305/ESVS01 (FMCD, CSIR-NAL, Bangalore), November 27, 2013. With permission.

TABLE 3.9

Design Cases for SBS IT2FLS

Cases	Antecedent Combiner	Inference	Defuzzification
1	min	prod	wtaver and (3.135)
2	min	prod	wtsum and (3.135)
3	prod	prod	wtaver and (3.135)
4	prod	prod	wtsum and (3.135)

Source: From Kashyap, S. K. Fusion of LWIR and EO images using fuzzy logic type 1 and type 2. MSDF (internal report/unclassified) Rep. 1305/ESVS01 (FMCD, CSIR-NAL, Bangalore), November 27, 2013. With permission.

fused images were ordered, best to worst, by capability of clearly identifying these things. The qualitative analysis is summarised as follows (in some tables important results are highlighted by grey shades) [16]:

1. SBS (Sugeno) T1FL or SBS IT2FL performed better than MBS (Mamdani) T1FL or MBS IT2FL.
2. IT2FLS performs better than T1FL.

(a)   (b)

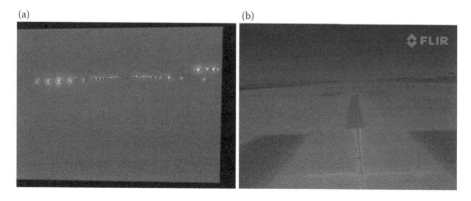

FIGURE 3.29
Input images of an airport runway: registered EOT (a), LWIR (b). (From Kashyap, S. K. Fusion of LWIR and EO images using fuzzy logic type 1 and type 2. MSDF (internal report/unclassified) Rep. 1305/ESVS01 (FMCD, CSIR-NAL, Bangalore), November 27, 2013. With permission.)

3. Within SBS/MBS IT2FLS, NST1 or NST2 based combination performs better than singleton fuzzifier-based combination.

4. In SBS T1FL, 'min' for consequent combining, 'prod' for inference and 'wtsum' for defuzzification perform very well (case 2 of Table 3.7; circle 2 in Figure 3.31).

5. In SBS IT2FLS (singleton, NST1 and NST2), 'min' for consequent combining, 'prod' for inference and 'wtsum' for defuzzification perform the best (case 2 in Table 3.9 and circle 2 in Figures 3.35 through 3.37).

6. In SBS, IT2FLS with NST1 is the best (circle 2 Figure 3.36) followed by IT2FLS with NST2 (circle 2 Figure 3.37), IT2FLS with singleton (circle 2, Figure 3.35) and last T1FLS (circle 2, Figure 3.31).

7. In MBS T1FL, except 'centroid' and 'bisector', application of all the other techniques for defuzzification result in unacceptable quality of the fused image.

8. In MBS IT2FLS and for singleton fuzzifier, 'min' for consequent combining, 'prod' for inference and 'center_of_sets' for TR perform very well (case 3 in Table 3.8 and circle 3 in Figure 3.32).

9. In MBS IT2FLS and for NST1 or NST2, 'prod' for consequent combining, 'prod' for inference and 'center_of_sets' for TR perform very well (case 7 in Table 3.8 and circle 7 in Figures 3.33 and 3.34).

10. In MBS, IT2FLS with NST1 is very good (circle 7 Figure 3.33) followed by IT2FLS with NST2 (circle 7 Figure 3.35), IT2FLS with singleton (circle 3, Figure 3.31) and last T1FL (white circle-patch 16 in Figure 3.30).

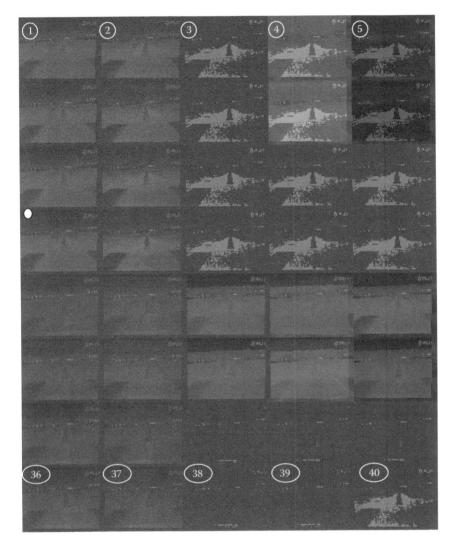

FIGURE 3.30
Fused images for the cases mentioned in Table 3.6. (From Kashyap, S. K. Fusion of LWIR and EO images using fuzzy logic type 1 and type 2. MSDF (internal report/unclassified) Rep. 1305/ESVS01 (FMCD, CSIR-NAL, Bangalore), November 27, 2013. With permission.)

3.9.5.2 Analytical Evaluation

Tables 3.10 through 3.17 show the numerical values of the performance metrics for the cases of Tables 3.6 through 3.9. For performance metrics, in general a good quality image should have a higher value of (a) average entropy, (b) average contrast, (c) average luminance, (d) signal to noise ratio, (e) image contrast, (f) correlation and (g) spatial frequency. The summary of

FIGURE 3.31
Fused images for the cases mentioned in Table 3.7. (From Kashyap, S. K. Fusion of LWIR and EO images using fuzzy logic type 1 and type 2. MSDF (internal report/unclassified) Rep. 1305/ESVS01 (FMCD, CSIR-NAL, Bangalore), November 27, 2013. With permission.)

FIGURE 3.32
Fused images for the cases mentioned in Table 3.8 – singleton fuzzifier. (From Kashyap, S. K. Fusion of LWIR and EO images using fuzzy logic type 1 and type 2. MSDF (internal report/unclassified) Rep. 1305/ESVS01 (FMCD, CSIR-NAL, Bangalore), November 27, 2013. With permission.)

the analytical evaluation is as follows (in some tables important results are highlighted by grey shades) [16]:

1. For the SBS FL, it can be seen from Tables 3.11, 3.15 through 3.17 that case 2 of each table satisfies the condition to qualify the best image within design cases of that table. This matches with qualitative

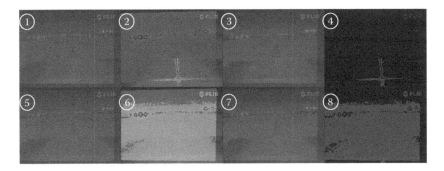

FIGURE 3.33
Fused images for the cases mentioned in Table 3.8 – non-singleton fuzzifier T1. (From Kashyap, S. K. Fusion of LWIR and EO images using fuzzy logic type 1 and type 2. MSDF (internal report/unclassified) Rep. 1305/ESVS01 (FMCD, CSIR-NAL, Bangalore), November 27, 2013. With permission.)

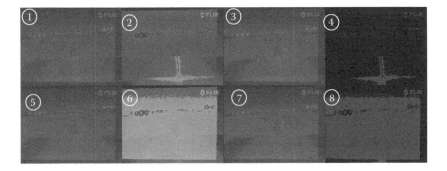

FIGURE 3.34
Fused images for the cases mentioned in Table 3.8 – non-singleton fuzzifier T2. (From Kashyap, S. K. Fusion of LWIR and EO images using fuzzy logic type 1 and type 2. MSDF (internal report/unclassified) Rep. 1305/ESVS01 (FMCD, CSIR-NAL, Bangalore), November 27, 2013. With permission.)

observation numbers 4 and 5. It is also observed that in general IT2FL (Tables 3.15 through 3.17) outperforms T1FL (Table 3.11), similar to observation number 2 in qualitative analysis.

2. Since, during qualitative analysis, it was found that for MBS T1FL design cases (Table 3.6) which uses bisector, lom, mom or som defuzzification methods yields a fused image of inferior quality therefore these cases are not considered in analytical evaluation. The left out cases are 1, 6, 11, 16, 21, 26, 31 and 36. During analytical evaluation (from Table 3.10) it was found that cases 1, 6, 11 and 16 stand out to be best ones producing a good quality image. In this, case 11 (against case 16, in case of qualitative observation number 7) seems to be the best among them.

FIGURE 3.35
Fused images for the cases mentioned in Table 3.9 – singleton fuzzifier. (From Kashyap, S. K. Fusion of LWIR and EO images using fuzzy logic type 1 and type 2. MSDF (internal report/unclassified) Rep. 1305/ESVS01 (FMCD, CSIR-NAL, Bangalore), November 27, 2013. With permission.)

FIGURE 3.36
Fused images for the cases mentioned in Table 3.9 – non-singleton fuzzifier T1. (From Kashyap, S. K. Fusion of LWIR and EO images using fuzzy logic type 1 and type 2. MSDF (internal report/unclassified) Rep. 1305/ESVS01 (FMCD, CSIR-NAL, Bangalore), November 27, 2013. With permission.)

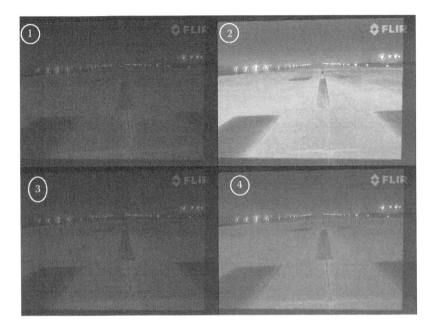

FIGURE 3.37
Fused images for the cases mentioned in Table 3.9 – non-singleton fuzzifier T2. (From Kashyap, S. K. Fusion of LWIR and EO images using fuzzy logic type 1 and type 2. MSDF (internal report/unclassified) Rep. 1305/ESVS01 (FMCD, CSIR-NAL, Bangalore), November 27, 2013. With permission.)

3. In case of MBS IT2FLSs, it is found from Tables 3.12 through 3.14 that case 6 produces the best quality images within the design cases of each table. In case of qualitative analysis case 7 was found to be the best (see observation numbers 10 and 11).

In the final analysis and summary from both qualitative and analytical evaluations we gather the following:

- IT2FLS performs better than T1FLS
- SBS, that is, Sugeno FL performs better than MBS, that is, Mamdani FL

3.10 DF Using Dempster-Shafer and Possibility Theory: Illustrative Example

In this example, we consider MSDF based on Dempster-Shafer (D-S) belief functions and possibility theory (FLS) for close range anti-personnel mine

TABLE 3.10

Performance Metrics for Fused Images for Cases of Table 3.6

Cases	a	b	c	d	e	f	g	h	i	j	k
1	5.0784	0.4518	76.2119	7.0493	5.5296	0.0215	0.9543	0.5133	0.9893	1.5731	1.3853
2	5.1289	0.5914	75.0730	6.6430	5.4565	0.0251	0.9363	0.5881	0.9876	1.7519	0.9301
3	2.0619	1.6410	67.3098	5.9729	3.1745	0.0926	0.8964	0.6884	0.9670	4.0070	0.0221
4	4.7535	2.0584	83.8466	6.1436	3.9438	0.1464	0.8708	0.3073	0.9362	4.2919	0.0365
5	4.7503	2.0386	46.1111	6.1680	2.1383	0.0879	0.9152	0.6436	0.9744	4.2720	0.0102
6	5.0597	0.4474	75.5893	7.1035	5.5867	0.0203	0.9538	0.5463	0.9900	1.5426	1.3521
7	4.9556	0.5857	74.5940	6.6454	5.4645	0.0249	0.9356	0.5946	0.9877	1.7559	0.8829
8	2.0620	1.6408	67.3098	5.9739	3.1746	0.0926	0.8965	0.6884	0.9670	4.0070	0.0221
9	5.6015	2.1074	84.0574	6.2719	3.8877	0.1095	0.9216	0.4450	0.9627	4.3821	0.0420
10	5.5741	2.0780	39.8452	6.2517	1.8285	0.0915	0.8279	0.6059	0.9639	4.3485	0.0077
11	5.3318	0.4906	75.3344	6.7161	5.3344	0.0222	0.9509	0.5460	0.9889	1.6051	1.2591
12	5.2236	0.6729	73.9910	6.3385	5.2412	0.0232	0.9366	0.6201	0.9885	1.9122	0.7166
13	2.0627	1.6483	67.3002	5.9727	3.1705	0.0928	0.8963	0.6884	0.9670	4.0150	0.0219
14	2.0627	1.6483	67.3002	5.9727	3.1705	0.0928	0.8963	0.6884	0.9670	4.0150	0.0219
15	2.0627	1.6483	67.3002	5.9727	3.1705	0.0928	0.8963	0.6884	0.9670	4.0150	0.0219
16	5.2580	0.4803	74.8775	6.7763	5.3874	0.0202	0.9544	0.5573	0.9901	1.5863	1.1796
17	5.0203	0.6698	73.4876	6.2946	5.2200	0.0228	0.9370	0.6240	0.9887	1.9375	0.6489
18	2.0627	1.6483	67.3002	5.9727	3.1705	0.0928	0.8963	0.6884	0.9670	4.0150	0.0219
19	2.0627	1.6483	67.3002	5.9727	3.1705	0.0928	0.8963	0.6884	0.9670	4.0150	0.0219
20	2.0627	1.6483	67.3002	5.9727	3.1705	0.0928	0.8963	0.6884	0.9670	4.0150	0.0219
21	4.9470	0.3776	72.8935	6.7159	7.1547	0.0099	0.9108	0.8912	0.9951	1.5214	1.3360
22	4.8858	0.4342	72.0230	7.1498	6.7501	0.0103	0.9153	0.8801	0.9949	1.5468	1.1775
23	5.3097	0.6264	73.4810	6.7522	3.8404	0.0309	0.9705	0.3441	0.9850	1.9972	0.4917
24	5.1838	0.6954	76.5133	6.8791	4.3811	0.0524	0.9487	0.3493	0.9745	2.1991	0.4582
25	5.5471	0.7086	69.3523	7.1806	3.1614	0.0337	0.9722	0.3176	0.9835	2.2496	0.4077
26	5.0890	0.4283	73.1088	7.0391	6.9615	0.0104	0.9156	0.8785	0.9948	1.4918	1.4058
27	5.0085	0.4727	72.5927	7.0232	6.4685	0.0100	0.9293	0.8622	0.9951	1.5579	1.1745
28	5.1630	0.5811	73.6201	6.7206	3.8234	0.0266	0.9761	0.3414	0.9871	1.9743	0.4978
29	5.5274	0.7249	77.7687	7.1316	4.3348	0.0281	0.9672	0.4868	0.9865	2.3172	0.4902
30	5.7352	0.7171	66.2865	7.2752	3.1144	0.0193	0.9626	0.5367	0.9906	2.3035	0.3466
31	5.1029	0.4236	72.6828	6.9643	7.0242	0.0107	0.9047	0.8896	0.9947	1.5120	1.2296
32	4.6523	0.4278	70.2348	7.0857	7.8664	0.0093	0.9049	0.9039	0.9955	1.5303	1.0027
33	0.4854	0.3344	64.1257	9.0533	11.2167	0.0146	0.6687	0.9742	0.9950	1.9318	0.0726
34	0.4854	0.3344	64.1257	9.0533	11.2167	0.0146	0.6687	0.9742	0.9950	1.9318	0.0726
35	0.4854	0.3344	64.1257	9.0533	11.2167	0.0146	0.6687	0.9742	0.9950	1.9318	0.0726
36	5.1580	0.4562	72.8508	7.2036	6.7845	0.0104	0.9182	0.8750	0.9949	1.5145	1.2289
37	4.6889	0.4628	70.7408	7.1616	7.5001	0.0099	0.9109	0.8905	0.9952	1.5359	1.0194
38	0.5232	0.3807	64.1571	8.5874	10.3288	0.0181	0.6554	0.9698	0.9940	2.0420	0.0607
39	0.5232	0.3807	64.1571	8.5874	10.3288	0.0181	0.6554	0.9698	0.9940	2.0420	0.0607
40	2.0627	1.6483	67.3002	5.9727	3.1705	0.0928	0.8963	0.6884	0.9670	4.0150	0.0219

Source: From Kashyap, S. K. Fusion of LWIR and EO images using fuzzy logic type 1 and type 2. MSDF (internal report/unclassified) Rep. 1305/ESVS01 (FMCD, CSIR-NAL, Bangalore), November 27, 2013. With permission.

TABLE 3.11

Performance Metrics for Fused Images for the Cases of Table 3.7

Cases	a	b	c	d	e	f	g	h	i	j	k
1	5.3571	0.4793	71.2951	7.0372	5.5700	0.0127	0.9513	0.7504	0.9937	1.6458	1.0746
2	6.5722	1.1215	78.9924	6.6828	3.1484	0.0324	0.9818	0.3078	0.9856	3.0982	0.3026
3	5.4402	0.5143	71.7121	7.1239	5.6216	0.0136	0.9323	0.8064	0.9933	1.6589	1.0661
4	5.4400	0.5142	71.7121	7.1258	5.6219	0.0136	0.9322	0.8065	0.9933	1.6589	1.0647

Source: From Kashyap, S. K. Fusion of LWIR and EO images using fuzzy logic type 1 and type 2. MSDF (internal report/unclassified) Rep. 1305/ESVS01 (FMCD, CSIR-NAL, Bangalore), November 27, 2013. With permission.

TABLE 3.12

Performance Metrics for Fused Images for the Cases of Table 3.8: Singleton Fuzzifier

Cases	a	b	c	d	e	f	g	h	i	j	k
1	5.4979	0.4653	76.0728	6.7886	6.3630	0.0181	0.9229	0.7725	0.9910	1.4848	1.6561
2	5.4230	1.3022	81.9386	6.4440	3.5258	0.0401	0.9622	0.4185	0.9819	3.2826	0.1310
3	5.6035	0.4872	74.8935	6.9017	5.6914	0.0210	0.9305	0.7180	0.9895	1.5370	1.5450
4	1.5395	0.6459	22.2335	7.0160	0.8081	0.0302	0.9802	0.6046	0.9932	2.5256	0.0138
5	5.4357	0.4586	76.6359	6.7551	5.9230	0.0247	0.9522	0.5270	0.9877	1.4916	1.7356
6	6.0795	3.4409	81.1502	6.1571	3.2549	0.2511	0.8783	0.2475	0.9234	5.6478	0.0236
7	5.4357	0.4586	76.6359	6.7551	5.9230	0.0247	0.9522	0.5270	0.9877	1.4916	1.7356
8	2.8075	2.3435	50.7249	6.4493	1.7804	0.1204	0.9256	0.5418	0.9728	4.9243	0.0078

Source: From Kashyap, S. K. Fusion of LWIR and EO images using fuzzy logic type 1 and type 2. MSDF (internal report/unclassified) Rep. 1305/ESVS01 (FMCD, CSIR-NAL, Bangalore), November 27, 2013. With permission.

TABLE 3.13

Performance Metrics for Fused Images for the Cases of Table 3.8: Non-Singleton Fuzzifier T1

Cases	a	b	c	d	e	f	g	h	i	j	k
1	5.4787	0.4696	76.8435	6.5704	6.4501	0.0197	0.9280	0.7383	0.9902	1.4996	1.7668
2	4.9812	0.8525	80.8959	7.2171	5.6331	0.0287	0.9453	0.5933	0.9861	2.5940	0.1920
3	5.5676	0.4878	75.8476	6.6943	5.8995	0.0199	0.9294	0.7326	0.9901	1.5501	1.6630
4	0.5790	0.4042	7.2564	9.4299	0.8309	0.0193	0.8728	0.9521	0.9954	2.0151	0.0020
5	5.4544	0.4740	77.0459	6.4975	6.1822	0.0264	0.9396	0.5905	0.9868	1.5039	1.7716
6	5.4389	1.4969	81.9233	6.6859	3.6852	0.0621	0.9572	0.4578	0.9767	3.5705	0.0935
7	5.5593	0.5117	75.7932	6.5888	5.5425	0.0283	0.9373	0.5833	0.9859	1.5683	1.5971
8	2.0401	1.1783	48.0712	7.0255	1.6224	0.0618	0.9649	0.5329	0.9863	3.4014	0.0147

Source: From Kashyap, S. K. Fusion of LWIR and EO images using fuzzy logic type 1 and type 2. MSDF (internal report/unclassified) Rep. 1305/ESVS01 (FMCD, CSIR-NAL, Bangalore), November 27, 2013. With permission.

TABLE 3.14

Performance Metrics for Fused Images for the Cases of Table 3.8: Non-Singleton Fuzzifier T2

Cases	a	b	c	d	e	f	g	h	i	j	k
1	5.4851	0.4682	77.0591	6.5768	6.4560	0.0205	0.9290	0.7228	0.9898	1.4955	1.8006
2	4.9818	0.9183	81.3706	7.0658	5.0475	0.0313	0.9551	0.5193	0.9850	2.6828	0.1825
3	5.5728	0.4912	75.9478	6.7195	5.8552	0.0207	0.9299	0.7201	0.9897	1.5549	1.6399
4	0.7870	0.4855	8.7145	8.2412	0.6850	0.0212	0.9346	0.9076	0.9949	2.1946	0.0037
5	5.4556	0.4679	77.2561	6.5652	6.1497	0.0238	0.9520	0.5449	0.9881	1.5012	1.8177
6	5.3157	1.4267	82.0157	6.8533	3.9013	0.0598	0.9560	0.5377	0.9794	3.4200	0.0997
7	5.5520	0.5040	75.8643	6.6604	5.4667	0.0266	0.9468	0.5475	0.9868	1.5628	1.6047
8	1.7868	0.9674	52.7140	7.4280	1.9142	0.0508	0.9666	0.5973	0.9886	3.0990	0.0198

Source: From Kashyap, S. K. Fusion of LWIR and EO images using fuzzy logic type 1 and type 2. MSDF (internal report/unclassified) Rep. 1305/ESVS01 (FMCD, CSIR-NAL, Bangalore), November 27, 2013. With permission.

TABLE 3.15

Performance Metric for Fused Images for the Cases of Table 3.9: Singleton Fuzzifier

Cases	a	b	c	d	e	f	g	h	i	j	k
1	5.6252	0.5340	71.8075	7.0098	5.5421	0.0143	0.9426	0.7643	0.9929	1.6499	1.1558
2	6.5431	1.0158	83.1011	6.5620	3.7740	0.0373	0.9796	0.2871	0.9822	2.7024	0.8168
3	5.5605	0.5481	71.9402	6.7619	5.5921	0.0146	0.9259	0.8093	0.9928	1.6650	1.1141
4	5.9259	0.6682	80.9896	6.9110	4.8759	0.0155	0.9775	0.4577	0.9925	1.9902	1.3486

Source: From Kashyap, S. K. Fusion of LWIR and EO images using fuzzy logic type 1 and type 2. MSDF (internal report/unclassified) Rep. 1305/ESVS01 (FMCD, CSIR-NAL, Bangalore), November 27, 2013. With permission.

TABLE 3.16

Performance Metrics for Fused Images for the Cases of Table 3.9: Non-Singleton Fuzzifier T1

Cases	a	b	c	d	e	f	g	h	i	j	k
1	5.6465	0.5419	72.8380	6.9327	5.3920	0.0153	0.9386	0.7660	0.9924	1.6918	1.2500
2	6.7703	1.2041	83.9468	6.5822	3.9211	0.0446	0.9823	0.2541	0.9793	2.8793	1.1378
3	5.6109	0.5572	72.5908	6.8237	5.4875	0.0153	0.9284	0.7953	0.9924	1.6812	1.2227
4	6.4223	0.9327	83.6440	6.8686	4.7287	0.0286	0.9749	0.3834	0.9862	2.3378	1.5732

Source: From Kashyap, S. K. Fusion of LWIR and EO images using fuzzy logic type 1 and type 2. MSDF (internal report/unclassified) Rep. 1305/ESVS01 (FMCD, CSIR-NAL, Bangalore), November 27, 2013. With permission.

detection [17]. The two main humanitarian-mine-detection/action-types/ tools (HMDAT) that could benefit from MSDF techniques are: (a) close-range anti-personnel mine detection and (b) mined area reduction. The first one consists of detection of (sub-) surface anomalies that may be related to the presence of mines, and/or detection of explosive materials. For the

TABLE 3.17

Performance Metrics for Fused Images for the Cases of Table 3.9: Non-Singleton Fuzzifier T2

Cases	a	b	c	d	e	f	g	h	i	j	k
1	5.6461	0.5372	72.7189	6.8851	5.4129	0.0157	0.9376	0.7630	0.9922	1.6805	1.2587
2	6.7947	1.5529	83.8918	6.9786	3.9198	0.0522	0.9773	0.2901	0.9755	3.3440	0.7917
3	5.6038	0.5542	72.5122	6.8823	5.5037	0.0151	0.9287	0.7964	0.9925	1.6809	1.1875
4	6.4149	1.3104	83.5556	7.1239	4.7235	0.0409	0.9655	0.3613	0.9801	2.8369	0.8691

Source: From Kashyap, S. K. Fusion of LWIR and EO images using fuzzy logic type 1 and type 2. MSDF (internal report/unclassified) Rep. 1305/ESVS01 (FMCD, CSIR-NAL, Bangalore), November 27, 2013. With permission.

close-range detection, efficient modelling and fusion of extracted features would improve the reliability and quality of single sensor-based processing, however, due to several situations within a minefield (such as specific moisture, depth, burial angles etc.) and between different minefields (such as types of mines, types of soil, minefield structure etc.), a good performance of HMDAT can only be obtained using MSDF approaches. Since the sensors detect different features, the combination of these complementary pockets of information would improve the results of detection and classification. In order to take into account the inter- and intra-minefield variability, uncertainty, ambiguity and partial knowledge, FLS or possibility theory and D-S/belief functions could prove to be very useful in this MSDF exercise of HMDAT. This application is illustrated using real data [17] coming from three complementary sensors: (a) metal detector, (b) ground-penetrating radar (GPR) and IR camera.

3.10.1 Information Fusion for Close-Range Mine Detection

Due to the types of mines and conditions, there is no single sensor used for HMD(AT) that can reach the high detection rate in many possible scenarios. Hence, one of the most promising sensor combinations consisting of an IR camera, an imaging metal detector (IMD) and GPR can be effectively used.

3.10.1.1 Belief Function Fusion

D-S evidence theory/belief function, that is, D-S theory allows representing both imprecision and uncertainty. This is done using plausibility and belief functions derived from a mass function. The mass of a proposition A is a part of the initial unitary amount of belief that signifies that the solution is exactly in A, and is specified defined as a function m from 2^Θ into [0, 1], where Θ is the decision space. This decision space is also called frame of discernment or full set. The constraints imposed are as follows, for null set and set A:

$$m(0) = 0; \quad \text{and} \quad \sum_{A \subseteq \Theta} m(A) = 1 \tag{3.136}$$

The D-S theory can be successfully used to include ignorance or partial ignorance, confusion between classes and partial reliability, and the masses are assigned by different sources/classifiers and governed by the orthogonal rule

$$m_{ij}(S) = \sum_{\substack{k,l \\ A_k \cap B_l = S}} m_i(A_k) \cdot m_j(B_l) \tag{3.137}$$

Here, S is any subset of the full set, and m_i and m_j are masses assigned by measures i and j, and their focal elements are $A_1, A_2, \dots , A_p$ and $B_1, B_2, \dots , B_q$. After the combination in this un-normalised form, the mass that is assigned to the empty set, is given as follows [17]:

$$m_{ij}(0) = \sum_{\substack{k,l \\ A_k \cap B_l = 0}} m_i(A_k) \cdot m_j(B_l) \tag{3.138}$$

is interpreted as a measure of conflict between the sources. All imprecision in the data should be introduced explicitly at the modelling level, especially in the choice of the focal elements: ambiguity between two classes in one source of information has to be modelled using a disjunction of hypotheses, so that conflict with other sources is limited and ambiguity can be possibly solved during the combination. We have a belief function as follows

$$Bel(A) = \sum_{B \subseteq A, B \neq 0} m(B); \quad \forall A 2^{\Theta} \tag{3.139}$$

and we have the plausibility function as follows:

$$Pls(A) = \sum_{B \cap A \neq 0} m(B); \quad \forall A 2^{\Theta} \tag{3.140}$$

Once the combination is done, the final decision is taken in favour of a simple hypothesis using: (i) the maximum of plausibility (generally over simple hypotheses), (ii) the maximum of belief or (iii) the pignistic decision rule. In case of HMDAT, it may be necessary to give more importance to some classes, for example, mines at the decision level; then maximum of plausibility can be used for the classes that should not be missed, and maximum of belief for some other situations.

3.10.1.2 *Fuzzy and Possibility Fusion*

FLSs and possibility theory (FLSPOS)-based models explicitly represent imprecision in the information and possible ambiguity between classes or decisions. In the fusion-combination step, the merits of FLSPOS rely on the operators that may deal with heterogeneous information [17]: t-norms, s-norms (T-conorms), mean operators, symmetrical sums and the operators which take into account any conflict between sources or reliability of the sources/sensors (Chapter 2). Unlike, the Bayesian or D-S-based fusion combination, FLSPOS provides a great flexibility in the choice of the operator, since each piece of information is converted in MFs of FLSs or PODI over the same decision space, and the decision is usually taken from the maximum of membership or possibility values after the fusion-combination step.

3.10.2 Close-Range Mine Detection Measures

A number of measures can be employed: (i) the area and the shape (elongation and ellipse fitting) of the object observed using the IR sensor, (ii) the size of the metallic area in IMD data and (iii) the propagation velocity (i.e. the type of material), the burial depth of the object observed using GPR, and the ratio between object size and its scattering function. Even as the semantics are different, similar information can be modelled in both possibilistic and belief function models. The aim is to design the possibility and mass functions as similarly as possible and concentrate on the comparison at the fusion-combination step. The ignorance is explicitly modelled in the D-S approach through a mass on the whole set (to guarantee the normalisation of the mass function over the power set). It is only expressed implicitly in the possibilistic model through the absence of the normalisation constraint.

3.10.2.1 *IR Measures*

The possibility degrees are derived from elongation and ellipse fitting measures and are given by π_{1l} and π_{2l}, respectively: due to relation to shape regularity, they are defined for a regular-shaped mine (RM), an irregular-shaped mine (IM), a regular-shaped non-dangerous (i.e. friendly) object (RF) and an irregularly shaped friendly object (IF). In the D-S framework, the full set is: $\Theta = \{RM, IM, RF, IF\}$. As elongation- and ellipse-fitting aim at distinguishing regular and irregular shapes, masses assigned by these two measures, m_{1l} and m_{2l}, are split between $RM \cup RF$, $IM \cup IF$ and Θ. For elongation, calculate r_1 as the ratio between minimum and maximum distance of bordering pixels from the CoG of (using thresholded images) and r_2 as the ratio of minor and major axis obtained from second moment calculation. Using these ratios, the possibility degrees derived are as follows [17]:

$$\pi_{1l}(RM) = \pi_{1l}(RF) = \min(r_1, r_2) \qquad (3.141)$$

$$\pi_{1I}(IM) = \pi_{1I}(IF) = 1 - \pi_{1I}(RM) \tag{3.142}$$

For D-S we have the following equations for the masses:

$$m_{1I}(RM \cup RF) = \min(r_1, r_2) \tag{3.143}$$

$$m_{1I}(IM \cup IF) = |r_1 - r_2| \tag{3.144}$$

For the full set we have

$$m_{1I}(\Theta) = 1 - \max(r_1, r_2) \tag{3.145}$$

For ellipse fitting, A_{oe} is the part of object area that belongs to the fitted ellipse as well, A_o is the object area and A_e is the ellipse area. Then, the following PODI is defined [17]

$$\pi_{2I}(RM) = \pi_{2I}(RF) = \max\left[0, \min\left(\frac{A_{oe}-5}{A_o}, \frac{A_{oe}-5}{A_e}\right)\right] \tag{3.146}$$

$$\pi_{2I}(IM) = \pi_{2I}(IF) = 1 - \pi_{2I}(RM) \tag{3.147}$$

The D-S masses for this IR measure are given as follows:

$$m_{2I}(RM \cup RF) = \max\left[0, \min\left(\frac{A_{oe}-5}{A_o}, \frac{A_{oe}-5}{A_e}\right)\right] \tag{3.148}$$

$$m_{2I}(IM \cup IF) = \max\left[\frac{A_{oe}-5}{A_o}, \frac{A_{oe}-5}{A_e}\right] \tag{3.149}$$

$$m_{2I}(\Theta) = 1 - m_{2I}(RM \cup RF) - m_{2I}(IM \cup IF) \tag{3.150}$$

When we are sure that all the mines have a regular-shaped mine (RM), the PODI degrees of RM can be reassigned to mines of any shape $M = RM \cup IM$, while the PODI degrees of IM can be reassigned to friendly objects of any $F = RF \cup IF$. Similarly, the masses given to $RM \cup RF$ can be reassigned to M, while masses given to $IM \cup IF$ can be reassigned to F. Now, the area directly provides a degree $\pi_{3I}(M)$ of being a mine, and since the range of possible antipersonnel mine sizes is approximately known, the degree of PODI being a mine is derived as a function of the measured size,

$$\pi_{31}(M) = \frac{a_I}{a_I + 0.1 a_{Imin}} \exp \frac{-[a_I - 0.5(a_{Imin} + a_{Imax})]^2}{0.5(a_{Imax} - a_{Imin})^2} \tag{3.151}$$

Here, a_I is the actual area of the object of the IR image, while the approximate range of expectable mine areas is between a_{Imin} and a_{Imax} (for anti-personnel mines, it is reasonable to set the values as 15 and 225 cm^2, respectively). Friendly objects can be of any size, so the PODI degree is set to one: $\pi_{31}(F) = 1$. Then, the area/size mass assignment is given by

$$m_{31}(\Theta) = \frac{a_I}{a_I + 0.1 a_{Imin}} \exp \frac{-[a_I - 0.5(a_{Imin} + a_{Imax})]^2}{0.5(a_{Imax} - a_{Imin})^2} \tag{3.152}$$

$$m_{31}(RF \cup IF) = 1 - m_{31}(\Theta)$$

3.10.2.2 IMD Measures

The metal detector, IMD data are usually saturated and data gathering resolution in the cross-scanning direction is typically very poor. Hence, the IMD information consists of only one measure, which is the width of the region in the scanning direction, w cm. The friendly objects can contain metal of any size, hence, we have $\pi_{MD}(F) = 1$. In case there is some knowledge on the expected sizes of metal in mines (for antipersonnel mines, the range is typically between 5 and 15 cm), we can assign possibilities to mines as [17]

$$\pi_{MD}(M) = \frac{w}{20}[1 - \exp(-0.2w)]\exp\left(1 - \frac{w}{20}\right) \tag{3.153}$$

and the corresponding mass functions are given as

$$m_{MD}(\Theta) = \frac{w}{20}[1 - \exp(-0.2w)]\exp\left(1 - \frac{w}{20}\right) \tag{3.154}$$

$$m_{MD}(RF \cup IF) = 1 - m_{MD}(\Theta) \tag{3.155}$$

3.10.2.3 GPR Measures

For mine-burial depth information, D, friendly objects can be found at any depth, while it is known that there is some maximum depth up to which anti-personnel mines can be expected. This is mainly due to their activation principles. But, due to soil perturbations, erosions and so on mines can, in process of time, go deeper or shallower than the depth at which they were initially buried; they can rarely be found buried below 25 cm (D_{max}). For this GPR measure, PODI π_{1G} for mines and friendly objects is modelled as follows [17]:

$$\pi_{1G}(M) = \frac{1}{\cosh(D/D_{max})^2}; \quad \pi_{1G}(F) = 1 \qquad (3.156)$$

The belief masses for this GPR measure are as follows:

$$m_{1G}(\Theta) = \frac{1}{\cosh(D/D_{max})^2}; \quad m_{1G}(RF \cup IF) = 1 - m_{1G}(\Theta) \qquad (3.157)$$

One additional GPR measure is the ratio of the object size to its scattering function, d/k. Friendly objects can have any value of this measure. For mines, there is a range of values that mines can have. Outside that range, the object is quite certainly not a mine:

$$\pi_{2G}(M) = \exp[-\frac{[(d/k) - m_d]^2}{2p^2}]; \quad \pi_{2G}(F) = 1 \qquad (3.158)$$

Here, m_d is the d/k value at which the PODI reaches its maximum value (here, $m_d = 700$, based on a priori information) and p is the width of the exponential function ($p = 400$). Similarly, the belief masses for GPR measure are

$$m_{2G}(\Theta) = \exp\left[-\frac{[(d/k) - m_d]^2}{2p^2}\right]; \quad m_{2G}(RF \cup IF) = 1 - m_{2G}(\Theta) \qquad (3.159)$$

The propagation velocity v can provide information about the object's identity, and we can extract depth information in a different way than in the case of the burial depth measure. We preserve the sign of the extracted depth. This information indicates if a potential object is above the surface. In that case, the extracted v should be close to $c = 3 \times 10^8$ m/s, the propagation velocity (of electromagnetic wave/speed of light) in vacuum; and if the sign indicates that the object is below the soil surface, the value of v should be around the values for the corresponding medium, (from 5.5×10^7 to 1.73×10^8 m/s in case of sand) [17]

$$\pi_{3G}(M) = \exp\left[-\frac{(v - v_{max})^2}{2h^2}\right] \qquad (3.160)$$

Here, v_{max} is the value of velocity which is the most typical for the medium (for sand, it is $0.5(5.5 \times 10^7 + 1.73 \times 10^8) = 1.14 \times 10^8$ m/s and for air, it is equal to *c*). The width of the exponential function *h* is 6×10^7 m/s [17]. The friendly objects can have any value of the velocity $\pi_{3G}(F) = 1$. The corresponding belief mass functions are given as

$$m_{3G}(\Theta) = \exp\left[-\frac{(v - v_{\max})^2}{2h^2}\right] \tag{3.161}$$

$$m_{3G}(RF \cup IF) = 1 - m_{3G}(\Theta) \tag{3.162}$$

3.10.3 Fusion Combination Evaluation

The fusion combination of PODI degrees, as well as of masses, is carried out in two steps: (i) apply to all the measures derived from one sensor and (ii) combine the results obtained in the first step for all three sensors. For PODI, only the combination rules related to mines are considered. For IR (here, the mines can be regular or irregular), the information about regularity on the level of each shape measure is combined using a disjunctive (max) operator as follows [17]:

$$\pi_{1IM} = \max\left(\pi_{1I}(RM), \pi_{1I}(IM)\right) \tag{3.163}$$

$$\pi_{2IM} = \max\left(\pi_{2I}(RM), \pi_{2I}(IM)\right) \tag{3.164}$$

The choice of the max operator, smallest disjunction and idempotent operator, as a T-conorm is related to the fact that the measures are not fully independent from each other. Hence, there is no reason to reinforce the measures by using a larger T-conorm, and the idempotent one is preferable in such situations and the two shape constraints should be both satisfied to have a high PODI degree of being a mine (they are combined in a conjunctive way using a product). The object is possibly a mine, if it has a size in the expected range or it satisfies the shape constraint, hence, the combination for IR measure is given as follows:

$$\pi_I(M) = \pi_{3I}(M) + [1 - \pi_{3I}(M)]\,\pi_{1IM}\pi_{2IM} \tag{3.165}$$

The conjunction of the second term ensures $\pi_I(M)$ is in [0,1]. For GPR, it is a mine if the object is at shallow depths and its dimensions resemble a mine and the extracted propagation velocity is appropriate for the medium. The fusion combination of the obtained possibilities for mines is performed using a t-norm, signifying the conjunction of all the criteria. The product t-norm is used as follows:

$$\pi_G(M) = \pi_{1G}(M)\pi_{2G}(M)\pi_{3G}(M) \tag{3.166}$$

For the metal detector, IMD (as one measure is used, there is no first combination step) the PODI is directly obtained (see Equation 3.153). For PODI, the second combination step is performed using the AS as follows:

$$\pi(M) = \pi_1(M) + \pi_{MD}(M) + \pi_G(M) - \pi_I(M)\pi_{MD}(M) - \pi_I(M)\pi_G(M)$$
$$- \pi_{MD}(M)\pi_G(M) + \pi_I(M)\ \pi_{MD}(M))\pi_G(M) \tag{3.167}$$

which leads to a strong disjunction, as the final possibility should be high if at least one sensor provides a high possibility. This means it is better to assign a friendly object to the mine class than it is to miss a mine. In the D-S framework, for IR and GPR sensors, the masses assigned by the measures of each of the two sensors are combined by the D-S rule in un-normalised form, Equation 3.137, the idea is to preserve conflict, that is, mass assigned to the empty set, Equation 3.138. The interest is in the possibility that sensors do not refer to the same object, as the unreliability can be modelled and resolved through discounting factors. After combining masses per sensor, the fusion of sensors/data is performed, using Equation 3.137, and if the mass of the empty set (after combination of sensors is high), they should be clustered as they do not sense the same object.

3.10.4 Comparison and Decision Results

For IR it can be shown that $Pl_I(M) \le \pi_I(M)$, and is in accordance with the least commitment principle used in the possibilistic model, however, as far as IMD is concerned, there is no difference, this is because it provides only one measure. In case of GPR, we have the following situation [17]:

$$\pi_G(M) = m_{1G}(\Theta)m_{2G}(\Theta)m_{3G}(\Theta) \tag{3.168}$$

Then, the application of the D-S theory to the mass assignments for the GPR measures yields the fused mass for the full set of the sensor as follows:

$$m_G(\Theta) = m_{1G}(\Theta)m_{2G}(\Theta)m_{3G}(\Theta) \tag{3.169}$$

From Equations 3.168 and 3.169, we can see the following equality:

$$\pi_G(M) = m_G(\Theta) \tag{3.170}$$

From the foregoing, we see that the ignorance is modelled as a mass on Θ in the D-S framework, while it prefers the class that should not be missed (M) in the possibilistic sense, for the ignorance would lead to safely decide in favour of mines. For the ultimate decision about the identity of the object, it should be left to the de-miner, not only because his life is in danger but also because of his experience; the fusion output is a suggested decision together with confidence degrees. In case of PODI-possibility theory, the final decision is obtained by thresholding the fusion result for M and providing the corresponding PODI degree as the confidence degree. Since, almost all PODI degrees obtained at the fusion centre are either very low or very high, the selected regions having very low values of $\pi(M)$ (below 0.1) are classified as F. The ones with very high values (>0.7) are classified as M. Only for a few regions at which the resulting PODI degree for M has an intermediary value and there (as mines must not be missed), the decision is M. This decision is referred to as *dec*1, and an alternative, *dec*2, for the final decision making is: (i) to derive the combination rule for F as well and (ii) compare the final values for M and F, and derive an adequate decision rule [17]. Based on the operation principles of GPR and IMD, the measures of these two sensors would only give information where mines are possibly not present. When they are non-informative with respect to F-objects, it is not useful to combine their PODI degrees for F. Hence, for deriving the final combination rule for F, $\pi(F)$, we can rely only on IR, and have $\pi(F) = \pi_1(F)$. For IR sensor, since F-objects can be regular or irregular, one applies a disjunctive 'max' operator for each of the shape constraints, then, when deciding F, one can combine the two shape constraints and the area measure using a conjunctive operator, resulting in the following equation:

$$\pi(F) = \max\left(\pi_{1I}(R_F), \pi_{1I}(I_F)\right) \max\left(\pi_{2I}(R_F), \pi_{2I}(I_F)\right) \qquad (3.171)$$

In the regions where IR gives an alarm, the decision rule chooses M or F depending on which one of the two has a higher possibility value. In other regions, at which IR does not give an alarm, although at least one of the two other sensors gives an alarm, the decision is based on the fusion result for M, as in *dec*1. In case of D-S, usual decision rules based on beliefs (plausibilities) do not give useful results because there are no focal elements containing mines alone. These usual decision rules would always favour F-objects, the reason being that the humanitarian demining sensors are anomaly detectors and not mine detectors, and in this sensitive situation, no mistakes are allowed. So, in case of any ambiguity, much more importance should be given to mines, as such then, guesses $G(A)$ are defined [17], where $A \in \{M,F,0\}$

$$G(M) = \sum_{M \cap B \neq 0} m(B) \qquad (3.172)$$

$$G(F) = \sum_{B \subseteq F, B \neq 0} m(B) \qquad (3.173)$$

$$G(0) = m(0) \qquad (3.174)$$

The guess value of a mine is the sum of masses of all the focal elements containing mines, regardless of their shape, and the guess of a F-object is the sum of masses of all the focal elements containing nothing else but F-objects of any shape. This means that the guesses are a cautious way to estimate confidence degrees. As the output of the D-S function fusion module, the three possible outputs (M, F, conflict) are provided together with the guesses, for each of the sensors and for their combination, and for GPR, the focal elements are only F and Θ. Hence, the guesses for this sensor become

$$G_G(M) = m_G(\Theta) \tag{3.175}$$

$$G_G(F) = m_G(F) \tag{3.176}$$

We, then conclude that for GPR, the possibility degree of a mine is equal to the guess of a mine

$$\pi_G(M) = G_G(M) \tag{3.177}$$

We also, observe that the guess of a mine is equal to its plausibility, and the guess of an F-object is equal to its belief, thus, actually, for IR we have the following inequality:

$$G_I(M) \leq \pi_I(M) \tag{3.178}$$

The approaches of this section were considered in Reference 17 for a set of known objects, buried in sand, leading to 36 alarmed regions in total of: 21 mines (M), 7 placed false alarms (PF, friendly objects) and 8 false alarms caused by clutter (FN, with no object). Since, all the mines were classified correctly, the results of the possibilistic fusion seem very promising [17] (see Table 3.18). The bracketed numbers indicate the number of regions selected in the pre-processing step for subsequent analysis, this being measure extraction and classification. In the second fusion step a decision is taken after the first step provides only 18 mines for IR, 9 for IMD and 13 for GPR, a result of combining heterogeneous sensors; the decision rules, *dec*1 and *dec*2, give the same results for mines and F-objects caused by clutter [17]. In case of placed false alarms, 2 are correctly classified in case of *dec*2, which is a slight improvement with respect to *dec*1 and the same result as for the D-S fusion. The placed false alarms are not well detected by any of the methods, since the model is designed in order to favour the detection of mines, also the type of result expected from de-miners. On the correct classification side of mines, the results of the possibilistic fusion are slightly better than those obtained using the D-S method (19 mines detected), this being due to the increased flexibility at the combination level. The false

TABLE 3.18

Correct Classification of Objects: Results of Fusion

Total	Sensor – IR		Sensor – MD		Sensor – GPR		Fusion – Possi		Fusion – D-S
	Possi	D-S	Possi	D-S	Possi	D-S	*dec1*	*dec2*	—
M = 21	18(18)	10(18)	9(9)	9(9)	13(13)	13(13)	21(21)	21(21)	19(21)
PF = 7	0(4)	3(4)	0(4)	0(4)	2(6)	1(6)	1(7)	2(7)	2(7)
FN = 8	0(1)	0(1)	0(0)	0(0)	6(7)	6(7)	6(8)	6(8)	6(8)

Source: Modified from Milisavljevic, N., Bloch, I. and Acheroy, M. 2008. Multi-sensor data fusion based on belief functions and possibility theory: Close range antipersonnel mine detection and remote sensing mined area reduction. In (Ed. Habib, M. K.), ISBN: 978-3-902613-11-0, In-Tech, available from: http://www.intechopen. com/books/humanitarian_demining/mltisensordata_fusion_based_on_belief_functions_and_possibility_theory__close_range_antipersonnel, accessed October 2014.

alarms with no objects are correctly identified by the D-S method (6 out of 8), and also is the case for the two possibilistic decision rules. The power of the methods studied in this section is in decreasing the number of clutter-caused false alarms without decreasing the result of mine detection. Although, the general shapes of the PODI are important and have been designed based on prior knowledge, they do not need to be estimated very precisely, and yet the results are robust to small changes in these functions. The functions are not crisp (the thresholding approach was not used) and that the rank is preserved (an object with a measure value outside of the usual range should have a lower possibility degree than an object with a typical measure value) [17]. The robustness is due to: (i) these PODIs are being used to model imprecise information, so they do not have to be precise themselves and (ii) each of them is combined in the fusion process with other pieces of information, which diminishes the importance and the influence of each of them.

Appendix 3A: Type 1 – Triangular MF-MATLAB Code

Source: Kashyap, S. K. Fusion of LWIR and EO images using fuzzy logic type 1 and type 2. MSDF (internal report/unclassified) Rep. 1305/ESVS01 (FMCD, CSIR-NAL, Bangalore), November 27, 2013.

```
[System]
Name = 'fis_25r_mamdani'
Type = 'mamdani'
Version = 2.0
```

```
NumInputs = 2
NumOutputs = 1
NumRules = 25
AndMethod = 'min'
OrMethod = 'max'
ImpMethod = 'min'
AggMethod = 'max'
DefuzzMethod = 'centroid'


[Input1]
Name = 'input1'
Range = [0 1]
NumMFs = 5
MF1 = 'VL':'trimf',[-0.25 0 0.25]
MF2 = 'L':'trimf',[0 0.25 0.5]
MF3 = 'M':'trimf',[0.25 0.5 0.75]
MF4 = 'H':'trimf',[0.5 0.75 1]
MF5 = 'VH':'trimf',[0.75 1 1.25]


[Input2]
Name = 'input2'
Range = [0 1]
NumMFs = 5
MF1 = 'VL':'trimf',[-0.25 0 0.25]
MF2 = 'L':'trimf',[0 0.25 0.5]
MF3 = 'M':'trimf',[0.25 0.5 0.75]
MF4 = 'H':'trimf',[0.5 0.75 1]
MF5 = 'VH':'trimf',[0.75 1 1.25]


[Output1]
Name = 'output1'
Range = [0 1]
NumMFs = 5
MF1 = 'VL':'trimf',[-0.25 0 0.25]
MF2 = 'L':'trimf',[0 0.25 0.5]
MF3 = 'M':'trimf',[0.25 0.5 0.75]
MF4 = 'H':'trimf',[0.5 0.75 1]
MF5 = 'VH':'trimf',[0.75 1 1.25]


[Rules]
11,1(1):1
12,2(1):1
13,2(1):1
14,3(1):1
15,3(1):1
21,1(1):1
22,2(1):1
23,2(1):1
```

```
24,3(1):1
25,3(1):1
31,2(1):1
32,2(1):1
33,3(1):1
34,4(1):1
35,4(1):1
41,3(1):1
42,3(1):1
43,4(1):1
44,4(1):1
45,5(1):1
51,3(1):1
52,3(1):1
53,4(1):1
54,4(1):1
55,5(1):1
```

Appendix 3B: Type 2 – Gaussian MF-MATLAB Code

Source: Kashyap, S. K. Fusion of LWIR and EO images using fuzzy logic type 1 and type 2. MSDF (internal report/unclassified) Rep. 1305/ESVS01 (FMCD, CSIR-NAL, Bangalore), November 27, 2013.

```
[System]
Name = 'fis_25r_mamdani_type2'
Type = 'mamdani'
Version = 2.0
NumInputs = 2
NumOutputs = 1
NumRules = 25
AndMethod = 'min'
OrMethod = 'max'
ImpMethod = 'min'
AggMethod = 'max'
ReducMethod = 'center_of_sets'
DefuzzMethod = 'centroid'

[Input1]
Name = 'input1'
Range = [0 1]
NumMFs = 5
MF1 = 'VL':'mgausstype2',[-0.05 0.05 0.1062]
MF2 = 'L':'mgausstype2',[0.2 0.3 0.1062]
```

```
MF3 = 'M' : 'mgausstype2', [0.45 0.55 0.1062]
MF4 = 'H' : 'mgausstype2', [0.7 0.8 0.1062]
MF5 = 'VH' : 'mgausstype2', [0.95 1.05 0.1062]

[Input2]
Name = 'input2'
Range = [0 1]
NumMFs = 5
MF1 = 'VL' : 'mgausstype2', [-0.05 0.05 0.1062]
MF2 = 'L' : 'mgausstype2', [0.2 0.3 0.1062]
MF3 = 'M' : 'mgausstype2', [0.45 0.55 0.1062]
MF4 = 'H' : 'mgausstype2', [0.7 0.8 0.1062]
MF5 = 'VH' : 'mgausstype2', [0.95 1.05 0.1062]

[Output1]
Name = 'output1'
Range = [0 1]
NumMFs = 5
MF1 = 'VL' : 'mgausstype2', [-0.05 0.05 0.1062]
MF2 = 'L' : 'mgausstype2', [0.2 0.3 0.1062]
MF3 = 'M' : 'mgausstype2', [0.45 0.55 0.1062]
MF4 = 'H' : 'mgausstype2', [0.7 0.8 0.1062]
MF5 = 'VH' : 'mgausstype2', [0.95 1.05 0.1062]

[Rules]
11,1(1):1
12,2(1):1
13,2(1):1
14,3(1):1
15,3(1):1
21,1(1):1
22,2(1):1
23,2(1):1
24,3(1):1
25,3(1):1
31,2(1):1
32,2(1):1
33,3(1):1
34,4(1):1
35,4(1):1
41,3(1):1
42,3(1):1
43,4(1):1
44,4(1):1
45,5(1):1
51,3(1):1
52,3(1):1
53,4(1):1
54,4(1):1
55,5(1):1
```

Appendix 3C: Fuzzy Inference Calculations – MATLAB Code

Source: Kashyap, S. K. Fusion of LWIR and EO images using fuzzy logic type 1 and type 2. MSDF (internal report/unclassified) Rep. 1305/ESVS01 (FMCD, CSIR-NAL, Bangalore), November 27, 2013.

```
function [output,firing,ofs,ARR] = evalfis2(input, fis, numofpoints)

% EVALFIS2 Perform fuzzy inference calculations of singleton,
% non-singleton type1, non-singleton type2 mamdani/sugeno Fuzzy interval type 2.
%
% Y = EVALFIS2(U,FIS) simulates the Fuzzy Inference System FIS for the
% input data U and returns the output data Y. For a system with N
% input variables and L = 1 output variables,
%     *U is a M = 1 by-N matrix, each row being a particular input vector
%     *Y is M = 1 by-L matrix, each row being a particular output vector.
%
% Y = EVALFIS2(U,FIS,NPts) further specifies number of sample points
% on which to evaluate the membership functions over the input or output
% range. If this argument is not used, the default value is 101 points.
%
% [Y,IRR,ORR,ARR] = EVALFIS2(U,FIS) also returns the following range
% variables when U is a row vector (only one set of inputs is applied):
%     *firing: the result of evaluating the input values through the membership
%      functions. This is a matrix of size Nr-by-2, where Nr is the number
%      of rules.
%     *Ofs: the result of evaluating the output values through the membership
%      functions. This is a matrix of size NPts-by-Nr-by-2. The first Nr
%      columns of this matrix correspond to the first output, the next Nr
%      columns correspond to the second output, and so forth.
%     *ARR: the NPts-by-2 matrix of the aggregate values sampled at NPts
%
% Dependency: fuzzy and fuzzy2 toolbox
% See also fuzzy2 readfis2.
% Sudesh K Kashyap, FMCD,CSIR-NAL, India, 18-10-2013.
% Copyright 1994-2005 The MathWorks, Inc.
% $Revision: 1 $ $Date: 2013/10/18 10:17:22 $

output = [];
ofs = [];
IRR = [];
ORR = [];
ARR = [];
ni = nargin;
if ni < 2
   disp('Need at least two inputs');
   output = [];
   IRR = [];
   ORR = [];
   ARR = [];
   return
end

if~isfis(fis)
   error('The second argument must be a FIS structure.')
elseif strcmpi(fis.type,'sugeno') & ~strcmpi(fis.impMethod,'prod')
   warning('Fuzzy:evalfis:ImplicationMethod','Implication method should be "prod"
for Sugeno systems.')
```

```
end
[M,N] = size(input);
Nin = length(fis.input);
if M==1 & N==1,
   input = input(:,ones(1,Nin));
elseif M==Nin & N~=Nin,
   input = input.';
elseif N~=Nin
   error(sprintf('%s\n%s',...
      'The first argument should have as many columns as input variables and',...
      'as many rows as independent sets of input values.'))
end

% Check the fis for empty values

checkfis(fis);

% Issue warning if inputs out of range
inRange = getfis(fis,'inRange');
InputMin = min(input,[],1);
InputMax = max(input,[],1);
if any(InputMin(:) < inRange(:,1)) | any(InputMax(:) > inRange(:,2))
   warning('Fuzzy:evalfis:InputOutOfRange','Some input values are outside of the
specified input range.')
end

% Compute output
if ni==2
   numofpoints = 101;
end
%% Fuzzification
if(strcmp(fis.mType,'singleton'))
   for i = 1:Nin
      for j = 1:length(fis.input(i).mf)
         % needs to check the membership function type
         fin(i,j,:) = mgausstype2(input(i),fis.input(i).mf(j).params);
      end
   end
end
if(strcmp(fis.mType,'non-singleton type1'))
   x = 0:1/numofpoints:1;
   for i = 1:Nin
   yst1 = gaussmf(x, [fis.mSig, input(i)]);
   for j = 1:length(fis.input(i).mf)
      y = mgausstype2(x,fis.input(i).mf(j).params);
      yu = y(1:length(x));
      yl = y(length(x)+1:end);
      yl_1 = min(yl,yst1');
      yu_1 = min(yu,yst1');
      fin(i,j,:) = [max(yu_1) max(yl_1)];
      end
   end
end
if(strcmp(fis.mType,'non-singleton type2'))
   x = 0:1/numofpoints:1;
   for i = 1:Nin
      yst_u = gaussmf(x, [fis.mSig(1), input(i)]);
      yst_l = gaussmf(x, [fis.mSig(2), input(i)]);
      for j = 1:length(fis.input(i).mf)
         y = mgausstype2(x,fis.input(i).mf(j).params);
```

```
            yu = y(1:length(x));
            yl = y(length(x) + 1:end);
            yl_1 = min(yl,yst_l');
            yu_1 = min(yu,yst_u');
            fin(i,j,:) = [max(yu_1)  max(yl_1)];
        end
      end
end
%% Combining antecedents
nr = length(fis.rule);
for i = 1:nr
  if(fis.rule(i).connection ==1)
      if(strcmp(fis.andMethod,'min'))

firing(i,:) = min(fin(1,fis.rule(i).antecedent(1),:),fin(2,fis.rule(i).
 antecedent(2),:));
      end
      if(strcmp(fis.andMethod,'prod'))
          firing(i,:) = fin(1,fis.rule(i).antecedent(1),:).*fin(2,fis.rule(i).
            antecedent(2),:);
      end
  end
  if(fis.rule(i).connection ==2)
      if(strcmp(fis.andMethod,'max'))

firing(i,:) = max(fin(1,fis.rule(i).antecedent(1),:),fin(2,fis.rule(i).
 antecedent(2),:));
      end
  end
end

if(strcmp(fis.type,'mamdani'))
    %% Implication
    Levels = fis.output.range(1):(fis.output.range(2)-fis.output.range(1))/
numofpoints:fis.output.range(2);
    Nop = length(fis.output);
    for i = 1:Nop
      for j = 1:length(fis.output(i).mf)
          tmp = mgausstype2(Levels,fis.output(i).mf(j).params);
          of(:,j,1) = tmp(1:length(Levels));
          of(:,j,2) = tmp(length(Levels) + 1:end);
      end
end

% based on rule fired
% threshold: firing_th
% rule fired if firing is greater than threshold
firing_th = 1.0e-4;
norf = 1;
for i = 1:nr
    if(strcmp(fis.impMethod,'min') && firing(i,1) > firing_th &&
firing(i,2) > firing_th)
        ofs(norf,:,1) = min(firing(i,1),of(:,fis.rule(i).consequent,1));
        ofs(norf,:,2) = min(firing(i,2),of(:,fis.rule(i).consequent,2));
        fiindx(norf) = i;
        norf = norf + 1;
      end
      if(strcmp(fis.impMethod,'prod') && firing(i,1) > firing_th &&
 firing(i,2) > firing_th)
        ofs(norf,:,1) = firing(i,1).*of(:,fis.rule(i).consequent,1);
```

```
         ofs(norf,:,2)=firing(i,2).*of(:,fis.rule(i).consequent,2);
         fiindx(norf)=i;
         norf=norf+1;
      end

   end

%% Aggregation
if(strcmp(fis.aggMethod,'max'))
   ARR(:,1)=max(ofs(:,:,1));
   ARR(:,2)=max(ofs(:,:,2));
end
if(strcmp(fis.aggMethod,'sum'))
  ARR1(:,1)=min(1,sum(ofs(:,:,1)));
  ARR1(:,2)=min(1,sum(ofs(:,:,2)));
end
%% Type reduction

if(strcmp(fis.reducMethod,'center_of_sets'))
   % Centre of sets
   % Step1: Centroid of each consequent set

   for i=1:norf-1
     c_u(i)=sum(Levels.*ofs(i,:,1))/sum(ofs(i,:,1));
     c_l(i)=sum(Levels.*ofs(i,:,2))/sum(ofs(i,:,2));
   end

   % Step2: Firing degree
   f_upper=firing(fiindx,1);
   f_lower=firing(fiindx,2);

   %% from fuzzy2 toolbox
   l_out=adapt(c_l,f_lower',f_upper',-1);
   r_out=adapt(c_u,f_lower',f_upper',1);
   %% Defuzzification: average
   output=(l_out+r_out)/2;
end

if(strcmp(fis.reducMethod,'height'))
   %Centre of sets
   %Step1: max height of each consequent set

   for i=1:norf-1
     c_u(i)=defuzz(Levels,ofs(1,:,1),'lom');
     c_l(i)=defuzz(Levels,ofs(1,:,2),'lom');
   end

   % Step2: Firing degree
   f_upper=firing(fiindx,1);
   f_lower=firing(fiindx,2);

   %% from fuzzy2 toolbox
   l_out=adapt(c_l,f_lower',f_upper',-1);
   r_out=adapt(c_u,f_lower',f_upper',1);

   %% Defuzzification: average
   output=(l_out+r_out)/2;
end

if(strcmp(fis.reducMethod,'centroid'))
   error(strcat(fis.reducMethod,' not available'));
```

```
      return
end

if(strcmp(fis.reducMethod,'modified_height'))
   error(strcat(fis.reducMethod,' not available'));
   return
end

if(strcmp(fis.reducMethod,'center_of_sums'))
   error(strcat(fis.reducMethod,' not available'));
   return
   end
end

if(strcmp(fis.type,'sugeno'))
   %based on rule fired
   %threshold: firing_th
   %rule fired if firing is greater than threshold
   firing_th=1.0e-4;
   norf=1;
   for i=1:nr
      if(firing(i,1)>firing_th && firing(i,2)>firing_th)
        if(strcmp(fis.output.mf(fis.rule(i).consequent).type,'linear'))
           a=fis.output.mf(fis.rule(i).consequent).params(1);
           b=fis.output.mf(fis.rule(i).consequent).params(2);
           c=fis.output.mf(fis.rule(i).consequent).params(3);
        end
        if(strcmp(fis.output.mf(fis.rule(i).consequent).type,'constant'))
           a=0;
           b=0;
           c=fis.output.mf(fis.rule(i).consequent).params;
         end
         c_u(norf)=a*input(1)+b*input(2)+c;
         c_l(norf)=a*input(1)+b*input(2)+c;
         fiindx(norf)=i;
         norf=norf+1;
      end
   end
   f_upper=firing(fiindx,1);
   f_lower=firing(fiindx,2);
   if(strcmp(fis.defuzzMethod,'wtaver'))
     y1=sum(c_u.*f_upper')/sum(f_upper);
     y2=sum(c_l.*f_lower')/sum(f_lower);
   end
   if(strcmp(fis.defuzzMethod,'wtsum'))
     y1=sum(c_u.*f_upper');
     y2=sum(c_l.*f_lower');
   end
   output=(y1+y2)/2;
end

end
```

EXERCISES

3.1 In T1FL, what is really fuzzified?

3.2 In T1FL, what is really meant by fuzzification?

3.3 What does T2FL exactly do? What is fuzzified here?

3.4 What is fuzzified in IT2FL and in which way it is different from T2FL?

3.5 After making a great fuss about FL and its use in analysis and development of FL-based control and intelligent systems, we finally end up carrying out defuzzification, why so?

3.6 What is the real meaning of FIS? What are we really inferring and from what information?

3.7 Show that the s-norm AS contains the t-norm AP.

3.8 Show that the s-norm bounded sum (BS) contains the s-norm AS and t-norm AP.

3.9 Why should the weights/coefficients in any DF rule be normalised to unity value?

3.10 Show that the t-norm bounded difference/bounded product contains the s-norm AS and t-norm AP.

3.11 Is a membership value/MF really fuzzy in T1FS?

3.12 Is a membership value/MF really fuzzy in T2FS and IT2FL?

3.13 Is the secondary MF in IT2FL really fuzzy (i.e. T1FL) or crisp?

References

1. Akhoundi, M. A. A. and Valavi, E. Multi-sensor data fusion using sensors with different characteristics. *The CSI Journal on Computer Science and Engineering (JCSE)*, Oct 28, 2010, pp. 1–2 (abstract, pp. 1–9).
2. Raol, J. R. *Multisensor Data Fusion with MATLAB*. CRC Press, FL, 2010.
3. Raol, J.R. and Gopal, A. K. (Eds.) *Mobile Intelligent Autonomous Systems*. CRC Press, FL, USA, 2013.
4. Hagras, H. and Wagner, C. Introduction to interval type-2 fuzzy logic controllers – Towards better uncertainty handling in real world applications. eNewsletter, *IEEE SMC Society*, No. 27, 2009. www.http://www.my-smc.org/news/back/-2009_06/SMC-Hagras.html, accessed November 2012.
5. Hagras, H. CE888: Type-2 fuzzy logic systems. Lecture ppts.courses.essex.ac.uk/ce/ce888/Lecture%20Notes/FuzzyLecture1.pdf, accessed November 2012.
6. Mendel, J. M., John, R. I. and Liu, F. Interval type-2 fuzzy logic systems made simple. *IEEE Transactions on Fuzzy Systems*, 14(6), 808–821, 2006.
7. Raol, J. R., Girija, G. and Singh, J. *Modelling and Parameter Estimation of Dynamic Systems, IEE/IET Control Series Book*, 65, IIE/IET Professional Society, London, UK, 2004.
8. Bouchon-Meunier, B., Dubois, D. and Prade, H. Chapter 1: Fuzzy sets and possibility theory in approximate and plausible reasoning. *Approximate Reasoning and Information Systems* (Eds. Bezdek, J. C., Dubois, D. and Prade, H.). The

handbooks of Fuzzy Sets Series Vol. 5, pp. 15–190, Springer, USA, ftp://www.irit.fr/pub/IRIT/ ADRIA/Chapter_1.pdf; and link.springer.com/chapter/10.1007/978-1-4615-5243-7_2, accessed May 2013.

9. Gebhardt, J. and Kruse, R. POSSINFER – A software tool for possibilistic inference. In *Fuzzy Set Methods in Information Engineering: A Guided Tour of Applications* (Eds. Dubois, D., Prade, H. and Yager, R., R.). Wiley, New York, 407–418, 1996.

10. Fuzzy logic tool box, MATLAB®, MathWorks. www.mathworks.com/products/fuzzylogic/

11. Castillo, O. and Melin, P. *Chapter 2: Fuzzy Logic Systems, Recent Advances in Interval Type-2 Fuzzy Systems, Springer Briefs in Computational Intelligence*, Springer-Verlag, Berlin, Heidelberg, Germany, 2012.

12. Qilian, L. and Jerry, M. M. Interval type-2 fuzzy logics systems: Theory and design. *IEEE Transaction on Fuzzy Systems*, 8(5), 535–550, 2000.

13. Dongrui, W. A brief tutorial on interval type-2 fuzzy sets and systems. A part of IT2FLS package available on www.mathworks.com, accessed on October 2013.

14. Ozek, M. B. and Akpolat, Z. K. A software tool: Type-2 fuzzy logic toolbox. In *Computer Applications in Engineering Education*, 16 (2), pp. 137–146, Wiley Periodicals, Inc., USA, 2008.

15. Garlin, L. D. and Naidu, V. P. S. Assessment of Color and Infrared images using No-reference image quality metrics. In *Proceedings of NCATC-2011*, Francis Xavier Engineering College, Tirunelveli, India, International Neural Network Society, India Regional Chapter, Paper No. IP05, pp. 29–35, April 6–7, 2011.

16. Kashyap, S. K. Fusion of LWIR and EO images using fuzzy logic type 1 and type 2. MSDF (internal report/unclassified) Rep. 1305/ESVS01 (FMCD, CSIR-NAL, Bangalore), November 27, 2013.

17. Multi-sensor data fusion based on belief functions and possibility theory: Close range antipersonnel mine detection and remote sensing mined area reduction. In *Humanitarian Demining: Innovative Solutions and the Challenges of Technology* (Ed. Habib, M. K.), ISBN: 978-3-902613-11-0, pp. 392, ITech Education and Publishing, Vienna, Austria, http://www. intechopen.com/books/humanitarian_demining/mltisensor_data_fusion_based_on_belief_functions_and_possibility_theory__close_range_antipersonnel, accessed October 2014.

18. Raol, J. R. and Singh, J. *Flight Mechanics Modeling and Analysis*. CRC Press, FL, 2009.

4

Filtering, Target Tracking and Kinematic Data Fusion

4.1 Introduction

The state estimation in the context of target tracking is the most important problem in sensor data fusion (DF). Estimation is the process of inferring the value of a quantity of interest from often indirect, inaccurate, uncertain and noisy observations. In other alternative situations if the knowledge regarding the noise process is not available, then the problem is solved as deterministic estimation with the models of the input disturbances regarded as the non-random processes. In the decision-making process (Chapter 8), we are interested in selecting one (alternative) out of a set of discrete (and/or seemingly looking distinct) alternatives – the best choice from a discrete (state) vector space. Now, tracking of a moving object is the estimation of the states (of that object) and is carried out using the measured data (observations) from one or more sensors (that are fixed in their locations or are on some moving platforms). Thus, in fact object tracking involves estimation and statistical decision theory. Filtering is the estimation of the current state of a dynamic system from noisy measurement data and it amounts to 'filtering out' the noise (of mainly the higher frequencies compared to the signal frequencies) from the signal that is contaminated with random noise. However, fundamentally, an estimator (or even the process of estimation) itself is considered as a decision process (rule) which takes (as an argument) a sequence of measurements and whose action is to compute and select (in some way, a decision aspect) values for parameters/states of interest. Almost all the DF problems involve this estimation process: we obtain a number of measurements from a set of sensors and using these data (in an estimator code) we want to find some estimate of the true state of the environment/object/target that we are observing [1].

Thus, state estimation encompasses many important aspects of the DF problem, including [1,2]: (i) the mathematical models of the dynamics of moving objects that are required to be specified in the state-space form, (ii) the sensor models that are required to understand what kind of information is provided by the actual sensors, (iii) environment (noise, disturbance) models are required to relate the observations made, to the parameters/states to be estimated and (iv) a criterion, a cost, or a loss function of information value (basically error in a state or a parameter) is needed to judge the performance of the state/parameter estimation/estimator. Thus, defining and solving an estimation problem is then the key to almost all successful engineering DF system. In this chapter (and/or Chapter 5), we consider three main approaches/algorithms [1]: (i) the group-sensor method, (ii) the sequential-sensor method and (iii) the inverse covariance form. The track-to-track fusion (TTTF) algorithm is also described. The problems of multiple-target tracking and data association (DA) are described. Some important algorithms for DA are discussed.

4.2 The Kalman Filter

Kalman filter (KF) has evolved to a very high state-of-the-art method of state estimation and filtering, and now as a sensor DF method, for dynamic systems, which are described by difference, or differential equations [1,2]. The impact of the KF approach is such that it has generated worldwide and globally extensive applications to aerospace engineering systems and many related problems that include [2–5]: (i) theoretical KF derivations, (ii) computational/numerical aspects, (iii) comparison of various versions of KF for non-linear systems, (iv) factorisation and square root filtering, (v) asymptotic results, (vi) applications to satellite orbit estimation, (vii) attitude determination, (viii) target tracking, (ix) sensor DF and (x) aircraft state/parameter estimation applications.

The KF has several features that make it very suitable for dealing with complex multi-sensor (and multi-target) estimation and DF situations. The explicit description of process/plant models and observations allows a large variety of different sensor models that can be incorporated within the basic structure of the KF algorithm. Also, since the consistent use of the statistical measures of uncertainty, via covariance matrices, is possible, the KF makes it possible to quantitatively evaluate the role of each sensor in overall DF system performance. The KF employs an explicit statistical model: (i) of how the parameter/state of interest, $x(t)$, evolves over time by way of mathematical model of the dynamic system and (ii) of how the measurements (via measurement or sensor model), $z(t)$, are related to these parameters/states. The KF gains ensures that, with certain assumptions about the measurement and process models (especially the statistics of state-, and measurement-noise process), the resulting estimate of the $x(t)$ minimises mean square (measurement/MSE) error. The KF and indeed any MSE estimator compute an estimate which is the conditional mean, that is, an average (defined by the mathematical expectation), rather than a most likely value (unlike maximum likelihood estimation).

4.2.1 State and Sensor Models

We consider first a linear dynamic system with process/state/plant model (basically a mathematical model of the system) and the sensor model in discrete-time domain, respectively, as

$$x(k + 1) = \phi x(k) + Gw(k) \tag{4.1}$$

$$z(k) = Hx(k) + v(k) \tag{4.2}$$

Here, x is $n \times 1$ state vector and z is $m \times 1$ measurement vector. w is a white Gaussian random process/state noise sequence with gain vector/matrix as G, of zero mean and covariance matrix Q. v is a white Gaussian random measurement noise sequence with zero mean and covariance matrix R, ϕ is the $n \times n$ transition matrix of the plant/system dynamics that takes states from (discrete-) time k to $k + 1$, and H is the $m \times n$ measurement model/sensor dynamics matrix as the case may be, for multi-sensory data fusion (MSDF) it might be a compounded matrix. The deterministic input u and associated matrices B and D are not included in Equations 4.1 and 4.2, but can be easily incorporated into the KF, because the input u is deterministic and stochastic aspects do not apply to it. We strongly feel that the discrete-time KF (DTKF) formulation is easy to understand and use for practical applications. Even if one uses the continuous-time formulation of KF, one needs to discretise it in order to implement it on a digital computer. If required one can convert the continuous-time system to discrete-time model and then use DTKF algorithm that can be easily implemented on a digital computer.

The filtering/state-estimation problem using KF is formulated as: given the model of the dynamic system, the initial condition of the state vector x, the statistics of the noise processes (Q, R; process noise and measurement noise covariance matrices, respectively) and the (noisy) measurement data (z), determine the optimal estimate of the state, x, of the system. If a few of the coefficients of the state/sensor models are not known, then these can be regarded as additional unknown states and appended to the state vector x yielding the augmented state vector, which then can be estimated – invariably this will result into a non-linear dynamic system for which an extended KF (EKF) can be used. The mathematical expression of Equation 4.1 is a Gauss–Markov model, since the plant noise is assumed to be Gaussian, the system described is linear, and the state x is a Markov chain, since the model state equation is of first-order discrete model.

4.2.2 The Kalman Filter Algorithm

We present a simple and straightforward derivation of the DTKF without invoking the H2/Hilbert vector space related and probabilistic arguments [2].

4.2.2.1 Time Propagation/Time Update Algorithm

If the continuous-time state equation is given as

$$\dot{x}(t) = A(t)x(t) \tag{4.3}$$

then the state vector x evolves according to the transitional equation as

$$x(t) = \phi(t, t_0)x(t_0) \tag{4.4}$$

where $x(t_0)$ is the initial state at time t_0. In formal discrete representation Equation 4.4 is written as

$$x(k + 1) = \phi(k, k + 1)x(k) \tag{4.5}$$

The transition matrix ϕ takes/moves/propagates the state $x(k)$ at time k to the state $x(k + 1)$ at time $k + 1$. The equation for the covariance matrix propagation, considering the state and its state errors as stochastic processes, can be easily derived based on the definition of covariance matrix and Equation 4.5. Let $P(k) = E\{\underline{x}(k)\,\underline{x}^T(k)\}$ be the covariance matrix of $x(k)$ at time index k, where $\underline{x}(k) = \hat{x}(k) - x(k)$. We have from Equation 4.5, the discrete-time state-estimator, ignoring the process noise term (in Equation 4.1), since it is not known anyway, and with $\tilde{x}$ as a predicted estimate of x

$$\tilde{x}(k + 1) = \phi\,\hat{x}(k) \tag{4.6}$$

Then, we have by simple computations, the covariance matrix of the state errors

$$
\begin{aligned}
\tilde{P}(k + 1) &= E\{(\tilde{x}(k + 1) - x(k + 1))(\tilde{x}(k + 1) - x(k + 1)^T\} \\
&= E\{(\phi\,\hat{x}(k) - \phi x(k) - Gw(k))(\phi\,\hat{x}(k) - \phi x(k) - Gw(k))^T\} \\
&= E\{(\phi\hat{x}(k) - \phi x(k))(\phi\,\hat{x}(k) - \phi x(k))^T\} \;+\; E\{Gw(k)w^T(k)G^T\}
\end{aligned}
$$

We assume that state errors and process noises are uncorrelated sequences of white processes and hence the cross-terms are neglected. With these simplifications and without any other approximations, we get the state-error covariance matrix propagation equation as follows:

$$\tilde{P}(k+1) = \phi \hat{P}(k)\phi^T + GQG^T \qquad (4.7)$$

Equations 4.6 and 4.7 give the two equations for the time propagation of state estimate and the time propagation of error covariance of the KF algorithm. Next, we derive the data-update part of the KF that is actually called the KF algorithm and it gives the estimate of the state when and after the measurement data are incorporated into the KF. The idea here is to know how the state estimate from Equation 4.6 gets upgraded when (more and further) information about the state is taken into account via the measurement Equation 4.2.

4.2.2.2 Measurement/Data-Update Algorithm

We now see that the state estimate and state-error covariance matrix at k have evolved/propagated to time $k+1$ using Equations 4.6 and 4.7. At this stage, at time instant $k+1$, a new measurement is available, and it contains new information regarding the state in the form of Equation 4.2. Then the following situation is imminent: H, R (measurement covariance matrix) and measurements z are known; we assume: $\tilde{x}(k) \rightarrow$ a priori estimate of state at time k (before the measurement data are incorporated); $\hat{x}(k) \rightarrow$ updated estimate of state at time k (after the measurement data are incorporated; $\tilde{P} \rightarrow$ a priori covariance matrix of state estimation error (that is available from the previous time-propagation cycle). The KF-measurement update algorithm is then given as

$$\hat{x}(k) = \tilde{x}(k) + K[z(k) - H\,\tilde{x}(k)] \quad \text{(filtering part)} \qquad (4.8)$$

$$\hat{P}(k) = (I - KH)\tilde{P}(k) \quad \text{(state-error covariance matrix update)} \qquad (4.9)$$

The KF equations (4.8) and (4.9) are based on the derivations presented next. We want an unbiased estimator (filter) in a recursive form, with overall (in MSE-sense) minimum errors in the estimates as measured by P. Let this recursive structure be given as

$$\hat{x}(k) = K_1\tilde{x}(k) + K_2 z(k) \qquad (4.10)$$

The expression of Equation 4.10 is a fair weighted and intuitively wise (and appealing) combination of the a priori estimate of state x, obtained by Equation 4.6 and the new measurement that provides direct/indirect information regarding the state. We have now two information sources about the state which should be properly combined to obtain the good and optimal estimate of the state. Interestingly Equation 4.10 sounds very much like a fusion rule, since, it is a weighted combination of two information sources, prior estimate and the new measurement. One can see that some fusion rule/formula is at the heart of the KF derivations, and for somewhat (slightly) different but similar reasons, the KF is a (measurement-level) DF algorithm [4]. The weighting factors (gains K_1 and K_2) should be optimally chosen. Let $\underline{x}(k) = \hat{x}(k) - x(k)$ and $\underline{x}^*(k) = \tilde{x}(k) - x(k)$ be the errors in the state estimates, then we have by simple substitution

$$\underline{x}(k) = [K_1\tilde{x} + K_2 z(k)] - x(k) = K_1\tilde{x} + K_2 H x(k) + K_2 v(k) - x(k) \quad (4.11)$$

Using the measurement Equation 4.2, we get

$$\begin{aligned}\underline{x}(k) &= K_1 [\underline{x}^*(k) + x(k)] + K_2 H x(k) + K_2 v(k) - x(k) \\ &= [K_1 + K_2 H - I]x(k) + K_2 v(k) + K_1\underline{x}^*(k)\end{aligned} \quad (4.12)$$

As $E\{v(k)\} = 0$ and if $E\{\underline{x}^*(k)\} = 0$, then, for the unbiased estimate, we have

$$E\{\underline{x}(k)\} = E\{(K_1 + K_2 H - I)\, x(k)\} \quad (4.13)$$

In order to have the unbiased estimate after the measurement is incorporated, we should have $E\{\underline{x}(k)\} = 0$, and hence we get the relations between two weighing factors as

$$K_1 = I - K_2 H \quad (4.14)$$

By substituting the gain of Equation 4.14 into Equation 4.10 and simplifying we obtain

$$\begin{aligned}\hat{x}(k) &= (I - K_2 H)\, \tilde{x}(k) + K_2 z(k) \\ &= \tilde{x}(k) + K_2[z(k) - H\, \tilde{x}(k)]\end{aligned} \quad (4.15)$$

Since, now we have one gain (vector/matrix), changing K_2 to K, we derive the measurement data-update algorithm of the KF, that is, Equations 4.8 and 4.9, however, we need to obtain the optimal expression for the gain, and covariance matrix update. The term $[z(k) - H\, \tilde{x}(k)]$ is called the measurement prediction error or the residual of the measurement; it is also called innovations sequence. Next, we consider the formulation of P (posterior covariance matrix) to determine the covariance of the state error after the measurement is incorporated as

$$\hat{P} = E\{\underline{x}(k)\underline{x}^T(k)\} = E\{(\hat{x}(k) - x(k))(\hat{x}(k) - x(k))^T\}$$
$$= E\{(\tilde{x}(k) - x(k) + K[Hx(k) + v(k) - H\tilde{x}(k)])(\cdot)^T\} \quad (4.16)$$
$$= E\{[(I - KH)\underline{x}* + Kv(k)][\underline{x}*^T(I - KH)^T + v(k)K^T]\}$$

$$\hat{P} = (I - KH)\tilde{P}(I - KH)^T + KRK^T \quad (4.17)$$

In the detailed expression for P, '$\cdot$' means that the second term, within the parentheses, is the same as the first term. In order to get optimal K, the error covariance matrix $\hat{p}$ is minimised in terms of some norm with respect to K. The appropriate cost function is chosen as (MSE)

$$J = E\{\underline{x}^T(k)\underline{x}(k)\} \quad (4.18)$$

This is equivalent to the following process:

$$J = \text{trace}\{\hat{P}\}$$
$$= \text{trace}\{(I - KH)\tilde{P}(I - KH)^T + KRK^T\} \quad (4.19)$$

$$\frac{\partial J}{\partial K} = -2(I - KH)\tilde{P}H^T + 2KR = 0$$

$$KR = \tilde{P}H^T - KH\tilde{P}H^T$$
$$KR + KH\tilde{P}H^T = \tilde{P}H^T$$

$$K = \tilde{P}H^T(H\tilde{P}H^T + R)^{-1} \quad (4.20)$$

By substituting the expression of K from Equation 4.20 into Equation 4.17 and simplifying, we obtain the short and the optimal form for the covariance matrix as

$$\hat{P} = (I - KH)\tilde{P} \quad (4.21)$$

The KF in the collective form is given as follows:

a. State propagation
   State estimate:

$$\tilde{x}(k + 1) = \phi \hat{x}(k) \quad (4.22)$$

Covariance:

$$\tilde{P}(k + 1) = \phi \hat{P}(k) \phi^T + GQG^T \tag{4.23}$$

b. Measurement update
   Residual/innovations:

$$r(k + 1) = z(k + 1) - H \tilde{x}(k + 1) \tag{4.24}$$

Filter gain:

$$K = \tilde{P}H^T (H\tilde{P}H^T + R)^{-1} \tag{4.25}$$

Filtered state:

$$\hat{x}(k + 1) = \tilde{x}(k + 1) + Kr(k + 1) \tag{4.26}$$

Covariance:

$$\hat{P} = (I - KH)\tilde{P} \tag{4.27}$$

The KF equations (4.22) through (4.27) can be easily coded in MATLAB and the simulated data can be generated using Equations 4.1 and 4.2 with desired signal-to-noise ratio (SNR) for the two noise processes: state and measurements.

4.2.3 The Innovations: Kalman Filter Residuals

We can make prediction as to what possible measurement will be made at a time $k + 1$ based on the measurements that have been made up to time k by simply taking expectations of the measurement, and the difference between the measurement $z(k + 1)$ and the predicted measurement using $H \tilde{x}(k + 1)$, is termed the innovation or residual, Equation 4.24. It is very important measure of the deviation of the filter/filtered estimates from the true states. It is a measure of how well the KF is performing, and it is particularly very important in DA and the DF process. In principle, the innovations are supposed to be an orthogonal, uncorrelated, white sequence [1,2], and can be utilised to monitor the KF performance. We see from Equation 4.25 that $K = \tilde{P}H^T S^{-1}$ with $S = H\tilde{P}H^T + R$, and the matrix S is the covariance matrix of the residuals or the innovations. We see that two important aspects in KF are covariance matrices Q and R. These matrices are to be properly chosen and specified in the filter for its good and adequate performance. Often these statistics are unknown, partially known, or they have very approximate values. In this case, some adaptive filtering approaches are required so that the elements of these matrices are adapted, in an optimal manner or even heuristically, while filter is recursively being run. Some methods are based on monitoring the residuals/innovations process online. The actual residuals (i.e. their numerical values) are computed from Equation 4.24 and are compared with their standard deviations obtained by taking the square root of the diagonal elements of the matrix S. This process of checking and tuning the filter to bring the computed residuals within the bound of at least two standard deviations is a very crucial tuning exercise for adequate performance of the KF [2]. Other possibilities are: (i) to check the whiteness of the measurement residuals, that is, of innovations; since the innovations are supposed to be a white stochastic process, this should be checked using the whiteness test [2] and (ii) to check if the computed covariance matrix (of states) matches the theoretical covariance matrix obtained from the covariance equations of the filter (Equations 4.23 and 4.27). The test (i) signifies the fact that the measurement residuals being white, no information is left out to be utilised in the filter, since the white process is an unpredictable process, and hence, the maximum amount of information has been utilised in obtaining the estimates. The test (ii) signifies that the computed state errors from the KF are within the bounds given by the filter predictions (theoretical estimates of the covariance), and hence, proper tuning has been achieved. The satisfaction of these tests means that the KF is performing properly as per the theoretical predictions, and the results of performance of the KF can be trusted. It is also very important to check the actual innovations (i.e. the residuals of the KF) for whiteness in case the EKF is used for state estimation of the non-linear dynamic systems.

4.2.4 Steady-State Filters

When the KF is used in the synchronous mode, and the state and measurement matrices are time invariant, the associated mathematical models are written as

$$x(k) = \phi x(k-1) + Bu(k) + w \tag{4.28}$$

$$z(k) = Hx(k) + v \tag{4.29}$$

Now, we see from Equations 4.23, 4.25 and 4.27 that the computations of the state estimate covariance matrices and KF gains (for linear systems) do not at all depend on the value of the measurements made, that is, on z. In fact, these quantities for a linear system (even if is a time-variant system) can be easily pre-computed, and stored for later use. Now, if ϕ, H, Q and R are time invariant, then the innovation covariance matrix S will be actually constant and the covariance/KF gain matrices/vector will also tend to some constant steady-state value [1]

$$P(k|k) \rightarrow P_\infty^+; \quad P(k|k-1) \rightarrow P_\infty^-; \quad K(k) \rightarrow K_\infty \qquad (4.30)$$

This convergence can often be very rapid. Thus, if the covariance/KF gain matrices tend to some constant steady-state values, say after only a few sampling intervals, it is very reasonable to avoid the extensive computations of these values online, and we can simply insert these constant values into the KF at the beginning. A constant gain vector/matrix is used in the formula $\hat{x}(k+1) = \tilde{x}(k+1) + K\, r(k+1)$, and the steady-state value of the gain K is most easily computed offline. If we are using a 'constant velocity' model (often encountered and used in target tracking [4]), the gain is described by two coefficients α and β [1]

$$\hat{x}(k+1) = \tilde{x}(k+1) + \begin{bmatrix} \alpha \\ \beta/T \end{bmatrix} r(k+1) \qquad (4.31)$$

This filter is well known as the classical α–β filter – constant gain filter in the literature on target tracking, T being the sampling-interval/time step. For a 'constant acceleration' target-tracking model, we use a filter with three gains as α, β and γ

$$\hat{x}(k+1) = \tilde{x}(k+1) + \begin{bmatrix} \alpha \\ \beta/T \\ \gamma/T^2 \end{bmatrix} r(k+1) \qquad (4.32)$$

These parameters are usually determined by performing a simulation and engineering judgment of the much reduction of computations. This is of considerable advantage for multi-target tracking applications where some thousands of filters must be working concurrently [1]. If the assumptions are valid, the errors, by using constant gain filters, are negligible. In many flight-testing ranges for aerospace vehicles (aircraft, helicopter and missiles), these steady-state filters are routinely used even if the assumptions of linear time-invariance and constant noise are not valid.

4.2.5 Asynchronous, Delayed and A-Sequent Measurements

In many DF systems, the measurement process, that is, receiving the data from the sensors/signal channels, might be often asynchronous. This would happen because the information/data/signals could come from many different sources/sensors (in un-coordinated manner, signals are not time-tagged). In addition to this each source/sensor might have different sampling rates and latencies/delays that are associated with the acquisition and communication of the respective measurements. In such a network (NW)-data acquisition systems, we need to address the three timing aspects, each one with increasing complexity [1]: (i) asynchronous data – information/data arrive at random intervals but in a timely manner (RIT – random intervals – in time) so that they are still processed, (ii) delayed data – information/data arrive both at random intervals and later on after an estimate of the state/parameter has already been computed (RIL) and (iii) A-sequent data – information/data arrive randomly, late and also out of time sequence (RILOS – temporal order). The asynchronous data are relatively easy to incorporate in the existing KF structure. For asynchronous measurements, the discrete-time state and measurement models of Equations 4.1 and 4.2 can be used directly. We should note that the process noise covariance matrix Q and other model matrices φ, G and H are functions of the time interval $(t_k - t_k-1)$, and hence the computed variances $P(k|k)$, $P(k|k-1)$, $S(k)$ and the gain matrix $K(k)$ would not remain constant and must be computed at every time slot.

The delayed measurement (delayed data) problem is handled by maintaining two estimates [1]: (i) one associated with the true time at which the last measurement was obtained and (ii) a prediction (based on this last estimate) that describes the state at the present time instant, when a new and the delayed measurement arrives, the current prediction is discarded and a new prediction (up to the time of this delayed measurement) is computed – this being done on the basis of the estimate at the time of the previous delayed measurement. This is then combined with the new measurement to produce a new estimate (that itself is predicted forward to the current time). Let t_c be the present time instant at which a new delayed measurement is available. Let t_p be the previous time-instant when a delayed measurement was available [1]. Let $t_c > t_k > t_{k-1}$ and $t_c > t_p > t_{k-1}$. Now, we assume that we already have an estimate $\hat{x}(t_{k-1}|t_{k-1})$ and an estimate (strictly a prediction) $\hat{x}(t_p|t_{k-1})$. We now acquire a measurement $z(t_k)$. We begin by simply discarding the estimate $\hat{x}(t_p|t_{k-1})$ and generating a new prediction $\hat{x}(t_k|t_{k-1})$ and prediction covariance $P(t_k|t_{k-1})$. These estimates and the delayed measurement are used to compute a new estimate $\hat{x}(t_k|t_k)$ at the time-instant the delayed measurement was made. This estimate is predicted forward to the present time-instant to produce an estimate $\hat{x}(t_c|t_k)$ and its covariance $P(t_c|t_k)$ according to the following equations:

$$\hat{x}(t_p|t_k) = \phi(t_p)\hat{x}(t_k|t_k) + B(t_p)u(t_p) \tag{4.33}$$

$$P(t_p|t_k) = \phi(t_p)P(t_k|t_k)\phi^T(t_p) + G(t_p)Q(t_p)G^T(t_p) \tag{4.34}$$

Both the estimates $\hat{x}(t_k|t_k)$ and $\hat{x}(t_p|t_k)$ with their associated covariance matrices should be maintained for the next measurement. We should note here that if the measurements are delayed, then the estimate provided at the present time will not be as good as the estimates obtained when the measurements are obtained in time. This happens because the additional prediction required injects additional process noise into the state estimate.

A-sequent data occur when the measurements are delayed and arrive at the filter out of time-order. This is a common situation in multi-sensory (MS) systems due to the pre-processing/communication delays that may be quite different between various sensors.

4.2.6 The Extended Kalman Filter

Most real-life dynamic systems are non-linear and estimation of states of such systems is often required, for example, satellite orbit determination. For target tracking and DF involving non-linear system models we need to use the EKF. Such a non-linear system can be expressed with the set of equations as [1,2]

$$\dot{x}(t) = f[x(t), u(t), \Theta] \tag{4.35}$$

$$y(t) = h[x(t), u(t), \Theta] \tag{4.36}$$

$$z(k) = y(k) + v(k) \tag{4.37}$$

Here, f and h are general non-linear vector-valued functions. Θ is the vector of unknown constants/parameters expressed as

$$\Theta = [x_0, b_u, b_y, \beta] \tag{4.38}$$

Here, x_0 is the state variable at time $t = 0$, b_u is the bias in control inputs (often called as nuisance parameters), b_y is the bias in model response y (again nuisance parameters) and β are some unknown parameters in the models of the system. We see that the KF recursions equations for linear system are not directly applicable for estimation of the states of the non-linear system. For applying a KF to the non-linear systems, these non-linear functions f and h need to be linearised around the current estimate of the state (predicted or filtered estimate). With this the KF recursions can be easily used for the estimation. This results into the EKF which is a sub-optimal solution to the non-linear filtering problem. If required simultaneous estimation of states and parameters can be achieved by augmenting the state vector with unknown parameters (regarding these as additional states) and using the filtering algorithm with the augmented non-linear model [2–5]. Thus, the augmented state vector can be given by

$$x_a^T = \left\lfloor x^T \ \Theta^T \right\rfloor \tag{4.39}$$

$$\dot{x} = \begin{bmatrix} f(x_a, u, t) \\ 0 \end{bmatrix} + \begin{bmatrix} G \\ 0 \end{bmatrix} w(t) \tag{4.40}$$

The augmented mathematical state and measurement models are given as

$$\dot{x} = f_a(x_a, u, t) + G_a w(t) \tag{4.41}$$

$$y(t) = h_a(x_a, u, t) \tag{4.42}$$

$$z_m(k) = y(k) + u(k), \quad k = 1, \dots, N \tag{4.43}$$

with

$$f_a^T(t) = \begin{bmatrix} f^T & 0^T \end{bmatrix}; \quad G_a^T = \begin{bmatrix} G^T & 0^T \end{bmatrix} \tag{4.44}$$

The necessary linearised system matrices are defined and obtained as

$$A(k) = \frac{\delta f_a}{\delta x_a}\bigg|_{x_a = \hat{x}_a(k), \, u = u(k)} \tag{4.45}$$

$$H(k) = \frac{\delta h_a}{\delta x_a}\bigg|_{x_a = \tilde{x}_a(k), \, u = u(k)} \tag{4.46}$$

with the appropriate state transition matrix given as

$$\phi(k) = \exp\left[-A(k)\,\Delta t\right] \quad \text{where} \quad \Delta t = t_{k+1} - t_k \tag{4.47}$$

4.2.6.1 Time Propagation of States/Covariance Matrix

The states are propagated from the present state to the next time instant, and the predicted state is given by

$$\tilde{x}_a(k + 1) = \hat{x}_a(k) + \int_{t_k}^{t_{k+1}} f_a[\tilde{x}_a(t), u(k), t]dt \tag{4.48}$$

The covariance matrix propagates from instant k to $k + 1$ as

$$\tilde{P}(k + 1) = \phi(k)\hat{P}(k)\phi^T(k) + G_a(k)QG_a^T(k) \tag{4.49}$$

Here $\tilde{P}(k+1)$ is the predicted covariance matrix at the instant $k+1$, G_a is the process noise coefficient matrix and Q is the process noise covariance matrix.

4.2.6.2 Measurement Update

This part of the EKF updates the predicted estimates obtained in the previous cycle by incorporating the measurements as and when they are available

$$\hat{x}_a(k+1) = \tilde{x}_a(k+1) + K(k+1)\left\{z_m(k+1) - h_a[\tilde{x}_a(k+1), u(k+1), t]\right\} \quad (4.50)$$

The gain formula is given as

$$K(k+1) = \tilde{P}(k+1)H^T(k+1)[H(k+1)\tilde{P}(k+1)H^T(k+1) + R]^{-1} \quad (4.51)$$

The covariance matrix is updated using the Kalman gain and the linearised measurement matrix from the predicted covariance matrix $\tilde{P}(k+1)$ and is given by

$$\hat{P}(k+1) = \left[I - K(k+1)H(k+1)\right]\tilde{P}(k+1) \quad (4.52)$$

The EKF is very suitable for target tracking if either the state model and/or the measurement model is non-linear.

4.2.7 Kalman Filter: A Natural Data-Level Fuser

It is very interesting to note that the KF can be regarded as a direct data fuser (the measurement/data-update part) and it is very useful in the kinematic DF process [4]. In general, three major ways to do the DF at kinematic level are: (i) fusion of the raw measurement data (of course these data should be in the form of engineering units, REUD and should not be absolutely raw) – called centralised fusion, or data-level fusion, (ii) fusion of estimated state vectors – state vector fusion (SVF) – after the original measurements have been processed by using KFs and (iii) hybrid approach which allows fusion of (raw) data and the processed state vector as desired.

4.2.7.1 Fusion-Measurement Update Algorithm

In order to use the KF as a natural data-fuser processing algorithm, we need to specify/have the following data sequences and quantities: (i) the measurements z at time $k+1$ (and further on), (ii) the measurement model H (and the models of other data/sensors that are to be fused here), (iii) assumed value of R (for all the measurement data channels), (iv) $\tilde{x}(k)$ as a priori estimate of the state at time k (before the measurement data are incorporated at time $k+1$) and (v) $\tilde{P}$ as a priori covariance matrix of state-estimation errors. Then, the measurement/data-update algorithm (i.e. the state vector filtering taking the measurement data/information from various channels into account) to obtain $\hat{x}(k+1)$ as the updated estimate of state at time $k+1$, that is, after the measurement data is incorporated, is given as

Residual equation:

$$r(k+1) = z_c(k+1) - H_c\,\tilde{x}_f(k) \tag{4.53}$$

Kalman Gain:

$$K = \tilde{P}H_c^T(H_c\tilde{P}H_c^T + R_c)^{-1} \tag{4.54}$$

Filtered state estimate:

$$\hat{x}_f(k+1) = \tilde{x}_f(k) + K\,r(k+1) \tag{4.55}$$

Covariance matrix (posterior):

$$\hat{P} = (I - KH_c)\tilde{P} \tag{4.56}$$

We note here that H and R are compound matrices (block matrices which are appropriately chosen for each combination of senor/measurements channels) since, we are fusing the data at the measurement level. We see clearly that via z, as a vector composed of several observables, like position, and angles, the KF in the form given above is itself a measurement data-level fusion algorithm. Then, KF as explained above, in principle as well as in practice accomplishes a DF process task. Also, in the similar way EKF or any version of KF and related filters (the so-called unscented KF, derivative-free filters and even including the filters based on H-infinity [HI] norms) can be used as data-level fuser. This is a very fundamental, very interesting and encouraging fact that the data-level fusion process and the filtering/estimation process are so intimately connected. This is a boon in application of sensor DF for multi-target tracking and estimation scenarios. Also, it is very interesting to note here that the basic probabilistic modelling and arguments discussed in Chapter 2, on which the KF development can be (alternatively) based, are equally applicable for the sensor DF as we have discussed in Chapter 2.

4.3 The Multi-Sensor Data Fusion and Kalman Filter

Several techniques and procedures developed and used for a single sensor KF are directly applicable to multi-sensor filtering/estimation and tracking situations [1]. Multi-target tracking and MSDF invariably involve a group of sensors and these sensors can be looked upon as a single sensor with a large (extended/expanded) and fairly complex measurement model, the H comprising of H_1, H_2 and so on in a suitable way. Though, in principle a good and feasible approach, but in practice, it is limited to relatively small numbers of sensors or a small group. In the second approach, we consider each measurement from each sensor as a separate/independent realisation (made according to a specific H model) and then we incorporate this measurement into the filter/estimator in a sequential manner. Single-sensor filter/estimation technique, applied sequentially, can be applied to the multi-sensor problem. This approach requires the computation of a new prediction and gain matrix for each measurement from each sensor at every time step, and hence is computationally expensive.

One more feasible approach is to explicitly derive equations for combining and integrating the multiple measurements (and their models) available at the same time into a common state estimate, here, the state model is the same if the target is only one (one-trajectory model). We can start from the formulation of the multi-sensor KF by utilising a single model for a group of sensors. Then a set of recursive formulae for integrating individual sensor measurements are obtained. These formulations and equations would be more naturally expressible in terms of 'information' (information-state/-matrix) rather than conventional state and covariance form, that is, the usual KF equations (often this KF formulation is called covariance form). The systems considered then are all 'centralised': the measurements from the sensors get reported back to a central processing station/unit (CPS) in a raw-REUD form where they are processed by a single filter in essentially the same way as single-sensor systems. One can also formulate the multi-sensor estimation problem in terms of a number of local sensor-filters, each generating the state estimates. These estimates are subsequently communicated in the processed form back to a CPS. The latter is a distributed processing structure and has merits in terms of modularity and flexibility of the resulting architecture. Then we need the algorithms to fuse estimates or track information at the central fusion centre and these algorithms might be complex.

4.3.1 Measurement Models

A common mathematical model of the true state is provided in the usual linear discrete-time form

$$x(k) = \phi x(k-1) + Bu(k) + Gw(k) \tag{4.57}$$

Here, $u(k)$ is the control input vector with associated gain vector/matrix and $w(k)$ is a random vector describing model/process uncertainty. The latter is assumed to be zero mean and time-uncorrelated. We see that here, since all the sensors are measuring the same state, that is, the same target, the process math-model should be common to all the sensors. The measurements of the state of this dynamic system/target, for example, an aircraft, are made synchronously by a number of different sensors according to a set of linear measurement models in the form given by

$$z_s(k) = H_s x(k) + v_s(k), \quad s = 1,...,S \tag{4.58}$$

Here, $z_s(k)$ is the measurement available at time k from the sensor 's' of a common state $x(k)$ according to the measurement model H_s with an additive noise $v_s(k)$. This noise is also zero mean Gaussian with covariance matrix R. If the measurements made are asynchronous (i.e. the different sensors make measurements at different rates), these can be handled by using the measurement model $H_s(t_k)$, by explicitly involving time-instant. It is further assumed that the noises $v_s(k)$ are all zero mean, uncorrelated between sensors and are time-uncorrelated (white), and that the process and measurement noises are also uncorrelated with each other (cross-covariance is null). However, one can include the correlated noise processes in the formulation, if so warranted.

4.3.2 Group-Sensor Method

The easy way of handling a multi-sensor estimation problem is to combine all the measurements and all the measurement models in a single composite 'group sensor'. Then the estimation problem uses an identical algorithm to that employed in single-sensor systems. Define a composite observation vector as [1]

$$z_c(k) = [z_1^T(k),...,z^T{}_s(k)]^T \tag{4.59}$$

The composite measurement model is given as

$$H_c(k) = [H_1^T(k),...,H_s^T(k)]^T \tag{4.60}$$

Similarly, the noise processes are also combined. Then the resulting measurement covariance matrix is a block diagonal as given by

$$R_c(k) = \text{block} - \text{diag}\{R_1(k),...,R_s(k)\} \tag{4.61}$$

Then the set of the measurement equations are rewritten as a single 'group' model

$$z_c(k) = H_c(k)x(k) + v(k) \tag{4.62}$$

Then Equations 4.59 to 4.62 are appropriately used in the standard KF to obtain the estimates of the states of the target, see Section 4.2.7.1. As the number of sensors incorporated in the group sensor increases, the dimension of the innovation vector/covariance matrix also increases, and this might cause a problem because of the inversion of the innovation covariance matrix required at each data-update cycle [1]. The 'combined-single' approach to the DF problem is of limited use if the number of sensors increases.

4.3.3 Sequential-Sensor Method

In this method to multi-sensor estimation problem, each sensor measurement is treated as an independent one, and the sequential updates to the state estimate are computed one by one. Here, the dimensions of the innovations/covariance matrix at each data-update stage remain of the same size as their single-sensor equivalents [1]. This means that the measurements from one sensor are used and the state estimate is updated, then this is repeated with the next sensor and so on until all the sensors are covered, then the prediction cycle of the KF is invoked. And the process then repeats with the measurements from the next sampling instant. It is assumed that every sensor gives measurements synchronously at every time step. What is to be kept in mind is that all the measurement/data updating stages and the prediction-stage state estimation should be completed in one sampling interval before the data from the next sampling instant arrive, this is important if the KF processing is done online/real time. Within a given sampling interval the measurements can be processed in any decided order. The state estimate determined from one sensor becomes the 'previous' state for the next sensor and so on. This approach is computationally more efficient than the one with 'group'-sensor method.

4.3.4 Inverse-Covariance Form

The group-sensor and the sequential-sensor algorithms are not very suitable when the number of sensors becomes very large. It is fundamentally difficult to integrate the multiple measurements at a 'single time' step because the innovations get correlated – this is due to the fact that the innovations at a single time step have a common prediction. The explicit set of equations for the MS estimation problem is derived in Reference 1. One can use Equations 4.22 and 4.23 for the prediction cycle and the following two equations for the measurement update [1]:

$$\hat{x}(k) = \tilde{P}(k)[\tilde{P}^{-1}(k)\tilde{x}(k) + \sum_{i=1}^{S} H_i^T(k)R_i^{-1}(k)z_i(k)] \tag{4.63}$$

$$\hat{P}(k) = [\tilde{P}^{-1}(k) + \sum_{i=1}^{S} H_i^T(k)R_i^{-1}(k)H_i(k)]^{-1} \tag{4.64}$$

One can easily see that in this formulation of the multi-sensor estimation problem the largest matrix inversion required is of the dimension of the state vector and the addition of new sensors is very simple.

4.3.5 Track-to-Track Fusion

In TTTF, also called the track fusion (TTF), the associated algorithms combine the estimates from several sensor sites and this is different from the algorithms that combine measurements from different sensors, that is, the 'scan' fusion (SF). In the former, the TTF, the algorithms and the local sensor sites generate local track estimates using a local KF [1]. Thus generated tracks/ estimates are then communicated to a central fusion site where they are combined to generate a global track-estimate. In some configurations, the global track-estimate is then communicated back to the local sensor sites, and it is called a feed-back configuration. The TTF algorithms offer certain merits [1]: (i) local track information is (anyway) available for use locally at each sensor site and (ii) track information (i.e. the state estimate) may be communicated at a lower and more compact rate to the central site for fusion, compared to the original measurements data that could be of very large dimension.

The global track estimate is obtained by the so-called SVF equations. In fact this is obtained by the weighted (by the covariance matrices, see Chapter 2) average of the individual states, and the corresponding fused covariance matrix is also easily obtained. However, in the TTF a common underlying state- (process) model is used. This also means that the KF prediction cycle uses the common predicted error resulting from a common process model. As a result of this any two tracks are correlated. The equations for the (track) fused state and the fused covariance matrix, taking into account the cross-covariance, are given as [1]

$$x_f = x_i + [P_i - P_{ij}][P_i + P_j - P_{ij} - P_{fij}]^{-1}[x_j - x_i] \tag{4.65}$$

$$P_f = P_i - [P_i - P_{ij}][P_i + P_j - P_{ij} - P_{fij}]^{-1}[P_i - P_{ij}]_f \tag{4.66}$$

A more formal approach to the development of the SVF formulae is presented in Section 5.5.

4.4 Non-Linear Data Fusion Methods

Real-life situations demand that we treat the MS DF problem in non-linear domain, due the presence of non-Gaussian, often multimodal pdfs, and/ or non-linear dynamic equations, the latter due to the fact that the process and/or measurement models are not linear, for example, as in several aerospace dynamical systems/models. In such cases, the conventional KF is not suitable, and we need an EKF, or any higher version of filters, like some non-linear filters, or the filters that deal with non-Gaussian noise processes.

4.4.1 Likelihood Estimation Methods

A solution to the recursive estimation problem can be derived entirely in probabilistic form using Bayesian arguments [1]. In general, the pdf (Chapter 2) needs to be computed for all the times k. This pdf describes the posterior pdf of target/vehicle state, at time k, given the measurements and control inputs up to and including time k together with the initial state $x(0)$. The posterior pdf, following a control $u(k)$ and observation $z(k)$, is then to be computed using Bayesian formula. The transition model for the state is defined in terms of a pdf on state transitions. The state transition is assumed to be a Markov process in which the state at x_k depends only on the preceding state at $x(k-1)$ and the control input $u(k)$. This is independent of the measurements made. With these models, Bayesian formula is utilised to define a recursive solution. For the recursive update rule for the posterior, the chain rule of conditional pdf is utilised to expand the joint distribution on the state and measurements in terms of the state and then in terms of the measurements.

4.4.2 Derivative-Free Filtering and Fusion

In the conventional EKF the non-linear system models of the given dynamic system are linearised to parameterise the underlying pdf in terms of its mean and covariance. In the derivative-free KF (DFKF), this linearisation of the process models is not carried out. Instead, the pdf is parameterised via the non-linear transformation with the chosen sigma points, and these points are chosen deterministically. We consider the time propagation of a random variable x (of dimension L, say $L = 2$) via a non-linear function $y = f(x)$. Assume that mean and covariance of (primarily) sigma points for the random variable are given as $\bar{x}$, P_x, respectively. Then these sigma points are computed as [4]

$$
\left.
\begin{aligned}
\chi_0 &= \bar{x} \\
\chi_i &= \bar{x} + \left(\sqrt{(L+\lambda)P_x}\right)_i & i &= 1,\dots,L \\
\chi_i &= \bar{x} - \left(\sqrt{(L+\lambda)P_x}\right)_{i-L} & i &= L+1,\dots,2L
\end{aligned}
\right\}
\qquad (4.67)
$$

The associated weights for these sigma points are computed as [4]

$$
\left.\begin{aligned}
w_0^{(m)} &= \frac{\lambda}{L + \lambda} \\
w_0^{(c)} &= \frac{\lambda}{L + \lambda} + (1 - \alpha^2 + \beta) \\
w_i^{(m)} &= w_i^{(c)} = \frac{1}{2(L + \lambda)} \qquad i = 1, \ldots, 2L
\end{aligned}\right\} \tag{4.68}
$$

To provide the unbiased transformation, the weights have to satisfy the condition $\sum_{i=1}^{2L} w_i^{(m \text{ or } c)} = 1$. The scaling parameters in DFKF [4] are: (i) α determines the spread of sigma points around $\bar{x}$, (ii) β incorporates a prior knowledge about distribution of $\bar{x}$, (iii) $\lambda = \alpha^2(L + \kappa) - L$ and (iv) κ is a secondary tuning parameter. These sigma points are propagated through the system's non-linear function, $y_i = f(\chi_i)$, where $i = 0, \ldots, 2L$. This then results in the transformed sigma points. The mean and covariance of these transformed points are computed as follows [4]:

$$
\bar{y} = \sum_{i=0}^{2L} w_i^{(m)} y_i \tag{4.69}
$$

$$
P_y = \sum_{i=0}^{2L} w_i^{(c)} \{y_i - \bar{y}\}\{y_i - \bar{y}\}^T \tag{4.70}
$$

Thus, as can be seen the DFKF is a straightforward and simple extension of the derivative-free transformation, a three-step process as described above. The complete state of the DFKF can be constructed by the augmented state vector consisting of the actual system states, process noise states and measurement noise states. Then, the dimension of augmented state vector would be $n_a = n + n + m = 2n + m$. More details and some results are presented in Reference 4. The fusion is either carried out using measurement level fusion (MLF) or state vector level fusion as per requirements.

4.4.3 Other Non-Linear Tracking Filters

There are several non-linear filtering algorithms that can be used for target tracking and sensor DF: (i) sum-of-Gaussians (SOG) method – in this SOG method, the pdfs of the general formulation are replaced with an approximation as a sum of Gaussians, this can be used to model any pdf, and there are efficient update rules for an SOG approximation (Appendix A.3); (ii) distribution approximation filter – the DAF provides an efficient method of approximating the non-linear transformation of Gaussian pdfs; (iii) particle filtering algorithm (Appendix A.4) and (iv) Pugachev filter

(Appendix A.5). The algorithmic details of some of these non-linear filters are given in Appendix A.

4.5 Data Association in MS Systems

In MS tracking-cum-estimation situations, the measurements being collected must be correctly associated with the state-space tracking models, since the states are being measured, often indirectly via these measurements (-sensors) and this aspect leads to the so-called DA problem [1] – that includes: (i) validating the measured data ensuring that these data are not erroneous or arising from clutter (other sources than the true sources), (ii) associating the correct measurements to the correct state-tracking models – this is important especially for multi-target (and MS) tracking situations and (iii) initialising/starting new tracks/states as and when needed. In the normal target-tracking-cum-filtering/estimation, one is concerned with uncertainty in measurement location. In DA one is concerned with uncertainty in the origin of the measurements. In DA one uses the normalised innovations or the so-called validation gate. In the target-tracking situations, the target tracks are updated by correlating, that is, by making the correspondence of the available measurements with the existing tracks or by initiating new tracks (using the measurements coming from different sensors). The methods/process of gating and DA enables a proper and accurate tracking in multi-sensory multi-target (MSMT) scenario. Thus, the gating helps in deciding if the measurement made (which could include clutter, false alarms and electronic counter measures) is a probable candidate for track maintenance or track update. DA is a step to associate these available measurements to the targets with certainty (and with high probability) when several targets are in the same neighbourhood/cluster. The measurements arriving from the sensors may not be true due to [1,4]: (i) (too much drowning by) the effect of clutter, (ii) false alarms, (iii) interference from other neighbouring targets, (iv) limited resolution capability (spatial coverage limitation of the sensor) and (v) clouding of/congestion by several targets in the neighbourhood of the target/track of interest. Thus, the process of gating is used to screen out spurious signal, as mentioned above, and the DA algorithms are used for [1,4]: (a) automatic track initiation (ATI), (b) measurement to track correlation (MTTC/DA) and (c) track-to-track correlation (TTC). In MTTC/DA, the sensor data are associated with existing number of tracks to determine and ascertain which sensor data belong to which target. Once such a determination has been made that there are more than one measurement for a particular target, then these measurements are combined/aggregated at raw-REUD level using process of the measurement fusion. The gating determines if a measurement belongs to a previously established target or to a new target. Then these gates are defined for one or more of the existing tracks. A measurement that satisfies the gate, meaning thereby that it falls within the size of the gate, then it becomes valid for association with that track. This region encompassed by the gate is then called the validation/confirmation region.

The various situations that could have occurred during the gating process are that [1,4]: (i) more than one measurement might have satisfied the gate of a single track, (ii) the measurement might have satisfied the gates of more than one track, (iii) the measurement might not be used at all for updating the existing track even if it fell within the validation region (it might be used to initiate a new track) and (iv) the measurement might have not fallen within the validation region of any of the existing tracks – in such a case the measurement is used to initiate a new track. If the probability of detection of the track/target is 1 or there are no expected extraneous returns (the responses from other sources) the gate size should be infinite – the ideal situation. In case the target state is observed with detection probability less than 1, in the presence of clutter, then this would give rise to false measurements.

A measurement is said to satisfy the gate if all the elements of the residual/innovations vector are less than 'the gate size times the STD of the residuals'. If $z(k)$ is the measurement at scan k given by [1,4]

$$z(k) = Hx(k) + v(k) \tag{4.71}$$

If $y = H\hat{x}(k|k-1)$ is the predicted value of the measurement with $\hat{x}(k|k-1)$ representing the predicted state at scan/time instant $k-1$, then the residual/innovation sequence is given by

$$v(k) = z(k) - y(k) \tag{4.72}$$

As we know from the set of the KF equations the innovations covariance matrix S is given by

$$S = HPH^T + R \tag{4.73}$$

Then a distance d^2 representing the norm of the residual vector is very easily specified as

$$d^2 = v^T S^{-1} v \tag{4.74}$$

Then a correlation between the measurement and a track is allowed if the distance d^2 is less than a certain gate-threshold value G

$$d^2 = v^T S^{-1} v \leq G \tag{4.75}$$

From Equation 4.75, we can ascertain that the measurement falling within thus defined gate is more likely to be from the track rather than from any other extraneous source. A method to choose the gate size G is based on chi-square distribution with m DOF (measurement vector dimension). In Equation 4.75, the distance d^2 is the sum of the squares of m (assumed to be) independent Gaussian variables with zero means and unit standard deviations. Then the quadratic quantity d^2 has a χ^2 distribution and a gate on d^2 can be determined using the chi-square tables (from open literature). This is because if the innovations are zero mean and white (random process), then the normalised innovation is a χ^2 random variable with m DOF. As such the normalised innovations-process serves as the basis for all, or many DA methods. This is because the innovations process is the only primary measure of divergence of the state estimates (in fact the predicted measurements obtained from the predicted state) from the measurement sequence, and it admits a precise probabilistic measure of correct association.

The DA methods are also extended to tracks, and these situations occur in distributed architectures and in TTTF algorithms [1]. Here, a gate defines an ellipsoidal region to the normal validation gate. In this case the test statistic is χ^2 distributed with nx DOF, the dimension of the state vector. When the validation gate is determined and established, the nearest-neighbour Kalman filter (NNKF) algorithm is usually used for associating these tracks. Of course one can use either probabilistic data association filter (PDAF) or multiple-hypothesis testing (MHT-filter) algorithm for the same purpose. Also, in multiple target problems, one needs to maintain a record of how these measurements and tracks are associated together, this will obtain the track file. The NNKF, PDAF and MHT are briefly discussed next.

4.5.1 Nearest-Neighbour Standard Filter

The NNKF selects the measurement that is nearest to the predicted measurement considering it as the validated one. The remaining measurements, even if within the gate, are not considered at all, of course thereby losing some information in practice (this is a basic limitation of NNKF). This only one measurement is chosen for determining/updating the state of the target. Thus, in the NNKF, the measurement that is nearest to the track is chosen, of course this measurement should be within the gate size. This chosen measurement is used for updating the track, by associating each measurement with one track. Hence, no two tracks share the same measurement. We presume that the probability of detection is very high. Here, the 'nearest' is based on the fact that the normalised innovations sequence is minimum using the predicted measurement and the newly made measurement. Thus, if a valid measurement exists, the track is updated using NNKF. The time propagation of the state and covariance follow the usual KF equations [1,4]

$$\tilde{X}(k/k - 1) = \Phi\hat{X}(k - 1/k - 1) \qquad (4.76)$$

$$\tilde{P}(k/k - 1) = \Phi\hat{P}(k - 1/k - 1)\Phi^T + GQG^T \qquad (4.77)$$

The updated state estimate is given as

$$\hat{X}(k/k) = \tilde{X}(k/k - 1) + Kv(k) \qquad (4.78)$$

$$\hat{P}(k/k) = (I - KH)\tilde{P}(k/k - 1) \qquad (4.79)$$

The Kalman gain is $K = \tilde{P}(k/k - 1)H^T S^{-1}$, and the innovations vector is $v(k) = z(k) - \tilde{z}(k/k - 1)$. The innovations covariance is computed as $S = H\tilde{P}(k/k - 1)H^T + R$. In case there is no valid measurement, the track will be just updated without using any measurement

$$\hat{X}(k/k) = \tilde{X}(k/k - 1) \qquad (4.80)$$

$$\hat{P}(k/k) = \tilde{P}(k/k - 1) \qquad (4.81)$$

The NNKF can give erroneous results if there is a high level of clutter and/ or several closely space targets, in that case correct association becomes too complex and is not feasible, and we need to use some more measurements, as is done in the PDAF.

4.5.2 Probabilistic Data Association Filter

In the PDAF, we use more than one measurement unlike in NNKF and thus, in the cases of high clutter and/or more number of closely spaced sensors the PDAF is more useful than NNKF, and is supposed to give a better tracking performance. All the 'valid' measurements within the gate (not only one closest measurement like NNKF) are used with certain association probabilities for each valid measurement at the current instant for the target. All the valid measurements within the gate and their associated probabilities are subsequently used in the PDAF, thus, it gives some importance to all the measurements which form a valid gate. It is, in some sense, like using the measurements/data-level fusion process within the structure of the PDAF and the valid gate. This inference regarding the PDAF further strengthens the concept of measurement level DF in the KF, that is, the KF is a natural data fuser. So, in PDAF this aspect of using the more number of observables within the validation gate (and not only using one nearest measurement) with associated weightings not only amounts to being like MLF but also amounts to using a DF rule with appropriate weights. We assume that '*m*' measurements have fallen within a particular gate and that there is only one track-target of interest and the track has been initialised, then we consider [1,4]

$$z_i = \begin{cases} \{y_i \text{ is the target originated measurement}\}, & i = 1, 2, \ldots, m, \\ \{\text{none of the measurements is target originated }\}, & i = 0 \end{cases}$$

(4.82)

as mutually exclusive and exhaustive events for $m \geq 1$. Then the conditional mean of the state is given as

$$\hat{X}(k/k) = \sum_{i=0}^{m} \hat{X}_i(k/k) p_i$$

(4.83)

In Equation 4.83, $\hat{X}_i(k/k)$ is the updated state and is conditioned on the event that the *i*th instant validated measurement is (assumed to be) correct. p_i is the conditional probability of the event, that is, the correct measurement has occurred. Then the estimates conditioned on the measurement '*i*' (being assumed to be correct) are computed by the update equations [1,4]

$$\hat{X}_i(k/k) = \tilde{X}(k/k - 1) + K v_i(k), \quad i = 1, 2, \ldots, m$$

(4.84)

The conditional innovations are computed using the standard equation

$$v_i(k) = z_i(k) - \hat{z}(k/k - 1)$$

(4.85)

For $i = 0$, if none of the measurements is valid ($m = 0$), then we have only the prediction

$$\hat{X}_0(k/k) = \tilde{X}(k/k - 1)$$

(4.86)

The standard state update equation is given as

$$\hat{X}(k/k) = \tilde{X}(k/k - 1) + K v(k)$$

(4.87)

The merged/combined innovations are computed by

$$v(k) = \sum_{i=1}^{m} p_i(k) v_i(k)$$

(4.88)

The covariance matrix corresponding to the computed state is given as

$$\hat{P}(k/k) = p_0(k)\tilde{P}(k/k-1) + (1 - p_0(k))P^c(k/k) + P^s(k)$$ (4.89)

The covariance of the state updated with the correct and associated measurement is then given as

$$P^c(k/k) = \tilde{P}(k/k-1) - KSK^T$$ (4.90)

Finally, the spread of the innovations is computed using the following equation:

$$P^s(k/k) = K\left(\sum_{i=1}^{m} p_i(k)v_i(k)v(k)_i^T - v(k)v(k)^T\right)K^T$$ (4.91)

The conditional probability for the Poisson clutter model is given as

$$p_i(k) = \frac{e^{-0.5v_i^T S^{-1}v_i}}{\lambda\sqrt{|2\Pi S|}((1-P_D)/P_D) + \sum_{j=1}^{m} e^{-0.5v_j^T S^{-1}v_j}}, \quad \text{for } i = 1,2,...,m$$ (4.92)

$$= \frac{\lambda\sqrt{|2\Pi S|}((1-P_D)/P_D)}{\lambda\sqrt{|2\Pi S|}((1-P_D)/P_D) + \sum_{j=1}^{m} e^{-0.5v_j^T S^{-1}v_j}}, \quad \text{for } i = 0$$ (4.93)

In Equation 4.92, we have λ as the false alarm probability, and P_D as the detection probability. Thus, we see that, although, the PDAF is a more complex target-tracking filter than the NNKF, we can surely infer from the foregoing theoretical arguments/developments that PDAF would perform better than NNKF, and this is again because it is a natural data fuser algorithm like the KF, and in addition to it, encompasses the data-fusion concept and probabilistic weightings to do the DA. This is a very important observation/inference on the basic characteristics of the PDAF. The PDAF might provide poor performance if the targets are nearly closed with each other, and can be incorrect if there are multiple targets. Also, track initialisation and deletion mechanisms need to be provided. In order to avoid certain limitations of PDAF, one can use Joint-PDAF [4].

4.5.3 Multiple-Hypothesis Filter

The MHT filter and the related track splitting filter (TSF)-based tracking algorithms maintain separate tracks for each possible associated measurement [1].

At each sampling instant/scan or time step, the predicted measurement is used to establish a validation gate. Then for each such measurement within the gate, a new hypothesised track is generated. A single track is split into n tracks [1]: (i) one track associated with each valid measurement and (ii) one track for the no-association hypothesis. Each of these tracks is treated independently and utilised to obtain the new predictions for the next time instant/step. Here, the number of branches into which the track is split would grow exponentially, and hence the likelihood function of each split track is computed and the unlikely ones are discarded. The MHT tracking algorithm works on the complete sequences of measurements.

4.6 Information Filtering

In the conventional KF, the state and its (error) covariance matrix are propagated in time-evolution cycle/time-propagation part of the KF, and in the data-update cycle the measurements are used and the state and its error covariance matrix are updated. In an information filter (IF), the state is propagated (and updated) along with the propagation and updating of the appropriate information matrices. This information matrix is defined in terms of Fisher's information matrix, and it is the inverse of the covariance matrix of the KF for the corresponding variables. The basic problem of the estimation of the states and/or the parameters using information concept is formulated next. The measurements are modelled using, say, the linear system, although the non-linear model can be used

$$z = Hx + v \tag{4.94}$$

Here, z, x and H have the usual meanings. Here, x can be regarded as the parameter vector also, depending on the problem at hand. Also, v is an m-vector of measurement noise with zero mean and identity covariance matrix for the sake of convenience. The usual least squares (LS) estimate of x is obtained by minimising the sum of the squares of the measurement errors and a cost function is formulated [2,4] as

$$J(x) = (z - Hx)^T (z - Hx) \tag{4.95}$$

Now, in addition to the above system of equations, we also have a priori unbiased estimate $\tilde{x}$ of x, and a priori information matrix both forming a priori state/parameter-information matrix pair: $(\tilde{x}, \tilde{\Lambda})$, here, the state/parameter is still in the conventional covariance form. By incorporating this a priori information pair in Equation 4.95, we obtain the modified cost function [2,4]

$$J_1(x) = (x - \tilde{x})^T \tilde{\Lambda}(x - \tilde{x}) + (z - Hx)^T(z - Hx) \tag{4.96}$$

Obtaining the partial differentiation of the cost function of Equation 4.96 with respect to the unknown vector x, and setting the resultant expression equal to zero would yield the formulae which when solved would give the optimal LS estimate of the state/parameter x. This optimal LS estimate is obtained with the fact that the a priori information matrix of the state/parameter was taken into account for the estimation. This is the fundamental approach of incorporating the information (matrix) into the problem of state/parameter estimation.

4.6.1 Square Root Information Filtering

In order to obtain the square root information filter-formulation (SRIF), we should factor the information matrix itself into its square roots to obtain the following form of the cost function [2,4]:

$$J_1(x) = (x - \tilde{x})^T \tilde{R}^T \tilde{R}(x - \tilde{x}) + (z - Hx)^T(z - Hx)$$

$$J_1(x) = (\tilde{y} - \tilde{R}x)^T(\tilde{y} - \tilde{R}x) + (z - Hx)^T(z - Hx) \tag{4.97}$$

The second equation of Equation 4.97 is obtained by the substitution, $\tilde{y} = \tilde{R}\tilde{x}$ (this being the information state [IS] as against the covariance state [CS] x, of the KF and the measurement z), the simplification and rearrangement of the first equation. The first parenthesis term is now written as $\tilde{y} = \tilde{R}x + \tilde{v}$ in the similar form as Equation 4.94 and we can readily see that the cost function J represents the following composite system [2,4]:

$$\begin{bmatrix} \tilde{y} \\ z \end{bmatrix} = \begin{bmatrix} \tilde{R} \\ H \end{bmatrix} x + \begin{bmatrix} \tilde{v} \\ v \end{bmatrix} \tag{4.98}$$

Thus, it can be seen very interestingly that the a priori information is considered as an additional measurement in the form of a data equation (DE), just like the measurement Equation 4.94, we can call the first-row term of Equation 4.98 the information state data equation (ISDE). Then the LS solution is obtained by applying the method of orthogonal transformation (by using Householder transformation) matrix T yielding the solution to the LS functional as follows:

$$-\begin{bmatrix} \tilde{v} \\ v \end{bmatrix} = \begin{bmatrix} \tilde{R} \\ H \end{bmatrix} x - \begin{bmatrix} \tilde{y} \\ z \end{bmatrix}$$

$$T\begin{bmatrix} \tilde{R}(k-1) & \tilde{y}(k-1) \\ H(k) & z(k) \end{bmatrix} = \begin{bmatrix} \hat{R}(k) & \tilde{y}(k) \\ 0 & e(k) \end{bmatrix}; \quad k = 1,...,N \qquad (4.99)$$

What we have essentially done is that Equation 4.98 is flipped over and the sum of squares of errors/noises is minimised in the form of the cost function, Equation 4.97, by applying the transformation to the composite known parts of Equation 4.99, the left-hand side block matrices. We see that e_j is the sequence of residuals, and it can be seen that the new information pair is generated ($\tilde{y}(k)$, $\hat{R}(k)$), after the incorporation of the current measurement data and the process of the orthogonal transformation. This process is then repeated with the incorporation of next/new measurement ($z(k + 1)$) (along with the information pair obtained from the previous cycle) and the new information pair is thus generated to obtain the recursive SRIF. This further forms the basis for decentralised SRIF which is discussed in Chapter 5. We hasten to add here that we still have not considered incorporation of the state-space model of dynamic system that would generate the state x. In order to explain the DF concept using the SRIF, the foregoing formulation is sufficient for the present.

4.6.2 DF Based on Square Root Information Filtering

Let us have a linear system with two sensors' models as H_1 and H_2. By using the formulation of Equation 4.99, we can fuse the sensor measurements at the data level itself as follows:

$$T\begin{bmatrix} \tilde{R}(k-1) & \tilde{y}(k-1) \\ H_1(k) & z_1(k) \\ H_2(k) & z_2(k) \end{bmatrix} = \begin{bmatrix} \hat{R}(K) & \tilde{y}(k) \\ 0 & e(k) \end{bmatrix}; \quad k = 1,...,N \qquad (4.100)$$

This results in the straightforward parameter estimates with the effect of two sensors' data taken into account for estimation and fusion in a composite manner. The resultant state/parameter estimate and the information matrix (in fact the square root of the information matrix) are a global effect of the processing of the two sensors' data. Alternatively, it is also feasible to process the measurements from each sensor individually to obtain the estimate of the IS-vectors and then fuse these estimated IS-vectors to obtain the combined IF-SVF, the equations for this being given as

$$\tilde{y}_f = \tilde{y}_1 + \tilde{y}_2 \quad \text{and} \quad \hat{R}_f = \hat{R}_1 + \hat{R}_2 \qquad (4.101)$$

Here, '$\tilde{y}$' is the fused state from the point-of-view-of the SRIF concept. It should be noted here that this IS-vector may not be the same as the one that would occur in the full information filtering algorithm. Interestingly and very easily the fused covariance-state can be obtained by

$$\hat{x}_f = \hat{R}_f^{-1}\tilde{y}_f \tag{4.102}$$

So, we observe here that the DE formulation of the information pair (IS and SRIM) and the orthogonal transformations obtain very simple, elegant, very powerful and useful solutions to the composite estimation-sensor DF problem with enhanced numerical reliability, efficiency and stability in the recursive manner.

4.7 HI Filtering-Based DF

The concept of HI is related to the optimal control synthesis in the frequency domain and it explicitly addresses the question of modelling the errors in the deterministic domain rather than in the usual stochastic domain. The idea is to treat the worst-case scenario, that is, minimise the maximum of the errors, and hence, it is also called the min–max problem also. The HI framework has certain simple and important properties [4]: (i) it is capable of dealing with system's modelling errors and unknown disturbances in the deterministic domain, (ii) it represents a natural extension to the existing theory, that is, the either one is a superset or sub-set of the other in some way, or to say in usual way, one is general or special case to the other, (iii) it is amenable to a meaningful optimisation process in the conventional sense of the term (min–max) and (iv) it is applicable to all the multi-variable control and estimation/filtering problems. Basically, the so-called HI norm is defined as

$$\frac{\sum_{k=0}^{N}(\hat{x}^f(k)-x(k))^t(\hat{x}^f(k)-x(k))}{(\hat{x}_0^f-x_0^f)^t P_0^f(\hat{x}_0^f-x_0^f)+\sum_{k=0}^{N}w^t(k)w(k)+\sum^m\sum_{k=0}^{N}v^{m^t}(k)v^m(k)} \tag{4.103}$$

The HI norm has several important terms (all the symbols/state/noise variables have usual meaning): (i) those in the denominator constitute the energies (equivalent to the power/variance/covariance and hence, the squared uncertainties) due to the errors in the initial conditions, state disturbances (deterministic discrepancies, equivalently generalising the process noise in the KF formulation), and the measurement disturbances (deterministic discrepancies, equivalently generalising the measurement noise in the KF formulation) for both the sensors (it can be easily extended to more number of sensors), and (ii) the output-numerator term as the energy (of the filter) due to the errors in fused state (considered as the deterministic error equivalently generalising the state errors that are considered in the KF as the stochastic errors-processes). Intuitively, then it means that any estimator/filtering algorithm derived based on this HI norm signifies that the corresponding filter minimises the effects of the input errors (energies) on the estimated states and it also, sees that the output error (energy) is minimum, since the maximum of the HI norm is minimised by the filter-derivation process. Equation 4.103 can be looked upon as the transfer function between output errors and the total input errors. In this case, the maximum gain of the transfer function is minimised by the ensuing filtering solution. It would be logical to term the filtering algorithm derived based on the HI norm as the HI estimator rather than the HI filter, however we can use the terms interchangeably for convenience and usual convention. In fact the HI norm, that is, I/O (input to output) gain of the norm, should be less than square of gamma, a scalar parameter that can be considered as an upper bound on the maximum energy gain from the input (errors/energies) to the output (error/energy). This HI norm is used in deriving the so-called robust estimation algorithms that can be easily used for target tracking and sensor DF. The studies and the results of DF using the HI filters are relatively few in the literature.

4.7.1 HI Posterior Filter

The estimates are obtained for each sensor ($i = 1,2$) using HI a posterior filter [6]. The basic mathematical model-problem formulation is the same as the state-space equations as for the KF, however, various noise processes are considered at the deterministic disturbances rather than the stochastic processes. We have the following composite R matrix:

$$R_i = \begin{bmatrix} I & 0 \\ 0 & -\gamma^2 I \end{bmatrix} + \begin{bmatrix} H_i \\ L_i \end{bmatrix} P_i(k)[H_i^t \; L_i^t] \tag{4.104}$$

The variable L specifies what linear combination of the state variables is desired to be estimated. The state covariance matrix time propagation is computed as

$$P_i(k+1) = \phi P_i(k)\phi' + GQG' - \phi P_i(k)[H_i^t \; L_i^t]R_i^{-1}\begin{bmatrix} H_i \\ L_i \end{bmatrix} P_i(k)\phi' \tag{4.105}$$

It must be emphasised here that the symbols/notations have the usual meaning, however, P and R are not usual covariance matrices, but are called Gramian matrices and the state/measurement variables are considered as the generalised random processes/variables, rather than the stochastic processes, since we are handling the deterministic disturbances. Hence, the usual stochastic theory is really not necessary to invoke for the derivation of the HI filter. We however, continue to use the usual notations/terminology for the sake of remaining connected with the conventional KF formulations.

The HI filter gain is given as

$$K_i = P_i(k+1)H_i^t(I + H_iP_i(k+1)H_i^t)^{-1} \tag{4.106}$$

The measurement update of states is given by

$$\hat{x}_i(k+1) = \phi\hat{x}_i(k) + K_i(y_i(k+1) - H_i\phi\hat{x}_i(k)) \tag{4.107}$$

If we have processed the measurement data from two sensors using the HI filter, then the fusion of the state estimates from the two HI estimators can be obtained by the conventional SVF formulae

$$\hat{x}_f(k+1) = \hat{x}_1(k+1) + \hat{P}_1(k+1)(\hat{P}_1(k+1) + \hat{P}_2(k+1))^{-1}(\hat{x}_2(k+1) - \hat{x}_1(k+1)) \tag{4.108}$$

$$\hat{P}_f(k+1) = \hat{P}_1(k+1) - \hat{P}_1(k+1)(\hat{P}_1(k+1) + \hat{P}_2(k+1))^{-1}\hat{P}_1^t(k+1) \tag{4.109}$$

We see that the fused state vector and Gramian matrix utilise the individual estimated state vectors of each sensor and the associated Gramian matrices. An extended HI filter for non-linear continuous-time dynamic system is given in Appendix A.6.

4.7.2 Risk-Sensitive HI Filter

This filter is based on the minimisation of the so-called risk-sensitive cost function, the scalar 'β' being called the risk-sensitivity parameter, thus, if $\beta = 0$, the filter is risk neutral and is equivalent to the conventional estimation. This risk-sensitive HI filter [4,6] is given by the following set of equations:

$$R_i = \begin{bmatrix} R_{vi} & 0 \\ 0 & -\beta^{-1}I \end{bmatrix} + \begin{bmatrix} H_i \\ L_i \end{bmatrix} P_i(k)[H_i^t \ L_i^t] \tag{4.110}$$

The time-propagation Gramian is computed as

$$P_i(k+1) = \phi P_i(k)\phi' + GQG' - \phi P_i(k)[H_i^t \ L_i^t]R_i^{-1}\begin{bmatrix} H_i \\ L_i \end{bmatrix}P_i(k)\phi' \tag{4.111}$$

The filter gain is given as

$$K_i = P_i(k+1)H_i^t(I + H_iP_i(k+1)H_i^t)^{-1} \tag{4.112}$$

The measurement update of states is given by

$$\hat{x}_i(k+1) = \phi\hat{x}_i(k) + K_i(z_i(k+1) - H_i\phi\hat{x}_i(k)) \tag{4.113}$$

4.7.3 Global HI Filter for DF

In this global HI filter, one can obtain the fusion of states and fusion of the Gramians iteratively like the measurement level DF KF. Hence, the global HI can be regarded as a natural data level fuser. For this global filtering algorithm [7,8], the local filters are given for each sensor ($i = 1,2 \ldots m$) as follows.

The state and Gramian time propagation are given by

$$\tilde{x}_i(k+1) = \phi\hat{x}_i(k) \tag{4.114}$$

$$\tilde{P}_i(k+1) = \phi\hat{P}_i(k)\phi + GQG' \tag{4.115}$$

The Gramian matrix update is obtained from

$$\hat{P}_i^{-1}(k+1) = \tilde{P}_i^{-1}(k+1) + [H_i^t \ L_i^t]\begin{bmatrix} I & 0 \\ 0 & -\gamma^2 I \end{bmatrix}^{-1}\begin{bmatrix} H_i \\ L_i \end{bmatrix} \tag{4.116}$$

The local filter gains are given as

$$A_i = I + 1/\gamma^2\hat{P}_i(k+1)L_i^t L_i; \quad K_i = A_i^{-1}\hat{P}_i(k+1)H_i^t \tag{4.117}$$

The measurement update of local states is obtained from

$$\hat{x}_i(k+1) = \tilde{x}_i(k+1) + K_i(y_i(k+1) - H_i\tilde{x}_i(k+1)) \tag{4.118}$$

The time propagation of fusion state/Gramian is given as

$$\tilde{x}_f(k+1) = \phi\hat{x}_f(k) \tag{4.119}$$

$$\tilde{P}_f(k+1) = \phi\hat{P}_f(k)\phi' + GQG' \tag{4.120}$$

The measurement update of the fusion states-Gramian is given as

$$\hat{P}_f^{-1}(k+1) = \tilde{P}_f^{-1}(k+1) + \sum_{i=1}^{m}(\hat{P}_i^{-1}(k+1) - \tilde{P}_i^{-1}(k+1)) + \frac{m-1}{\gamma^2}L^tL \quad (4.121)$$

The global gain is given as

$$A_f = I + 1/\gamma^2 \hat{P}_f(k+1)L^tL \quad (4.122)$$

The global measurement update, that is, the fused state is given as

$$\hat{x}_f(k+1) = [I - A_f^{-1}\hat{P}_f(k+1)H_f^t H_f]\tilde{x}_f(k+1)$$
$$+ A_f^{-1}\hat{P}_f(k+1)\sum_{i=1}^{m}\{\hat{P}_i^{-1}(k+1)A_i\hat{x}_i(k+1) - (\hat{P}_i^{-1}(k+1)A_i + H_i^t H_i)\phi\hat{x}_i(k)\}$$

$$(4.123)$$

Some results of the application of these HI filters to sensor DF problems can be found in Reference 4. The results of the application of the HI posterior filter for simultaneous localisation and mapping problem (SLAM) for a mobile vehicle/robot can be found in Chapter 15 of Reference 5.

4.7.4 Hybrid H₂ and HI Filter

As we know the KF is an estimation/filtering algorithm that minimises the average estimation errors (minimisation of the mean squares errors) whereas any min–max (HI) filter minimises the worst-case estimation error as discussed in Section 4.7. It is important to note here certain points for both filters [9]: (i) the KF filter assumes that the dynamic system's noise properties (Q, R) are known exactly (if these are not known, they are guessed or estimated and then used, (ii) the mini-max/HI filters assume that nothing is known of these noise properties, in fact these noises are not considered as the random processes at all, but as the unknown deterministic disturbances; and as we have seen in Section 4.7, these variables then are considered as the generalised random variables, (iii) the KF minimises the average estimation error (MSE) and (iv) the mini-max/HI filter minimises the worst-case (maximum of the) estimation errors. The item (ii) keeps the link with the conventional KF, but there is nothing stochastic about the HI/min–max filters. There are some structural similarities between KF and many HI filters, but the conditions for their existence and uniqueness are different. If in a situation both the conditions of (i) and (ii) exist, then one can use the hybrid filter, that is, hybrid KF–HI filter (HF). The HF concept is heuristic but seems powerful.

Assume that we design a KF for the system and find the steady-state gain denoted by K^2. Then we design a min–max filter for the same system with the gain denoted by K^1. Then we can choose the hybrid filter gain as

$$K = wK^2 + (1 - w)K^1 \qquad (4.124)$$

Here, $w = 0$ or 1 to be decided by the user and is considered as a switching parameter. If $w = 0$, then the HF is equivalent to the min–max filter, and if $w = 1$, then the HF is equivalent to the KF. One can use two aspects while choosing the parameter w: (i) although the KF and the HI filters both might be stable, a combination of the two filters might be unstable, so, w must be chosen such that the HF is stable; and (ii) w can be chosen based on the relative weights given by the filter designer, his/her confidence in the a priori noise statistics, in case of the requirement specified by the component KF. It might be possible to use the value of w based on the norms (or some function of the norm) of the information matrices (inverse of the covariance matrix of the KF, and the inverse of the Gramian matrix of the HI filter – or could be the norm of the covariance/Gramian matrix itself) provided by the respective filters. This suggestion remains so far unexplored for HF. In case of the KF, we have the information matrix as the inverse of the covariance matrix, whereas in the case of the HI estimator we can take the inverse of the Gramian matrix as the information matrix (we can call this as the infogramian matrix to signify the fact the information comes from the basic Gramian matrix of the HI filters). However, these ideas need to be explored further. The preliminary results of one application of the HF can be found in Reference 9.

4.8 Optimal Filtering for Data Fusion with Missing Measurements

As we have seen MSDF/DF is very important for several civilian and military applications for target tracking and automatic target recognition (ATR). The main idea is that we want to ascertain the status and/or an identity of the object under investigation/observation wherefore the measurements are available from more than one sensor channel. Then more complete information about the status of the object/target is obtained by fusing or combining the information/data from these sensors in some appropriate way. Sometimes, in a data-communications channel a few or many measurements might be missing (from one or more sensors). It then becomes very important to study and evaluate the performance of the data processing algorithms in the presence (despite) of missing measurements (missing data, MD), that is, in the absence of certain data during certain times; we might have only the measurement/random noises present for these channels. The aspect of MD is also very important in MSDF, and has gained tremendous impetus recently in various fields of science and engineering. We discuss certain algorithms based on fundamental KF and evaluate the performance of these algorithms when some measurement data are randomly missing (RMD) and are not available for further processing in the filters meaning thereby, these data are gone forever. We use MATLAB-based numerical simulations for evaluating the performances of such algorithms.

As we have seen earlier typically there are two basic methods of fusion of information/data from two (or more) sets of data: (i) measurement data level and (ii) SVF. In the data-level fusion, the data sets from the two (or more) sensor channels are directly combined and used in the conventional Kalman filter (KF, or in any other filter) and after processing (by the filter), we obtain one automatically fused state of the target. In the SVF, the data sets coming from each sensor are initially processed by individual KFs and then by a well-known formula of SVF (Equations 4.65 and 4.66), these individual esti- mates of the states are fused. This formula obtains a weighted average of the individual state estimates, and the weights are derived from the covariance matrices of the individual KFs which have been used for the processing of original data sets. As we have seen above the major requirement in DF is the availability of measurements from more than one sensor and the appropri- ate signal processing algorithm, that is, the most suitable algorithm for the linear system is the KF. If some measurements are missing in the processing of the data by KF, then the performance of the filter would not be as/so good, since the KF does not have so much of inherent robustness to handle the situations when the data are missing for a very long time. What would hap- pen is that the filter would continue to predict the state estimates based on the time-propagation equations which do not take any measurements into account anyway, and the measurement data-update would not be very effec- tive, since some data are missing anyway. Hence, it is very important to study the problem of missing measurements in the filter processing cycle, and yet this aspect has not gained much attention in the context of MSDF and target tracking (though some work has been done in certain special cases [10,11]). The data may be missing due to one or more of these reasons [11]: (a) the sensor has failed, (b) there is a problem is a communication channel and (c) the received data might be only noise and the signal is missing. So, it is very important to handle the situation of the missing data in KF in some formal way. In recent work in this direction [11], two formal approaches of handling the missing data in an optimal filter for DF have been presented.

So, first we discuss the optimal filters (OFs) [11] for the sake of completion and to set a stage for the development of variants of these filters. We note here that the conventional discrete-time KF is usually given in two parts: (i) time propagation and (ii) the measurement data update. Whereas the OFs of Reference 11 are the composite ones meaning that the basic (Kalman-type) fil- ter in the OFs is in one part: the time propagation and the measurement data update are composite. We discuss these filters [11]: (i) the one that takes care of the randomly missing data in any sensor (and hence in both the sensors), and then we can use SVF, we name this OFSVF1, and (ii) the other that takes care of the randomly missing data and then uses MLF, we name this OFMLF1.

Then, we present some ramifications of the OFs: (a) optimal filter in two parts for state vector fusion (OFSVF2) with the KF-type filter in two parts (and data can randomly miss in any or both the sensors) and then uses SVF, and (b) OFMLF22 with the KF-type filter in two parts for data randomly missing only in the second sensor, whereas the first sensor provides all the data, without a miss (alternatively the MD could have occurred in the first sensor only). We do this because conventionally the KF is presented in two parts: (i) time propagation (prediction) and (ii) measurement/data update (correction). Thus, any ramification of handling the MD need to be applied to the measurement part only, and the prediction part always remains the same.

4.8.1 Basic Filter for Missing Measurements: SVF

In this section, we give the filtering equations that handle the randomly missing measurements. Let the linear system be described as [11]

$$
\begin{aligned}
x(k) &= \phi x(k-1) + Gw(k) \\
z^i(k) &= \gamma^i(k)H^i(k)x(k) + v^i(k)
\end{aligned}
\tag{4.125}
$$

Here, z represents the measurement vector, with data from, say, any of the two sensors ($i = 1,2$). The process and measurement noise covariance matrices are Q and R, and these white Gaussian noise processes are zero mean and uncorrelated with each other. The scalar quantity γ^i is a Bernoulli sequence which takes values 0 and 1 randomly; thus, we have $E\{\gamma^i(k) = 1\} = b^i(k)$ and $E\{\gamma^i(k) = 0\} = 1 - b^i(k)$, with b as the percentage of measurements that arrive truthfully to the sensor fusion node. This also means that some measurement data are randomly missing. The constant b is assumed to be known and pre-specified. Then, the optimal KF-like filtering algorithm to handle and incorporate the missing measurements in the composite-filter cycle is given by the set of equations [11], (4.126) to (4.130).

The state estimation (combined state prediction/time propagation and the measurement data-update cycle) is given as

$$
\hat{x}(k) = \phi \tilde{x}(k) + K(z(k) - b(k)H\tilde{x})
\tag{4.126}
$$

State-error covariance matrix (composite) is given as follows:

$$
\begin{aligned}
\hat{P} &= [\phi - bKH\phi]\tilde{P}[\phi - bKH\phi]^t + [I - bKH]GQG' \\
&\quad [I - bKH]^t + b(1-b)KHX H^t K^t + KRK^t
\end{aligned}
\tag{4.127}
$$

For simplicity, we have dropped the time index k from certain variables, since from the context it will clear, and if required it can be easily reintroduced. The optimal gain is obtained from the following two expressions as $K = C*$inverse (D):

$$C = b\phi\tilde{P}\phi^t H^t + bGQG'H^t \qquad (4.128)$$

$$D = b(1-b)H\phi X\phi^t H^t + b^2 H\phi\tilde{P}\phi^t H^t + bHGQG^t H^t + R \qquad (4.129)$$

We also have, as required in Equations 4.127 and 4.129, the following:

$$X_k = \phi(k)X_{k-1}\phi'(k) + Q \qquad (4.130)$$

It is emphasised here that the optimal filter equations as given above are put in the utmost simplified form (compared to the ones given in the original source [11]) so that the implementation in MATLAB is rendered very easy and intuitively appealing. Thus, the estimated state and covariance matrix are marked with '^' and the a priori/previous estimates of the state/ covariance matrix with '~'. We name this filter OFSVF1.

4.8.2 Optimal Filter for Missing Measurements: Measurement Level Fusion

We have again the following composite model for the dynamics of the tracking system [11]:

$$\begin{aligned} x(k) &= \phi x(k-1) + Gw(k) \\ z_{12}(k) &= \gamma(k)H_{12}(k)x(k) + v_{12}(k) \end{aligned} \qquad (4.131)$$

Here, we have the following details for two sensors:

$$z_{12} = \begin{bmatrix} z_1 \\ z_2 \end{bmatrix}; \quad H_{12} = \begin{bmatrix} H_1 \\ H_2 \end{bmatrix}; \quad v_{12} = \begin{bmatrix} v_1 \\ v_2 \end{bmatrix} \qquad (4.132)$$

Then the composite state estimator is given as follows [11]:

$$\hat{x}(k) = \phi\tilde{x}(k) + K(z_{12}(k) - bH_{12}\tilde{x}) \qquad (4.133)$$

In Equation 4.133, we ascertain that the new and the previous state estimates are the fused state estimates automatically obtained by this optimal filter, since the fusion takes place at the measurement data level itself as is specified in Equation 4.132. The gain matrix/vector is also the composite one. We drop the time index k from certain variables, since from the context of the use it will clear where it is implied, and can be re-inserted if required. Then the composite state-error covariance matrix for this MLF filter is given as

$$
\begin{aligned}
\hat{P} = [\phi - bKH_{12}\phi]\tilde{P}[\phi - bKH_{12}\phi]^t + [I - bKH_{12}]GQG' \\
[I - bKH_{12}]^t + b(1 - b)KH_{12}XH_{12}{}'K^t + KR_{12}K^t
\end{aligned}
\tag{4.134}
$$

We note here that the optimal gain, the covariance matrix P and the covariance matrix R are the composite ones. Hence, we have

$$
R_{12} = \begin{bmatrix} R_1 & 0 \\ 0 & R_2 \end{bmatrix}
\tag{4.135}
$$

The optimal gain of the fusion filter is obtained from the following two expressions as $K = C*inverse\ (D)$. We have

$$
C = \begin{bmatrix} C_1 & C_2 \end{bmatrix}
\tag{4.136}
$$

$$
D = \begin{bmatrix} D_{11} & D_{12} \\ D_{21} & D_{22} \end{bmatrix}
\tag{4.137}
$$

The components of the C and D are given as follows [11]:

$$
C_1 = b\phi\tilde{P}\phi^t H_1^t + bGQG'H_1^t
\tag{4.138}
$$

$$
C_2 = b\phi\tilde{P}\phi^t H_2^t + bGQG'H_2^t
\tag{4.139}
$$

$$
D_{11} = b(1 - b)H_1\phi X\phi^t H_1^t + b^2 H_1\phi\tilde{P}\phi^t H_1^t + bH_1GQG'H_1^t + R_1
\tag{4.140}
$$

$$
D_{12} = b(1 - b)H_1\phi X\phi^t H_2^t + b^2 H_1\phi\tilde{P}\phi^t H_2^t + bH_1GQG'H_2^t
\tag{4.141}
$$

$$
D_{21} = b(1 - b)H_2\phi X\phi^t H_1^t + b^2 H_2\phi\tilde{P}\phi^t H_1^t + bH_2GQG'H_1^t
\tag{4.142}
$$

$$
D_{22} = b(1 - b)H_2\phi X\phi^t H_2^t + b^2 H_2\phi\tilde{P}\phi^t H_2^t + bH_2GQG'H_2^t + R_2
\tag{4.143}
$$

We name this filter OFMLF1.

4.8.3 Optimal Filter in Two Parts for SVF

We now describe a simple optimal filter following the basic KF that is generally given in two parts. We state that the data can randomly miss in any of the sensors (in fusion theory usually two sensors are considered for the purpose of illustration). We consider the measurement data missing randomly as described in Section 4.8.1. Since, the data are missing at the measurement (data-) level the time-propagation part of the filter remains the same as in the conventional KF. Also, since this is an individual filter applicable to any sensor channel, for the fusion we need to run two such filters and then the individual state estimates are fused using the SVF fusion formulae. We again use very compact notations. The time-propagation part to be used for each sensor channel for state estimation and covariance propagation is given as follows:

$$\tilde{x} = \phi\hat{x} \tag{4.144}$$

$$\tilde{P} = \phi\hat{P}\phi^T + GQG^T \tag{4.145}$$

The measurement/data-update part is given as follows.
State estimation

$$\hat{x} = \tilde{x} + K(z - bH\tilde{x}) \tag{4.146}$$

The optimal filter gain for the randomly missing data filter is given as

$$K = b\tilde{P}H^T(b^2H\tilde{P}H^T + R)^{-1} \tag{4.147}$$

The data-update state-error covariance matrix is given as

$$\hat{P} = [I - bKH]\tilde{P}[I - bKH]^t + KRK^t \tag{4.148}$$

We note that the z is the measurement data vector and where some measurements might be missing randomly. The derivation of this filter is given in Appendix A. We name this filter OFSVF2.

4.8.4 Optimal Filter in Two Parts for MLF

Here, we describe again a simple optimal filter following the basic KF that is generally given in two parts, and we state that the data can randomly miss in only one sensor whereas the data from the other sensor are not missing. We consider the measurement data missing as described in Section 4.8.1. Since, the data are missing at the measurement (data) level the time-propagation part of the filter remains the same as in the conventional KF, and the filter equations are similar to Equations 4.144 and 4.145:

The time-propagation part to be used is given as follows:

$$\tilde{x}_f = \phi \hat{x}_f \tag{4.149}$$

$$\tilde{P}_f = \phi \hat{P}_f \phi^T + GQG^T \tag{4.150}$$

We note that in Equations 4.149 and 4.150 'f 'stands for the fused state vector and state-error covariance matrix, since we are considering the MLF of data from sensor number 1, and the data from sensor number 2, where for the latter some data are randomly missing in the sense described in Section 4.8.1. We state here that data are randomly missing from sensor two only (i.e. in z_2, denoted also as z_m in Appendixes A.1 and A.2) and hence, the factor b appears appropriately in the corresponding part of the equations. We note here that the derivation of this filter is given in Appendix A.2.

The measurement data-update part is given as

$$\hat{x}_f = \tilde{x}_f + K_2(z_1 - H_1\tilde{x}_f) + K_3(z_2 - bH_2\tilde{x}_f) \tag{4.151}$$

The optimal gains in Equation 4.151 can be obtained from the following set of the equations:

$$K_2(H_1\tilde{P}H_1^t + R_1) + K_3(bH_2\tilde{P}H_1^t) = \tilde{P}H_1^t \tag{4.152}$$

$$K_2(bH_1\tilde{P}H_2^t) + K_3(b^2H_2\tilde{P}H_2^t + R_2) = b\tilde{P}H_2^t \tag{4.153}$$

The fused state-error covariance matrix (of the measurement data-update cycle) is given as

$$\hat{P}_f = [I - K_2H_1 - bK_3H_2]\tilde{P}[I - K_2H_1 - bK_3H_2]^t + K_2R_2K_2^t + K_3R_2K_3^t \tag{4.154}$$

We name the filter OFMLF22. The filter gains are obtained by simultaneously solving Equations 4.152 and 4.153. We state here, that all the filters presented in Section 4.8 are optimal for linear systems, and are unbiased state estimators, because the filters' structures are obtained based on the condition of unbiasedness, and the filter gains are derived using minimisation of appropriate MSE for each filter.

4.8.5 Performance Evaluation of the Filters for Handling Missing Data: Illustrative Examples

In this section, we evaluate the performance of the four filters studied in the previous sub-sections. We have simulated the data generation using

MATLAB and the dynamic model given in Equation 4.125. The target model has position and velocity as states, and the sensor model has only position as the measurement: $F = [1\ T; 0\ 1]$; $G = [T^2/2; T]$; $H1 = H2 = [1\ 0]$; Two sensors are considered. The sampling interval is taken as 0.5 s, and the total simulation scan is for 500 data points. The variance of the measurement noise in sensor number 1 is 1 unit and for the sensor number 2 is 1.44 units. The process noise covariances are $Q1 = 0.002$, $Q2 = 0.001$. The randomly missing measurements situation has been simulated based on the discussion in Section 4.8.1, and the presence of parameter b has been heuristically implemented. We have studied the performance of these filters and the discussion follows. The percentage fit errors (PFE for the measurement time histories) and the percentage state errors (PSE) metrics obtained for all the cases are given in Tables 4.1 (cases 1 to 3) and 4.2 (cases 4 and 5). The results of the five filters studied are discussed next.

Case 1: Basic Kalman filter when the data are missing non-randomly, KF. The conventional KF is used where the data are missing non-randomly, that is, the data are missing for 100 points (in between) in any one sensor, whereas the data in the other sensor are not missing. The fusion rule is SVF using Equations 4.65 and 4.66, neglecting the cross-covariance matrices in the fusion equations. The performances of the KF with and without missing data are almost similar.

Case 2: The optimal filter for SVF – OFSVF1. The optimal filter discussed in Section 4.8.1 [11] was implemented for two sensors each individually. The data are missing randomly (DRM) in both the sensors at either levels 0.9 or levels 0.75. Also, the results are presented when the missing data fact is not taken in the account in the filters. The results for the latter aspect are slightly worse, as shown by the metrics' numerical values marked in bold and italics.

Case 3: The optimal fusion filter for missing measurements – OFMLE1. The optimal filter discussed in Section 4.8.2 [11] was implemented for two sensors in the form of the measurement (data-) level fusion. The data are missing randomly in both the sensors at either levels 0.9 or levels 0.75. Also, the results are presented when the missing data fact is not taken in the account in the filters. The results for the latter aspect are slightly worse, as shown by the metrics' numerical values marked in bold and italics. Also, we see that the performance of the OFMLE1 filter is slightly better (marked in only bold values) compared to the equivalent entries of the filter OFSVF1 (i.e. compared to the SVF as is normally the situation).

Case 4: The optimal filter in two parts for state vector fusion – OFSVF2. The optimal filter presented in Section 4.8.3 (and derived in Appendix A.1) was implemented for two sensors in the form of SVF. The data are missing randomly in both the sensors at either level 0.9 or 0.75. Also, the results (Table 4.2) are presented when the missing data fact is not taken in the account in the filter. The results for the latter aspect are slightly worse, as shown by the metrics' numerical values marked in bold and italics.

TABLE 4.1

Percentage Fit Errors and Percentage State Errors for Optimal Filters for Missing Data

Filter	Sensor	No Missing Data				Missing Data (MD)				Remarks
		HI Norm	PFE-position	PSE-position	PSE-velocity	HI Norm	PFE-position	PSE-position	PSE-velocity	
Case 1 KF	S1	**0.0296**	**0.3993**	0.1154	4.5188	0.0299	0.4016	0.1125	4.5069	DRM (data randomly missing) Data non-randomly missing in S1
	S2		0.4793	0.1156	4.7524		0.4793	0.1156	4.7524	
	SF			**0.1015**	**3.9769**			0.1020	3.9788	
Case 2 OFSVF1 [11]	S1	**0.0296**	0.3993	0.1154	4.5188					
	S2		0.4793	0.1156	4.7524					
	SF			0.1015	3.9769					DRM in S1 and S2 (state vector fusion)
		DRM – missing level = 0.9, 0.9; Bk1 = Bk2 = 0.9				DRM – missing level = 0.9, 0.9; Bk1 = Bk2 = 1 (MD not accounted in the filter)				
	S1	0.0318	0.4030	0.1218	4.5808	*0.0328*	0.4027	0.1206	*4.8043*	
	S2		0.4820	0.1457	5.4583		*0.4830*	*0.1472*	*6.091*	
	SF			0.1053	4.7220			*0.1069*	*5.145*	
		DRM – missing level = 0.75, 0.75; Bk1 = Bk2 = 0.75				DRM – missing level = 0.75, 0.75; Bk1, Bk2 = 1 (MD not accounted in the filter)				
	S1	0.0352	0.4094	0.1481	4.5860	*0.0435*	0.4073	0.1369	5.2845	
	S2		0.4886	0.2041	8.0580		*0.4973*	*0.2226*	*10.9911*	
	SF			0.1107	5.3217			*0.1230*	*7.1716*	

(Continued)

TABLE 4.1 (Continued)

Percentage Fit Errors and Percentage State Errors for Optimal Filters for Missing Data

Filter		No Missing Data			Missing Data (MD)			Remarks
Case 3								
OFMLE1 [11]	0.0050	No missing data						DRM in S1 and S2 (data level fusion)
		S1	0.3948					
		S2	0.4772					
		SF	0.1034	4.1972				
	0.0059	DRM – missing level = 0.9, 0.9; Bk = 0.9			0.0054	DRM – missing level = 0.9, 0.9; Bk = 1 (MD not accounted in the filter)		
		S1	0.3971			S1	0.3975	
		S2	0.4727			S2	0.4736	
		SF	0.1062	5.0087		SF	0.1076	5.6211
	0.0076	DRM – missing level = 0.75, 0.75; Bk = 0.75			0.0067	DRM – missing level = 0.75, 0.75; Bk = 1 (MD not accounted in the filter)		
		S1	3.3406			S1	3.3407	
		S2	0.4720			S2	0.4755	
		SF	0.1108	5.1012		SF	0.1193	7.7153

TABLE 4.2

Percentage Fit Errors and Percentage State Errors for Optimal Filters (in Two Parts: OFSVF2 and OFMLF22[a]) for Missing Data

Filter	Sensor	No Missing Data				Missing Data (MD)				Remarks
		HI Norm	PFE-position	PSE-position	PSE-velocity	HI Norm	PFE-position	PSE-position	PSE-velocity	
Case 4 OFSVF2		0.0296								DRM (Data randomly missing)
	S1		0.3993	0.1154	4.5188					
	S2		0.4793	0.1156	4.7524					
	SF			0.1015	3.9769					
		DRM – missing level = 0.9, 0.9; Bk1 = Bk2 = 0.9				DRM – missing level = 0.9, 0.9; Bk1 = Bk2 = 1 (MD not accounted in the filter)				DRM in S1 and S2 (SVF)
		0.0381				0.0328				
	S1		0.5571	0.1218	4.5808		0.4027	0.1206	*4.8043*	
	S2		0.4732	0.1457	5.4808		*0.4734*	*0.1472*	*6.0909*	
	SF			0.1053	4.7221			*0.1069*	*5.1449*	
		DRM – missing level = 0.75, 0.75; Bk1 = Bk2 = 0.75				DRM – missing level = 0.75, 0.75; Bk1, Bk2 = 1 (MD not accounted in the filter)				
		0.0352				*0.0435*				
	S1		2.5267	0.1481	4.5859		3.3416	0.1369	5.2845	
	S2		0.4791	0.2041	8.0583		0.4879	0.2226	10.9911	
	SF			0.1107	5.3217			0.1230	7.1716	
Case 5 OFMLF22[a]		DRM – missing level = 0.9 (for sensor 2 only); Bk = 0.9				DRM – missing level = 0.9 (for S2 only); Bk = 1 (MD not accounted in the filter)				DRM in S2 only (data-level fusion)
		0.0053				0.0052				
	S1		0.3954				0.3956			
	S1		0.4723				0.4727			
	SF			0.1055	5.1272			0.1066	5.4522	
		DRM – missing level = 0.75 (for S2 only); Bk = 0.75				DRM – missing level = 0.75 (for S2 only); Bk = 1 (MD not accounted in the filter)				
		0.0056				0.0054				
	S1		**0.3957**				0.3966			
	S2		**0.4701**				0.4712			
	SF			**0.1062**	**5.0735**			0.1094	6.1395	

[a] Missing data randomly simulated in sensor number 2 only.

Case 5: *The optimal filter in two parts for MLF – OFMLF22.* The optimal filter presented in Section 4.8.4 (and derived in Appendix A.2) was implemented for two sensors in the form of the MLF. The data are missing randomly in only sensor number 2 at either level 0.9 or 0.75, whereas sensor number 1 has no missing data. Also, the results (Table 4.2) are presented when the missing data fact is not taken in the account in the filter. The results for the latter aspect are slightly worse, as shown by the metrics' numerical values marked in bold and italics. Also, we see that the performance of the OPMLE22 filter (MLF) is slightly better (marked in only bold values) compared to the equivalent entries of the filter OFSVF2 (i.e. compared to the SVF as is normally the situation). We infer from the results of Tables 4.1 and 4.2 that

i. The results of the cases where the data are not missing are almost similar across all the filters thereby establishing the general good and acceptable performance of the OFs in the normal situation.

ii. The results when the data missing is not accounted for in the filters are slightly worse (most of the times) or almost similar to the fact when this aspect is taken care of in the filters.

iii. The results of the filters with the MLF are similar or better (in many cases) than the corresponding filter with SVF.

iv. It is interesting to note that the results of the two-step filter OFSVF2 (derived in Appendix A.1) are almost similar to the one-step filter OFSVF1 [11].

We see from the plots of Figures 4.1 through 4.4 for OFSVF1, that the performances of these filters are acceptable as well as very encouraging. The plots of Figures 4.5 and 4.6 for the new filter OFMLF22 also show that the performance of the new filter is acceptable as well as very encouraging. For all the results presented in Tables 4.1 and 4.2, the corresponding plots of the time histories of the state errors, residuals and the covariance norms were found to be very satisfactory.

The results of the examples presented in this section are only illustrative of the basic use of the OFs for missing measurements data and sensor DF. However, further analytical as well as practical work requires to be done with applications to dynamic systems with more DOF, and more and different practical ways of handling randomly missing data artifice in the filtering algorithms.

4.9 Factorisation Filtering and Sensor DF: Illustrative Example

Target tracking is the process of obtaining the values of those parameters (and states) that completely specify the motion of the vehicle (it could be a target) based on the measurements. These measurements are obtained from radar,

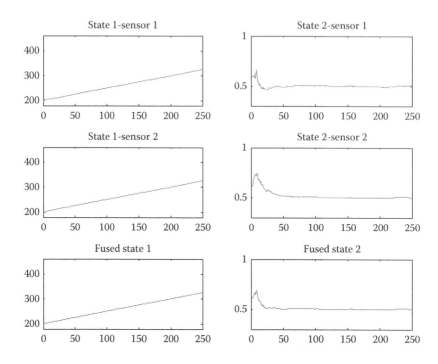

FIGURE 4.1
Time histories of the individual states and fused state with OFSVF1 case 1 with DRM missing
levels 0.9, 0.9 and Bk1 = Bk2 = 0.9.

sonar or electro-optical devices (EOT/Ds). A precision and accurate track determination algorithm also takes into account the measurement noise, and even
the correlated process/measurements noises. This latter aspect is often built in
KF (as correlated KF), the algebraic equivalent of which is the celebrated UD
factorisation (UDF) filter [12]. For a (vehicle) flight test range, the tracking of
a flight vehicle (or a target) and sensor DF are of great importance. Here, we
describe the UD filter and its variant that handles bias parameters and correlated process noise. The algorithms were implemented in PC MATLAB, and
C language (in an alpha DEC computer), and were validated using simulated
as well as real data (in post-processing offline mode), and using socket-pro-
gramming features, the UDP (used-defined) protocol for data communication
between the two alpha DEC machines was established [13].

   Although, KF has found very wide and numerous applications in tracking
problems, because of its optimal and recursive features, the conventional KF
algorithm is not numerically robust due to round-off errors, and numerical
accuracy can degrade to the point where results could cease to be useful.
Typical problems that could occur in the KF are loss of positive definiteness
of the covariance matrix (P) resulting from numerical errors such as finite
word length computations and cancellation errors due to the subtraction
term in the covariance update, Equation 4.9.

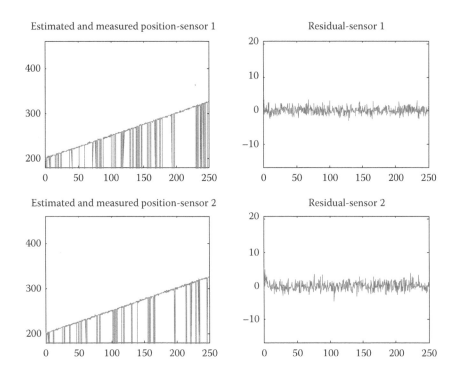

FIGURE 4.2
Measurement time histories and the residuals of the individual sensors OFSVF1 case 1 with DRM missing levels 0.9, 0.9 and Bk1 = Bk2 = 0.9.

4.9.1 Kalman UD Factorisation Filter

Square root-type filtering formulations (like factorisation filters) offer a solution to this problem of numerical accuracy and hence stability of the filter. The improved numerical behaviour of square root algorithms is due to reduction of the numerical ranges of the variables (square root of P is propagated and updated, instead). In the UD filter [12], the covariance update formulae and the estimation recursions are reformulated so that the covariance matrix does not appear explicitly, this avoids numerical ill-conditioning; since, specific recursions are used for U and D factors of the covariance matrix, $P = UDC^T$. Thus, computing and updating with triangular (U is unit upper triangular matrix factor) and diagonal (D, the diagonal factor) matrices involve fewer arithmetic operations, and the algorithm processes measurements, one component at a time. The major advantage from UD formulation comes from the fact that since, the square root-type algorithms process square roots of the covariance matrices (P), they essentially use half the word length normally required by the conventional KF. This is very advantageous for multi-sensor DF applications, since, we need to process several data sets, from several sensor channels, for several hours, and that these algorithms might have been

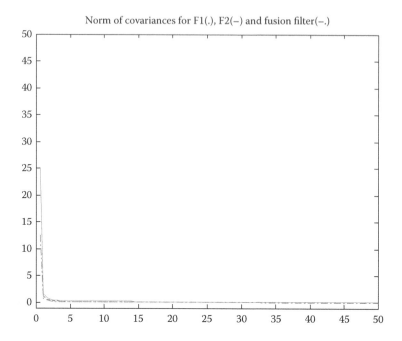

FIGURE 4.3
Covariance comparison for the OFSVF1 case 1 with DRM missing levels 0.9, 0.9 and Bk1 = Bk2 = 0.9.

implemented on onboard/real-time computers (in the tracking vehicle) or on the tracking mobile/fixed site. These computers might be operating with 8-bit or 16-bit registers, memory units. In such situations, the factorisation filtering algorithms provide numerically stable, efficient and accurate trajectories.

The UD filtering algorithm for linear system is given in two parts [2,12]:

Time Propagation

We have for the covariance propagation/evolution part as follows:

$$\tilde{P}(k+1|k) = \phi\hat{P}(k)\phi^T + GQG^T \tag{4.155}$$

We are given $\hat{P} = \hat{U}\hat{D}\hat{U}^T$ and Q as the process noise covariance matrix, then, the time update factors $\tilde{U}$ and $\tilde{D}$ are obtained through modified Gram–Schmidt orthogonalisation process as follows: define $V = \lfloor \phi\hat{U}|G \rfloor$ and $\bar{D} = $ diag$[\hat{D},Q]$, and $V^T = [v_1, v_2, \ldots, v_n]$; here, P is formulated as $\tilde{P} = \tilde{V}\tilde{D}\tilde{V}^T$, and the U and D factors of $\tilde{V}\tilde{D}\tilde{V}^T$ are computed as follows.

For $j = 1, \ldots, n$ the following equations are recursively processed [2,12]:

$$\tilde{D}_j = <v_j, v_j>_{\bar{D}} \tag{4.156}$$

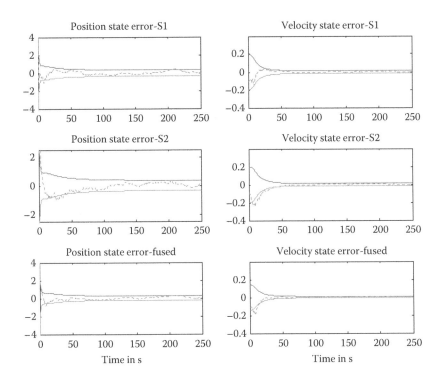

FIGURE 4.4
State-errors' time histories for the OFSVF1 case 1 with DRM missing levels 0.9, 0.9 and Bk1 = Bk2 = 0.9.

$$\tilde{U}_{ij} = (1/\tilde{D}_j) < v_i, v_j >_{\bar{D}} \quad i = 1, \dots, j - 1 \tag{4.157}$$

$$v_i = v_i - \tilde{U}_{ij}v_j \tag{4.158}$$

Here, $< v_i, v_j >_{\bar{D}} = v_i^T \bar{D} v_j$ signifies the weighted inner product between v_i and v_j. Thus, the time-propagation algorithm efficiently produces the required U, D factors, taking the effect of previous U, D factors and the process noise, and it preserves the symmetry of the (original) P matrix.

Measurement Update

As we know the measurement update in KF combines a priori estimates $\tilde{x}$ and error covariance $\tilde{P}$ with a scalar observation $z = cx + v$ to construct an updated estimate and covariance given as follows [12]:

$$K = \tilde{P}c^T / s$$

$$\hat{x} = \tilde{x} + K(z - c\tilde{x})$$

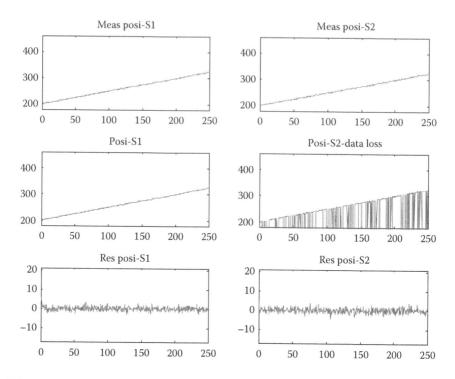

FIGURE 4.5
Measurement time histories and the residuals of the individual sensors OFSVF22 case 5; DRM – missing level = 0.75 (for S2 only); Bk = 0.75.

$$s = c\hat{P}c^T + R$$

$$\hat{P} = \hat{P} - Kc\tilde{P} \tag{4.159}$$

Here, we take $\tilde{P} = \tilde{U}\tilde{D}\tilde{U}^T$; c as the measurement matrix, R as the measurement noise covariance and z as the vector of noisy measurements. Based on Equation 4.159, and factorisation process, we obtain Kalman gain K and updated covariance factors $\hat{U}$ and $\hat{D}$ from the following equations [2,12]:

$$g = \tilde{U}^T c^T; \quad g^T = (g_1, \ldots, g_n) \quad w = \tilde{D}g;$$

$$\hat{d}_1 = \tilde{d}_1 R/s_1, \quad s_1 = R + w_1 g_1 \tag{4.160}$$

Then, for $j = 2, \ldots, n$ the following equations are processed [2,12]:

$$s_j = s_{j-1} + w_j g_j$$

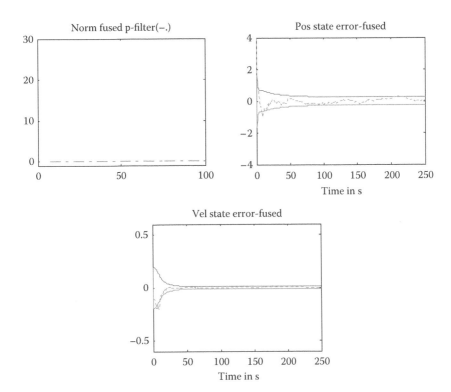

FIGURE 4.6
Fused covariance norm and the state-errors' time histories for OFSVF22 case 5; DRM – missing level = 0.75 (for S2 only); Bk = 0.75.

$$\hat{d}_j = \tilde{d}_j s_{j-1}/s_j$$

$$\hat{u}_j = \tilde{u}_j + \lambda_j K_j, \quad \lambda_j = -g_j/s_{j-1}$$

$$K_{j+1} = K_j + w_j \tilde{u}_j; \quad \tilde{U} = [\tilde{u}_1, \dots, \tilde{u}_n]$$

Finally, the Kalman gain is given by

$$K = K_{n+1}/s_n \qquad (4.161)$$

We see that $\tilde{d}$ is the predicted diagonal element, and $\hat{d}_j$ is the updated diagonal element of the D matrix. The time propagation and measurement update of the state vector (state equations) is similar to KF and hence, not repeated here (see Equations 4.22 and 4.26). Here, the measurement update/ data processing is done sequentially, meaning thereby that each observable (measurement) is processed in turn, and the state estimate updated, this avoids the matrix inversion in the Kalman gain formulation.

4.9.2 UD Factorisation Filter for Correlated Process Noise and Bias Parameters

A mathematical model that represents a track trajectory is usually not exact (even in a statistical sense) and the process noise might not be a strictly white noise process. For this case the state model can be given as follows:

$$
\begin{bmatrix} x \\ p \\ y \end{bmatrix}_{k+1} = \begin{bmatrix} V_x & V_p & V_y \\ 0 & M & 0 \\ 0 & 0 & I \end{bmatrix} \begin{bmatrix} x \\ p \\ y \end{bmatrix}_k + \begin{bmatrix} 0 \\ w_k \\ 0 \end{bmatrix} \tag{4.162}
$$

In Equation 4.162, x is the state vector, p is the state variable representing the correlated noise and y is the bias vector, and we see that the transition matrix is almost triangular. The mapping, that is, the time propagation/evolution of the UD factors is carried out using the following equivalence [12]:

$$
\begin{bmatrix} \hat{U}_x & \hat{U}_{xp} & \hat{U}_{xy} \\ 0 & \hat{U}_p & \hat{U}_{py} \\ 0 & 0 & \hat{U}_y \end{bmatrix}; \quad \hat{D} = \text{Diagonal}\,(\hat{D}_x, \hat{D}_p, \hat{D}_y)\ \text{all at time}\ k. \tag{4.163}
$$

$$
[\bar{U}_p\ \bar{U}_{py}\ \bar{U}_y\ \bar{D}_p\ \bar{D}_y] = [\hat{U}_p\ \hat{U}_{py}\ \hat{U}_y\ \hat{D}_p\hat{D}_y] \tag{4.164}
$$

$$
\bar{U}_{xp} = V_x\hat{U}_{xp} + V_p\hat{U}_p; \quad \bar{U}_{xy} = V_x\hat{U}_{xy} + V_p\hat{U}_{py} + V_y\hat{U}_y \tag{4.165}
$$

$$
\bar{U}_x\bar{D}_x\bar{U}_x^T = (V_x\hat{U}_x)\hat{D}_x(V_x\hat{U}_x)^T \tag{4.166}
$$

$$
\tilde{U}_y = \bar{U}_y = \hat{U}_y; \quad \tilde{D}_y = \bar{D}_y = \hat{D}_y \tag{4.167}
$$

$$
\tilde{U}_{py} = M\bar{U}_{py} = M\hat{U}_{py}; \quad \tilde{U}_{xy} = \bar{U}_{xy} \tag{4.168}
$$

$$
\begin{bmatrix} \tilde{U}_x & \tilde{U}_{xp} \\ 0 & \tilde{U}_p \end{bmatrix}\begin{bmatrix} \tilde{D}_x & 0 \\ 0 & \tilde{D}_p \end{bmatrix}\begin{bmatrix} \tilde{U}_x^T & 0 \\ \tilde{U}_{xp} & \tilde{U}_p^T \end{bmatrix} = \begin{bmatrix} \bar{U}_x & \bar{U}_{xp} \\ 0 & M\bar{U}_p \end{bmatrix}\begin{bmatrix} \bar{D}_x & 0 \\ 0 & \bar{D}_p \end{bmatrix}\begin{bmatrix} \bar{U}_x^T & 0 \\ \bar{U}_{xp} & \bar{U}_p^TM^T \end{bmatrix} + \begin{bmatrix} 0 & 0 \\ 0 & Q \end{bmatrix}
$$

$$
\tag{4.169}
$$

The updating, Equation 4.169 is mechanised by using the modified Gram–Schmidt orthogonalisation algorithm. The factors related to the correlated process noise and the bias are mapped as ultimate factors using the modified Gram–Schmidt algorithm to obtain

$$\tilde{U}\tilde{D}\tilde{U}^T = W\,\mathrm{Diagonal}(D,Q)W^T \tag{4.170}$$

$$W = \begin{bmatrix} \bar{U}_x\bar{U}_{xp} & 0 \\ 0 & M\bar{U}_p I \end{bmatrix} \tag{4.171}$$

The merit of this UD filter expressed by Equations 4.163 to 4.171 (UDCBF-filter) is that it has all the properties of the basic UD filter, and in addition handles the process noise correlation and bias parameters in the state-space model.

4.9.3 Sensor Fusion Scheme

For the sensor channels: EOT, PCMC, S-Band radars, TM, RADAR 1, RADAR 2, INS, and GPS for fusion, it is necessary to develop (sensor/data) a fusion logic so that we can utilise the available information from these sensors/channels [13] in a proper way, as shown in Figure 4.7. The priority logic is determined based on the sensor accuracy within the range of the individual sensor's capability, hence, the following sequence could be given to the sensors (of the first module) within the range limit of, say RL km:

EOT
PCMC
TM and S-Band fusion
S-Band
TM
Track loss

For range more than RL km PCMC radar tracks the vehicle, and for the second module which contains the RADARS, the sequence followed is

RADAR 1 and RADAR 2 fusion
RADAR1
RADAR2
Track loss

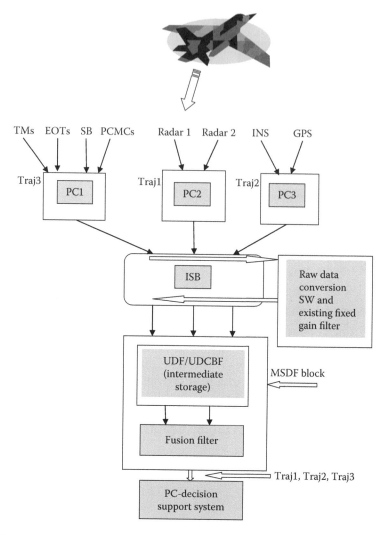

FIGURE 4.7
Typical multi-channel-data acquisition and fusion scheme for target tracking (PC – personal computer, TM – Telemetry, SB – S-band radar, ISB – intelligent switch board, Traj – tracking trajectories).

And, for the third module that contains the INS and GPS, the following sequence is followed:

INS and GPS fusion (GPS data replaced by INS)

INS

GPS (GPS data replaced by INS)

Track loss

In the third module, if the GPS data are not available, these are replaced by INS data for the sake of demonstration of the scheme. Depending on the situation, either measurement level of state level fusion is employed. Also, all the channel data are processed sequentially: position (velocity, acceleration or any other angular/bearing data), from one sensor channel, then from the next sensor channel and so on (until all the channels are covered) within one sampling interval, and then again the processing is started from the first sensor channel (to all the channels) for the next sampling interval, and so on. We see from Figure 4.7 that three trajectories are to be formulated and tracked for a given single moving object/target. The PC1 can combine the available or derived information to formulated Traj1, PC2 can do similarly to have Traj2, and so the PC3. These three trajectories are processed by fixed gain (existing, and there was no intention to remove it!) tracking filter in real time, and then these are further processed by UD/UDCB filters, so ultimately the three trajectories and/or the final fused trajectory can be sent to the DSS.

4.9.4 Performance Evaluation of UD and UDCB Filters for Tracking and Fusion

The mathematical model of the vehicle state includes position, velocity and sometimes acceleration as state variables, and the measurement model relates state variables to available measurement variables, as in Equations 4.1 and 4.2. For simplicity the states of target motion are defined in the spherical coordinates such that the state equation can be decoupled into three independent channels, so that the tracking filter can work independently on each channel. The simulated data with correlated process noise were generated. The model-transition matrices used are given as [13]

$$V = \begin{bmatrix} 1 & \Delta t \\ 0 & 1 \end{bmatrix}; \quad M = \exp(-\Delta t/\tau) \tag{4.172}$$

The target states are position and velocity, and a constant bias has been added to the position data so that $V_y = 1$. Δt is the sampling interval of 0.5 s and M is 0.9, indicating high correlation. The UDF, without accounting for correlated noise and bias, and the UDCBF, accounting for correlated process noise and bias, are used to process these simulated data. Vehicle position is used as the only observable. From Table 4.3 and Figure 4.8, we see that the UDCB filter performs better than the UD filter when the data are contaminated by the correlated noise and affected by a constant bias [13]. The real (flight) data of a flight vehicle (in Cartesian coordinates X,Y,Z) were processed (in post-processing/offline mode obtained from the country's flight

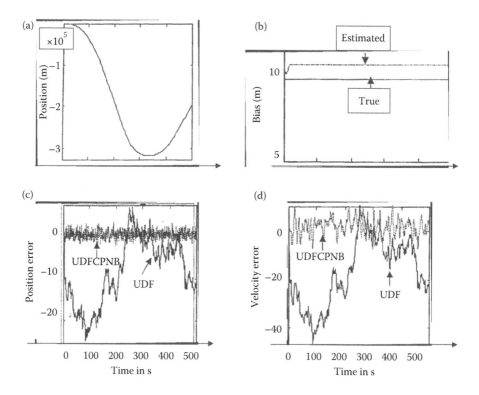

TABLE 4.3

Results of the Simulated and Real Data for Tracking and Fusion Using
UD Filters

Filter Used	Mean Values of the State Errors-Simulated Data		
	Position	Velocity	—
UDF	−8.4	−17.211	—
UDCBF	0.8489	1.68	—
	Mean values of the residuals-real data		
	x	Y	Z
UDF	0.88	1.51	1.75
UDCBF	0.03	0.18	0.2

FIGURE 4.8
State-estimation using UD and UDFCPNB–(UDCB) filters for tracking and sensor DF (simulated data): (a) position estimate, (b) bias estimate, (c) position error (m), (d) velocity error (m/s).

test range) using both the UD filters, and the transition matrix of the state used is given as follows [13]:

$$V_x = \begin{bmatrix} 1 & \Delta t & \Delta t^2/2 \\ 0 & 1 & \Delta t \\ 0 & 0 & 1 \end{bmatrix} \tag{4.173}$$

Here, $H = [1\ 0\ 0]$; the state vector x has position, velocity and acceleration as its components. From Table 4.3 it is seen that the performance of UDCB filter is better than the UD filter [13]. Similar inferences were obtained for the real data from the trajectories-matches/bias estimation results, the plots are not produced here.

Acknowledgement

Mrs. M. Spoorthy's work (under the guidance of Dr. J. R. Raol) on missing data for Chapter 4 is appreciated.

EXERCISES

4.1 Given I_{m1} and I_{m2} as the information matrices (for one tracked object with two sensors' measurement information) obtain the formula for P_f, the fused covariance matrix neglecting the cross-covariance matrices.

4.2 In a sensor NW, one sensor's mathematical model is incomplete, and this sensor's measurements are to be used for processing by a KF. How will you determine these parameters of the sensor's mathematical model along with the states of the dynamic system?

4.3 Why should the innovations be white stochastic process?

4.4 Compare the salient features of the filtering algorithms discussed in this chapter.

4.5 In an EKF what is the advantage of using function f in Equation 4.48 and h in Equation 4.50?

4.6 What are the differences in the HI norm, Equation 4.103; the cost function J, Equations 4.18 used in the derivation of the KF; and the cost function used in the derivation of SRIF, Equation 4.96?

4.7 The KF and the SRIF are fundamentally the DF rules in themselves, apart from being used (later on) as the measurement level data fuser. Comment on this aspect.

4.8 Try to include more than one measurement model in the posterior HI filter and modify the equations of the filter appropriately.

4.9 Can your stretch your imagination and come out with some (conceptual) schemes for adaptive filtering (for tuning of KF) as suggested in Section 4.2.3.

4.10 Obtain KF equations following Section 4.8.2 and Appendix A.2 when $b \to 0$.

References

1. Durrant-Whyte, H. Multi sensor data fusion – Lecture Notes. Australian Centre for Field Robotics, University of Sydney, NSW 2006, Australia, January 2001.
2. Raol, J. R., Girija, G. and Singh, J. *Modelling and Parameter Estimation of Dynamic Systems*, IET/IEE Control Series Book, Vol. 65. IET/IEE, London, 2004.
3. Raol, J. R. and Singh, J. *Flight Mechanics Modeling and Analysis*. CRC Press, FL, USA, 2010.
4. Raol, J. R. *Multisensor Data Fusion with MATLAB*. CRC Press, FL, USA, 2010.
5. Raol, J. R. and Gopal, A. K. *Mobile Intelligent Automation Systems*. CRC Press, FL, USA, 2012.
6. Hassibi, B., Sayad, A. H. and Kailath, T. Linear estimation in Krein spaces – Part II: Applications. *IEEE Transactions on Automatic Control*, 41(1), 34–49, January 1996.
7. Jin, S. H. Park, J. B., Kim, K. K. and Yoon, T.S. Krein space approach to decentralized H_∞ state estimation. *IEE Proceedings on the Control Theory and Applications*, 148(6), 502–508, November 2001.
8. Lee, T. H., Ra, W. S., Yoon, T. S. and Park, J. B. Robust Kalman filtering via Krein space estimation. *IEE Proceedings Control Theory and Applications*, 151(1), 59–63, January 2004.
9. Simon, D. and El-Sherief, H. Hybrid Kalman/minimax filtering in phase-locked loops. In *Control Engineering Practice Jl*. Editor, Kugi, A. International Federation of Automatic Control, Elsevier, 4(5), pp. 615–623, May 1996.
10. Shanthakumar, N., Girija, G. and Raol, J. R. Performance of Kalman filter and gain fusion algorithms for sensor data fusion with measurements loss. *International Radar Symposium India (IRSI)*, Bangalore, December 2001.
11. Mohamed, S. M. and Nahavandi, S. Optimal multisensor data fusion for linear systems with missing measurements. *SOSE 2008, IEEE International Conference on System of Systems Engineering, IEEE*, Monterey Bay, CA, 2–5 June, 1–4, 2008.
12. Bierman, G. J. *Factorisation Methods for Discrete Sequential Estimation*. Academic Press, NY, 1977.
13. Shanthakumar, N., Girija, G. and Raol, J. R. Factorisation filtering algorithm with colored noise for tacking. Presented at the *International Radar Symposium India (IRSI/Organized by Institution of Electronics and Telecommunications Engineers, India, Bangalore Chapter)*, Bangalore, December 1999. (http://nal-ir.nal.res.in/4574/1/Factorisation_filtering__Algorithm.PDF).

5

Decentralised Data Fusion Systems

5.1 Introduction

In any multi-sensory (MS) sensor data fusion (DF) system, there are a number of sensors physically distributed around a region and/or an environment of interest. In a centralised system (CDF), the raw engineering unit data (REUD)-sensor information/data are communicated to a central processor (CP). Here at the centre, the information/data are combined to produce a single picture of the environment, entity, object, scene or scenario. In a distributed DF (DDF) system each sensor has its own local processing unit that extracts the useful information from these sensor data prior to communication to other nodes or any global node if required. This DDF: (i) requires less information to be communicated, (ii) the computational load on the CP is considerably less and (iii) the sensors can be arranged in a reasonably modular manner [1]. The degree of local processing/computation that is to be carried out at a sensor/node site varies from simple validation and data compression to the full determination of tracks. This can even extend to the interpretation of information at the local node. In many systems only CDF is adequate, however, in many practical situations, several forms of distributed processing (DDF) are required due to [1]: (i) the enhanced sophistication of the systems/sensor configurations/algorithms for processing, (ii) more functional requirements due to more assigned tasks and (iii) complexity and size of the DF system itself. This is also coupled with the reducing cost of computing power. The design of DDF system requires the development of appropriate algorithms that can operate at a number of distributed sites in an adequate, reliable and consistent manner. Based on the development of the information filter (IF) and square root information filter (SRIF) presented in Chapter 4, one can easily see that the IF, and more generally the log-likelihood (LL) implementations of Bayesian theorem are very suitable to DDF systems. In this chapter it is shown how these can be mapped to many distributed (DDF) and decentralised data fusion (DCF) systems [1]. The major issue of communication in a DDF system is due to limited communication bandwidth (BW), transmission time delays (TD) and communication (-channel) failures. These occur between sensing and fusion processes/stations. Some more issues in the distributed data decentralised fusion (DDCF) (DDF/DCF) systems are [1]: (i) the problem of model distribution, where each local sensor/node maintains a different model of the environment; (ii) sensor management, in which a limited set of sensor and their resources should be used cooperatively; and (iii) system organisation, in which the optimal design of sensor networks is required. Several aspects of DCF are treated in References 1–3.

5.2 Data Fusion Architectures

In DDF at the simplest level, sensors could communicate information and data directly to a CP where these data are combined [1]. The local processing of information/data is not required. Of course, the merit of having several sources of information/data is lost due to having a complete centralised control over the processing/interpretation of these data. When more processing occurs at local nodes, the computational and communication burden is removed from the fusion centre. But, this is at the cost of reduced direct control of lower level sensor information/data. Then the increase in the intelligence of the local sensor nodes results in a hierarchical structure for the fusion architecture. This has a merit of imposing some order in the fusion process. In yet some more distributed architectures, the sensor nodes have significant local ability to generate tracks and to do the fusion tasks, that is, blackboard and agent-based systems. Fully DCF have no CP and no common communication system, here, the nodes operate in a fully autonomous manner. They only coordinate via the anonymous communication information.

5.2.1 Hierarchical Data Fusion Architectures

In the hierarchical structure (hierarchical data fusion architecture [HDFA]), the processing elements/nodes at the lowest level transmit information upwards (forward in the information flow direction) via successive/progressive levels [1]. The information flows (in parallel) from several sensor/devices to their respective tracking systems from which the state estimates are available. For state estimation, any of the suitable filtering algorithms from Chapter 4, may be utilized. These estimates are then fed to the track fusion algorithm which then output a combined track. Thus, in the intermediate levels the information/data are combined and refined in such a way that at the top level a global view of the state of the system is obtainable. Also, it is possible to have a multiple level hierarchical multiple tracking systems wherein the components are the HDFAs and after the group-picture tracks (previously called track fusion) we combine these picture tracks to obtain the final fusion. The HDFAs are quite common in many organisations and data fusion systems, and have many advantages over the fully CDF systems. The HDFA reduces the load on the

CP, and yet maintains strict control over sub-processor operations. This architecture has been employed in a number of DF systems resulting in a variety of useful algorithms for combining information/data at different levels of an HDFA. The general hierarchical Bayesian algorithms are based on the independent likelihood pool architectures, or on the LL opinion pools. The focus is on hierarchical estimation and tracking algorithms. We assume that all the sensors observe a common state/track $x(k)$. The measurements are collected at local sites according to a local measurement equation

$$z_i(k) = H_i(k)x(k) + v_i(k), \quad i = 1,\dots,S \tag{5.1}$$

Each site then has a Kalman filter (KF) to provide local state estimates based only on local observations

$$\hat{x}_i = \tilde{x}_i + K_i r_i \quad i = 1,\dots,S \tag{5.2}$$

These local estimates are then passed on further up the hierarchy, that is, in the forward cycle, to an intermediate or central processor (CP). This CP then fuses the tracks to form a global estimate based on all measurements

$$\hat{x} = \sum_i^S w_i(k)\hat{x}_i(k) \tag{5.3}$$

In Equation 5.3, $w_i(k)$ are the local site weighting factors/matrices. Each sensor/node should be observing a common true state. The local process models are related via a common global model. And hence, the predictions made at the local sites/nodes are correlated and so the updated local estimates in Equation 5.2 would also be correlated. This is true despite the fact that the measurements at each node are different. So, in a true sense, the local estimates generated at each node cannot be combined in an independent manner per Equation 5.2. One should explicitly account for these correlations in the calculation of the node weighting matrices. This can be done by the track to track fusion (TTF) algorithms described in the form of Equations 4.65 and 4.66.

Interestingly the HDFA has several demerits [1]: (i) it relies on the CP, and the controlling level within the hierarchy compromises the reliability/flexibility, (ii) failure of the CP → the failure of the entire system, (iii) changes in the system → the changes also occur in both the CP and all the sub-units, (iv) the burden on the CP in terms of combining data can still be very high; this would make the design methodology unable to be extended to incorporate a large number of sources of data and (v) the inability of data sources to communicate (other than through some higher level in the hierarchy) eliminates the possibility of any synergy that can be developed between two or more complimentary sources, this allows the system designer to use only some rigid/predetermined combinations of data.

5.2.2 Distributed DF Architectures

The desire to make the system more modular and flexible and the recognition of the fact that a CDF/HDFA would put large overheads on the communication of the data and the computation in the central node, drives one to go for the DDF [1]. A well-known DDF is one that is called blackboard data fusion architecture (BBDF/blackboard architecture (BBA)/BB medium) consisting of [1]: (i) a few individual tracking systems (after the sensor data are available and processed), (ii) tracking fusion algorithms, (iii) track identity knowledge base, (iv) situation knowledge base and (v) remote sensing knowledge base. In fact all these five components bilaterally interact with the blackboard medium (BBM). It consists of a number of independent autonomous agents, where each agent is a source of expert knowledge and/or has a special information processing capability. These agents can exchange information via a shared memory called blackboard (BB). Every agent in the system is also able to write information and/or local knowledge to this resource and is able to read from this resource and that too in an unrestricted manner. Every agent is made modular and new agents can be added to the system as and when required. This is done without changing the underlying architecture. BBA is most widely used for knowledge-based DF systems. This is true, especially in data interpretation and situation assessment (SA). However, the BBA uses a common communication/memory resource, and it practically becomes a one level hierarchy and lacks flexibility.

5.2.3 Decentralised Data Fusion Architectures

A DCF system consists of a network (NW) of sensor nodes wherein each node has its own processing arrangement. It consists of mainly three components: (i) sensor/s with their own fusion processor/s, (ii) sensor node/s receiving the data from, say, a radar and (iii) communications medium (mediums) connecting all the components of (i) and (ii); such a DCF system is implemented with a point-to-point communication architecture [1]. In structure, this DCF architecture is almost similar or the same to the one in Figure 1.6b. Hence, these nodes as such do not need any central fusion/central communication (CFCC) provision. The DF occurs locally at each node on the basis of local measurements and the information/data communicated from the neighbouring nodes/devices/sensors. There is no common place for fusion or global decision. However, a DCF system has some constraints [1]: (i) no single central fusion (CDF) exists; (ii) no common communication (CC) provision and (iii) the sensor nodes do not know NW topology. The main point is that the DCF system has no central fusion centre (CF/C). This is not true of many so called 'decentralised' systems that are often described in the literature on DF – actually those DF systems are typically distributed or hierarchical (DDF/HDFA) in nature, rather than truly decentralised (DCF). The DCF systems have several useful features [1]: (a) it eliminates the CFCC – hence, the system is scalable, (b) since no node is central and no global knowledge of the NW topology is required for fusion means – hence, the system is fault tolerant to the online loss/addition of sensing nodes as well as to any dynamic change in the NW and (c) since all the fusion processes take place locally at each sensor site, the nodes are constructed and programmed in a modular fashion. Thus, a DCF differs from a DDF system since it does not have CFCC facility, and hence each sensor node is completely self-sufficient. Each node operates completely independent of any other component in the system. The communication between nodes is truly and strictly one-to-one. It does not need any remote knowledge of any node capability. In this chapter, we particularly distinguish between DCF system with no common resources, and the DDF system wherein some small centralised arrangement is maintained. Also, a DCF system can be implemented with a broadcast, fully connected, communication architecture, although such a common communication facility violates DDF restriction – however a broadcast medium is a good model of real communication NWs [1]. Also one can have DCF fusion system with a hybrid, broadcast and point-to-point, communication arrangements.

5.3 Decentralised Estimation and Fusion

The DCF architecture/system/concept is based on the notion of using information measures for quantifying, communicating and assimilating sensor data. It is perhaps very natural then to use decentralised estimation concepts (DCE) which are best implemented in the form of an IF [1].

5.3.1 Information Filter

In contrast to the conventional KF (Chapter 4) [4–7] the IF deals with the information state (IS) and the information matrix (IM) [2]. The conventional KF is often known as the covariance filter, since it utilises the covariance matrices in its time-propagation/data update cycles. The linear model of the dynamic-tracking system is given as

$$x(k + 1) = \phi x(k) + w(k)$$
$$z(k) = Hx(k) + v(k)$$
(5.4)

We define and form the IM as the inverse of the state-error covariance matrix P, then the IS and IM are defined as

$$\hat{y} = P^{-1}\hat{x}$$
$$\hat{Y} = P^{-1}$$
$$\hat{y} = \hat{Y}\hat{x}$$
(5.5)

The state x in the KF is called the covariance state that is associated with the covariance matrix P, and obviously the IS of the IF can be obtained by pre-multiplying the covariance state with the IM, see Equation 5.5. This process is intuitively appealing since the covariance matrix represents the uncertainty in the variable/state estimation, and if the covariance is large, the information is less and vice versa. Then the time-propagation/prediction equations are given by the following equations [1].

First compute the following intermediate quantities:

$$M = \phi^{-t}Y(k - 1/k - 1)\phi^{-1} \tag{5.6}$$

$$S = G^t MG + Q^{-1} \tag{5.7}$$

$$L = MGS^{-1} \tag{5.8}$$

Then the prediction equations for the IM and the IS are given as

$$Y(k/k - 1) = M - LSL^t \tag{5.9}$$

$$\tilde{y}(k/k - 1) = [I - LG^t]\phi^{-t}\hat{y}(k - 1/k - 1) + Y(k/k - 1)Q^t G^t u \tag{5.10}$$

The measurement data-update or the estimation equations are given as

$$\hat{y}(k/k) = \tilde{y}(k/k - 1) + i(k) \tag{5.11}$$

$$\hat{Y}(k/k) = \tilde{Y}(k/k - 1) + I(k) \tag{5.12}$$

where the second terms in the above equations are given, respectively, as

$$i(k) = H^T(k)R^{-1}z(k) \tag{5.13}$$

$$I(k) = H^T(k)R^{-1}H(k) \tag{5.14}$$

In the above equations the variables G and Q have the usual meaning from the prediction cycle of the KF. From Equation 5.13, $i(k)$ provides the incremental update of the IS due to the incorporation of the current available measurement. Similarly Equation 5.14 provides the incremental update of the IM itself due to taking the effect of the current measurement in the IF via the measurement vector/matrix $H(k)$ and the measurement noise covariance matrix $R(k)$. One should remember that in the IF all the quantities do not have IS/IM interpretations, as we can see that some covariance matrices are also used. From the measurement data update part of the IF one can see that this part has intuitively appealing features when the state estimates are handled in the information domain, the two equations are very straight forward. The new information for the IS and the new IM are just added to the previously available quantities from the prediction cycle. Hence, for the DCF systems this is very advantageous. Only that the prediction part, Equation 5.10 is more complex. The complete development of the IFs for linear as well as non-linear dynamic systems can be found in References 1–3. From Equations 5.11 and 5.12 it is obvious that for the DCF, the IF is and should be the most natural choice.

5.3.2 Information Filter and Bayesian Theorem

There is a strong connection between the underlying pdf of the Bayesian theorem and the IS with the assumption that the likelihoods are Gaussian [1]. The first derivative of the LL is called the score function and the second derivative is the Fisher information equation. Thus, the IF is the implementation of the Bayesian theorem. The first derivative is the centre of mass and the second derivative is the moment of inertia, and hence, these derivatives are truly the moment generating functions. Thus, in essence the IF is a means of recursive computation of the sufficient statistics when the underlying distributions are Gaussian.

5.3.3 Information Filter in Multi-Sensor Estimation

Considering the application of the KF to two sensor-data fusion for the single target, it was seen that although the innovation vectors at different times are uncorrelated, the innovations generated by different sensors at the same time are not, meaning these are correlated [1]. This is due to the fact that they use a common prediction equation (because the state model is one for one target and two-sensor data fusion situations). As a result of this the innovation covariance matrix cannot be made diagonal. Hence, it cannot be partitioned and inverted to yield a gain matrix for each individual measurement. This problem is not faced in the IF. As we have seen the information contributions can be simply added to obtain the fused information. As we see from the measurement update cycle of the IF, the information contributions made by the measurements are directly related to the underlying likelihood functions for the states rather than to the state estimates themselves. This can be seen by considering the IF as directly an implementation of Bayesian theorem in terms of LL (rather than in terms of states). Often, because of this observation, the IF is called as the likelihood filter. The IF provides a more natural means of assimilating information than does the conventional KF and hence, the IF is a much simpler method of handling complex multi-sensor data fusion situations. The linear addition of information in the IF is rendered by direct algebraic manipulation. The total information provided to the filter at any time step is just the sum of the information contributions from the individual sensors, that is, the IS contributions as well as the IM contributions. The single sensor information updates are extended to multiple sensor updates in a simple manner and are given as follows with 'i' as the sensor number index [1]:

$$\hat{y}(k/k) = \tilde{y}(k/k-1) + \sum_{i-1}^{m} i_i(k) \tag{5.15}$$

$$Y(k/k) = Y(k/k-1) + \sum_{i-1}^{m} I_i(k) \tag{5.16}$$

5.3.4 Hierarchical Information Filter

The IF can also be partitioned to provide a simple hierarchical estimation arrangement based on the communication of [1]: (i) the incremental information terms $i(.)$ and $I(.)$ from sensor nodes to a CP and (ii) partial information-state estimates from nodes to a central fusion node. Thus, the IF can also be used for hierarchical data fusion systems. It was suggested in Section 2.4.4, that due to a simple information summation, the Bayesian LL theorem can easily be rendered to a number of different architectural forms. Also, the information additions in Equations 5.15 and 5.16 can be distributed in a simple manner.

In an HDFA each sensor incorporates a full state model and takes measurements according to a system comprising 'm' number of sensors. In such an HDFA the ISs are computed at each sensor node and then transmitted to a CP. Here, a common estimate is obtained by simple summation, thus, all the state predictions are undertaken at the CP. Then the IS contributions from their measurements are computed in incremental terms $i_i(k)$ and $I_i(k)$. These terms are communicated further to the CP and incorporated into the global estimate via Equations 5.15 and 5.16. Then the IS prediction is computed centrally using Equations 5.9 and 5.10. The true state estimate (i.e. the covariance state of KF) itself can be found as and when required by using Equation 5.5. Any validation/data association can take place at the CP in order to avoid communicating predictions-estimates to nodes.

Another hierarchical system allows local tracks to be maintained at local sensor sites, and each sensing node produces local IS (y) estimates on the basis of its own measurements. The system then communicates these back to a CFC. At this CP they are assimilated to provide a global estimate of IS. Let $y_i(.|.)$ be the IS estimate arrived at by each sensor site based only on its own measurements. Let the local information-state prediction be $y_i(k|k-1)$. This local estimate can be determined from the following set of equations [1]:

$$\begin{aligned} \hat{y}_i(k/k) &= \tilde{y}_i(k/k-1) + i_i(k) \\ Y_i(k/k) &= Y_i(k/k-1) + I_i(k) \end{aligned} \tag{5.17}$$

Subsequently these (partial) IS/IM estimates, y and Y are communicated to a CP. Here, these estimates are assimilated using the following set of equations [1]:

$$Y(k/k) = Y(k/k-1) + \sum_{i-1}^{m} [\hat{Y}_i(k/k) - Y(k/k-1)]$$

$$\hat{y}(k/k) = \tilde{y}(k/k-1) + \sum_{i-1}^{m} [\hat{y}_i(k/k) - \tilde{y}(k/k-1)] \tag{5.18}$$

If we assume that the local predictions at each node are the same as the prediction generated by a CP, one can easily see that the assimilation equations are identical to Equations 5.15 and 5.16.

In yet another hierarchical fusion system, the global tracks are maintained at local sensor nodes/sites [1]. In this case once a global estimate is generated at the CP, it is communicated back to the local site. Here, the estimate is assimilated to form a global estimate. In this architecture each site acts in an 'autonomous' fashion and has access to the global track information.

5.3.5 Decentralised Information Filter

It is very easy to decentralise assimilation Equation 5.18 in systems where there is a fully connected network of sensing nodes [1]. Here, each node generates a prediction, takes a measurement and computes a local estimate. This estimate is communicated to all the neighbouring nodes. Then each node receives all the local estimates and performs a local form of the assimilation to produce a global estimate of IS. This global estimate is equivalent to that obtained by a CP. This is the same as reproducing/repeating the central assimilation Equation 5.18 at every local sensor node. Then the resulting equations are simplified. In this case it is assumed that each local sensor node maintains a state model identical to an equivalent centralised model. Each node then starts by computing a local estimate based on a local prediction and the measured local information. If each node starts with a common initial IS estimate and the NW is fully connected, then the estimates generated by each node are identical. The set of the nine IF equations for the decentralised case is given as [1]

a. Prediction equations:

$$M_i = \phi^{-t} Y_i(k-1/k-1)\phi^{-1} \tag{5.19}$$

$$S_i = G^t M_i G + Q^{-1} \tag{5.20}$$

$$L = M_i G S_i^{-1} \tag{5.21}$$

$$Y_i(k/k - 1) = M_i - L_i S_i L_i^t \tag{5.22}$$

$$\tilde{y}_i(k/k - 1) = [I - L_i G^t]\phi^{-t}\hat{y}_i(k - 1/k - 1) + Y_i(k/k - 1)Q^t G^t u \tag{5.23}$$

b. Estimate-incorporation of measurement data/information:

$$\hat{y}_i(k/k) = \tilde{y}_i(k/k - 1) + i_i(k) \tag{5.24}$$

$$\hat{Y}_i(k/k) = \tilde{Y}_i(k/k - 1) + I_i(k) \tag{5.25}$$

These estimates are communicated to the neighbouring nodes and then assimilated using the following equations [1]:

c. Assimilation:

$$Y_i(k/k) = Y_i(k/k - 1) + \sum_{j=1}^{m} [\hat{Y}_j(k/k) - Y_j(k/k - 1)] \tag{5.26}$$

$$\hat{y}_i(k/k) = \tilde{y}_i(k/k - 1) + \sum_{j=1}^{m} [\hat{y}_j(k/k) - \tilde{y}_j(k/k - 1)] \tag{5.27}$$

From the above development one can see that the quantities communicated between various sensor nodes consist of the difference between the local information at a time k and the prediction based on the information only up to time $k - 1$, and this is interpreted as the new piece of information gained by that node at the current time instant. Actually the communicated terms are equivalent to $i_j(k)$ and $I_j(k)$. Logically the new information available at a time k is just the information obtained via measurement at that time. The operation of the sensor network (SNW) can be envisioned as a group of local estimators. These estimators communicate new and independent information between each other. They assimilate both local and communicated information to obtain a globally optimal local estimate.

There are some interesting aspects about these decentralised equations [1]: (i) the additional computation required of each node is small – this is due to the use of the information form of the filter (in IF the computational burden is on the generation of predictions); (ii) the amount of communication is actually less than in an HDFA, this is because each node individually computes a global estimate, and hence the estimates/predictions are not communicated prior to an estimation cycle, and the communication BW is halved and (iii) the assimilation equations are identical to the ones in a system with distributed sensing nodes and a broadcast communication system.

5.3.5.1 Square Root Information Filter and Fusion

A scheme of decentralised square root information (DCSRIF) consists of an NW of nodes with its own SRIF-processing facility. Here, the fusion occurs locally at each node on the basis of local measurements and the information communicated from the neighbouring nodes. This processing node is a sensor-fusion node and takes local measurements and shares information with other fusion nodes. It thus assimilates the communicated information and computes the estimate [4]. The details of these equations are given below. Let a linear system be given as

$$x(k + 1) = \phi x(k) + Gw(k) \tag{5.28}$$

Here, it is assumed that the a priori information given about x_0 and w_0 can be put in data equation as

$$z_w = R_w w_0 + v_w \tag{5.29}$$

$$\tilde{y}_0 = \tilde{R}_0 x_0 + \tilde{v}_0 \tag{5.30}$$

The noise processes/initial conditions v_0, $\tilde{v}_0$ and v_w are assumed to be zero mean, independent and are with unity covariance matrices (for the sake of convenience). Here, $\hat{y}, \tilde{y}$ represent the ISs (Section 4.6.1). Then by incorporating a priori information, the time-propagation part of the DCSRIF yields the following form. The local mapping/time propagation is given as [4]

$$T(j+1)\begin{bmatrix} \tilde{R}_w(k) & 0 & \tilde{y}_w(k) \\ -R^d(k+1)G & R^d(k+1) & \tilde{y}(k) \end{bmatrix}$$
$$= \begin{bmatrix} \hat{R}_w(k+1) & \hat{R}_{wx}(k+1) & \hat{y}_w(k+1) \\ 0 & \hat{R}(k+1) & \hat{y}(k+1) \end{bmatrix} \tag{5.31}$$

Here, T is the Householder transformation matrix that will automatically result into the factorisation of Equation 5.31, that is, we obtain the right-hand side block matrix. That is, given a priori (square root of) IM (square root of the information matrix [SRIM]) and the IS vector, the transformation by the matrix T yields the new IS/SRIM. The subscript w signifies those variables related to the process noise. We next have the following important relation with transition matrix:

$$R^d(k+1) = \hat{R}(k)\phi^{-1}(k+1) \tag{5.32}$$

Then the local estimates are generated using measurement update of DCSRIF [4]

$$T(k + 1/k + 1)\begin{bmatrix} \hat{R}(k + 1/k) & \hat{y}(k + 1/k) \\ H(k + 1) & z(k + 1) \end{bmatrix} = \begin{bmatrix} R^*(k + 1) & y^*(k + 1) \\ 0 & e(k + 1) \end{bmatrix} \quad (5.33)$$

The local updated estimates are denoted with '*'. Here, the simple variable z represents the measurements, whereas variable y as the IS, and we note here that this IS may be/not be exactly identical to the IS of the IF. These local estimates are communicated between all nodes in a fully connected network. Then, at each node these estimates are assimilated to produce global DCSRIF estimates. The assimilation to produce global DSRI estimates at the ith node (with $i = 1, ..., N - 1$ representing the remaining nodes) is processed by the following equations [4]:

$$\hat{y}_i(k + 1/k + 1) = y^*(k + 1/k + 1) + \sum_{i=1}^{N-1} y_i^*(k + 1/k + 1) \quad (5.34)$$

$$\hat{R}_i(k + 1/k + 1) = R^*(k + 1/k + 1) + \sum_{i=1}^{N-1} R_k^*(k + 1/k + 1) \quad (5.35)$$

As we have seen in Chapter 4, the formulation of the information pair (IS–IM) and the use of the orthogonal transformation matrix T yield very elegant, simple, useful and intuitively appealing equations for the decentralized data fusion (DCDF). This DCDF estimation algorithm would possess better numerical reliability, and stability properties due to the use of the square root filtering formulation compared to the normal IF-based schemes, especially for any large dimension problem or when the estimation scheme is implemented on a finite word length computer. This aspect is very important if such fusion process is to be implemented for online/real-time applications using digital signal processing (DSP) hardware/field programmable gate array (FPGA), and for very large wireless sensor network/data fusion (WSN/DF) systems. In SRIF structure, it is easy to derive algorithmic modification and/or approximations with the data equation in a straightforward manner. In the decentralised IF for fusion requires communication of the local ISs and information matrices to all the neighbouring nodes for computing the global estimates, whereas in the case of the DCSRIF for fusion the ISs and the SRIM are estimated together in a compact form and in recursive manner with the help of the Householder orthogonal transformation matrix. The handling of the reduced and smaller range of numbers in the DCSRIF formulation enables the results to be represented by fewer bits, thereby saving on the communication overheads. This is very advantageous in the large wireless sensor networks (WSN), the topic of Chapter 9, and for health monitoring of large aerospace structures.

5.4 Decentralised Multi-Target Tracking

In this section, the data association for multi-target tracking (MTT) is described. The aspect of data association in a distributed system is a complex problem. This is because the firm and hard association decisions already made locally (in an optimal manner with respect to local measurements) may not be optimal at the global level [1]. This is true especially if and when all the sensor information is provided. It is almost impossible to undo an incorrect association decision once the data have been fused into a track.

5.4.1 Decentralised Data Association

The usual approach to overcome this problem is to maintain both a local and a global track file and periodically to synchronise the two. Other approaches involve using either probabilistic data association filter (PDAF) or multiple hypothesis testing (MHT). This is done to avoid the need to make hard local association decisions. Most data association methods require that the normalised innovation is readily available at each of the local processor nodes for decisions to be taken. In a DCF, the information transmitted from one node to another is in the form of the information incremental terms $i(k)$ and $I(k)$. In order to implement a local validation procedure it is necessary to have an expression from the prediction and communication terms that would allow further computation of a normalised innovation by every node on the basis of local IS–IM estimates $y(k|k-1)$ and $Y(k|k-1)$. This normalised information innovation would yield an information gate (equivalent to the innovation gate). It is found that the normalised information innovation is identically the conventional (measurement) residual [1]. The data association or a gating strategy for a decentralised sensing NW (DCSNW) using IF is implemented using a normalised gate, or modified LL and the following equations [1]:

$$v(k) = i(k) - I(k)Y^{-1}(k/k-1)\hat{y}(k/k-1) \tag{5.36}$$

$$\Gamma(k) = E\{v(k)v(k)^t | Z^{k-1}(k)\} \tag{5.37}$$

Equation 5.36 is the information residual vector which is in fact the innovations (of the conventional KF) projected onto the information space. Equation 5.37 is the information innovations covariance. The gate obtained from

Equations 5.36 and 5.37 is identical to the gate normally used in conventional MTT situations. With the gate established thus using Equations 5.36 and 5.37 one can use the data aggregation (DAG) methods described in Chapter 4.

5.4.2 Decentralised Identification and Bayesian Theorem

The DCF concepts are easily extended to DF systems and situations for which the underlying pdfs are not necessarily Gaussian and also if the pdfs are of discrete type [1]. This facilitates implementation of decentralised/discrete system identification/state estimation algorithms. The approach is based on the use of LLs and also extends the hierarchical LL architectures described in Section 2.4.4. We have the following recursive form of the Bayesian formulation [1]:

$$p(x \mid Z(k)) = p(z(k)|x)p(x|Z(k-1))/p(z(k) \mid Z(k-1)) \qquad (5.38)$$

We rewrite Equation 5.38 in the following form:

$$\ln p(x \mid Z(k)) = \ln p(x|Z(k-1)) + \ln \frac{p(z(k) \mid x)}{p(z(k) \mid Z(k-1))} \qquad (5.39)$$

Here, x is the state to be estimated and Z is the set of measurements up to the kth time step. In Equation 5.39, the first term is the information accumulated about the state up to time $k-1$, and the second term is the new information generated at the time step k. Equation 5.39 represents the communication needs of the fully connected DCSNW, wherein each sensor node communicates a LL based on its own measurement to all others and receives the LLs of all its neighbouring nodes. These nodes are then fused to form a global estimate. Equation 5.39 is re-written in terms of an individual sensor node 'i', as follows [1]:

$$\ln p(x_i \mid Z(k)) = \ln p(x_i|Z(k-1)) + \sum_j \ln \frac{p(z_j(k)|x_j)}{p(z_j(k)|Z(k-1))} \qquad (5.40)$$

Equation 5.40 gives the global estimate and the second term is the summation of the communicated terms.

5.5 Millman's Formulae in Sensor Data Fusion

As we have seen in Chapter 2 (and also in Chapter 4), the theory and techniques of statistics and probability can be very well utilised for the sensor data fusion problems. We also emphasise here that in fact until the arrival and renaissance of the soft computing paradigms (artificial neural networks, fuzzy logic and genetic algorithms) the major estimation work and approaches were developed based on the basis of probabilistic models, except the inception of the H-infinity theory. In that sense the modern estimation theory is well entrenched in the notion of modelling the uncertainty in measurements (and states), by using probability theory and more so by using Bayesian rule. This theory of estimation, including system identification and filtering, has been so well developed that scores of estimators are available and have been very successfully utilised to solve tens of hundreds of practical problems in industry, target tracking and satellite orbit estimation. These techniques are also the main body and concepts for developing DF processes and approaches. In recent times there is increasing use of soft computing approaches for the DF tasks as discussed in Chapters 3 and 10. The point of this discussion is that a lot of theoretical work would be required to bring the new approaches to the level of theoretical development/sophistication of the (classical) and modern estimation theory and practice. Then the new estimation theory will be labelled as Intelligent Estimation, which is already evolving.

As we have seen in Chapter 4 and in many previous sections of this chapter, the MS problems are divided into [8]: (i) measurement-to-measurement, (ii) measurement-to-track and (iii) track-to-track fusion situations. These aspects are coupled with the fusion architectures: (a) centralised (CDF), (b) hierarchical (HDFA), (c) decentralised (DCF) or (d) mixed ones. In this sense the interactions among local estimators, the presence of a central estimator, communications of data types (measurements or estimates) and communication speeds define the DF architecture. We discuss Millman's formula (MF) which has applications to various data fusion situations [8] discussed in Chapter 4 and some previous sections of the present chapter.

5.5.1 Generalised Millman's Formula

We consider that there are N number of local estimators of the state x in the Euclidean vector space and we also have the associated covariance matrices specified as $P_{ij} = \text{cov}(\tilde{x}_i, \tilde{x}_j)$, with $i,j = 1, 2, ..., N$. Then our aim is to determine the optimal linear estimate of x possibly of the following form:

$$\hat{x} = \sum_{i=1}^{N} w_i \hat{x}_i \tag{5.41}$$

The weights sum up to 1. The state-error covariance matrix is given by the following formula:

$$P = \sum_{i=1}^{N} \sum_{j=1}^{N} w_i P_{i,j} w_j^T \tag{5.42}$$

The weighting factors/matrices are determined by minimising the following mean square error (MSE) criterion:

$$J(w_1, w_2, \ldots, w_N) = E\left(\left\| x - \sum_{i=1}^{N} w_i \hat{x}_i \right\|^2 \right) \tag{5.43}$$

The process of minimisation leads to the following equations [8]:

$$\sum_{i=1}^{N-1} w_i (P_{ij} - P_{iN}) + w_N (P_{Nj} - P_{NN}) = 0 \tag{5.44}$$

$$\sum_{i=1}^{N} w_i = I \tag{5.45}$$

with $j = 1, 2, \ldots, N-1$, in Equation 5.44. Equations 5.41, 5.42, 5.44 and 5.45 represent the Millman's data/estimates fusion in the generalised sense for $N > 2$. For $N = 2$, we have the following formulae (also known as the Bar-Shalom–Campo formula) for fused state/covariance when the two estimates are correlated [8]:

$$\begin{aligned} \hat{x} &= \hat{x}_1 + (P_{11} - P_{12})(P_{11} + P_{22} - P_{12} - P_{21})^{-1}(\hat{x}_2 - \hat{x}_1) \\ &= (P_{22} - P_{21})(P_{11} + P_{22} - P_{12} - P_{21})^{-1}\hat{x}_1 \\ &\quad + (P_{11} - P_{12})(P_{11} + P_{22} - P_{12} - P_{21})^{-1}\hat{x}_2 \\ &= w_1 \hat{x}_1 + w_2 \hat{x}_2 \end{aligned} \tag{5.46}$$

$$\begin{aligned} P &= w_1 P_{11} w_1^T + w_1 P_{12} w_2^T + w_2 P_{21} w_1^T + w_2 P_{22} w_2^T \\ &= P_{11} - (P_{11} - P_{12})(P_{11} + P_{22} - P_{12} - P_{21})^{-1}(P_{11} - P_{21}) \end{aligned} \tag{5.47}$$

From Equation 5.46, we clearly see the explicit formulae for the weights. The Millman's formulae for the two uncorrelated estimates are given by the following expressions [8]:

$$\begin{aligned} \hat{x} &= P_{22}(P_{11} + P_{22})^{-1}\hat{x}_1 + P_{11}(P_{11} + P_{22})^{-1}\hat{x}_2 \\ &= w_1 \hat{x}_1 + w_2 \hat{x}_2 \end{aligned} \tag{5.48}$$

$$\begin{aligned} P &= w_1 P_{11} w_1^T + w_2 P_{22} w_2^T \\ &= P_{11}(P_{11} + P_{22})^{-1} P_{22} \end{aligned} \tag{5.49}$$

Again the formulae for the weights are explicitly seen in Equation 5.48.

5.5.2 Millman's Fusion Formulae in Filtering Algorithms

We have seen in Section 4.2.7, that KF is a natural measurement data level fusion algorithm itself. We have the following maximum likelihood state estimator for the KF [8]:

$$\hat{x}^{ml}(k) = (H^T R^{-1} H)^{-1} H^T R^{-1} z_k \text{ with } P_{ml} = (H^T R^{-1} H)^{-1} \tag{5.50}$$

Using Equation 5.50 and the standard KF equations from Section 4.2.2, we can write the following state-estimator equation:

$$\hat{x}(k) = P(k)P^{-1}(k-1)\hat{x}(k-1) + P(k)P_{ml}^{-1}\hat{x}^{ml}(k) \tag{5.51}$$

We see from Equation 5.51 that the two estimates are independent and hence can be expressed by Millman's formula

$$\hat{x}(k) = w_1 \hat{x}_1 + w_2 \hat{x}_2 = w_1 \hat{x}(k-1) + w_2 \hat{x}^{ml}(k) \tag{5.52}$$

With the weights given by the following expressions:

$$
\begin{aligned}
w_1 &= P_{22}(P_{11} + P_{22})^{-1} = (P_{11}^{-1} + P_{22}^{-1})^{-1} P_{11}^{-1} \\
&= (P^{-1}(k-1) + H^T R^{-1} H)^{-1} P^{-1}(k-1) \\
&= P(k)P^{-1}(k-1)
\end{aligned} \tag{5.53}
$$

$$
\begin{aligned}
w_2 &= P_{11}(P_{11} + P_{22})^{-1} = (P_{11}^{-1} + P_{22}^{-1})^{-1} P_{22}^{-1} \\
&= (P^{-1}(k-1) + H^T R^{-1} H)^{-1} P_{ml}^{-1} \\
&= P(k)P_{ml}^{-1}
\end{aligned} \tag{5.54}
$$

In Equations 5.53 and 5.54, we have $P_{11} = P(k-1)$ and $P_{22} = P_{ml}$. Also, the weight constraint Equation 5.45 is satisfied

$$w_1 + w_2 = P(k)(P^{-1}(k-1) + P_{ml}^{-1}) = P(k)P^{-1}(k) = I \tag{5.55}$$

The associated posterior covariance matrix is now written in the form of MF:

$$
\begin{aligned}
P(k) &= w_1 P_{11} w_1^T + w_2 P_{22} w_2^T \\
&= (P_{11}^{-1} + P_{22}^{-1})^{-1}(P_{11}^{-1} + P_{22}^{-1})(P_{11}^{-1} + P_{22}^{-1})^{-1} \\
&= (P^{-1}(k) + P_{ml}^{-1})^{-1} = (P^{-1}(k) + H^T R^{-1} H)^{-1}
\end{aligned} \tag{5.56}
$$

5.5.3 Millman's Fusion Formulae in Smoothing Algorithms

The well-known merit of the smoothed estimate is that, not only is the estimate of state better than the filtered estimate, but also, the smoother uses more data. In fact it is the latter reason why the smoother gives such better results. We assume that the KF has been run in the forward direction and we have obtained the filtered state estimate as well as the associated covariance matrices. Then using more measurement data and, perhaps some IF-type algorithm or any other suitable filtering algorithm, these data are processed in the backward direction. Then, the backward estimate is given by the following expression:

$$\hat{x}(k/l) = Y^{-1}(k/l)\hat{y}(k/l) \tag{5.57}$$

In Equation 5.57, in general Y is the IM and is the inverse of the conventional covariance matrix P of the KF. In that sense y is the IS (estimate) vector and is the result of the backward data processing cycle using, say IF. The fusion equations based on the Millman's fusion concept and Equations 5.48 and 5.49 are given by the following equations [8]:

$$K(k) = P(k)S(k/l)(I + P(k)S(k/l))^{-1}$$
$$P(k) = (I - K(k))P(k) \tag{5.58}$$
$$\hat{x}(k) = (I - K(k))\hat{x}(k) + P(k)\hat{y}(k/l)$$

It must be noted here that in Equation 5.58 the covariance matrix P from the second sub-equation feeds into the third sub-equation. It is interesting to add here that though the smoothing problem is different from the filtering one, the equation is obtained from the concept of KF and the MF.

5.5.4 Generalised Millman's Formula in State Estimation

In Section 5.5.2, we have considered only one sensor as such. Now, we consider more than one sensor and obtain the expressions of fusion of state estimates using the generalised MF as per Equations 5.44 and 5.45. We consider three sensor-data-fusion configurations (SDFC): (i) centralised, (ii) MS fusion and (iii) hierarchical fusion.

5.5.4.1 Optimal Centralised Fusion

In this SDFC, we presume that all the measurements are provided at the CP, which processes the data, say by using KF. In this case, the measurements/ and associated equations are merged into a block as follows:

$$z(k) = Hx(k) + v(k) \tag{5.59}$$

In Equation 5.59, we have vector z as the collection of all the measurement variables, that is, the observables, for example, position, velocity and acceleration of a target, similarly for a satellite orbit estimation, z would be the vector that contains the observables as the satellite elevation, azimuth and range (i.e. slant distance of the satellite from the observation site). Likewise H is the matrix that contains all the associated measurement (sensor) mathematical models, and v is the vector of the measurement noise processes affecting all the observables. We assume that the measurement noise matrix is diagonal with its elements as the individual covariance matrix of the measurement.

5.5.4.2 Multi-sensory Fusion

In this SDFC, the measurements from each sensor are processed by sensor's own estimator, say by KF, this is termed as the local KF (LKF). Then at the fusion centre all these estimates are fused. This is like SVF discussed in Chapter 4. We assume that each sensor provides an estimate of the state and its associated covariance matrix, for $j = 1, 2, ..., S$ (as the number of individual sensors). We further assume that the process noise is quite small, and hence can use the generalized Millman's formula (GMF) because we regard the local estimates as independent from each other. In that case the cross-covariance matrices are null. Then, we have the following closed form as the solution of Equations 5.44 and 5.45:

$$w_i = \left(\sum_{j=1}^{N} P_{jj}^{-1} \right)^{-1} P_{ii}^{-1} \tag{5.60}$$

Then using Equation 5.60 in Equations 5.41 and 5.42, we obtain the following expressions for the fused state and covariance matrix [8]:

$$\hat{x}(k) = P(k) \sum_{j=1}^{N} P^{(j)^{-1}}(k) \hat{x}^{(j)}(k) \tag{5.61}$$

$$P^{-1}(k) = \sum_{j=1}^{N} P^{(j)^{-1}}(k) \tag{5.62}$$

If the process noise is not zero then we need to compute the cross-covariance matrices as follows:

$$P^{(ij)}(k) = [I - K^{(i)}(k)H^{(i)}][\phi P^{(ij)}(k-1)\phi^T + GQG^T][I - K^{(j)}(k)H^{(j)}]^T \tag{5.63}$$

In Equation 5.63, K is the ith LKF gain.

5.5.4.3 Hierarchical Data Fusion

In this SDFC, there is a feedback from the CP to all the local estimators. This CP derives the measurement information from the estimates available to the CP. In this case, the equivalent measurements are obtained by the reverse use of the MF in Equation 5.51. Then the information independently coming from each sensor is fused with the prior state/covariance estimates. This is done by using the GMF closed form solution of Equation 5.60 and is given as follows [8]:

$$P^{-1}(k)\hat{x}(k) = P^{-1}(k-1)\hat{x}(k-1) + \sum_{j=1}^{S} \{P^{(j)^{-1}}(k)\hat{x}^{(j)}(k)$$

$$-P^{(j)}(k-1)\hat{x}^{(j)}(k-1)\} \tag{5.64}$$

$$P^{-1}(k) = P^{-1}(k-1) + \sum_{j=1}^{S} \{P^{(j)^{-1}}(k) - P^{(j)^{-1}}(k-1)\}$$

Then in order to restart the local filter (by using the feedback from the CP), the one-step predictions are replaced by the fused estimate predictions as follows:

$$\hat{x}^{(j)}(k+1/k) = \hat{x}(k+1/k) \quad \text{and} \quad P^{(j)}(k+1/k) = P(k+1/k) \tag{5.65}$$

In Equation 5.65, we have the respective one-step central predictions as follows:

$$\hat{x}(k+1/k) = \phi\hat{x}(k)$$
$$P(k+1/k) = \phi P(k)\phi^T + GQG^T \tag{5.66}$$

The local filters are the conventional and standard KFs.

In Equation 5.64, we use the following differences as equivalent measurements:

$$P^{(j)^{-1}}(k)\hat{x}^{(j)}(k) - P^{(j)^{-1}}(k-1)\hat{x}(k-1)$$
$$P^{(j)^{-1}}(k) - P^{(j)^{-1}}(k-1) \tag{5.67}$$

A numerical simulation example for target tracking using 2DOF state-space model (for $S = 3$ sensors, with only position as the measurement, and with measurement noise $R = 1$) that evaluated the approaches presented in Section 5.5.3 is given in Reference 8. It compares: (i) centralized Kalman filter (CKF), (ii) hierarchical KF, (iii) MS case, (iv) MS with GMF and (v) the single KF. The overall MSEs and the overall Mahalanobis distances (MD i.e. square

TABLE 5.1

Performance Metrics for the Comparison of Multi-sensory Fusion Using MF

Estimator/Metric	CKF	HKF	MSC-GMF	MSC	KF
MSE	0.101	0.101	0.114	0.114	0.153
MD	1.251	1.251	1.254	1.520	1.255

Source: Adapted from Ajgl, J., Simandl, M. and Dunik, J. Millman's formula in data fusion. *10th International PhD Workshop on Systems and Control*, Institute of Information Theory and Automation AS CR, Hluboka nad Vltavou, 1–6, September 2009. http://www.as.utia.cz/files/121.pdf, accessed April 2013.

root of the normalised square errors of the states estimates) were evaluated and are reproduced in Table 5.1. The normalisation in MD is done using the covariance matrix *P*. Overall satisfactory results have been obtained using MF in estimation/DF.

5.6 SRIF for Data Fusion in Decentralised Network with Four Sensor Nodes: Illustrative Example

Simulated data of the position of a target moving with constant acceleration are generated using MATLAB-based code. The SRIF and the decentralised SRISFA (SRIF-sensor fusion algorithm) are validated for four-node interconnection SNW [9]. The system states are position, velocity and acceleration: $x^T = [x \, \dot{x} \, \ddot{x}]$ and the transition matrix is given by

$$\phi = \begin{bmatrix} 1 & \Delta t & \Delta t^2/2 \\ 0 & 1 & \Delta t \\ 0 & 0 & 1 \end{bmatrix}; \quad G = \begin{bmatrix} \Delta t^3/6 \\ \Delta t^2/2 \\ \Delta t \end{bmatrix} \quad (5.68)$$

Here, Δt is the sampling interval. The measurement model for each sensor is given by

$$z_m(k+1) = Hx(k+1) + v_m(k+1) \quad (5.69)$$

where $H = [1 \ 0 \ 0]$ for each sensor. The process noise covariance strength is 0.0001. The vector v is the measurement noise which is assumed to be white and Gaussian with zero mean. The position data pertaining to the four nodes/measuring sensors are generated by adding random noise with STDs as $\sigma_{v_1} = 1$, $\sigma_{v_2} = 3$, $\sigma_{v_3} = 5$, $\sigma_{v_4} = 10$ to the true measurements with 2000 data points. The sampling interval is 0.1 unit/s. We use the whitened measurements with appropriate measurement models for *H*. The initial

conditions for the state are: $x_0 = [200\ 0.5\ 0.05]$ in SI units. The SRIF generates the state estimates. The fused global estimates at the nodes are obtained
utilising the local estimates at the node and the communicated information
from the neighbouring nodes by using SRISFA. Also, some results are generated with the IF, and compared with some results of the SRIF for the same
state-space target model and the measurement models [10].

Figure 5.1 shows the states and their estimates obtained using SRIF. Figure
5.2 shows the state-errors with their bounds using SRIF. Figure 5.3 compares
the states and their estimates using SRIF and IF. Figure 5.4 shows the position
state-errors for two sensors and fused (SRIF) and comparison with IF (second
sub-plot). Figure 5.5 shows the comparison of the original states, each senor-
side estimated states (four sensors) and the fused states – SRIF. The fourth sub-
plot of Figure 5.5 shows the comparison of the norm of the SRIF-information
matrices (at four sensor nodes and the fused one). Table 5.2 gives shows the
percentage fit errors at the four local sensor nodes and the global fused (estimate) with SRIF and IF. The percentage fit error is calculated using the relation

$$\textit{Fit error} = 100 * norm(\hat{x} - x_t)/norm(x_t) \tag{5.70}$$

where x_t is the true state.

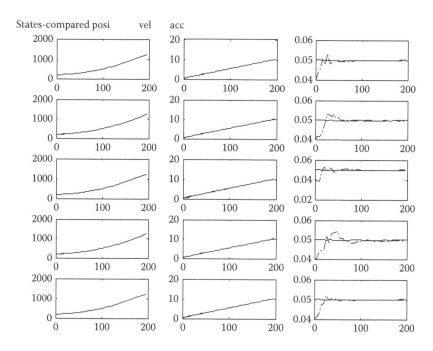

FIGURE 5.1
States and their estimates for each sensor node, first four rows of plots; fused, the last row
plot – SRIF.

State-errors with bounds->posi   vel   acc

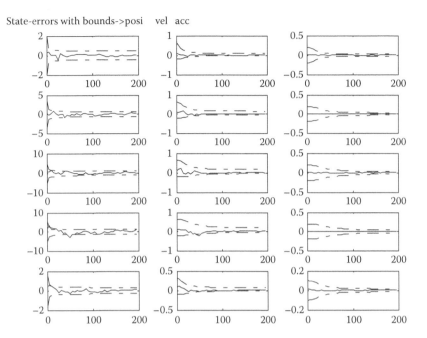

FIGURE 5.2
State-errors for each sensor node, first 4 rows of plots; fused, the last row plot – SRIF.

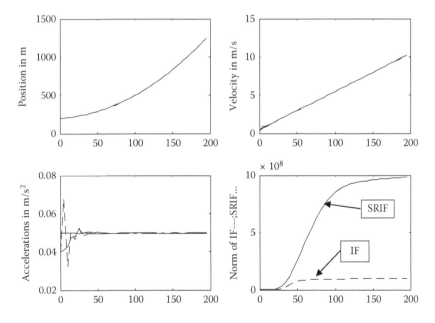

FIGURE 5.3
States compared with state estimates obtained by using SRIF and IF.

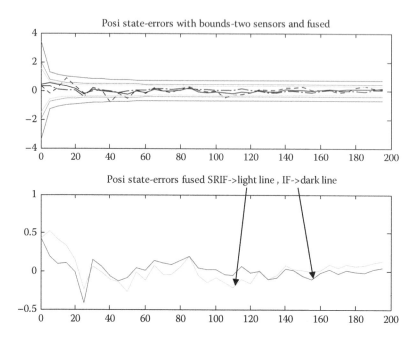

FIGURE 5.4
Position state-errors for two sensors and fused (SRIF) and comparison with IF (second sub-plot).

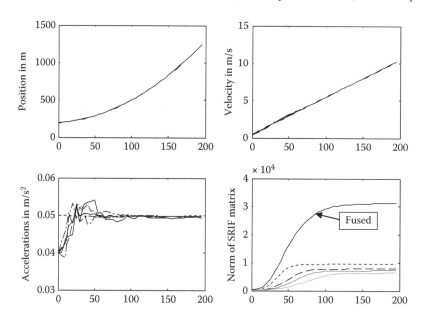

FIGURE 5.5
Comparison of the original states, each sensor-side estimated states (four sensors) and the fused states – SRIF.

TABLE 5.2

Fit Errors (%) at the Four Local Sensor Nodes and the Global Fused
(Estimate) – SRIF and IF

Sensor Node	SRIF			IF		
	Position	Velocity	Acceleration	Position	Velocity	Acceleration
1	0.0176	0.3044	3.7954	0.0182	0.4510	10.2016
2	0.0401	0.6321	5.3134	0.0524	1.4794	22.3175
3	0.0723	0.8071	5.2932	0.0899	1.2204	11.9736
4	0.1675	1.0076	5.4856	0.2280	4.2994	42.9865
Fused	0.0247	0.4336	4.4633	0.0163	0.4252	9.4430

From the presented results (plots and Table 5.2) we infer that the performance of the DSRISF algorithm (SRIF) in terms of numerical accuracy when compared with the decentralised IF algorithm is relatively better. However, the fused results with the IF are not far away from the SRIF.

EXERCISES

5.1 Compare salient features of various DF architectures discussed in this chapter.

5.2 Compare salient features of the non-linear filters discussed in this chapter.

5.3 What is the major difference between optimal filter based on conditional pdf (of the states given the measurements) and Pugachev's filter?

5.4 Even though Pugachev's filter is not based on the conditional pdf, why is it called a conditionally optimal filter?

5.5 Compare the structure of Pugachev's estimator with the structure of KF, and see under what conditions the latter can be obtained from e Pugachev's estimator.

5.6 What is the fundamental (philosophical/conceptual) difference between derivative-free KF (DFKF) and the EKF?

5.7 What is the process of whitening the measurements? Explain by taking a simple example.

5.8 Download any Householder transformation code (in MATLAB) from the net and run to verify the transformation Equations 4.99 and 5.31.

5.9 Equation 5.60 seems odd due to there being three inversions, check and verify that it is unit-wise correct as such

$$w_i = \left(\sum_{j=1}^{N} P_{jj}^{-1} \right)^{-1} P_{ii}^{-1}.$$

5.10 Can you extend Equation 5.33 to incorporate more than one sensor (measurements/data channels)

$$T(k+1/k+1)\begin{bmatrix} \hat{R}(k+1/k) & \hat{y}(k+1/k) \\ H(k+1) & z(k+1) \end{bmatrix} = \begin{bmatrix} R^*(k+1) & y^*(k+1) \\ 0 & e(k+1) \end{bmatrix}?$$

5.11 Draw a block diagram (schematic diagram) for a centralised data fusion system.

5.12 Draw a block diagram for a DCF system.

5.13 Draw a block diagram for a distributed data fusion system.

5.14 Draw a block diagram for a hybrid data fusion system.

References

1. Durrant-Whyte, H. Multi sensor data fusion – Lecture Notes. Australian Centre for Field Robotics, University of Sydney, NSW 2006, Australia, January, 2001.
2. Mutambra, A. G. O. *Decentralized Estimation and Control for Multisensor Systems*. CRC Press, FL, 1998.
3. Manyika, J. and Durrant-White, H. *Data Fusion and Sensor Management – A Decentralized Information – Theoretic Approach*. Ellis Horwood Series, Prentice Hall, Upper Saddle River, NJ, 1994.
4. Raol, J. R. *Multisensor Data Fusion with MATLAB*. CRC Press, FL, 2010.
5. Raol, J. R., Girija, G. and Singh, J. *Modelling and Parameter Estimation of Dynamic Systems*. IET/IEE Control Series Book Volume 65, IET/IEE London, 2004.
6. Raol, J. R. and Singh, J. *Flight Mechanics Modelling and Analysis*. CRC Press, FL, 2010.
7. Raol, J. R. and Gopal A. K. *Mobile Intelligent Automation Systems*. CRC Press, FL, 2012.
8. Ajgl, J., Simandl, M. and Dunik, J. Millman's formula in data fusion. *10th International PhD Workshop on Systems and Control*, Institute of Information Theory and Automation AS CR, Hluboka nad Vltavou, 1–6, September 2009. http://www.as.utia.cz/files/121.pdf, accessed April 2013.
9. Girija, G. and Raol, J. R. Sensor data fusion algorithms using square-root information filtering. *IEE Proceedings on Radar, Sonar and Navigation (UK)*, 149(2), 89–96, 2002.
10. Girija, G. and Raol, J. R. Application of information filter for sensor data fusion. *38th Aerospace Sciences Meeting and Exhibit* (AIAA), Reno, NV, USA, paper No. 2000-0894, Jan 10–13, 2000.

6

Component Analysis and Data Fusion

6.1 Introduction

The theory and concepts of independent component analysis (ICA), principal component analysis (PCA), wavelet transforms (WTs), curvelet transform and discrete-cosine transform (DCT) are being increasingly used for data fusion (DF) and especially for image fusion in recent times [1–8]. Although, several aspects of image algebra and image fusion are further discussed in Chapter 7, we study component analysis (CA) in this chapter, since some of these analyses and transform methods are also being increasingly used for image separation and image restoration, in addition to image fusion, and also since CA is intimately connected with signal processing and kinematic DF. Hence, after treating CA in this chapter, CA in general and in particular for image processing/fusion, we formally study image algebra and image fusion further in Chapter 7. The images of moving objects are often tracked via centroid tracking algorithms. In centroid tracking, mainly the $x - y$ coordinates of the object image (centroid) are being tracked using Kalman filter (Chapters 4 and 8). This is the tracking of the geometric centre of the object image, whereas one can use the concept of tracking of the centroid of image intensities. This also opens up the scope of using any such tracking-estimation algorithm and augmentation of such algorithm with fuzzy logic to improve that tracking adaptation. One such application is illustrated in Chapter 10, in the context of using fuzzy logic as a soft computing paradigm.

The main idea in the CA is to transform the original data signal to a level of reduced dimensions by way of finding the most important and relevant information from the original data. The huge amount of data might contain a small amount of useful information. Thus, any CA is a way of identifying patterns in data (searching patterns and hidden aspects, hidden Markov models, etc.), and expressing these data in such a particular way as to highlight their similarities and differences. In many data analyses, problem representation is often sought to have linear transformation of the original data set that gives a new data set that is called an output, and this new set is often of reduced dimension [1]. The original data set is called the source input. In this case each component of the output data is a linear combination of the original source variables. These methods for analysis are: (i) ICA – non-Gaussianness/non-Gaussianity (i.e. non-Gaussian [NG] property of the components is important here), (ii) PCA – Gaussian property of the components is important here, (iii) factor analysis (FA) and (iv) projection [1]. Quite often the (hidden) patterns in the data might be hard to find, especially, in data of higher dimensions, and where graphical representation is not available. In such cases the PCA is a very powerful method/decomposition for analysing such data. Certain mathematical aspects of CA are presented in this chapter and further aspects of image DF are discussed in Chapter 7.

The CA procedures (especially ICA) are generally very powerful approaches and are able to separate independent sources linearly and originally mixed in several sensors' situation. According to us the problem of any CA looks like the inverse of the DF problem. In DF we have the data of individual sensors and then seek adequate and proper methods to fuse the information from these sensor data to obtain one representation of the combined information. In a CA problem, we have a mixer of the data (the combined information that has already happened due to some reason from several sensors – at this stage the fusion was not intentional, but was incidental!) and then we are seeking to operate over this (output) mixer data and try to separate it into its individual components. Surprisingly and despite this inverse relationship between DF and CA, the latter can still be used for DF. This aspect is explained next. For example, in case of the EEG (electroencephalograms) signal, the ICA separates the features hidden in the EEG data and when this is done on other sets of EEG data, then one can separate out and collect such distinct features (from various equivalent sets of EEG data, but mostly taken independently as much as possible for the ICA to be very effective). These separated-out features are then combined/fused (disregarding other insignificant features) in some appropriate way to obtain the enhanced information of the chosen distinct feature. Thus, first the ICA is carried out on EEG data of each individual set and then the DF process is carried out to improve the prediction of the important and distinct features that were extracted from each ICA analysis. Thus, first the CA – the separation process, and then the fusion process – the combination operation are performed, keeping in mind that the CA process operates over the original signals and the DF process operates over the most significant features extracted. In DF one uses the weighted combination of these extracted features/attributes of the underlying phenomena. This would be further clear when we study the methods of PCA, WT and DCT for image fusion. Thus, we see that there is an interesting relationship between CA and the DF process. Similar observations/inferences can be made about other CA methods. We also discuss the singular value decomposition (SVD) based procedure for image fusion in Section 6.10.

In general signal processing problems one often needs to find suitable representations for an image, or audio data for tasks like compression and de-noising. Data representations are then based on some discrete linear transformations. The standard linear transformations used in image processing are the Fourier, Haar and Cosine transforms. This process can be considered as a part of CA. The basic signal/image is then separated in terms of its hidden components such as Fourier coefficients. Then most important and significant components are chosen for further analysis, interpretations and fusion.

6.2 Independent Component Analysis

The ICA performs a linear projection (of the output signals/data) into independent components (ICs/independent sources) and the usual assumptions are that of: (i) the linearity – the mixture has happened linearly, (ii) no delays in the data and (iii) statistical independence of sources – uncorrelatedness [1]. In the ICA method, our goal is to find a linear representation of NG data so that the inherent components are statistically independent (or as independent as possible). This representation is expected to capture the essential structure of the data in many applications, including feature/attribute extraction and signal separation. Assume that the two recorded/measured/observed signals are each a weighted sum of the speech (or any other types of signals: audio signal from a car, human speech, etc.) emitted by the two speakers. This is expressed as a linear equation [1]

$$x_1(t) = a_{11}s_1 + a_{12}s_2 \tag{6.1}$$

$$x_2(t) = a_{21}s_1 + a_{22}s_2 \tag{6.2}$$

Here, a_{11}, a_{12}, a_{21} and a_{22} (that constitute a mixing model or called a mixer model) are the unknown parameters/constants that might depend on the distances of the microphones from the speakers or other parameters for other types of signals. The main idea in the ICA method is to estimate these two original signals $s_1(t)$ and $s_2(t)$ (which in the first place have got mixed in a liner manner) using only the recorded signals $x_1(t)$ and $x_2(t)$. Thus, we have a simple mixing model of the original signals which seem to have come from two (or more) independent sources. The CA problem is a complex one, because as such we do not know the original signals (was it an audio from a car or human speech?) as well as the coefficients in the mixing model. It is a blind-source identification problem. A similar situation occurs when there is an image that consists of several independent images and one needs to separate the original source components (e.g. images). Thus, it is imperative that in order to make the problem analytically tractable, we should have some information (assumptions) either on the original signals and/or the mixing matrix (of these coefficients) [1]. One straightforward approach is to use some statistical properties of the signals $s_i(t)$ to estimate the unknown coefficients. It might suffice to assume that $s_1(t)$ and $s_2(t)$ (at each time instant t) are statistically independent, however, more knowledge is always useful. The problem of CA seems to be like an 'inverse fusion problem', that is, to determine or extract the original signals (source signals/images) and the mixing matrix from the (unintended) 'fused' signal/image! A statistical 'latent variables' model (LVM) is used by assuming n linear mixtures $x_1, x_2, \ldots, x_n$ of n ICs as [1]

$$x_j = a_{j1}s_1 + a_{j2}s_2 + \cdots + a_{jn}s_n, \quad \text{for all } j \tag{6.3}$$

Here, we assume that each mixture x_j and each IC s_k is a random variable signal. The observed values $x_j(t)$ are a sample of this random variable. We assume that both the mixture variables (MVs) and the ICs have zero mean. Thus, x is the random vector whose elements are the mixtures $x_1, \ldots, x_n$ and s is the random vector with elements $s_1, \ldots, s_n$. Let A be the matrix of elements a_{ij}, then the mixing model is written as

$$x = As \tag{6.4}$$

The model of Equation 6.4 is the ICA model, a generative model, meaning that it describes how the observed data are generated by a process of mixing the ICs s_i. These components are latent variables since they are not directly observable. In addition, the mixing matrix A is as such unknown. In fact, we observe only the random vector x, and we must estimate both A and s using x. However, this should be done under as general assumptions as possible, since we have least knowledge of these unknowns. Yet, any knowledge of the mode of mixing and the source signals/images, if available would be good, and can be utilised in determining the coefficients of the mixing matrix A as well as the ICs.

The first step in most ICA algorithms is to whiten the data to remove any correlations in the data, that is, the covariance matrix of the data should be diagonal and all the diagonal elements should be equal to 1. Thus, the process of whitening is a linear change of the mixer data, and in fact, the whitening of the data restores the initial shape of the data when represented in some probability/vector space. The ICA then rotates the resulting matrix back to the original axes. It performs the rotation by minimising the Gaussian nature of the data projected on both the axes. Thus, the ICA is recovering the original sources that are statistically independent. The ICA can also handle the data of higher than two DOF. Some important features of ICA are [1]: (i) it can only separate the linear mixer of the source data, (ii) because it deals with the collection of the data (the points/data could be scattered in a vector space and forming a structure like clouds), the order in which these data are processed generally has no effect on the algorithm, (iii) also the channel-order change would have no or little effect on the outcome of the ICA algorithm,

(iv) because the ICA separates the original sources from the output by maximising their NG nature, the perfectly Gaussian sources cannot be separated and (v) if the sources are not independent, the ICA would find a space where the they are independent maximally.

6.2.1 Independence

We have said earlier that the fundamental restriction in ICA is that the ICs should have NG nature. Assume that the mixing matrix is orthogonal and the s_i are Gaussian, then x_1 and x_2 are Gaussian, uncorrelated and of unit variance with their joint probability density function (pdf) given by [1]

$$p(x_1, x_2) = \frac{1}{2\pi} e^{\left\{-\left(x_1^2 + x_2^2\right)/2\right\}}$$ (6.5)

It can be easily inferred that the pdf given by Equation 6.5 is symmetric. Hence, it does not have any information on the directions of the columns of A and hence A cannot be estimated. Also, we can establish that the distribution of any orthogonal transformation of the Gaussian (x_1, x_2) has the same distribution as (x_1, x_2), and that x_1 and x_2 are independent, that is, in the case of Gaussian variables, the ICA model up to an orthogonal transformation can be determined [1]. Matrix A is not identifiable for Gaussian ICs, however, if one of the components is Gaussian, the ICA model can still be determined. As, we have seen in Chapter 2, the independence is defined by the pdfs as follows:

$$p(s_1, s_2) = p(s_1)p(s_2)$$ (6.6)

Equation 6.6 states that the joint pdf is the product of individual pdfs of the component variables. We also have as an extension of Equation 6.6 the following property:

$$E\{g(s_1)g(s_2)\} = E(g(s_1)) \times E(g(s_2))$$ (6.7)

Equation 6.7 can be easily established by using the definition of E, the mathematical expectation, that contains the pdfs and then we apply the property of Equation 6.6 again to obtain the expression of Equation 6.7. A weaker form of independence is uncorrelatedness, that is, the two random variables s_1 and s_2 are said to be uncorrelated, if their covariance is zero, and if the variables are independent, they are uncorrelated, which follows from Equation 6.7 immediately, however, uncorrelatedness does not imply independence. We also ascertain here that since independence implies uncorrelatedness, ICA methods can constrain the estimation process to always give uncorrelated estimates of the ICs, thereby reducing the number of free parameters and also simplifying the problem.

6.2.2 NG Property

We know from the probability theory's result of the central limit theorem (CLT) that the pdf of the sum of independent random variables tends towards a Gaussian pdf, under certain conditions. This CLT states that a sum of two independent random variables usually has a distribution that is closer to Gaussian than any (distributions) of the two original random variables. If there are more ICs then the pdf of the sum is very Gaussian. We assume that the data vector x is distributed according to the model of Equation 6.4, thereby ascertaining that it is a mixture of ICs, with identical distributions. Now, we consider a linear combination of the x_i (to estimate one of the ICs), say, $y = w^T x = \sum_i w_i x_i$, here, w being a vector to be estimated by ICA. If w were one of the rows of the inverse of A, then this linear combination (i.e. $\sum_i w_i x_i$) itself would be one IC. Now, if we make a change of variables, as $z = A^T w$, then we have $y = w^T x = w^T As = z^T s$, implying that y is a linear combination of s_i (weights given by z_i). We know from the CLT that a sum of even two independent random variables is more Gaussian than the original variables, $z^T s$ is more Gaussian than any of the s_i and becomes least Gaussian (more NG) when it equals one of the s_i. Obviously only one of the elements z_i of z is non-zero. We note here that the s_i were assumed to have the identical distributions. So, we should take w as a vector that maximises the non-Gaussianity of $w^T x$. This vector would necessarily correspond, in the transformed domain, to z that has only one non-zero component, this implies that $w^T x = z^T s$ is equals to one IC, that is, maximising the non-Gaussianity of $w^T x$ would really give one of the ICs. The optimisation landscape for non-Gaussianity in the n-dimensional space/domain of vectors w has $2n$ local maxima. This is for two ICs, corresponding to s_i and $-s_i$ (the ICs can be estimated only up to a multiplicative sign). To estimate more ICs, we require to find all the local maxima. This is made easier since the different ICs are uncorrelated, that is, we can constrain the search to the space that gives estimates uncorrelated with the previous ones, and so on. This is the orthogonalisation (process) in a suitably transformed domain, that is, whitened space.

We summarise the foregoing development as follows, for intuitive understanding: the ICs $\rightarrow$ sum of the ICs $\rightarrow$ CLT $\rightarrow$ Gaussian. So, we revert the argument and say: Gaussianness = (obtained by the) sum of the ICs $\rightarrow$ sum of the sources; then, we say that the *non*-Gaussianness (would imply that) $\rightarrow$ the sources are *not* the sum $\rightarrow$ but then the sources *might be the ICs*. Hence, the key point in ICA estimation is the non-Gaussianness or non-Gaussianity property (of the ICs).

6.2.3 Determination of NG Property

There are several methods for determining the NG property of the variables. A few of these are discussed here.

6.2.3.1 Kurtosis

One measure of NG property is the kurtosis or the fourth-order cumulant, defined by

$$kurt(y) = E\{y^4\} - 3(E\{y^2\})^2 \tag{6.8}$$

Since, y is of unit variance, $kurt(y) = E\{y^4\} - 3$, that is, kurtosis is a normalised version of the fourth-order moment $E\{y^4\}$. For a Gaussian y, the fourth moment (the first term in Equation 6.8) equals $3(E\{y^2\})^2$, thereby implying that the kurtosis (kurt) is zero. For most (if not for all) NG random variables, kurt is non-zero, being either positive or negative. Random variables with a −ve kurt are called sub-Gaussian (a typical example is that of the uniform pdf) and those with positive value are super-Gaussian, the latter random variables have typically a spiky pdf with heavy tails, the example is the Laplace distribution with its pdf as $p(y) = (1/\sqrt{2})e^{(\sqrt{2}|y|)}$.

6.2.3.2 Neg-Entropy

Another important measure of non-Gaussianity is neg-entropy that is based on the quantity of (differential) entropy. The more 'random', that is, unpredictable and unstructured the variable is, the larger its entropy, the covariance matrix of such a variable being very large (Section 2.6.1). A result from information theory tells that a Gaussian variable has the largest entropy among all random variables of equal variance. This signifies that the Gaussian distribution is the most random or the least structured of all distributions. In that case the entropy is small for distributions that are clearly concentrated on certain values. This is true if the variable is clearly has a pdf that is very spiky. Neg-entropy is defined as follows:

$$J(y) = H(y_g) - H(y) \tag{6.9}$$

In Equation 6.9, y_g is a Gaussian random variable of the same covariance matrix as y (see Equation 2.43 for the definition of the entropy, H). The neg-entropy in Equation 6.9 is always non-negative. Neg-entropy is invariant for invertible linear transformations. The merit of neg-entropy is that it is well justified by statistical theory. Neg-entropy is in some sense the optimal estimator of non-Gaussianity, however, it is computationally difficult. The method of approximating neg-entropy is to use higher-order moments [1]

$$J(y) \approx \frac{1}{12}\{E(y^3)\}^2 + \frac{1}{48}\{kurt(y)\}^2 \tag{6.10}$$

The variable y is assumed to be of zero mean and unit variance, and the approximation carries some limitation from the non-robustness with kurtosis. Another approximation is based on the maximum-entropy principle

$$J(y) \approx \sum_{i=1}^{n} a_i [E\{f_i(y)\} - E\{f_i(v)\}]^2 \tag{6.11}$$

In Equation 6.11: (i) a_i are some positive constants, (ii) v is a Gaussian variable with zero mean and unit variance, (iii) y is also a zero mean/unit variance process and (iv) the f_i are some non-quadratic functions. Even if this approximation is not very accurate, it can be used as a measure of non-Gaussianity since it is consistent in the sense that it is always non-negative and is also equal to zero if y is Gaussian. Also, one can use only one function, choosing it wisely, in Equation 6.11, as long as a good approximation to neg-entropy is obtained and some of such functions are given as follows:

$$f(y) = \frac{1}{\alpha} \log\{\cosh(\alpha y)\}; \quad f(y) = -e^{(-y^2/2)} \tag{6.12}$$

with $1 \le \alpha \le 2$ as a suitable constant. These approximations are conceptually simple, fast to compute and have appealing statistical properties.

6.2.4 Determination of ICs Based on Information Theory

We can use the concept of differential entropy (i.e. neg-entropy) to define the mutual information (MI) I between n (scalar) random variables, y_i, $i = 1$, 2, ..., n as follows:

$$I(y_1, y_2, \ldots, y_n) = \sum_{i=1}^{n} H(y_i) - H(y) \tag{6.13}$$

MI gives a measure of the dependence between random variables and it is always non-negative. MI is zero if and only if the variables are statistically independent and takes into account the whole dependence structure of the variables, and not only the covariance. For an invertible linear transformation $y = Wx$, we have the following expression:

$$I(y_1, y_2, \ldots, y_n) = \sum_{i=1}^{n} H(y_i) - H(x) - \log|det(W)| \tag{6.14}$$

If we constrain the y_i to be uncorrelated variables with unit variance, then we have the following expression:

$$E\{yy^T\} = WE\{xx^T\}W^T = I \qquad (6.15)$$

Equation 6.15 implies the following:

$$det(I) = 1 = det(WE\{xx^T\}W^T) = det(W)det(E\{xx^T\})det(W^T) \qquad (6.16)$$

Equation 6.16 implies that $det(W)$ must be constant. Also, for y_i with unit variance, entropy and neg-entropy differ only by a constant, and the sign, as seen from the following equation:

$$I(y_1, y_2, \ldots, y_n) = C - \sum_{i=1}^{n} J(y_i) \qquad (6.17)$$

Equation 6.17 shows the relation between neg-entropy and MI. Now, since MI is the natural measure of the independence of random variables, we can use MI as the criterion for determining the ICs transform. In this approach, we define the ICs of a random vector x as an invertible transformation, where the matrix W is determined so that the MI information of the transformed components s_i is minimised. We see from Equation 6.17 that determining an invertible transformation W (that minimises the MI) is approximately/ roughly equivalent to determining the directions in which the neg-entropy is maximised. Also, it is roughly equivalent to determining one-dimensional (1D) subspaces such that the projections in those subspaces have maximum neg-entropy. Equation 6.17 clearly shows that estimation of ICs by minimising MI is really equivalent to maximising the sum of non-Gaussianities of the estimates (when the estimates are constrained to be uncorrelated).

6.2.5 Maximum Likelihood Estimation

A well known and very popular method for estimating the ICs model is maximum likelihood (ML) estimation (Chapter 2). It is essentially equivalent to minimisation of MI. It is possible to formulate directly the LL in ICs model without noise. Then, ML method is used for estimation of IC model. If we have $W = (w_1, \ldots, w_n)^T$ $(= A^{-1})$, the LL is given as follows [1]:

$$L = \sum_{t=1}^{T} \sum_{i=1}^{n} \log\{p_i(w_i^T x(t))\} + T \log\{|det(W)|\} \qquad (6.18)$$

In Equation 6.18, p_i are the pdfs of the s_i (assumed known) and the $x(t)$, $t = 1, \ldots, T$ are the realisations of x. Also, for any random vector x (with pdf $p(x)$) and for any W, the pdf of $y = Wx$ is given by $p_x(Wx)|detW|$. We can establish the connection between LL and MI by taking the expectation of the LL, Equation 6.18 [1]

$$\frac{1}{T}E\{L\} = \sum_{i=1}^{n} E[\log\{p_i(w_i^T x(t))\}] + \log\{|det(W)|\} \qquad (6.19)$$

In Equation 6.19, if p_i were equal to the actual distributions of $(w_i^T x(t))$, the first term would be equal to $-\sum_{i=1}^{n} H(w_i^T x)$. Hence, LL would be equal (up to an additive constant) to the negative of MI. In applying the ML method to ICA, a reasonable approach would be to estimate the pdf of $(w_i^T x(t))$ (as part of the ML estimation method), and then use this as an approximation of the pdf of s_i.

6.2.6 Demonstration of FastICA Code: Illustrative Example

The results of a particular run of the GUI/MATLAB (7.x and 6.x) based (version 2.5) FastICA tool box (developed by © Hugo Gävert, Jarmo Hurri, Jaakko Särelä, and Aapo Hyvärinen; see Appendix D.20) to perform ICA of the mixer of signals are given here. Run the tool as per the instruction given in Appendix D.20. Figure 6.1 shows the original data, Figure 6.2 shows the whitened data and Figure 6.3 depicts the ICs. Further details on FastICA are given in Reference 1.

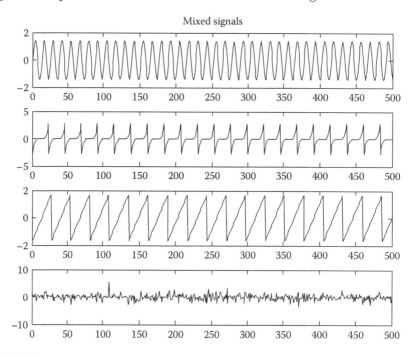

FIGURE 6.1
Mixed signals used for ICA analysis.

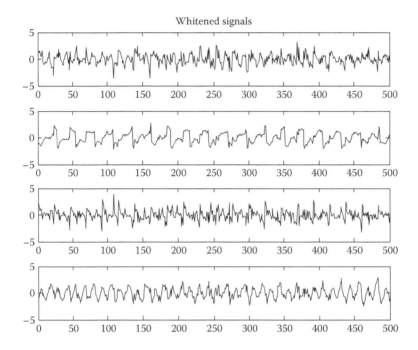

FIGURE 6.2
The whitened signals for ICA analysis.

6.3 An Approach to Image Fusion Using ICA Bases

Image fusion (as we will discuss in Chapter 7 in more detail) is a process of obtaining a single/one image from a set of input images with more complete information [2,3]. In fact, the process of combining the important features from the input images to form a single enhanced image can be called image fusion. Image fusion can be performed at three different levels: (i) pixel level, (ii) feature level and (iii) decision–decision level. The fused image is supposed to improve image content, so that it becomes easier for the user to detect, recognise and identify objects/targets. Image fusion has applications in computer vision, remote sensing, robotics and medical systems. Hence, image fusion is the task of enhancing the perception of a scene/object by combining information captured by different sensors. Often, the pyramid decomposition (PD) and the dual-tree wavelet transform (DTWT) are applied in image fusion as analysis and synthesis tools [3]. One uses different types of pixel-/region-based fusion rules to combine the important features of the input-candidate images in the transform domain (TD) to obtain an enhanced image – this is a fused image. We discuss procedures of image fusion using transform/s constructed using ICA and topographic ICA (TICA) bases. These bases are obtained by offline training with images that are of similar

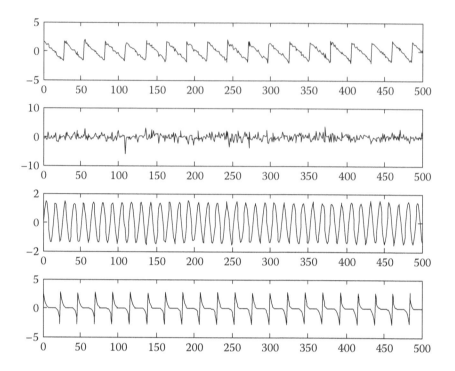

FIGURE 6.3
ICs obtained by the FastICA GUI/MATLAB tool box (see Appendix D.20).

context to the observed scene. Then, the images are fused in the TD using pixel-/region-based fusion rules.

6.3.1 Fusion Preliminaries

Let $I_1(x, y)$, $I_2(x, y)$, ..., $I_T(x, y)$ be the T images of size M_1 and M_2 that capture the same scene. We also assume that each image has been acquired using different modalities/sensors, thus each image has different characteristics: degradation and thermal and visual features [3]. The idea is to employ multiple sensors/image capturing devices that are placed relatively close to observe the same scene. Thus, the acquired images, although they are nearly similar, might have some translational motion, that is, mis-correspondence between several points of the scene. In such case, image registration process of establishing point-by-point correspondence between a number of images is utilised (Chapter 7). Here, we assume that the images are properly registered. The fusion methods can be categorised as spatial domain (SD) and TD techniques. In the SD approach, the images are fused in the SD, that is, we use the localised spatial features. Assuming that $f(.)$ represents the fusion rule, then the SD technique can be stated as follows as a general fusion rule:

$$I_f(x,y) = f\{I_1(x,y),\ldots,I_T(x,y)\} \tag{6.20}$$

The idea of using a TD is to capture the salient features of the input images very clearly and then fuse these features. Here, the choice of the transform is very crucial. We use $\Gamma(.)$ as transform operator and $f(.)$ as some fusion rule, then transform-domain fusion techniques can then be outlined, as follows [3]:

$$I_f(x,y) = \Gamma^{-1}\{f(\Gamma\{I_1(x,y)\},\ldots,\Gamma\{I_T(x,y)\})\} \tag{6.21}$$

6.3.2 Major Fusion Strategies

The fusion operator $f(.)$ describes the merging of information from various input images, and finally it gives out one fused image (Equations 6.20 and 6.21). Several fusion rules are available in the open literature and these fusion strategies are categorised as: (i) pixel-based and (ii) region-based [3,9].

6.3.2.1 Pixel-Based Fusion

The fusion is performed pixel-by-pixel, either in the TD or SD. Each pixel (x, y) of the input images is combined by using a fusion rule and a corresponding pixel (x, y) of the fused image is formed. We have the following basic TD schemes:

i. Mean rule: in fusion by averaging method, the corresponding coefficients in each image are averaged as follows:

$$\Gamma\{I_f(x,y)\} = \frac{1}{T}\sum_{i=1}^{T}\Gamma\{I_i(x,y)\} \tag{6.22}$$

ii. Max-abs rule: in fusion by absolute maximum method, we select the greatest in absolute value of the corresponding coefficients in each image as follows [3]:

$$\Gamma\{I_f(x,y)\} = sgn(\Gamma\{I_i(x,y)\})\max_i\left|\Gamma\{I_i(x,y)\}\right| \tag{6.23}$$

iii. In fusion by de-noising (hard/soft thresholding) simultaneously fusion and de-noising is performed by thresholding the TD coefficients.

iv. In high/low fusion process, we combine the high-frequency parts of some images with the low-frequency parts of some other images.

6.3.2.2 Region-Based Fusion

In these schemes image pixels are grouped to form contiguous regions, and the different fusion rules to each image region are applied.

More aspects of these major fusion strategies are discussed in Chapter 7.

6.3.3 ICA and TICA Bases

We assume an image $I(x, y)$ of size M_1 by M_2. Also, assume that a window W of size N by N is centred around a particular pixel (x_0, y_0). Next, define an image patch as the product between an N by N neighbourhood centred around the pixel (x_0, y_0) and the window W as follows [3]:

$$I_w(k,l) = W(k,l)I(x_0 - \lfloor N/2 \rfloor + k, y_0 - \lfloor N/2 \rfloor + l); \quad \forall k,l \in [0,N] \quad (6.24)$$

Alternatively, we can assume the window as follows:

$$W(k,l) = 1; \quad \forall k,l \in [0,N-1] \quad (6.25)$$

Thus, in Equation 6.25 it is assumed that N is odd, and hence $\lfloor \_.\_ \rfloor$ represents a lower integer part of the resultant number. In that case in Equation 6.26, we have $N - 1$, for we now assume a rectangular window.

6.3.3.1 Bases

It is a usual practice in image analysis to express it as the synthesis of several other basis images, like it is assumed that a periodic signal is composed of its Fourier components, as in the latter case the bases are orthogonal functions. In image analysis these bases are chosen according to the image properties and features that we want to highlight and extract for the analysis. There are several bases: cosine, complex cosine, Hadamard and WT, and these are well defined in order to serve some specific purpose of analysis. However, it is possible to have arbitrary bases. This is achieved by training with a population of similar content images, and these bases are estimated by optimising a cost function. This cost function is chosen such that the bases have desirable vproperties. The N by N image patch $I_w(k, l)$ is expressed as a linear combination of a set of K basis images $b_j(k, l)$ as follows:

$$I_w(k,l) = \sum_{j=1}^{K} u_j b_j(k,l) \quad (6.26)$$

In Equation 6.26, we have u as the scalar parameters that determine the combinations of the bases images. As is often done, a two-dimensional (2D) representation is simplified to 1D representation. This is done by lexico-graphic ordering to facilitate the image analysis. The image patch $I_w(k, l)$ is arranged into a vector I_w, by taking all elements from matrix I_w in a row-wise fashion and arranging these elements in successive columns, one below another → a first row is flipped from the horizontal position to a vertical column, and then the second row is made vertical and put directly under the first column, and so on. These image patches are expressed in lexicographic ordering as follows [3]:

$$\bar{I}_w(t) = \sum_{j=1}^{K} u_j(t)\bar{b}_j = [\bar{b}_1\ \bar{b}_2 \dots \bar{b}_K] \begin{bmatrix} u_1(t) \\ u_2(t) \\ \cdot \\ \cdot \\ \cdot \\ u_K(t) \end{bmatrix} \tag{6.27}$$

In Equation 6.27, we have tth image patch from the original image. We also have $B = [\bar{b}_1\ \bar{b}_2 \dots \bar{b}_K]$ and $\bar{u}(t) = [u_1(t)u_2(t)\dots u_K(t)]^T$. Then, from Equation 6.29 we obtain the following compact equations:

$$\bar{I}_w(t) = B\bar{u}(t); \quad \bar{u}(t) = B^{-1}\bar{I}_w(t) = A\bar{I}_w(t) \tag{6.28}$$

In Equation 6.28, A represents the analysis kernel and B the synthesis ker-nel. We also see that in the transformations of Equation 6.28, the observed image signal (I_w) is projected onto a set of bases vectors, b, wherefore our aim is to estimate a finite set of basis vectors. These vectors should be able to cap-ture most of the signal's structure/energy. We need N^2 bases for a complete representation of the N^2-D signals of image $I_w(t)$. In certain situation one can use only $K < N^2$ bases. Determination of these K vectors can be performed using a bundle of training image patches $I_w(t)$ and optimisation of a cost function.

6.3.3.2 ICA Bases

ICA can identify statistically independent basis vectors in a linear generative model. Several approaches are available to analyse the generative model of Equation 6.29. In this case, we assume statistical independence between the coefficients u_i in the TD. The principles that we studied in the ICA, also hold for any linear combination I_w of these independent random variables u_j. Some methods estimate u_j by minimising the Kullback–Leibler (KL) divergence between the estimated coefficients u_i and probabilistic priors on the same coefficients [3]. Some other methods minimise the MI conveyed by the esti-mated coefficients. Yet other methods perform approximate diagonalisation of a cumulant tensor of I_w. Some more methods, as we have seen earlier, estimate u_i by estimating the directions of the most NG components using kurtosis or neg-entropy. In the approach used for this purpose, PCA is used as a pre-processing step to select the K most important vectors and the data are orthonormalised. Then, ICs are identified using orthogonal projections $a^T_i z$, here, a are the elements of the matrix A and z are the orthogonal components. Then, the projecting vectors a_i are estimated by minimising the following non-quadratic approximation of neg-entropy [3]:

$$J(\bar{a}_i) = [E\{f(\bar{a}^T_i, \bar{z})\} - E\{f(v)\}]^2 \tag{6.29}$$

Compare Equation 6.29 with Equation 6.11, and the variables have similar meanings as explained below. Equation 6.11, and f is the same function as given in Equation 6.12. One estimator/update rule is given as follows [3]:

$$\bar{a}_i = E\{\bar{a}_i\phi(\bar{a}^T_i \bar{z})\} - E\{\phi'(\bar{a}^T_i \bar{z})\}\bar{a}_i; \quad 1 \le i \le K \tag{6.30}$$

$$A = A(A^T A)^{-0.5} \tag{6.31}$$

In Equation 6.30, we have $\phi(.) = -\partial f(.)/\partial(.)$. The update rule in Equation 6.30 is randomly initialised and iterations are performed until the convergence is obtained.

6.3.3.3 Topographic ICA: TICA Bases

In many practical situations, the assumption of independence does not remain valid. Then, this dependency/structure being very informative could be utilised somehow to define a topographic order between the (non-) ICs [3]. One can therefore modify the original ICA model to include a topographic order between the components. The components that are near to each other in the topographic representation are relatively strongly dependent in the sense of higher-order correlations/MI. The new model is known as the topographic ICA model (TICA). This topography feature is introduced using a neighbourhood function $h(i, k)$ that expresses the proximity between the ith and the kth component. The neighbourhood model is given as

$$h(i,k) = \begin{cases} 1, & \text{if } |i - k| \le L \\ 0, & \text{otherwise} \end{cases} \tag{6.32}$$

In Equation 6.32, L defines the width of the neighbourhood. As a result the estimated coefficients u_i are no longer independent/ICs, but can be modelled by some generative random variables d_k, g_i that are controlled by the neighbourhood function and are shaped by a non-linear function $\phi(.)$. The T-source model is given as [3]

$$u_i = \phi\left(\sum_{k=1}^{K} h(i,k)d_k)g_i\right) \tag{6.33}$$

We assume a fixed-width neighbourhood L by L and the input data preprocessed by PCA. Then, ML estimation of the synthesis kernel B is performed using the linear model in Equation 6.30 and the T-source model of Equation 6.33. Several assumptions for the generative random variables d_k and f_i are made. Then, a gradient-based TICA update rule is obtained by optimising an approximation of the derived log-likelihood as follows [3]:

$$\bar{a}_i = \bar{a}_i + \eta E\{\bar{z}(\bar{a}_i^T \bar{z})\bar{r}_i\}; \ 1 \leq i \leq K \tag{6.34}$$

$$A = A(A^T A)^{-0.5}$$

In Equation 6.34, η is the learning rate parameter and r_i are given by the following expression:

$$r_i = \sum_{k=1}^{K} h(i,k)\phi\left(\sum_{j=1}^{K} h(j,k)(\bar{a}_i^T z)^2\right) \tag{6.35}$$

6.3.4 Training and Properties of ICA Bases

In this section, we discuss properties and training aspects of ICA bases.

6.3.4.1 Training

The training procedure needs to be completed only once, after the successful training, the estimated transform is used for fusion of similar content images. The procedure is as follows [3]: (i) select a set of images with similar content to the ones that are intended to be used for image fusion, (ii) a number of N by N patches (usually ~10,000) are randomly selected from the training set images, (iii) apply lexicographic ordering to the selected images patches, (iv) perform PCA on the selected patches and select the $K < N^2$ most important bases, according to the eigenvalues corresponding to the bases, (v) iterate the ICA update rule of Equation 6.31 or the TICA rule of Equation 6.35 for a chosen L by L neighbourhood until convergence.

6.3.4.2 Properties

Both the transforms ICA and TICA are invertible and hence they guarantee perfect reconstruction. Using the symmetric orthogonalisation step of Equation 6.32, we assure that the bases remain orthogonal. The ICA/TICA basis vectors seem to be related to WTs and Gabor functions, as they represent similar features in different scales. However, these bases have more DOF than WTs; the discrete WT (DWT) has only two orientations and the DTWT can give six distinct sub-bands at each level with orientation, whereas the ICA bases can get arbitrary orientations to fit the training patches. One drawback of these transforms is that they are not shift invariant, however, if we assume that the observed images are all registered then this is not a major limitation. TICA provide an ordered representation of the data, compared to the unordered representation of the ICA bases. In an image fusion, topography can identify groups of features that can characterise certain objects in the image. The TCA can offer a more comprehensive representation compared to the general ICA model.

6.3.5 Image Fusion Using ICA Bases

In this section, we discuss the process of image fusion using ICA/TICA bases. It is assumed that ICA/TICA transform $\Gamma\{.\}$ is already available. Also, the registered sensor input images $I_k(x, y)$ are available for the fusion. From each input image every possible N by N patch is isolated and these images are lexicographically ordered yielding it a vector $I_k(t)$. The size N of the patches should be the same as that used in the transform estimation. These representations $I_k(t)$ are transformed to the ICA/TICA domain as $u_k(t)$ (once we have A as the estimated analysis kernel) as follows:

$$\bar{u}_k(t) = \Gamma\{\bar{I}_k(t)\} = A\bar{I}_k(t) \tag{6.36}$$

Then, one can apply a hard threshold on the coefficients and perform optional de-noising (sparse code shrinkage). One can perform image fusion in the ICA/TICA domain in the same way as is done in the WT/DTWT domain, that is, the corresponding coefficients $u_k(t)$ from each image are combined in the ICA domain to construct a new image $u_f(t)$. The method $f(.)$ that combines the coefficients in the ICA domain is termed fusion rule

$$\bar{u}_f(t) = f(\bar{u}_1(t), \ldots, \bar{u}_k(t), \ldots, \bar{u}_T(t)) \tag{6.37}$$

Once the $u_f(t)$ is constructed in the ICA domain, we go back to the SD, using the synthesis kernel B, that is, we synthesise the image $I_f(x, y)$ by averaging the (overlapping) image patches $I_f(t)$ in the same order as they were selected during the analysis step.

6.3.6 Pixel- and Region-Based Fusion Rules Using ICA Bases

We describe two rules [3]: (i) an extension of the max-abs pixel-based rule referred as the weight combination (WC) rule and (ii) a combination of the WC and the mean rule in a region-based scenario.

6.3.6.1 WC Pixel-Based Method

This rule is given by the following expression:

$$\Gamma\{\bar{I}_f(t)\} = \sum_{k=1}^{T} w_k(t)\Gamma\{\bar{I}_k(t)\} \tag{6.38}$$

One can use the so-called activity measures in the estimation of the contribution $w_k(t)$ of each image to the fused one. When we process each image in N by N patches, we can use the mean absolute value of each patch (arranged as a vector) in the TD as an activity indicator in each patch

$$E_k(t) = \|\bar{u}_k(t)\|_1; \quad k = 1, 2, \ldots, T \tag{6.39}$$

These weights in Equation 6.38 should emphasise sources that feature more intense activity, as represented by $E_k(t)$. Then, the weights for each patch can be determined by the contribution of the kth source image $u_k(t)$ over the entire contribution of all the T-source images at patch t, in terms of activity, and the weights are chosen as

$$w_k(t) = E_k(t) \bigg/ \sum_{k=1}^{T} E_k(t) \tag{6.40}$$

In case the denominator in Equation 6.40 is very small, then one can use the max-abs or mean fusion rule for those patches.

6.3.6.2 Region-Based Image Fusion Using ICA Bases

As we have seen earlier, we have already divided the image in small N by N patches/regions. Then, using the splitting/merging concept of region-based segmentation, one can find a criterion to merge the pixels corresponding to each patch. This is done in order to form contiguous areas of interest. One can use the energy activity measurement as discussed above to infer the existence of edges in the corresponding frame. Since, the ICA tends to focus on the edge information, it is clear that large values for $E_k(t)$, correspond to large activity in the frame (implying that edges exist), whereas the small values for $E_k(t)$ signify the existence of almost constant background in the frame. With this notion, one can segment the image in: (i) active regions containing details and (ii) non-active regions containing background information. The threshold equal to $2\text{mean}_t\{E_k(t)\}$ is used to characterise a region as active or non-active. Then, a following segmentation map is formed from each input image [3]:

$$m_k(t) = \begin{bmatrix} 1, & \text{if } E_k(t) > 2\text{mean}_t\{E_k(t)\} \\ 0, & \text{otherwise} \end{bmatrix} \tag{6.41}$$

These segmentation maps of input images are combined to form a single segmentation map, using the logical OR operator as follows:

$$m_k(t) = \text{OR}\{m_1(t), m_2(t), \dots, m_T(t)\} \tag{6.42}$$

Once, the images are segmented into active and non-active regions, one can fuse these regions using different pixel-based fusion schemes: (i) for the active region, one can use a fusion scheme that preserves the edges (the max-abs scheme or the weighted combination scheme) and (ii) for the non-active region, one can use a scheme that preserves the background information (the mean or median scheme).

6.3.6.3 Performance Evaluation Metrics

For the performance evaluation of the image-fusion schemes based on ICA bases, one can use the absolute maximum (max-abs), and the averaging (mean), the weighted combination (weighted) and the region-based schemes/rules of fusion as applicable. To evaluate the scheme's performance, one can use the signal-to-noise ratio (SNR) expression to compare the truth image with the fused image as given below [3]

$$SNR = 10\log_{10} \frac{\sum_x \sum_y I_t(x,y)^2}{\sum_x \sum_y [I_t(x,y) - I_f(x,y)]^2} \tag{6.43}$$

It is obvious from Equation 6.43 that if the quality of fusion is very good, then the SNR as specified will be very high, since, then the error between the true image and the fused image would be very low, and hence, the denominator will be very low. The SNR actually should be called an IEIR (power/energy/variance of) image to (power/energy/variance of) error-in-image ratio, since that is actually what Equation 6.43 represents, and the SNR name for this equation, though well entrenched in the literature on image fusion, is actually a misnomer. This metric is applicable if the ground truth image available for comparison. Further details on image fusion and performance metrics can be found in Reference 3.

6.4 Principal Component Analysis

The PCA is a simple and non-parametric method to extract some important features from the data signal. It is also an approach to reduce the complex data into simplified data structures that were hidden in the original data [4]. The advantage of PCA is that once we have found these patterns/hidden structures/features from the original data, the resultant data are in the compressed form with reduced dimensions [5]. This happens without much loss of information that was contained in the original data. This approach is used in image compression and image fusion. It involves an eigen analysis (eigenvalue/eigenvector analysis) of the correlation or covariance matrix of the given data set and the procedure can also be called an eigen system (ES) analysis. The main merit of PCA is that the data dimensionality (of the original data) is reduced by performing an analysis of covariance. It is a simple fact that if we have a set of data signals and if we obtain the covariance matrix of/from these signals, then the dimension of the original signals is drastically reduced, with the fact that the underlying information is now captured in the form of the covariance matrix, which when further transformed using the eigen analysis can reduce the dimension the data. Thus, PCA can be used to uncover (hidden/unknown) trends in the data (could be image intensities/features/pdfs, etc.) and will reveal the relevant components and patterns in an image scene, etc. Once the pattern is unearthed the data can be compressed by removing unimportant aspects (there could be more sub-patterns that might not be very important) and this method is used in image compression and image fusion [1,4,5]. In case of PCA, we have a situation in which at least some of the variables in a given data set might be correlated with each other. This signifies that some redundancy in the information provided by these variables is present in the data. The PCA exploits this underlying redundancy in such multi-variate data sets and then picks out patterns/relationships in the variables, in the compact form of ES components and hence reduces the dimensionality of the data set.

The important steps involved in performing the PCA analysis are [1,4,5]: (i) begin with the test data and remove the mean (of the data) from the test data, (ii) calculate/compute the covariance matrix of the data set, using MATLAB tool, (iii) compute the eigenvalues and eigenvectors of the covariance matrix and (iv) choose the important components, the principal ones, and form a feature/attribute vector/s. We then get the data reductions. Thus, the PCA is a numerical algorithm to transform a number of correlated variables into a number of uncorrelated variables called principal components (PCs), such that [9]: (i) the first principal component (PC) accounts for much of the variance/variation in the data, each succeeding component (not PCs) accounts for much of the remaining variance/variation, and the first PC along the direction with the maximum variance; (ii) the second component is constrained to be in the sub-space perpendicular to the first component, within the subspace this component points to the direction of maximum variance and (iii) the third component is taken in the maximum variance direction in the sub-space perpendicular to the first two, and so on. The PC-basis vectors depend on the given test data set.

Let X be a n-dimensional random/or some data vector with zero mean, and let the orthonormal projection matrix V be such that $Y = V^T X$, the data vector X is transformed to Y, by operation of V onto X. With the property of V, the covariance of Y, $cov(Y)$, is a diagonal matrix. Thus, using simple matrix/vector algebra, we obtain the following set [9]:

$$
\begin{aligned}
cov(Y) = E\{YY^T\} &= E\{(V^T X)(V^T X)^T\} \\
&= E\{(V^T X)(X^T V)\} = V^T E\{XX^T\}V \\
&= V^T cov(X)V
\end{aligned}
\tag{6.44}
$$

Since, the matrix V is known, the expectation operation can be taken inside and applied to the variables X in Equation 6.44. Then, by pre-multiplying both the sides of Equation 6.44 with V we obtain

$$
\begin{aligned}
V cov(Y) &= VV^T cov(X)V \\
&= cov(X)V
\end{aligned}
\tag{6.45}
$$

Here, VV^T is an identity matrix. Now, we can write V as $V = [V_1, V_2, \ldots, V_n]$ and $cov(Y)$ in the diagonal form as

$$
\begin{bmatrix}
\lambda_1 & 0 & \cdots & 0 & 0 \\
0 & \lambda_2 & \cdots & 0 & 0 \\
\vdots & \vdots & \ddots & \vdots & \vdots \\
0 & 0 & \cdots & \lambda_{n-1} & 0 \\
0 & 0 & \cdots & 0 & \lambda_n
\end{bmatrix}
\tag{6.46}
$$

Then, by substituting Equation 6.46 into Equation 6.45, and using the component form of V we obtain

$$
[\lambda_1 V_1, \lambda_2 V_2, \ldots, \lambda_n V_n] = [cov(X)V_1, cov(X)V_2, \ldots, cov(X)V_n]
\tag{6.47}
$$

Equation 6.47 can be written as follows:

$$
\lambda_i V_i = cov(X)V_i
\tag{6.48}
$$

Here λ_i are the eigenvalues and V_i are the eigenvectors (of $cov(X)$) for $i = 1$, $2, \ldots, n$. Thus, we see that the PCA in fact obtains the ES analysis of the given original data set. The eigenvalues have further deeper and very important interpretations as given in References 10, 11. Such interpretations can be further exploited for the scale invariant feature transform (SIFT) in the domain of image analysis/fusion (IAF) [12]. In such a process of IAF, the SIFs of a desired object (of investigation) are extracted and tracked in real time by extended KF (EKF) in a stream of video scenes.

6.4.1 Image Fusion Using PCA Coefficients

First, the source images are arranged in two column vectors. Then the data are organised into column vectors yielding a matrix Z of $n \times 2$ dimension. Then empirical mean for each column (the mean vector M has dimension of 2×1) is computed and subtracted from each column of the data matrix Z, resulting into X of dimension $n \times 2$. The covariance matrix C is computed as $C = X^T X$. The eigenvectors V and eigenvalues (D) are computed and vector V is sorted in the decreasing order, with V and D being of dimension 2×2. Then we consider the first column of V that corresponds to the largest eigenvalue to calculate the PCs pc_1 and pc_2 as

$$pc_1 = \frac{V(1)}{\Sigma V} \quad \text{and} \quad pc_2 = \frac{V(2)}{\Sigma V} \tag{6.49}$$

We see that $pc_1 + pc_2 = 1$. Then, the following fusion rule/formulae is formulated and fused image is given by

$$I_f = pc_1 I_{mage1} + pc_2 I_{mage2} \tag{6.50}$$

This simply means that the PCs are used as weights for image fusion, and thus we have very simple weighted-average fusion rule.

6.4.2 Image Fusion of Blurred Aircraft Images Using PCA Coefficients: Illustrative Example

In this section, we consider the pairs of images of a light transport aircraft (LTA). We then use their blurred images (one in each pair) for consideration for image fusion using PCA coefficients. The results of the MATLAB code of image fusion using PCA (*PCAimfuse_dem.m*, Appendix D.13) are shown in Figures 6.4 and 6.5. In Figure 6.4, we have an image pair 1 (saras51.jpg) of two LTA, of which the top one is a blurred image. Also, in Figure 6.4, we have image pair 2 (saras52.jpg) of the same aircraft of which the bottom one is a blurred image. Then, the PCA program is run and the fused image pair is shown in Figure 6.5. We see that the fused image pair is quite good and the blurring has almost disappeared. The PCA has extracted the most important and significant information/features from the original image pairs (image-data matrices) and the fusion is quite accurate.

Aircraft image – top one blurred   Aircraft image – bottom one blurred

Image pair 1 – fusion candidate   Image pair 2 – fusion candidate

FIGURE 6.4
Original sets of two image pairs to be fused using PCA.

6.5 Discrete-Cosine Transform

Another approach for image analysis/image fusion is based on DCT, and is a well-used method for image fusion. The large DCT coefficients are concentrated in the low-frequency region having good energy compaction properties. The DCT $X(k)$ of a 1D signal $x(n)$ of length N is represented as follows [8]:

$$X(k) = \alpha(k) \sum_{n=0}^{N-1} x(n) \cos\left(\frac{\pi(2n+1)k}{2N}\right), \quad 0 \le k \le N-1 \tag{6.51}$$

with

$$\alpha(k) = \begin{cases} \sqrt{\dfrac{1}{N}} & k = 0 \\ \sqrt{\dfrac{2}{N}} & k \ne 0 \end{cases} \tag{6.52}$$

Here n is sample index and k the frequency index (normalised). For $k = 0$, from Equation 6.51 we have $X(0) = \sqrt{(1/N)} \sum_{n=0}^{N-1} x(n)$, and it is the mean of the image signal $x(n)$ and is known as DC (direct coefficient). The other coefficients ($X(k)$, $k \ne 0$) are known as AC (alternative coefficients). The inverse DCT is obtained as

$$x(n) = \sum_{k=0}^{N-1} \alpha(k) X(k) \cos\left(\frac{\pi(2n+1)k}{2N}\right), \quad 0 \le n \le N-1 \tag{6.53}$$

Equation 6.51 is an analysis formula or the forward DCT transform and Equation 6.53 is the synthesis formula or inverse transform (IDCT). The orthogonal basis sequence $\cos((\pi(2n+1)k)/2N)$ is real and displays discrete time (co-) sinusoids. The 2D DCT $X(k_1, k_2)$ of an image signal $x(n_1, n_2)$ of size $N_1 \times N_2$ is obtained by [7]

$$
X(k_1, k_2) = \alpha(k_1)\alpha(k_2) \sum_{n_1=0}^{N_1-1} \sum_{n_2=0}^{N_2-1} x(n_1, n_2) \cos\left(\frac{\pi(2n_1+1)k_1}{2N_1}\right) \times
$$
$$
\cos\left(\frac{\pi(2n_2+1)k_2}{2N_2}\right), \quad \begin{array}{l} 0 \leq k_1 \leq N_1 - 1 \\ 0 \leq k_2 \leq N_2 - 1 \end{array} \tag{6.54}
$$

and the 2D IDCT is given by

$$
x(n_1, n_2) = \sum_{k_1=0}^{N_1-1} \sum_{k_2=0}^{N_2-1} \alpha(k_1)\alpha(k_2) X(k_1, k_2) \cos\left(\frac{\pi(2n_1+1)k_1}{2N_1}\right) \times
$$
$$
\cos\left(\frac{\pi(2n_2+1)k_2}{2N_2}\right), \quad \begin{array}{l} 0 \leq n_1 \leq N_1 - 1 \\ 0 \leq n_2 \leq N_2 - 1 \end{array} \tag{6.55}
$$

Here, $\alpha(k_1)$ and $\alpha(k_2)$ are defined as in Equation 6.52.

6.5.1 Multi-Resolution DCT

The multi-resolution (MR) DCT (MDCT) is similar to the wavelet transforms (WTs). The image is transformed into frequency domain by applying DCT column wise [8,9]: (i) perform IDCT on first 50% of points, 0 to 0.5π to obtain the low passed image L; (ii) perform IDCT on second 50% of points, 0.5π to π to obtain the high passed image H; (iii) the sub-image L is transformed into frequency domain by applying DCT row wise; (iv) perform IDCT on first 50% of points to obtain low passed image LL; (v) and perform IDCT on the remaining 50% to get the high passed image LH; (vi) the sub-image H is transformed into frequency domain by applying DCT row wise; (vii) perform IDCT on first 50% of points to obtain low passed image HL; and (viii) perform IDCT on the remaining 50% to obtain the high passed image HH. The sub-image LL contains the average image information corresponding to low-frequency band of multi-scale decomposition and is considered as smoothed and sub-sampled version of the source image. It is an approximation of the source image. The sub-images LH, HL and HH are detailed sub-images which contain directional (horizontal, vertical and diagonal) information of the source image due to the spatial orientation. The MR could be achieved by recursively applying the same algorithm to low pass (LP) coefficients (LL) from the previous decomposition. The MATLAB code of

Blurring is almost gone

Fused image of aircraft by PCA

FIGURE 6.5
Fused pair of the original candidate-pairs using PCA.

MDCT used for image decomposition (one level) is given in Appendix D.16. The image can be reconstructed by reversing the previously described procedure. The MATLAB code of IMDCT used for image reconstruction is given in Appendix D.17.

6.5.2 Multi-Sensor Image Fusion

The MDCT-based pixel level image fusion can be performed as follows [8,9]: (i) the registered source images I_{mage1} and I_{mage2} are decomposed into D ($d = 1, 2, \ldots, D$) levels using MDCT, (ii) the resultant decomposed images are $I_{mage1} \rightarrow \{^1LL_D, \{^1LH_d, {}^1HH_d, {}^1HL_d\}_{d=1,2,\ldots,D}\}$ and are $I_{mage2} \rightarrow \{^2LL_D, \{^2LH_d, {}^2HH_d, {}^2HL_d\}_{d=1,2,\ldots,D}\}$ and (iii) at each decomposition level($d = 1, 2, \ldots, D$), the fusion rule selects the larger absolute value of the two MDCT detailed coefficients, because these coefficients correspond to sharper brightness changes in the images such as edges and object boundaries. At the most coarse level ($d = D$), the fusion rule uses the average of the MDCT approximation coefficients since these at the coarser level are the smoothed and sub-sampled version of the original image. The image fusion rules are given as follows [7]:

$$
{}^f LH_d = \begin{cases} {}^1 LH_d & \left|{}^1 LH_d\right| \geq \left|{}^2 LH_d\right| \\ {}^2 LH_d & \left|{}^1 LH_d\right| < \left|{}^2 LH_d\right| \end{cases} \tag{6.56}
$$

$$
{}^f HH_d = \begin{cases} {}^1 HH_d & \left|{}^1 HH_d\right| \geq \left|{}^2 HH_d\right| \\ {}^2 HH_d & \left|{}^1 HH_d\right| < \left|{}^2 HH_d\right| \end{cases} \tag{6.57}
$$

$$
{}^f HL_d = \begin{cases} {}^1 HL_d & \left|{}^1 HL_d\right| \geq \left|{}^2 HL_d\right| \\ {}^2 HL_d & \left|{}^1 HL_d\right| < \left|{}^2 HL_d\right| \end{cases} \tag{6.58}
$$

$$
{}^f LL_D = 0.5({}^1 LL_D + {}^2 LL_D) \tag{6.59}
$$

The fused image I_{magef} can be obtained using IMDCT

$$
I_{magef} \leftarrow \{{}^f LL_D, \{{}^f LH_d, {}^f HH_d, {}^f HL_d\}_{d=1,2,\ldots,D}\} \tag{6.60}
$$

The MATLAB code for the image fusion is given in Appendix D.18.

6.6 WT: A Brief Theory

Most of the time-domain signals arising as responses from any dynamic system or as standalone signals such as time series signals contain information hidden in the frequency content of the signal. The frequency content of a signal is then fundamentally and basically the frequency components (the spectral components) of that signal. This is traditionally revealed by Fourier transform (FT) of the time-domain signal. Thus, the FT is a mathematical transformation applied to a signal (a REUD signal) to obtain further information from that signal that is not readily seen in the original signal [7,9]. The WT that was developed as an extension or an alternative to the FT provides the time–frequency representation. In FT data, the time information is not revealed or not provided, whereas the WT data provide the time and frequency information simultaneously. A time-domain original signal can be from various high pass (HP) and LP filters, thereby filtering out either high-frequency or low-frequency portions of the signal. This procedure is repeated every time a portion of the signal corresponding to some frequencies is removed from the signal, when we repeat the process of LP/HP we take the sub-set of the original signal. This process would result into to a

bunch of sub-signals which actually represent the same signal but correspond to different frequency bands. The, we can get a three-dimensional
(3D) plot with time on one axis, frequency on the second and amplitude on
the third axis. This 3D plot will reveal to us what frequencies exist at which
time, or more accurately what frequency bands exist at what time intervals.
If we do finer division we get better resolution in the frequency-transformed
signal. The so-called STFT (short-term Fourier transform) gives a fixed resolution at all times.

The WT is devised to give a variable resolution: (i) the higher frequencies
are better resolved in time and (ii) the lower frequencies are better resolved
in frequency. This would mean that a certain high-frequency component can
be located better in time with less relative error than a low-frequency component, whereas a low-frequency component can be located better in frequency
compared to high-frequency component [9]. In the WT analysis, the signal is
multiplied with a function, the wavelet (similar to the window function in
the STFT) and the transform is computed separately for different segments
of the time-domain signal. The WT-transformed signal itself is a function
of two variables: (i) the translation parameter and (ii) the scale parameter
's'. The transforming function is called the mother wavelet meaning a small
wave and it signifies that this function is of a finite length. The term 'mother'
signifies the functions with different regions of support that are used in the
WT process are derived from one main function. The term translation signifies the location of the window, and the window is shifted through the
signal. It corresponds to time information in the TD. The scale parameter is
defined as 1/frequency. The WTs (i.e. the components of the decomposition
of the original signal) are extensively used in image processing, since these
provide a MR decomposition of an image in a bi-orthogonal basis, and these
bases are wavelets. By the WT analysis a signal is decomposed into scaled
(dilated/expanded) and shifted (translated) versions of the chosen mother
wavelet or function $\psi(t)$. Then the wavelet is a small wave that grows and
decays in a limited time period, and should satisfy the properties [9]:

i. Time-integral property

$$\int_{-\infty}^{\infty} \psi(t)\,dt = 0 \tag{6.61}$$

ii. Square of wavelet integrated over time property

$$\int_{-\infty}^{\infty} \psi^2(t)\,dt = 1 \tag{6.62}$$

The WT of 1D signal $f(x)$ onto a basis of wavelet functions is obtained as

$$W_{a,b}(f(x)) = \int_{x=-\infty}^{\infty} f(x)\,\psi_{a,b}(x)dx \tag{6.63}$$

Then by the translation and dilation operations of the mother wavelet we have

$$\psi_{a,b}(x) = \frac{1}{\sqrt{a}}\,\psi\left(\frac{x-b}{a}\right) \tag{6.64}$$

For a DWT, the dilation factor is $a = 2^m$, the translation factor is $b = n2^m$ with m and n as integers.

6.6.1 Image Analysis for Image Fusion by WT

The WT separately filters and down-samples the 2D image in the vertical and the horizontal directions. For example, the input image $I_{image}(x,y)$ is filtered by LP filter L and HP filter H in horizontal direction [9]. Then the image signal is down-sampled by a factor of two (while keeping the alternative samples) to create the coefficient matrices $I_{imageL}(x,y)$ and $I_{imageH}(x,y)$. These coefficient matrices are further LP and HP filtered in vertical direction and down-sampled by a factor of two to create sub-bands (sub-images): $I_{LL}(x,y)$, $I_{LH}(x,y)$, $I_{HL}(x,y)$ and $I_{HH}(x,y)$. The $I_{LL}(x,y)$ sub-band has the average image information related to the low-frequency band of multi-scale decomposition and is considered the smoothed and sub-sampled version of the source image $I_{image}(x,y)$. It represents the approximation of source image $I_{image}(x,y)$. The $I_{LH}(x,y)$, $I_{HL}(x,y)$ and $I_{HH}(x,y)$ are detailed sub-images containing the directional (horizontal, vertical and diagonal) information of the source image $I_{image}(x,y)$ due to the spatial orientation. Inverse 2D WT (IWT) process to restore the image $I_{image}(x,y)$ from sub-images $I_{LL}(x,y)$, $I_{LH}(x,y)$, $I_{HL}(x,y)$ and $I_{HH}(x,y)$ is carried out by column-up sampling and filtering using LP/HP filters for each of the sub-images [4]. Row-up sampling and filtering with LP/HP filters of the resulting images and summation of all matrices is used for restoring the image $I(x, y)$. The FIR (finite impulse response) filter coefficients $(H(n)/L(n))$ of LP/HP filters for forward and reverse processes of decomposition/analysis and image restoration/synthesis should satisfy the conditions as follows [9]:

$$\sum_{n=1}^{m} H(n) = \sum_{n=1}^{m} \tilde{H}(n) = 0$$
$$\sum_{n=1}^{m} L(n) = \sum_{n=1}^{m} \tilde{L}(n) = \sqrt{2} \tag{6.65}$$

$$\tilde{H}(n) = (-1)^{n+1}L(n)$$

$$\tilde{L}(n) = (-1)^{n}H(n)$$

$$H(n) = (-1)^{n}L(m - n + 1)$$

We have m as the number of coefficients in the filter and n as the index of the filter coefficient. L and H are the vectors of numerator coefficients of LP/HP filters, respectively, used in image decomposition and $\tilde{L}$ and $\tilde{H}$ are those for the image reconstruction. For the image fusion using the WT, the source images $I_{mage1}(x, y)$ and $I_{mage2}(x, y)$ are decomposed into approximation and detailed coefficients (of the WT) at required level using DWT. Then the coefficients of both the images are combined using an appropriate fusion rule. Finally, the fused image is obtained by the IDWT as follows:

$$I_{magef}(x, y) = IDWT[f\{DWT(I_{mage1}(x, y)), DWT(I_{mage2}(x, y))\}] \quad (6.66)$$

In Equation 6.66, f represents the fusion rule. A simple fusion rule is (i) to average the approximation coefficients and pick the detailed coefficient in each sub-band with the largest magnitude and then (ii) to perform the IDWT operation on the resultant-fused coefficients to obtain the fused image.

6.6.2 Image Fusion of Blurred Aircraft Images Using WT Coefficients: Illustrative Example

In this section, we consider again the same pairs of images (as considered in Section 6.2.1) of the LTA. We then use their blurred images (one in each pair) for consideration for image fusion using WT. The results of the MATLAB program of image fusion using WT (*vpnwtfuseL1demo*, Appendix D.14) **vpnwtfuseL1demo** changed to *vpnwtfuseL1demo* are shown in Figures 6.6 and 6.9. In Figure 6.6, we have image pair 1 (saras91.jpg) of two LTA with the top one highly blurred. Also, in Figure 6.7 we have image pair 2 (saras92. jpg) of the same aircraft with the bottom one highly blurred. Then the WT program is run and resultant fused image pair is shown in Figure 6.8. We see that the fused image pair is quite good and that the blurring has almost disappeared. The WT has extracted the most important and significant information/features from the original image pairs (image-data matrices) and the fusion is quite accurate. Figure 6.9 shows the error-image pair, the

FIGURE 6.6
Image pair 1 as a candidate for fusion using WT.

FIGURE 6.7
Image pair 2 as a candidate for fusion using WT.

FIGURE 6.8
Fused image pair using WT analysis.

FIGURE 6.9
Error-image pair – WT.

difference between the true image pair, saras9t.jpg, and the fused image pair (from Figure 6.8).

6.7 An Approach to Image Fusion Using ICA and Wavelets

The idea of image fusion using ICA (and WT) is to merge the ICA decompositions of the images that are derived from WT [2]. The WT is very popular MR analysis procedure, especially applicable to image fusion. First, 2D DWT can be used in order to extract multiple sub-band images. These sub-bands contain coarse approximations of the input image/s as well as (i) horizontal, (ii) vertical and (iii) diagonal details of images at various scales. Subsequently, one extracts ICs features from these sub-bands. As we have seen in previous sections, we have an alternative method for (data/image) source separation and it is called ICA. ICA is a tool based on higher-order statistics: it de-correlates the input signals, as well as reduces higher-order statistical dependencies. The ICA method can be applied for detecting edges in natural images containing scenes. As such, ICA is can be regarded as a statistical computational model that uses linear transformations on multi-dimensional data to interpret the spectral signatures of signals. It can also be used to decipher hidden factors underlying signals/measurements/time series data. As we have seen in the ICA model, the data variables are assumed to be linear or non-linear mixtures of some unknown latent variables, and the mixing system/mixer model is also unknown. A simple procedure of image fusion using WT and ICA is described as follows [2]:

   i. Decompose each original input image (the candidate images for the fusion) and we obtain four resolution sub-bands images as: A, H, V and D. 2D DWT is used in order to decompose images of every channel into some sub-images at different frequency range. These are the groups of approximation (A), horizontal (H), vertical (V) and diagonal (D) details.

   ii. ICA method is used these image sub-spaces to extract ICs. The ICA is carried out in the sub-images individually to extract some ICs.

   iii. Then, the ICs are analysed to cancel the ICs that are related to noise.

   iv. The ICs related to the uncorrupted images are used to obtain/reconstruct the sub-images.

   v. Next, the matrix A is treated as the weighting coefficients of the ICs image bases.

   vi. Then, the images-bases are fused according to A matrix.

   vii. Finally, inverse DWT is used to obtain the fused image.

6.8 Non-Linear ICA and PCA

We discuss some ideas of non-linear independent component analysis (NICA), and non-linear principal components (NPCA). There are many definitions of NICs and NPCs and these open up several possibilities [13] for data-image-analysis and fusion. Under the situations where complicated information is to be captured from the data/images, then certain non-linear functions might do a better job than any linear function/modes.

6.8.1 Non-Linear ICA

The NICA is more sophisticated and rather involved. A non-linear mixer model is given as follows:

$$X_j = f_j\left\{\sum_{i=1}^{n} a_{ji}S_i\right\} = f_j(AS) \qquad (6.67)$$

Here, f is a non-linear function over and above the mixer model A. We assume that the class of system operators of which $\mathcal{F}$ is a member constrained to be a group exists. In that case the separable ICA model would exist. Then we write $\mathcal{F}$ as a function $f(S) = \{f_1(S), f_2(S), ..., f_p(S)\}$. Here, each function f' is a group action of S. The group action is built up from continuous and increasing functions. These functions are tied up by several steps of composition and inversion. More general non-linear functions would lead to multiple solutions. There are several other non-linear dimension reduction methods using [13]: (i) Shannon information, (ii) SVD, (iii) multi-dimensional scaling and (iv) self-organising maps.

6.8.2 Non-Linear PCA

As we know the PCs are based on obtaining eigenvalues (EVAs and eigenvectors (EVEs) of variance of X (the data matrix). For non-linear case one should choose real-valued non-linear functions f_i of x $j = 1, 2, ..., p$. Then carry out PCA on the covariance matrix of $f(x) = \{f_1(x), f_2(x), ..., f_p(x)\}$. So, now we have the covariance of f as $\mathrm{cov}(f, f)$ and the corresponding ES as $(\lambda_1, v_1),...,(\lambda_p, v_p)$ with the EVAs as the decreasing order set. We then have the non-linear PC set as $U = \{u_1, u_2, ..., u_p\}$, with the jth NPC component is given by

$$u_j = \sum_{i=1}^{p} v_{j,i} f_i(x) = v_j^T f(x) \qquad (6.68)$$

If $f(x) = x$ we get the linear PCs. One can have quadratic non-linear functions and the corresponding PCs. We can also have the orthogonal transformation matrix T such that $f(Tx) = Wf(x)$ to see the effects of this transformation on the results. This artifice is often useful when we use the transformed images for analysis and fusion. One can also define the following objective function by L-2 norm:

$$J(W) = E\|x - Wf(W^T x)\| \tag{6.69}$$

Then in order to determine W, the cost function J is minimised with respect to W, and we can obtain the recursive rule to compute the Ws. If the sequential optimisation is used then we can get the following formula for the first PC:

$$w_1 = \arg\max_{\|w\|=1} \text{Var}(f(w^T x)) \tag{6.70}$$

Subsequently other components can be determined by using orthogonality conditions.

6.9 Curvelet Transform for Image Fusion

Often we need to fuse high-spectral/low-spatial resolution multi-spectral and low-spectral/high-spatial resolution images in various applications of remote sensing [6,7]. A WT-based image fusion provides high-quality spectral content in fused images, but some WT-based methods give fused results with spatial resolution lower than obtained using the Brovey, intensity–hue–saturation (IHS) and PCA fusion methods [6]. On the other hand, one can use curvelet transforms (CT) that represent edges better than WTs. Since, the edges play an important role in image (feature), an effective means to enhance spatial resolution is to enhance the edges, CT-based image fusion provides richer information in the spatial and spectral domains simultaneously. In several remote sensing applications, the spatial as well as spectral information of a fused image is a very important factor. The approaches of multi-scale systems based on ridgelet transform (RT) and CT are different from WT methods. CT and RT methods take the form of base elements that exhibit very high directional sensitivity. The CTs represent edges better than WTs and are well suited for multi-scale edge enhancement. CT also can give the same colour as the original multi-spectral images because we also use the WT-based image fusion method.

The CT is based on the segmentation of the entire image into several small overlapping tiles and then the RT transform is applied to each tile, the purpose being of approximating the curved lines by small straight lines, the overlapping tiles avoiding edge effects [7]. The RT transform itself is a 1D WT transform applied on the Radon transform (RnT) of each tile, which is a tool of shape detection, hence the process is and the flow is as follows: start with additive WT → image tiles → RnT → 1D WT (the latter two aspects i.e. RnT and 1D WT being the RT). Because of CT's ability to deal with curved shapes it is also very useful in medical image fusion. In sub-band, filtering the image is decomposed into additive components, each being a sub-band of that image. This is to isolate the different frequency components of the image into different planes without down sampling. In the tiling process, the image is divided into overlapping tiles and these tiles are small in dimension to transform curved lines into small straight lines in the sub-bands, thereby improving the ability of the CT to handle such edges.

The RT belongs to the family of discrete transforms employing certain base functions and it can be viewed as a wavelet analysis in the Radon domain (RnD) (RnT for shape detection). Hence, the RT is a tool of ridge detection/shape detection of the objects in an image. The RT basis function is given by [7]

$$\Psi_{a,b,\theta}(x_1, x_2) = a^{-0.5}\psi\{(x_1 \cos\theta + x_2 \sin\theta - b)/a\}; \quad a > 0, b \in R, \theta \in [0, 2\pi \quad (6.71)$$

The function of Equation 6.71 is constant along with lines: $(x_1 \cos\theta + x_2 \sin\theta) = $ constant, the ridge-let-coefficients (RC) of an image, $I(x, y)$ are given by the following expression [7]:

$$R_I(a,b,\theta) = \int_{-\infty}^{\infty}\int_{-\infty}^{\infty} \Psi_{a,b,\theta}(x,y)I(x,y)dx\,dy \qquad (6.72)$$

The RT transform of Equation 6.72 is invertible and the image can be reconstructed by the following inverse transformation [7]:

$$I(x,y) = \int_{0}^{2\pi}\int_{-\infty}^{\infty}\int_{0}^{\infty} \{R_I(a,b,\theta)\Psi_{a,b,\theta}(x,y)da\,db\,d\theta\}/(4\pi a) \qquad (6.73)$$

The RnT is given by the following equation:

$$R_I(\theta,t) = \int_{-\infty}^{\infty}\int_{-\infty}^{\infty} I(x,y)\delta(x\cos\theta + y\sin\theta - t)dx\,dy \qquad (6.74)$$

The RT can be represented in the terms of the RnT by the following expression:

$$R_I(a,b,\theta) = \int_{-\infty}^{\infty} R_I(\theta,t)a^{-0.5}((t-b)/a)dt \qquad (6.75)$$

Hence, the RT is the application of the 1D WT to the slices of the RnT with the angular variable being constant and t being varying. One can make the RT discrete by making the RnT and WT as discrete transforms.

We divide images into sub-images which are differently scaled by WT. Then, local CT of every sub-image is taken. Its sub-blocks are different from each other on account of changes in the scales. The steps of using CT to fuse two images are: (i) resample and register the original images to correct any distortion so that they have similar pdf, then the WT coefficient of a similar component will stay in the same magnitude; (ii) use WT to decompose the original images into proper levels: (a) one low-frequency approximate component and (b) three high-frequency detail components in/for each level; (iii) obtain the CTs of these components and (iv) local area variance is chosen to measure definition for low-frequency component.

6.10 Image Fusion Using MR Singular Value Decomposition

As we have seen earlier, multi-sensory image fusion (MSIF) is a technique to merge the information content from several images (or acquired from different imaging sensors) taken from the same scene in order to accomplish a combined and integrated image that contains, perhaps, the finest information coming from these different original source images. The basic aspect in MSIF is to take the mean of the grey level registered images pixel by pixel, however, this technique would produce several undesired effects and reduced feature contrast. In this section, we discuss the MSIF technique based on MR singular value decomposition (MRSVD) [14].

6.10.1 Multi-Resolution SVD

Let $X = [x(1), x(2), \ldots, x(N)]$ represent some image data/a 1D signal of length N and it is assumed that N is divisible by 2^K for $K \geq 1$. We rearrange the samples in such a way that the top row contains the odd number indexed samples and the bottom row contains the even number indexed samples. Then the resultant matrix/data matrix be given as [14]

$$X_1 = \begin{bmatrix} x(1) & x(3) & \cdots & x(N-1) \\ x(2) & x(4) & \cdots & x(N) \end{bmatrix} \tag{6.76}$$

Let us denote the matrix

$$T_1 = X_1 X_1^T \tag{6.77}$$

Also, let V_1 be the eigenvector matrix that brings T_1 into diagonal matrix as follows:

$$V_1^T T_1 V_1 = S_1^2 \tag{6.78}$$

The matrix $S_1^2 = \begin{bmatrix} s_1(1)^2 & 0 \\ 0 & s_2(1)^2 \end{bmatrix}$ contains the squares of the singular values, with $s_1(1) > s_2(2)$. Basically, the diagonal elements are the eigenvalues. Let $\hat{X}_1 = V_1^T X_1$ so that

$$X_1 = V_1 \hat{X}_1 \tag{6.79}$$

The top row of $\hat{X}_1$, denoted by $\hat{X}_1(1,:)$ contains the approximation component that corresponds to the largest eigenvalue, and the bottom row of $\hat{X}_1$, denoted by $\hat{X}_1(2,:)$ contains the detail component that corresponds to the smallest eigenvalue. Then, we denote $\Phi_1 = \hat{X}_1(1,:)$ and $\Psi_1 = \hat{X}_1(2,:)$ to represent the approximation and detail components, respectively. The successive levels of decomposition repeat this procedure by placing the approximation component Φ_1 in place of X, the procedure is repeated recursively K times. Let $\Phi_0(1,:) = X$ so that the initial approximation component is the original data-signal. For each level l, the approximation component vector Φ_l has $N_l = N/2^l$ elements and these are represented as follows:

$$\Phi_l = \left[\phi_l(1), \phi_l(2), \ldots, \phi_l(N_l) \right] \tag{6.80}$$

The K level MRSVD for $l = 1, 2, \ldots, K-1$ is as follows [14]:

$$X_l = \begin{bmatrix} \phi_{l-1}(1) & \phi_{l-1}(3) & \cdots & \phi_{l-1}(2N_l - 1) \\ \phi_{l-1}(2) & \phi_{l-1}(4) & \cdots & \phi_{l-1}(2N_l) \end{bmatrix} \tag{6.81}$$

$$T_l = X_l X_l^T = V_l S_l^2 V_l^T \tag{6.82}$$

where the singular values are to be arranged as $s_l(1) \geq s_l(2)$

$$\hat{X}_l = V_l^T X_l \tag{6.83}$$

$$\Phi_l = \hat{X}_l(1,:) \tag{6.84}$$

$$\Psi_l = \hat{X}_l(2,:) \tag{6.85}$$

In general, it is sufficient to store the lowest resolution approximation component vector Φ_L, the detail component vectors Ψ_l for $l = 1, 2, \ldots, L$ and the eigenvector matrices V_l for $l = 1, 2, \ldots, L$.

Then, the MRSVD can be written as

$$X \rightarrow \{\Phi_L, \{\Psi_l\}_{l=1}^{L}, \{V_l\}_{l=1}^{L}\} \tag{6.86}$$

Since the steps are reversible, the original signal X can be reconstructed from the right-hand side.

The 1D MRSVD described above can be easily extended to 2D MRSVD and even for higher dimensions. First, divide the $M \times N$ image X into the non-overlapping 2×2 blocks and arrange each block in a 4×1 vector by stacking columns to form the data matrix X_1. These blocks may be taken in transpose raster-scan manner or proceeding downwards first and then to right. The eigen decomposition of the 4×4 scatter matrix is as follows:

$$T_1 = X_1 X_1^T = V_1 S_1^2 V_1^T \tag{6.87}$$

where the singular values are arranged in decreasing order as $s_1(1) \geq s_2(2) \geq s_3(3) \geq s_4(4)$.

Let $\hat{X}_1 = V_1^T X_1$ and the first row of $\hat{X}_1$ correspond to the largest eigenvalue and be considered as the approximation component. The remaining ones contain the detail component that may correspond to edges/texture in the image. The elements in each row are rearranged to form $M/2 \times N/2$ matrix. For the next level of decomposition, let Φ_1 denote $M/2 \times N/2$ matrix formed by rearranging the row $\hat{X}_1(1, :)$ into a matrix by first filling in the columns and then rows. Also, each of the three rows $\hat{X}_1(2, :)$, $\hat{X}_1(3, :)$ and $\hat{X}_1(4, :)$ may be arranged into $M/2 \times N/2$ matrices that are denoted as Ψ_1^V, Ψ_1^H and Ψ_1^D, respectively. Then, the next level of decomposition proceeds as above where X is replaced by Φ_1. The complete L level decompositions are represented as

$$X \rightarrow \{\Phi_L, \{\Psi_l^V, \Psi_l^H, \Psi_l^D\}_{l=1}^{L}, \{U_l\}_{l=1}^{L}\} \tag{6.88}$$

Then, the original image X is reconstructed from the right-hand side, since the steps are reversible. Figure 6.10 shows the structure of the 2D MRSVD with three decomposition levels [14].

6.10.2 Image Fusion Using MRSVD: Illustrative Example

The process of MRSVD image fusion is shown in Figure 10.11. The images I_1 and I_2 are decomposed into L ($l = 1, 2, \ldots, L$) levels using MRSVD. At each decomposition level ($l = 1, 2, \ldots, L$), the fusion rule will select the larger absolute value of the two MRSVD detailed coefficients [14]. This is because the

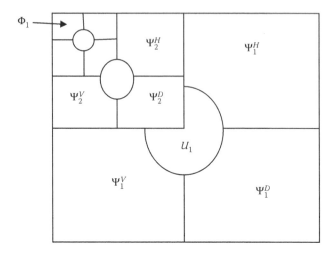

FIGURE 6.10
MRSVD process for image fusion. (Adapted and modified from Naidu, V. P. S. *Defense Science Journal*, 61(5), 479–484, 2011, DESDOC. With permission.)

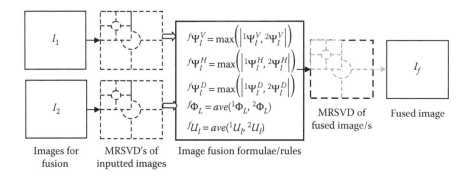

FIGURE 6.11
Image fusion using MRSVD – some steps. (Adapted and modified from Naidu, V. P. S. *Defense Science Journal*, 61(5), 479–484, 2011, DESDOC. With permission.)

detailed coefficients correspond to sharper brightness changes in the images such as edges and object boundaries and these coefficients fluctuate around zero. At the coarsest level ($l = L$), the fusion rule takes the average of the MRSVD approximation coefficients since the approximation coefficients at this level are the smoothed and sub-sampled version of the original image. Similalrly, at each decomposition level ($l = 1, 2, ..., L$), the fusion rule takes the average of the two MRSVD eigen matrices. The fused image I_f can be obtained using

$$I_f \leftarrow \{^f\Phi_L, \{^f\Psi_l^V, {}^f\Psi_l^H, {}^f\Psi_l^D\}_{l=1}^L, \{^fU_l\}_{l=1}^L\} \tag{6.89}$$

The performance of the image fusion scheme is evaluated when the reference image is available using the following metrics [14]:

1. Root mean square error (RMSE)

   It is computed as the root mean square error of the corresponding pixels in the reference image I_r and the fused image I_f. It will be almost zero when the reference and fused images are very alike and it will increase when the dissimilarity increases

$$RMSE = \sqrt{\frac{1}{MN}\sum_{x=1}^{M}\sum_{y=1}^{N}(I_r(x,y) - I_f(x,y))^2} \tag{6.90}$$

2. Inverse fusion error square (IFES)

$$IFES = 20\log_{10}\left(\frac{L^2}{\dfrac{1}{MN}\displaystyle\sum_{x=1}^{M}\sum_{y=1}^{N}(I_r(x,y) - I_f(x,y))^2}\right) \tag{6.91}$$

   where L is the number of grey levels in the image. The higher value of IFES implies better fusion. In the literature of image fusion this metric is also known as PSNR (peak signal to noise ratio), though it is a misnomer.

3. Spectral information divergence (SID)

$$SID(I_t, I_f) = I_t\log(I_t/I_f) + I_f\log(I_f/I_t) \tag{6.92}$$

   The values of I_t and I_f should be in between 0 and 1. The ideal value is zero and it will increase when there is a spectral divergence.

The following metrics are used when the reference image is not available [14]:

1. Standard deviation (SD)

$$\sigma = \sqrt{\sum_{i=0}^{L}(i - \bar{i})^2 h_{I_f}(i)}, \quad \bar{i} = \sum_{i=0}^{L} i h_{I_f} \tag{6.93}$$

TABLE 6.1

Numerical Values of Performance Metrics: Data Set 1

Levels of Decomposition	With Reference Image			Without Reference Image	
	RMSE	SID	IFES	SD	SF
$L = 1$	10.134	~0.0	38.107	52.454	15.123
$L = 2$	8.584	0.04	38.828	53.560	18.059

where $h_{I_f}(i)$ is the normalised histogram of the fused image $I_f(x,y)$. This metric is more efficient in the absence of noise in the images. It measures the contrast in the fused image. A fused image with high contrast would have a high SD.

2. Spatial frequency

Spatial frequency (SF) can be computed as follows:

$$SF = \sqrt{RF^2 + CF^2} \tag{6.94}$$

where the row frequency is given as

$$RF = \sqrt{\frac{1}{MN} \sum_{x=1}^{M} \sum_{y=2}^{N} [I_f(x,y) - I_f(x,y-1)]^2}$$

and the column frequency is given as

$$CF = \sqrt{\frac{1}{MN} \sum_{y=1}^{N} \sum_{x=2}^{M} [I_f(x,y) - I_f(x-1,y)]^2}$$

The SF indicates the overall activity level in the fused image, and the fused image with high activity level would have high SF.

An LTA image is considered as a reference image I_r to evaluate the performance of the proposed fusion algorithm. The complementary pair of the aircraft images (as data set 1) input images I_1 and I_2 are taken, see Figure 6.4. The numerical values of the performance metrics for evaluating the image fusion MRSVD algorithm are shown in Table 6.1, and we can say that satisfactory results have been obtained with MRSVD fusion procedure.

EXERCISES

6.1 Why is CA called blind-source identification problem?

6.2 In terms of projection, what does ICA do?

6.3 What are the applications of ICA?

6.4 Can ICA be considered a system identification/parameter estimation and optimisation problem?

6.5 What is the meaning of ICs?

6.6 How is central limit theorem (CLT) utilised in ICA?

6.7 How is NG property determined?

6.8 What are we actually looking for in ICA?

6.9 In ICA, we want to find the local maxima of non-Gaussianity of a linear combination $y = Wx$ under the constraint that the variance of x is constant. Then what does each local maximum give?

6.10 For ICA, we seek such directions that are most independent from each other, what are we looking for in PCA?

6.11 Is the assumption of ICs required for PCA?

6.12 What are the important differences between FT, short time FT and WT?

6.13 What are the merits of CT over other transforms for image fusion?

References

1. Aapo, H. and Oja, E. Independent component analysis: Algorithms and applications. *Neural Networks*, 13(4–5), 411–430, 2000.
2. Maokuan, Li. and Jian, G. Pixel level image fusion based on ICA and wavelet transform. http://read.pudn.com/downloads135/sourcecode/others/577277/ICA%20and%20Wavelet%20Transform.pdf, accessed April 2013.
3. Mitianoudis, N. and Stathaki, T. Pixel-based and region-based image fusion schemes using ICA bases. *Information Fusion*, 8(2), 131–142, 2007.
4. Shlens, J. A tutorial on principal component analysis. http://www.snl.salk.edu/ ~ shlens/pca.pdf, accessed April 2013.
5. Smith, L. I. A tutorial on principal components analysis. 26 February 2002, http://www.cs.otago.ac.nz/cosc453/student_tutorials/principal_components.pdf, accessed November 2012.
6. Choi, M., Young Kim, R., NAM, M.-R. and Kim, H. O. The curvelet transform for image fusion. http://www.isprs.org/proceedings/XXXV/congress/yf/papers/931.pdf, accessed April 2013.
7. Mamatha, G. and Gayatri, L. An image fusion using wavelet and curvelet transforms. *Global Journal of Advanced Engineering Technologies*, 1(2), 2012 (ISSN: 2277-6370).
8. Raol, J. R. and Gopal, A. K. (Eds.) *Mobile Intelligent Autonomous Systems*. CRC Press, FL, 2012.
9. Raol, J. R. *Multisensor Data Fusion with MATLAB*. CRC Press, FL, 2010.

10. Raol, J. R., Girija, G. and Singh, J. *Modelling and Parameter Estimation of Dynamic Systems*. IET/IEE Control Series Book, Vol. 65. IET/IEE, London, 2004.

11. Raol, J. R. and Singh, J. *Flight Mechanics Modelling and Analysis*. CRC Press, FL, 2010.

12. Lowe, D. Distinctive image features from scale invariant keypoints. *International Journal of Computer Vision*, 60(2), 91–110, 2004.

13. Clarke, B. Fokoue, E. and Zhang, H. H. *Principles and Theory for Data Mining and Machine Learning, Springer Series in Statistics*. Springer, Heidelberg, 2009.

14. Naidu, V. P. S. Image fusion technique using multi-resolution singular value decomposition. *Defense Science Journal*, 61(5), 479–484, 2011, DESDOC.

7

Image Algebra and Image Fusion

S. Sethu Selvi*

7.1 Introduction

Humans rely on vision to make sense of the world around. They not only look at things to identify and classify them, but they also can scan for differences and obtain an overall rough feeling for a scene with a quick glance. They also have evolved very precise visual skills. Human beings can identify a face in an instant, differentiate colours (unless they are colour blind) and can process a large amount of visual information very quickly. However, the objects around us change their appearance depending on the time of the day, the amount of sunlight or a variety of shadows. The image processing techniques deal with single image or snapshots of a visual scene. An image is a single picture that represents something of the observed object. It may be a picture of a person, animal, outdoor scene, microphotograph of a component or the result of medical imaging. A major portion of information received by a human from the environment is visual. The process of receiving and analysing visual information by humans is referred to as perception. The same process if performed by a digital computer is called digital image processing (DIP). The term image is used because the computer stores and processes the numerical image of a scene. There are three basic components of image formation: (i) illumination, (ii) reflectance models of surfaces which are imaged and (iii) the process of image formation in the retina of the human eye or in the sensor of a camera. Once the images (which are two-dimensional [2D] analogue signals) are formed, the next process involves sampling and digitisation of the analogue image.

7.1.1 A Digital Image

A digital image can be defined by a finite valued function over a discrete domain Z^2 and is a rectangular array of size $M \times N$, that is, $D = \{(x,y) \mid x = 0,1,...,M-1; y = 0,1,...,N-1\}$ obtained by sampling with step

* Professor and Head, Department of Electronics and Communications Engineering, M.S. Ramaiah Institute of Technology (MSRIT), Bangalore, India.

size Δx and Δy along x and y directions. So, a digital image $\{f(x, y)\}$, Equation 7.1, can be represented by an $M \times N$ matrix whose elements are integer numbers ranging from 0 to $L - 1$. Each element of the matrix is called pixel (picture element), pel or point. If an image is digitised to an $M \times N$ sampled array and L intensity levels, then the number of bits required to store this image is MNb where $L = 2^b$

$$\{f(x,y)\} = \begin{bmatrix} f(0,0) & f(0,1) & \cdots & f(0, N-1) \\ f(1,0) & f(1,1) & \cdots & f(1, N-1) \\ \vdots & \vdots & \vdots & \vdots \\ f(M-1,0) & f(M-1,1) & \cdots & f(M-1, N-1) \end{bmatrix} \quad (7.1)$$

7.1.2 Needs of Image Fusion

With the recent rapid developments in the field of sensing technologies multi-sensor systems have become a reality in a growing number of fields such as remote sensing, medical imaging, machine vision and the military applications for which they were first developed. Image fusion provides an effective way of reducing the increasing volume of information received while at the same time extracting all the useful information from the source images. The data fusion (DF) is variously defined: (i) '…is a process by which data from a multitude of sensors are used to yield an optimal estimate of a specified state vector pertaining to the observed system' [1]; and (ii) '…deals with the synergistic combination of information made available by various knowledge sources such as sensors, in order to provide a better understanding of a given scene' [2].

The aim of image fusion, apart from reducing the amount of data, is to create the fused images with enhanced information content that are more suitable for the purposes of human/machine interaction and perception, and for further image-processing tasks such as segmentation, object detection or target recognition in applications such as remote sensing and medical imaging. For example, visible band and infrared (IR) images may be fused to aid pilots landing aircraft in poor visibility. Complementary information about the same observed scene can be available in the following cases: (i) data recorded by different sensors (multi-sensor image fusion), (ii) data recorded by the same sensor scanning the same scene at different dates (multi-temporal image fusion), (iii) data recorded by the same sensor operating in different spectral bands (multi-frequency image fusion), (iv) data recorded by the same sensor at different polarisations (multi-polarisation image fusion) and (v) data recorded by the same sensor located on platforms flying at different heights (multiresolution [MS] image fusion). Multi-sensor images often have different geometric representations, which have to be transformed to a common representation for fusion. This representation should retain at least the best resolution of either sensor. A pre-requisite for successful image fusion is the alignment of multi-sensor images. An illustration of a multi-sensor image fusion system

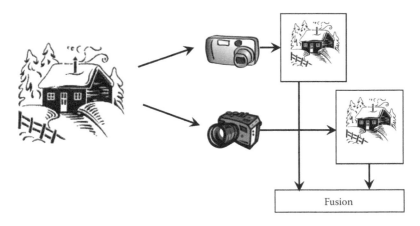

FIGURE 7.1
Multi-sensor image fusion system.

is provided in Figure 7.1, wherein an IR camera is supplementing the digital visual band camera and their individual images are fused to obtain a fused image. While the digital camera is appropriate for daylight scenes, the IR camera is suitable in poorly illuminated ones, including during the night, because the IR camera provides a thermal profile/image of the object.

Image fusion gives a single image from a set of input images. The fused image should have more complete information which is more useful for human or machine perception. The objectives of image fusion are: (i) to extract all the useful information from the source images, (ii) not to introduce artefacts or inconsistencies which will distract the human observer or computer processing and (iii) to be reliable and robust to imperfections such as misregistration. The advantage of image fusion is pictorially represented in Figure 7.2. The redundant information from two different sensors A and B improves the reliability of the fused image and the complementary information from

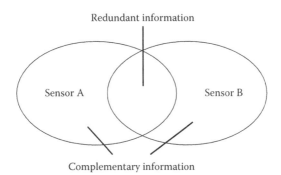

FIGURE 7.2
Advantage of image fusion.

the sensors improves the capability of image fusion. For example, if a television (TV) device is used as sensor A and IR waves/camera are used as sensor B, then the fused image has improved penetration and better visibility in day/night conditions. Similarly, if IR is sensor A and ultraviolet band/device is used as sensor B, the fused image provides better background discrimination than the individual images. The benefits of multi-sensor image fusion include [3]: (i) extended range of operation: multiple sensors that operate under different operating conditions can be deployed to extend the effective range of operation. For example, different sensors can be used for day/night operation, (ii) extended spatial and temporal coverage – joint information from sensors that differ in spatial resolution can increase the spatial coverage, (iii) reduced uncertainty – joint information from multiple sensors can reduce the uncertainty associated with the sensing or decision process, (iv) increased reliability – fusion of multiple measurements can reduce noise and therefore improve the reliability of the measured quantity, (v) robust system performance – redundancy in multiple measurements can help in system's robustness. In case one or more sensors fail or the performance of a particular sensor deteriorates, the system can depend on the other sensors, (vi) compact representation of information by the fused image – for example, in remote sensing, instead of storing imagery from several spectral bands, it is comparatively more efficient to store the fused information.

7.2 Image Algebra

The goal of image algebra is to develop a complete, unified algebraic structure that provides a common mathematical environment for image processing algorithm development, optimisation, comparison, coding and performance evaluation [4,5]. It also helps image fusion in an overall sense. The theory of image algebra provides a language/framework that if properly implemented as a standard image processing environment, can greatly reduce research and development costs. Since, the foundation of this language is purely mathematical and independent of any future computer architecture or language, the longevity of an image algebra standard is assured. Although commonality of language and cost savings are two major reasons for considering image algebra as a standard language for image processing, there are multitude of other reasons for desiring the broad acceptance of image algebra as a component of all image processing development systems. Premier among these is the predictable influence of an image algebra standard on future image processing technology. Image algebra is a mathematical theory concerned with the transformation and analysis of images. The main goal is the establishment of a comprehensive and unifying theory of image transformations, image analysis and image understanding in the discrete as well as the continuous domains. Sternberg [4] was the first to use the term 'image algebra'. Additional benefits provided by the use of image algebra are: (a) elemental operations are small in number, translucent, simple and provide a method of transforming images that are easily learned and used, (b) operations and operands provide the capability of expressing all image-to-image transformations, (c) theorems make computer programs amenable to both machine dependent and machine independent optimisation techniques, (d) the algebraic notation provides a deeper understanding of image manipulation operations due to conciseness and brevity of code and is capable of suggesting new techniques, (e) the notational adaptability to programming languages allows the substitution of extremely short and concise image algebra expressions for equivalent blocks of code, and therefore increases programmer productivity, (f) provides a rich mathematical structure that can be exploited to relate image processing problems to other mathematical areas and (g) without image algebra, a programmer will never benefit from the bridge that exists between an image algebra programming language and the multitude of mathematical structures, theorems and identities that are related to image algebra. Image algebra is a heterogeneous or many-valued algebra with multiple sets of operands and operators. Manipulation of images for purposes of image enhancement, analysis and understanding involves operations not only on images, but also on different types of values and quantities associated with these images. Thus, the basic operands of image algebra are images and the values and quantities associated with these images. Roughly speaking, an image consists of two things, a collection of points and a set of values associated with these points. Images are therefore endowed with two types of information, namely the spatial relationship of the points, and also some type of numeric or other descriptive information associated with these points. Consequently, the field of image algebra bridges two broad mathematical areas, the theory of point sets and the algebra of value sets, and investigates their interrelationship.

7.2.1 Point and Value Sets

A point set is simply a topological space. Thus, a point set consists of two things, a collection of objects called points and a topology which provides for such notions as nearness of two points, the connectivity of a subset of the point set, the neighbourhood of a point, boundary points and curves and arcs. Point sets are typically denoted by capital bold letters and points which are the elements of point sets are typically denoted by lower case bold letters. If $z \in R^2$, then z is of the form $z = (x, y)$ where x and y denote real numbers called the coordinates of z. The most common point sets occurring in image processing are discrete subsets of $n–D$ Euclidean space R^n with $n = 1,2,3$ together with the discrete topology. Point sets can assume arbitrary shapes, in particular, shapes can be rectangular or circular. Some of the more pertinent point sets are the set of integer points Z, the n-dimensional lattice $Z^n \subset R^n$ with $n = 2$ or $n = 3$ and rectangular subsets of Z^2. The most often encountered point set is of the form

$$X = Z^M \times Z^N = \{(x,y) \in Z^2 : 0 \le x \le M - 1, 0 \le y \le N - 1\} \qquad (7.2)$$

These point sets inherit the elementary vector space operations such as sum of two points, point subtraction, multiplication and addition of a scalar. In addition to these standard vector space operations, image algebra also incorporates three types of point multiplication. These are the Hadamard product (element wise multiplication), cross product and the dot product. The sum of two points, Hadamard product and the cross product are binary operations that take as input two points and produce another point. These operations can be viewed as mappings $Z \times Z \to Z$. The binary operation of dot product is a scalar and not another vector. This provides an example of a mapping $Z \times Z \to F$, where F denotes an appropriate field of scalars. Another mapping is the distance function $Z \times Z \to R$ which assigns to each pair of points x and y the distance from x to y. The most common distance functions occurring in image processing are the Euclidean distance, city block or diamond distance, and the chess board distance which are defined by

$$d(x,y) = \left[\sum_{k=1}^{n} (x_k - y_k)^2 \right]^{1/2} \qquad (7.3)$$

$$\rho(x,y) = \sum_{k=1}^{n} | x_k - y_k | \qquad (7.4)$$

$$\delta(x,y) = \max\{ | x_k - y_k | : 1 \le k \le n \} \qquad (7.5)$$

For a pair of point sets X and Z, a neighbourhood function from X to Z is a function $N : X \to 2^Z$ where $2^Z = \{X : X \subset Z\}$ which is the power set of a set Z defined as the set of all subsets of Z. For each point $x \in X$, $N(x) \subset Z$. The set $N(x)$ is called a neighbourhood of x. There are two neighbourhood functions on subsets of Z^2 which are of particular importance in image processing. The four neighbourhoods or Von Neumann neighbourhood $N : X \to 2^{Z^2}$ is defined by

$$N(x) = \{y : y = (x_1 \pm j, x_2) \quad \text{or} \quad (x_1, x_2 \pm k), j, k \in (0,1)\} \qquad (7.6)$$

and the eight neighbourhoods or Moore neighbourhood is defined by

$$M(x) = \{y : y = (x_1 \pm j, x_2 \pm k), j, k \in (0,1)\} \qquad (7.7)$$

where $x = (x_1, x_2) \in X \subset Z^2$. They are local neighbourhoods since they only include the directly adjacent points of a given point. Distances can be conveniently computed in terms of the norm of a point. The three norms of interest are derived from the standard L^p norms as

$$\|x\|_p = \left(\sum_{i=1}^{n} |x_i|^p \right)^{1/p}$$
(7.8)

The L^∞ norm is given by

$$\|x\|_\infty = \bigvee_{i=1}^{n} |x_i| = \max\left\{ |x_1|, |x_2|, \ldots, |x_n| \right\}$$
(7.9)

The Euclidean norm is given by

$$\|x\|_2 = \sqrt{x_1^2 + x_2^2 + \cdots + x_n^2}$$
(7.10)

Thus, we have

$$d(x, y) = \|x - y\|_2$$
$$\rho(x, y) = \|x - y\|_1$$
$$\delta(x, y) = \|x - y\|_\infty$$
(7.11)

A heterogeneous algebra is a collection of non-empty sets of different types of elements together with a set of finite operations which provide the rules of combining various elements in order to form a new element. A homogeneous algebra is a heterogeneous algebra with only one set of operands. This is simply a set together with a finite number of operations. These will be referred to as value sets. The operations on and between elements of a given value set F are the elementary operations associated with F. Thus, if $F \in \{Z, R, Z^{2k}\}$, then the binary operations are the usual arithmetic and logic operations of addition, multiplication and maximum, and the complementary operations of subtraction, division and minimum. Also, elementary unary operations associated with these sets such as absolute value, conjugation, trigonometric, logarithmic and exponential operations are possible.

7.2.2 Images and Templates

The primary operands in image algebra are images, templates and neighbourhoods. Images are the most fundamental entities since templates and neighbourhoods are special cases of the general concept of an image. Let F be a value set and X a point set. An F valued image on X is any element of F^X which denotes the set of all functions $X \to F$, that is, $F^X = \{f : f$ is a function from X to F}. Given an F valued image $g \in F^X$, then F is called the set of possible range values of g and X the spatial domain of g. It is convenient to let the graph of an image $g \in F^X$ represent g. The graph of an image is also referred to as the data structure representation of the image. Given the data structure representation $g = \{(x, g(x)) : x \in X\}$, then an element $(x, g(x))$ of the data structure is called a picture element or pixel. The first coordinate x of a pixel is called the pixel location or image point, and the second coordinate $g(x)$ is called the pixel value of g at location x. This definition of an image covers all mathematical images on topological spaces with range in an algebraic system. Operations on and between F valued images are the natural induced operations of the algebraic system F. For example, if γ is a binary operation on F, then γ induces a binary operation on F^X defined as follows: Let $a, b \in F^X$. Then $a\gamma b = \{(x, c(x)) : c(x) = a(x)\gamma b(x), x \in X\}$. Replacing γ by the binary operations $+$, $\cdot$, $\vee$ and $\wedge$, the basic binary operations on real valued images are obtained. All four operations are commutative and associative. The binary operation γ on F also induces scalar multiplication and addition of real valued images.

Although much of image processing is accomplished using real, integer or binary valued images, many higher level vision tasks require manipulation of vector and set valued images. A set valued image is of form $a : X \rightarrow 2^F$. The operations on set valued images are those induced by the Boolean algebra of the value set. The operation of complementation is a unary operation. A particularly useful unary operation on images which is induced by a binary operation on a value set is known as the global reduce operation. If γ is an associative and commutative binary operation on F and X is finite, then γ induces a unary operation $\Gamma : F^X \rightarrow F$ called the global reduce operation induced by γ which is defined as

$$\Gamma a = \mathop{\Gamma}_{x \in X} a(x) = \prod_{i=1}^{n} a(x_i) = a(x_1)\gamma a(x_2)\gamma \ldots a(x_n) \tag{7.12}$$

This provides four basic global reduce operations namely Σa, πa, $\vee a$ and $\wedge a$. Given an unary operation $f : F \rightarrow F$, then the induced unary operation $F^X \rightarrow F^X$ is denoted by f and is defined by

$$f(a) = \left\{(x, a(x)) : c(x) = f(a(x)), x \in X\right\} \tag{7.13}$$

The composition $f \circ a$ is viewed as a unary operation on F^X with operand a. An example is the characteristic function defined as

$$\chi_{\geq k}(r) = \begin{cases} 1, & \text{if } r \geq k \\ 0, & \text{otherwise} \end{cases} \tag{7.14}$$

Then for any $a \in R^x$, this is the two valued image on X, with value 1 at location x if $a(x) \geq k$ and value 0 if $a(x) < k$. An application of this operation is the thresholding of an image. Given an image a and using the characteristic function:

$$\chi_{[j,k]}(r) = \begin{cases} 1, & \text{if } j \leq r \geq k \\ 0, & \text{otherwise} \end{cases} \tag{7.15}$$

then the image b in the image algebra expression $b := a \cdot \chi_{[j,k]}(a)$ is given by

$$b = \{(x, b(x)) : b(x) = a(x) \quad \text{if } j \leq a(x) \leq k, \text{ otherwise } b(x) = 0\} \tag{7.16}$$

The unary operations on an image $a \in F^X$ have resulted either in a scalar by use of global reduction operation, or another F valued image by use of the composition $f \circ a = f(a)$. Given a function $f : F \to G$, then the composition $f \circ a$ provides for a unary operation which changes an F valued image into a G valued image $f(a)$. Using a function f between spatial domains provides a scheme for realising naturally induced operations for spatial manipulation of image data. In particular, if $f : Y \to X$, and $a \in F^X$, then the induced image $a \circ f \in F^Y$ is given by

$$a \circ f = \{(y, a(f(y))) : y \in Y\} \tag{7.17}$$

This operation transforms an F valued image defined over the space X into an F valued image over the space Y. Examples of spatial-based image transformations are affine and perspective transforms. Simple shifts of an image can be obtained by using either a spatial transformation or point addition. In particular, given $a \in F^X$, $X \subset Z^2$ and $y \in Z^2$, a shift of a by y is defined as

$$a + y = \{(z, b(z)) : b(z) = a(z - y), z - y \in X\} \tag{7.18}$$

This is an image on $X + y$ as $z - y \in X \Leftrightarrow z \in X + y$, which provides for the equivalent information as follows:

$$a + y = \{(z, b(z)) : b(z) = a(z - y), z \in X + y\} \tag{7.19}$$

The spatial transformation $f : X + y \to X$ can be defined as $f(z) = z - y$, to obtain the identical shifted image $a + y = a \circ f$. Templates are images whose values are images. The notion of a template, as used in image algebra, unifies and generalises the concepts of templates, masks, windows and neighbourhood functions into one general mathematical entity. A template is an image whose pixel values are images. An F valued template from Y to X is a function $t : Y \rightarrow F^X$. Thus $t \in (F^X)^Y$ and t is an F^X valued image on Y. For convenience this is defined as $t_y \equiv t(y)$, $\forall y \in Y$. The image t_y has representation as

$$t_y = \{(x, t_y(x)) : x \in X\} \tag{7.20}$$

The pixel values $t_y(x)$ of this image are called the weights of the template at point y. If t is a real valued template from Y to X, then the support of t_y is denoted by $S(t_y)$ and is defined as

$$S(t_y) = \{x \in X, t_y(x) \neq 0\} \tag{7.21}$$

For extended real valued templates, supports at infinity are also defined. If X is a space with an operation $+$ such that $(X,+)$ is a group, then a template $t \in (F^X)^X$ is said to be translation invariant if and only if for each triple x, y, $z \in X$, $t_y(x) = t_{y+z}(x+z)$. For example, let $X = Z^2$ and $z = (x, y)$ be an arbitrary point of X. If $x_1 = (x, y-1)$, $x_2 = (x+1, y)$, $x_3 = (x+1, y-1)$ and $t \in (R^X)^X$ defined by the weights $t_z(z) = 1$, $t_z(x_1) = 3$, $t_z(x_2) = 2$, $t_z(x_3) = 4$ and $t_z(x) = 0$ whenever x is not an element of $\{y, x_1, x_2, x_3\}$. The support is $S(t_z) = z, x_1, x_2, x_3$.

The definition of an image template product provides the rules for combining images with templates and templates with templates. The definition of this product includes correlation and convolution products used in image processing. If F is value set with two binary operations $\cdot$ and γ, where $\cdot$ distributes over γ, γ is associative and commutative, and if $t \in (F^X)^Y$, then for each $y \in Y$, $t_y \in F^X$. Thus if $a \in F^X$, where X is finite, then $a \cdot t_y \in F^X$; $\Gamma(a \cdot t_y) \in F$. It follows that the binary operations $\cdot$ and γ induce a binary operation defined by

$$b(y) = \Gamma(a \cdot t_y) = \underset{x \in X}{\Gamma}(a(x) \cdot t_y(x)) \tag{7.22}$$

If $X = x_1, x_2, \ldots, x_n$, then we have

$$b(y) = (a(x_1) \cdot t_y(x_1))\gamma(a(x_2) \cdot t_y(x_2))\gamma \cdots (a(x_n) \cdot t_y(x_n)) \tag{7.23}$$

This expression is the right convolution product of a with t. While a is an image on X, this is an image on Y. Thus, templates allow for the transformation of an image from one type of domain to an entirely different domain type. If $b = a \oplus t$, this is the linear image template product or convolution of a with t, where

$$b(y) = \sum_{x \in X}(a(x) \cdot t_y(x)), \quad a \in R^X \text{ and } t \in (R^X)^Y) \tag{7.24}$$

The utility of template products stems from the fact that

$$a \oplus (s \oplus t) = (a \oplus s) \oplus t \tag{7.25}$$

This equation can be utilised in order to reduce the computational burden associated with typical convolution problems. For example, if $r \in (R^{Z^2})^{Z^2}$ is defined by $\forall y \in Z^2$, then

$$r_y = \begin{array}{|c|c|c|} \hline 4 & 6 & -4 \\ \hline 6 & 9 & -6 \\ \hline -4 & -6 & 4 \\ \hline \end{array} \tag{7.26}$$

$$a \oplus r = a \oplus (s \oplus t) = (a \oplus s) \oplus t \tag{7.27}$$

where

$$s_y = \begin{array}{|c|c|c|} \hline 2 & 3 & -2 \\ \hline \end{array} \qquad t_y = \begin{array}{|c|} \hline 2 \\ \hline -3 \\ \hline 2 \\ \hline \end{array} \tag{7.28}$$

The construction of the new image $b := a \oplus r$ requires nine multiplications and eight additions per pixel. In contrast, the computation of the image $b := (a \oplus s) \oplus t$ requires only six multiplications and four additions per pixel. In general, for an image of size $N \times N$, direct convolution requires N^2 multiplications and $N^2 - 1$ additions per pixel but separable templates require $2N$ multiplications and $2(N - 1)$ additions per pixel. For large images of size 1024×1024, this amounts to significant savings in computation.

7.2.3 Recursive Templates

A recursive template is defined in terms of a regular template from some point set X to another point set Y with some partial order imposed on Y. A partially ordered set $(P, \prec)$ is a set P together with a binary relation $\prec$, satisfying the following three axioms for arbitrary $x, y, z \in P$: (i) reflexive: $x \prec x$, (ii) anti-symmetric: $x \prec y$ and $y \prec x \Rightarrow x = y$ and (iii) transitive: $x \prec y$ and $y \prec z \Rightarrow x \prec z$. The recursive template operation computes a new pixel value based on both the pixel values of the source image and some previously calculated new pixel values which are determined by the partial order and region of support of the participating template. Some partial orders that are commonly used in 2D recursive transforms are forward and backward raster scanning and serpentine scanning. The computation of a new pixel can be done only after all its predecessors have been computed. This in contrast to non-recursive template operations, recursive template operations are not computed in a globally parallel fashion. Let X and Y be finite subsets of R^n with Y partially ordered by $\prec$. If $a \in R^X$; $t \in (R^X, R^Y)^{(Y, \prec)}$, then the recursive linear convolution product $a \oplus_{\prec} t$ is defined by

$$a \oplus_{\prec} t = \left\{ (y, b(y)) : y \in Y, b(y) = \sum_{x \in S(t_{\neq y})} (a(x) \cdot t_{\neq y}(x)) + \sum_{z \in S(t_{\prec y})} (b(z) \cdot t_{\prec y}(z)) \right\} \quad (7.29)$$

If the recursive template t is defined such that $S(t_{\prec y}) = \emptyset$ for all $y \in Y$, then the non-recursive template operation is given by

$$a \oplus_{\prec} t = \left\{ (y, b(y)) : y \in Y, b(y) = \sum_{x \in S(t_{\neq y})} (a(x) \cdot t_{\neq y}(x)) \right\} \quad (7.30)$$

Hence, recursive template operations are natural extensions of non-recursive template operations.

7.2.4 Neighbourhoods and the p-Product: Illustrative Examples

There are several types of template operations that are more easily implemented in terms of neighbourhood operations. Typically, neighbourhood operations replace template operations whenever the values in the support of a template consist only of the unit elements of the value set associated with the template. A template $t \in (F^X)^Y$ with the property that for each $y \in Y$, the values in the support of t_y consist only of the unit of F is called a unit template. A more general form of neighbourhood reduction is a function

$$\Gamma : F^X \mid_N \rightarrow F \quad (7.31)$$

where $N \in (2^X)^Y$ and $F^X \mid_N = \{a \mid_{N(y)} : a \in F^X, y \in Y\}$
   An example is to define

$$\Gamma : R^X \mid_N \rightarrow R \quad \text{as} \quad (7.32)$$

$$\Gamma(a \mid_{N(y)}) = \frac{1}{\text{card}(N(y))} \sum_{x \in N(y)} a(x) \quad (7.33)$$

then Γ implements the averaging function. Similarly, for integer valued images, the median reduction

$$\text{median} : N^X \mid_N \rightarrow N \quad (7.34)$$

is defined as

$$\text{median}(a\,|_{N(y)}) = \text{median}\{a(x_{i_1}), a(x_{i_2}), \ldots, a(x_{i_k})\} \tag{7.35}$$

where $N(y) = x_{i1}, x_{i2}, \ldots, x_{ik}$

For reduce operation of summation, the unit values in the support must be 1, while for the maximum reduce operation, the values in the support must all be 0. Therefore, a single template operand cannot be defined to characterise a neighbourhood for reduction without regard to the image template operation to be used to reduce the values within the neighbourhood. But, the information of interest in unit templates can be captured exactly with the neighbourhood function. Thus, for example, the Moore neighbourhood M can be used to add the values in every 3×3 neighbourhood as well as to find the maximum or minimum in such a neighbourhood. Consider a simple example of image smoothing by local averaging. Suppose $a \in R^X$, where $X \subset Z^2$ is an $m \times n$ array of points, and $t \in (R^{Z^2})^{Z^2}$ is the 3×3 Moore template with unit values 1. The image b obtained from $b: = (a \oplus t)/9$ represents the image obtained by a by local averaging since the new pixel value $b(y)$ is given by

$$b(y) = \frac{1}{9} \sum_{x \in X \cap S(t_y)} a(x) \cdot t_y(x) = \frac{1}{9} \sum_{x \in X \cap S(t_y)} a(x) \tag{7.36}$$

In the linear domain, template convolution products and image template convolution products are equivalent to matrix products and vector matrix products, respectively. The notion of a generalised matrix product was developed to provide a general matrix theory approach to image template products and template convolution products. This provides for a transformation that combines the same or different types of values or objects into values of a possibly different type from those initially used in the combining operation. This p-product can be applied to express various image processing transforms in computing form. Let $F \in R$ and the set of all $m \times n$ matrices with entries from F is denoted by $F_{m \times n}$. F^n is the set of all n-dimensional row vectors with entries from F and the set of all m-dimensional column vectors with entries from F is given by $(F^m)'$. Let m, n and p be positive integers with p dividing both m and n. The following correspondences are defined as

$$c_p : Z_p^+ \times Z_{n/p}^+ \rightarrow Z_n^+$$
$$\text{by } c_p(k, j) = (k - 1)\frac{n}{p} + j$$

where

$$1 \le j \le \frac{n}{p} \quad \text{and} \quad 1 \le k \le p$$

and

$$r_p : Z^+_{m/p} \times Z^+_p \to Z^+_m$$
$$\text{by } r_p(i,k) = (i-1)p + k$$

where

$$1 \le k \le p \quad \text{and} \quad 1 \le i \le \frac{m}{p} \tag{7.37}$$

Since $r_p(i,k) < r_p(i',k') \Leftrightarrow i < i'$ or $i = i'$ and $k < k'$, r_p linearises the array $Z^+_{m/p} \times Z^+_p$ using the row scanning order. The entries of the matrix is in terms of a triple index $a_{s,(i,k)}$ by using the convention given by

$$a_{s,(i,k)} = a_{s,t} \Leftrightarrow r_p(i,k) = t$$

where

$$1 \le i \le m/p \quad \text{and} \quad 1 \le k \le p \tag{7.38}$$

EXAMPLE 7.1

Suppose $l = 2$, $m = 6$ and $p = 2$. Then $m/p = 3$, $1 \le k \le p = 2$ and $1 \le i \le m/p = 3$. Hence for $A = (a_{s,t}) \in F_{2 \times 6}$

$$A = \begin{pmatrix} a_{11} & a_{12} & a_{13} & a_{14} & a_{15} & a_{16} \\ a_{21} & a_{22} & a_{23} & a_{24} & a_{25} & a_{26} \end{pmatrix}$$

$$= \begin{pmatrix} a_{1,(1,1)} & a_{1,(1,2)} & a_{1,(2,1)} & a_{1,(2,2)} & a_{1,(3,1)} & a_{1,(3,2)} \\ a_{2,(1,1)} & a_{2,(1,2)} & a_{2,(2,1)} & a_{2,(2,2)} & a_{2,(3,1)} & a_{2,(3,2)} \end{pmatrix} \tag{7.39}$$

The factor Z^+_n of the Cartesian product $Z^+_n \times Z^+_q$ is decomposed in a similar fashion. The entries of the matrix $B = (b_{s,t}) \in M_{n \times q}(F)$ is in terms of a triple index $b_{(k,j),t}$ by using the convention given by

$$b_{(k,j),t} = b_{s,t} \Leftrightarrow c_p(k,j) = s,$$
$$\text{where } 1 \le k \le p \quad \text{and} \quad 1 \le j \le n/p \tag{7.40}$$

EXAMPLE 7.2

Suppose $n = 4$, $q = 3$ and $p = 2$. Then $n/p = 2$, $1 \le k \le p = 2$ and $1 \le j \le n/p = 2$. Hence for $B = (b_{s,t}) \in F_{4 \times 3}$, we have

$$
B = \begin{pmatrix} b_{11} & b_{12} & b_{13} \\ b_{21} & b_{22} & b_{23} \\ b_{31} & b_{32} & b_{33} \\ b_{41} & b_{42} & b_{43} \end{pmatrix}
$$

$$
= \begin{pmatrix} b_{(1,1),1} & b_{(1,1),2} & b_{(1,1),3} \\ b_{(1,2),1} & b_{(1,2),2} & b_{(1,2),3} \\ b_{(2,1),1} & b_{(2,1),2} & b_{(2,1),3} \\ b_{(2,2),1} & b_{(2,2),2} & b_{(2,2),3} \end{pmatrix} \tag{7.41}
$$

Now, let $A = (a_{s,j'}) \in F_{l \times m}$ and $B = (b_{i',t}) \in F_{n \times q}$. Using the maps r_p and $c_{p'}$ A and B can be rewritten as

$$
A = (a_{s,(i,k)})_{l \times m'}, \quad \text{where } 1 \le s \le l,\ 1 \le r_p(i,k) = j' \le m \text{ and}
$$
$$
B = (b_{(k,j),t})_{n \times q'}, \quad \text{where } 1 \le c_p(k,j) = i' \le n \quad \text{and} \quad 1 \le t \le q \tag{7.42}
$$

The p-product or generalised matrix product of A and B is denoted by $A \oplus_p B$ and is the matrix

$$
C = A \oplus_p B = F_{l(n/p) \times (m/p)q} \tag{7.43}
$$

defined by

$$
c_{(s,j),(i,t)} = \sum_{k=1}^{p} (a_{s,(i,k)} b_{(k,j),t}) = (a_{s,(i,1)} b_{(i,j),t}) + \cdots + (a_{s,(i,p)} b_{(p,j),t}) \tag{7.44}
$$

where $c_{(s,j),(i,t)}$ denotes the (s, j)th row and (i, t)th column entry of C. The lexicographical order is used here as $(s, j) \prec (s', j') \Leftrightarrow s \prec s'$ or if $s = s',\ j \prec j'$. For the above example, the $(2, 1)$ row and $(2, 3)$ column element $c_{(2,1),(2,3)}$ of the matrix $C = A \oplus_2 B \in R_{l(n/p) \times (m/p)q} = R_{4 \times 9}$ is given by

$$
c_{(2,1),(2,3)} = \sum_{k=1}^{2} a_{2,r2(2,k)} \cdot b_{c2(k,1),3}
$$
$$
= a_{2,r2(2,1)} \cdot b_{c2(1,1),3} + a_{2,r2(2,2)} \cdot b_{c2(2,1),3}
$$
$$
= a_{23} \cdot b_{13} + a_{24} \cdot b_{33} \tag{7.45}
$$

In particular,

$$
\begin{pmatrix} 1 & 2 & 0 & 5 & 4 & 3 \\ 2 & 3 & 4 & 1 & 0 & 6 \end{pmatrix} \oplus_2 \begin{pmatrix} 2 & 6 & 1 \\ 1 & 3 & 2 \\ 2 & 2 & 5 \\ 3 & 0 & 4 \end{pmatrix}
$$

$$
= \begin{pmatrix} 6 & 10 & 11 & 10 & 10 & 25 & 14 & 30 & 19 \\ 7 & 3 & 10 & 15 & 0 & 20 & 13 & 12 & 20 \\ 10 & 18 & 17 & 10 & 26 & 9 & 12 & 12 & 30 \\ 11 & 6 & 16 & 7 & 12 & 12 & 18 & 0 & 24 \end{pmatrix} \tag{7.46}
$$

If

$$A = \begin{pmatrix} 1 & 0 \\ -1 & 1 \end{pmatrix}, \quad B = \begin{pmatrix} 4 \\ 2 \\ 6 \\ 3 \end{pmatrix},$$

$$(A \oplus_2 B)' = (4 \quad 2 \quad 2 \quad 1) \neq (4 \quad -2 \quad 6 \quad -3) = B' \oplus_2 A' \tag{7.47}$$

This shows that the transpose property, which holds for the regular matrix product, is generally false for the p-product. p-product is not a dual operation in the transpose domain. In order to make the transpose property hold, the dual operation $\oplus'_p$ of $\oplus_p$ is defined as

$$A \oplus'_p B = (B' \oplus_p A')' \tag{7.48}$$

It follows that

$$A \oplus_p B = (B' \oplus'_p A')' \tag{7.49}$$

The transpose property is

$$(A \oplus_p B)' = B' \oplus'_p A' \tag{7.50}$$

7.3 Pixels and Features of an Image

We discuss here certain aspects of image fusion based on image pixels and features. Certain other aspects like probabilistic model-based image fusion are discussed in Reference 6. One common property that can be observed in nearly all images is that radiation is emitted from some source, then interacts with some material, and is sensed and ultimately transduced into an electrical signal, which is then digitised. The resulting images can be used to extract information about the radiation source, and about the objects with which the radiation interacts. The images can be classified according to the way in which the interaction occurs, and that images are of multiple types, Figure 7.3 [7]. Reflection images sense radiation that has been reflected from the surfaces of objects. The radiation may be ambient or artificial, and it may be from a localised source, or from multiple or extended sources. Most of our daily experience of optical imaging through the eye is of reflection images. Common non-visible examples include radar images, sonar images and some types of electron microscope images. The type of information that can be extracted from reflection images is primarily about object surfaces, that

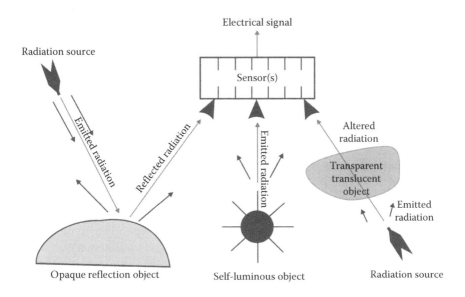

FIGURE 7.3
Image generation and different types of images.

is, their shape, texture, colour, reflectivity and so on. Emission images are even simpler, since in this case the objects being imaged are self-luminous. Examples include thermal or IR images, which are commonly encountered in medical, astronomical and military applications, self-luminous visible light objects, such as light bulbs and stars, and MRI images, which sense particle emissions. In images of this type, the information to be had is often primarily internal to the object; the image may reveal how the object creates radiation and thence something of the internal structure of the object being imaged. However, it may also be external; for example, a thermal camera can be used in low-light situations to produce useful images of a scene containing warm objects, such as people. Finally, absorption images yield information about the internal structure of objects. In this case, the radiation passes through objects and is partially absorbed or attenuated by the material composing them. The degree of absorption dictates the level of the sensed radiation in the recorded image. Examples include x-ray images, transmission microscopic images and certain types of sonic images.

Light is the predominant energy source for images, as human beings can observe it directly and it has the advantage of being safe, cheap, easily detectable, and readily processed with suitable hardware. A digital camera uses charge coupled device (CCD) which is an array of light sensitive cells, each of which produces a voltage proportional to the intensity of light falling on them. CCDs are used as they produce very good results, can be made with high resolution and are robust against noise. A complementary technology is the use of complementary metal oxide semiconductor (CMOS) chips. These have the advantage of being cheaper to produce and require less power to run. However, they are more susceptible to noise and used in low end cameras, such as in webcams. The output of CCD or CMOS is an array of values, each representing a sampled point from the original scene. The elements of this array are called picture elements, or more simply pixels. Although light is popular and easy to use, other energy sources are also used to create a digital image. For microscopy, x-rays or electron beams are used. Because of the shorter wavelength of x-rays than visible light, they are used to resolve smaller objects and determine the structure of objects hidden from view such as bones that are not possible with visible light.

Consider a monochromatic image or a photo which has no colour. This image is a 2D function $f(x, y)$, where the function values give the brightness of the image at any given spatial location (x, y). In such an image brightness values can be any real numbers in the range 0.0 corresponding to black to 1.0 corresponding to white. The ranges of x and y depend on the image. A digital image differs from a photo in that the x, y and $f(x, y)$ values are all discrete. Often the $f(x, y)$ values take only integer values, and the brightness values ranging from 0 (black) to 255 (white). A digital image can be considered as a large array of sampled points from the continuous image, each point with a particular quantised brightness. These points are the pixels which constitute the digital image.

7.4 Inverse Image

The inverse of an image is obtained by complementing the pixels values in an image. For binary images, the pixels with value 0 are replaced with 1 and the pixels with value 1 are replaced with 0. For an intensity image, the inverse image is obtained by subtracting each pixel value from the maximum pixel value that can be represented by the input. For example, if the input pixel values are given by $f(x, y)$ and the output pixel values of the inverse image are given by $g(x, y)$, then for a grey scale image $g(x, y) = 255 - f(x, y)$. The complement of a grey scale image is its photographic negative. Grey scale image complement operation is useful for enhancing the visibility of subtle brightness variations among grey levels in regions of a digital image where fine details are obscured. The grey scale complement operation belongs to the class of image processing algorithms often referred to as point operations. These functions are utilised to transform each input pixel in an image to a modified output pixel in a manner that is dependent only on the grey level value of the input pixel. A function that is used to map input brightness values to output brightness values is known as a grey scale transformation function. Due to the logarithmic response of the human eye to illumination level differences, subtle brightness variations among the grey levels in bright regions of an image may be difficult to detect. Performing a full complement operation on a grey scale digital image reverses the brightness range of the image and produces the equivalent of a photographic negative, which can improve the visibility of grey level variations in bright regions. In a similar manner, low contrast images sometimes contain regions where specimen detail is obscured due to inadequate illumination. Applying a complement operation to images having poor contrast due to improper illumination can often improve the visibility of darker details by making them brighter. An effect known as solarisation can be achieved by reversing only part of the input brightness range of an image. This type of complement operation is referred to as a partial complement, and can be used to improve the visibility of heavily shadowed regions in a digital image.

7.5 Red, Green and Blue, Grey Images and Histograms

The different types of digital images are as follows: (a) binary images: each pixel is black or white. As there are only two possible values for each pixel, it requires only 1 bit per pixel. This representation is suitable for printed or handwritten text, fingerprints and architectural plans; (b) grey scale images: each pixel is a shade of grey, normally from 0 (black) to 255 (white). Each pixel is represented by 8 bits or 1 byte. Other grey scale ranges are used, but generally they are a power of 2. Such images are used in x-rays, and images of printed works. A total of 256 different grey levels are sufficient for the recognition of most natural objects; (c) true colour or red, green and blue (RGB) images: each pixel has a particular colour, that colour being described by the amount of RGB in it. If each of these components has a range 0–255, this gives a total of $255^3 = 16,777,216$ different possible colours in the image. Such images are also called 24-bit colour images. These images are also considered as consisting of a stack of three matrices, representing RGB values for each pixel. For every pixel there are three values; (d) indexed images: most colour images have only a small subset of the more than 16 million possible colours. For convenience of storage and file handling, the image has an associated colour map, or colour palette, which is a list of all colours used in that image. Each pixel has a value which is an index to the colour in the map. It is convenient if an image has 256 colours or fewer, for then the index values will require only 1 byte each to store.

Given a grey scale image, its histogram consists of the distribution of its grey levels. It is a graph indicating the number of times each grey level occurs in the image. The histogram is simply the count of the number of pixels at each grey level value. For an 8-bit image, an accumulation table with 256 entries is needed. For higher bit depths, a table with the appropriate number of entries should be used. The appearance of an image can be deduced from its histogram. In a dark image, the grey levels are clustered at the lower end. In a uniformly bright image, the grey levels are clustered at the upper end and in a well contrasted image, the grey levels are spread out over the entire range. In a poorly contrasted image, the grey values are all clustered together in the centre of the histogram. To automatically determine the best values for brightness and gain control, one approach is to look at the darkest and brightest pixel values in an image and map them to pure black and white. Another approach is to find the average value in the image, map it to grey, and expand the range so that it more closely fills the values that can be displayed [8]. To simultaneously brighten some dark values and darken some light values, while still using the full extent of the available dynamic range, one popular algorithm is the histogram equalisation [9]. This gives an intensity mapping function which uses the cumulative distribution function such that the resulting histogram is flat. The resulting image may lack contrast and have a muddy look. One way to compensate for this is to partially compensate for the histogram unevenness by using a mapping function $\alpha g(I) + (1 - \alpha)f(I)$, which is a linear blend between the cumulative distribution function $g(I)$ and the identity transform. The resulting image maintains the original grey scale distribution while having a more appealing balance.

While global histogram equalisation can be useful, for some images it might be preferable to apply different kinds of equalisation in different regions. Instead of computing a single histogram equalisation curve, the image can be sub-divided into $M \times M$ pixel blocks and separate histogram equalisation can be performed in each sub-block. The resulting image exhibits a lot of blocking artefacts which is due to intensity discontinuities at block boundaries. A more efficient approach is to compute non-overlapped block-based equalisation functions, but to then smoothly interpolate the transfer functions between blocks. This technique is known as adaptive histogram equalisation (AHE) and its contrast limited version is known as CLAHE [10]. The weighting function for a given pixel can be computed as a function of its horizontal and vertical position within a block.

7.6 Image Segmentation

Segmentation checks each individual pixel to see whether it belongs to an object of interest or not. This generally produces a binary image. A pixel has the value one if it belongs to the object; otherwise it is zero. Segmentation is the operation at the threshold between low-level image processing and image analysis. After segmentation, it is known which pixel belongs to which object. The image is parted into regions and the discontinuities as the boundaries between the regions. After segmentation, the shape of objects is analysed with different operations. Pixel-based methods use the grey values of the individual pixels. Region-based methods analyse the grey values in larger areas. Finally, edge-based methods detect edges and then try to follow them. The common limitation of all these approaches is that they are based only on local information. Even then they use this information only partly. Pixel-based techniques do not even consider the local neighbourhood. Edge-based techniques look only for discontinuities while region-based techniques analyse homogeneous regions. In situations where the geometric shape of an object is known, model-based segmentation can be applied. Image segmentation is one of the most important steps leading to the analysis of processed image data. The main goal is to divide an image into parts that have a strong relation with objects or areas contained in the image. This processing is context independent; no object-related model is used and no knowledge about expected segmentation results contribute to the final segmentation. These methods can be divided into three groups according to the dominant features they employ: (i) global knowledge about an image or its part, (ii) edge-based segmentation and (iii) region-based segmentation.

7.6.1 Thresholding

Many objects or regions are characterised by constant reflectivity or light absorption of their surfaces. A brightness constant or threshold can be determined to segment objects and background. This is computationally inexpensive and fast. Segmenting an image R gives a finite set of regions $R_1, R_2,...,$ R_n, where $R = \bigcup_{i=1}^{n} R_i$; $R_i \cap R_j = \phi, i \neq j$. Thresholding is the transformation of an input image f to an output segmented binary image

$$g(x,y) = \begin{cases} 1, & \text{for } f(x,y) \geq T \\ 0, & \text{for } f(x,y) < T \end{cases}$$

where T is the threshold, $g(x, y) = 1$ for the objects and $g(x, y) = 0$ for the background or vice versa. This is a suitable segmentation method if objects do not touch each other, and if their grey levels are clearly distinct from background grey levels. Global thresholding or using a single threshold for the whole image may be successful under very unusual circumstances as even in very simple images there are likely to be grey-level variations in objects and background due to non-uniform lighting. So, adaptive thresholding or variable thresholds are used, in which threshold value varies over an image as a function of local image characteristics. One way is to divide the image into sub-images and determine a threshold independently in each sub-image and each sub-image is then processed with respect to its local threshold. Semi-thresholding which makes human-assisted analysis easier is given by

$$g(x,y) = \begin{cases} f(x,y) & \text{for } f(x,y) \geq T \\ 0 & \text{for } f(x,y) < T \end{cases} \tag{7.51}$$

This aims to mask out the image background, leaving grey level information present in the objects. Threshold can also be applied if the values $f(x, y)$ do not represent grey levels, but represents gradient, a local texture property or the value of any other image decomposition criterion. Threshold detection is based on histograms. If an image consists of objects of approximately the same grey level that differs from the grey level of the background, the resulting histogram is bi-modal. Object pixels form one of its peaks, while background pixels form the second peak. Threshold is determined as the grey value that has a minimum histogram value between the two maxima. If the histogram is multi-modal, more thresholds are determined at minima between any two maxima. Optimal thresholding is based on approximation of the histogram using a weighted sum of two or more probability densities with normal distribution. The threshold is set as the closest grey value corresponding to the minimum probability between the maxima of two or more normal distributions, which results in minimum error segmentation. This works well even if the histogram is not bi-modal. This assumes that regions of two grey values are present in the image. The mean grey value of the object and the background are computed using a global threshold T. The threshold is updated as the average of the means and the process continued till the threshold does not change in successive iterations.

7.6.2 Edge-Based Segmentation

Edge-based segmentation represents a large group of methods based on information about edges in the image. The edges mark image locations of discontinuities in grey values, colour and texture. Simple thresholding of an edge image can be applied to remove small edge values corresponding to non-significant grey level changes resulting from noise and small illumination irregularities. A problem with this is the thickening where there should be a simple thin edge. This is rectified if edges carry directional information by performing non-maximal suppression to suppress multiple responses in the neighbourhood of single boundaries. In Canny edge detector [11], the hysteresis approach is used to filter output of the edge detector. Edge magnitudes greater than a threshold T_1 are marked as edges and edge magnitudes less than T_0 are assumed to be noise induced. Edge magnitudes in the range $[T_0, T_1]$ are marked as an edge if such a pixel borders another already marked as an edge.

7.6.3 Region-Based Segmentation

Edge-based segmentation methods find borders between regions but region-based segmentation methods construct regions directly. Homogeneity is an important property of regions and is used as the main segmentation criterion, where the basic idea is to divide an image into zones of maximum homogeneity. The criterion for homogeneity is based on grey values, colour,

texture and shape. Region growing segmentation should satisfy the following conditions of complete segmentation:

$$R = \bigcup_{i=1}^{n} R_i \quad \text{and} \quad R_i \cap R_j = \phi, \quad i \neq j \tag{7.52}$$

and the maximum region homogeneity conditions

$$H(R_i) = \text{TRUE}, i = 1, 2, \ldots, n \quad \text{and} \quad H(R_i \cup R_j) = \text{FALSE}, i \neq j, R_i \text{ adjacent to } R_j \tag{7.53}$$

Region merging starts with an over segmented image in which regions satisfy
the homogeneity criteria. Region splitting begins with an under segmented
image which does not satisfy the criteria and the regions are sequentially
split. A combination of split and merge use pyramid image representation.
In watershed segmentation, the algorithm starts with finding a downstream
path for each pixel to local minima of image surface. A region or catchment
basin is defined as the set of pixels for which their respective downstream
paths all end up in the same altitude minimum. In another approach, each
grey value minimum represents one region and the strategy is to start filling
the regions from the bottom.

7.7 Noise Processes in an Observed/Acquired Image

Real images are often degraded by some random errors called noise. Noise
can occur during image capture, transmission, or processing and may be
dependent on, or independent of image content. Noise is described by its
probabilistic characteristics. Idealised noise, called white noise, has a constant
power spectrum, meaning that its intensity does not decrease with increased
frequency. This is frequently applied as the worst approximation of degradation and its advantage is that it uses simplified calculations. A special case
of noise is Gaussian noise. This noise is a very good approximation to noise
that occurs in many practical cases. When an image is transmitted through
some channel, noise which is independent of the image occurs. This signal independent degradation is called additive noise. The noise magnitude
depends in many cases on the signal magnitude. This is multiplicative noise.
Quantisation noise occurs when insufficient quantisation levels are used and
false contours appear. Impulsive noise means that an image is corrupted with
individual noisy pixels whose intensity differs significantly from that of the
neighbourhood. Salt and pepper noise describes saturated impulsive noise
and this refers to an image corrupted with white and/or black pixels.

Noise is any degradation in the image caused by external disturbance and is defined as an unwanted component of the image. Since image sensors count photons, especially in low light situations and the number of photons counted is a random quantity, images often have photon counting noise. The grain noise in photographic films is sometimes modelled as Gaussian and sometimes as Poisson. Many images are corrupted by salt and pepper noise, as if someone had sprinkled black and white dots on the image. Other noises include quantisation noise and speckle noise in coherent light situations. If an image is sent electronically from one place to another, via satellite or wireless transmission or through cables, errors occur in the transmitted images. These errors appear on the image in different ways depending on the type of disturbance.

Let $f(x, y)$ denote an image. The image is decomposed into a desired component $I(x, y)$, and a noise component $v(x, y)$. The most common decomposition is additive

$$f(x,y) = I(x,y) + v(x,y) \tag{7.54}$$

For example, Gaussian noise is considered to be an additive component. The second most common decomposition is multiplicative

$$f(x,y) = I(x,y)v(x,y) \tag{7.55}$$

An example of noise often modelled as multiplicative is speckle. The multiplicative model can be transformed into the additive model by taking logarithms and the additive model into the multiplicative one by exponentiation. The additive model is most appropriate when the noise in that model is independent of f. There are many applications of the additive model. Thermal noise, photographic noise and quantisation noise obey the additive model well. The multiplicative model is most appropriate when the noise in that model is independent of f. One common situation in which the multiplicative model is used is for speckle in coherent imagery. Finally, there are important situations in which neither the additive nor the multiplicative model fits the noise well. Poisson counting noise and salt and pepper noise fit neither model well.

7.7.1 Salt and Pepper Noise

This is also called impulse noise, shot noise or binary noise. This is caused by sharp, sudden disturbances in the image. Its appearance is randomly scattered white or black or both pixels over the image. Salt and pepper noise refers to a wide variety of processes that result in image degradation: only a few pixels are noisy, but they are very noisy. The effect is similar to sprinkling white and black dots – salt and pepper – on the image. One example where salt and pepper noise arises is in transmitting images over noisy digital links. Let each pixel be quantised to B bits in the usual fashion. The contribution to the error from the most significant bit is approximately three times that of all the other bits. The pixels whose most significant bits are changed will likely appear as black or white dots. Salt and pepper noise is an example of heavy-tailed noise. Let $f(x, y)$ be the original image and $v(x, y)$ be the image after it has been altered by salt and pepper noise, then we have

$$Pr(v = f) = 1 - \alpha, \quad Pr(v = \text{max}) = \alpha/2, \quad Pr(v = \text{min}) = \alpha/2 \quad (7.56)$$

where max and min are the maximum and minimum image values, respectively. For 8-bit images, max $= 255$ and min $= 0$. With probability $1 - \alpha$ the pixels are unaltered and with probability α the pixels are changed to the largest or smallest values. The altered pixels look like black and white dots sprinkled over the image.

7.7.2 Gaussian Noise

This is an idealised form of white noise, which is caused by random fluctuations in the image. This can be observed by watching a TV slightly mistuned to a particular channel. Gaussian noise can be modelled by random values added to an image. It is widely used to model thermal noise and, under some often reasonable conditions, is the limiting behaviour of other noises, for example, photon counting noise and film grain noise. The density function of univariate Gaussian noise q, with mean μ and variance σ^2 is given as

$$p_q(x) = \frac{1}{\sqrt{2\pi\sigma^2}} e^{-(x-\mu)^2/2\sigma^2} \quad \text{for} - \infty < x < \infty \quad (7.57)$$

The range of values of x where the probability density is non-zero, is infinite in both the positive and negative directions. But, if the image is considered as an intensity map, then the values must be non-negative. In other words, the noise cannot be strictly Gaussian. If it were, there would be some non-zero probability of having negative values. In practice, however, the range of values of the Gaussian noise is limited to approximately $\pm 3\sigma$ and the Gaussian density is a useful and accurate model for many processes. If necessary, the noise values can be truncated to keep $f > 0$. The effect of a small amount of Gaussian noise introduces a fuzzy appearance to the overall image. The filtered image is usually visually less pleasing than the original noisy one. When the noise is increased the degradation is much more objectionable. Various filtering techniques can improve the quality, though usually at the expense of some loss of sharpness.

7.7.3 Speckle Noise

This can be modelled by random values multiplied by pixel values. This is also called multiplicative noise. This is a major problem in some radar applications. Speckle is one of the more complex image noise models. It is signal dependent, non-Gaussian and spatially dependent. In coherent light imaging, an object is illuminated by a coherent source, usually a laser or a radar transmitter. When coherent light strikes a surface, it is reflected back. Because of the microscopic variations in the surface roughness within one pixel, the received signal is subjected to random variations in phase and amplitude. Some of these variations in phase add constructively, resulting in strong intensities, and others add deconstructively, resulting in low intensities. This variation is called speckle. The exponential density is tailed heavier than the Gaussian density, meaning that much greater excursions from the mean occur. In particular, the standard deviation of f equals the mean of f, that is, the typical deviation in the reflected intensity is equal to the typical intensity. It is this large variation that causes speckle to be so objectionable to human observers.

7.7.4 Quantisation and Uniform Noise

Quantisation noise results when a continuous random variable is converted to a discrete one or when a discrete random variable is converted to one with fewer levels. In images, quantisation noise often occurs in the acquisition process. The image may be continuous initially, but to be processed it must be converted to a digital representation. Quantisation noise is usually modelled as uniform. Uniform noise is the opposite of the heavy-tailed noises. Its tails are very light, that is, zero. When the number of quantisation levels is small, the quantisation noise becomes signal dependent. The noise is correlated on a pixel-by-pixel basis and is not uniformly distributed. The general appearance of an image with too few quantisation levels may be described as scalloped. Fine gradations in intensities are lost. There are large areas of constant colour separated by clear boundaries. The effect is similar to transforming a smooth ramp into a set of discrete steps. The smooth gradations are replaced by large constant regions separated by noticeable discontinuities visualised as false contours.

7.7.5 Photon Counting Noise

Fundamentally, most image acquisition devices are photon counters. Let a denote the number of photons counted at some location in an image. Then, the distribution of a is usually modelled as Poisson process with parameter λ given by

$$P(a = k) = e^{-\lambda}\lambda^k/k!, \quad k = 0, 1, 2, \ldots \tag{7.58}$$

This noise is also called Poisson noise or Photon counting noise.

7.7.6 Photographic Grain Noise

Photographic grain noise is a characteristic of photographic films. It limits the effective magnification one can obtain from a photograph. Slow film has a large number of small fine grains, whereas fast film has a smaller number of larger grains. The small grains give slow film a better, less grainy picture; the large grains in fast film cause a grainier picture.

7.7.7 Periodic Noise

In this, the image is subject to a periodic, rather than a random disturbance. The effect is of bars over the image.

Salt and pepper noise, Gaussian noise and speckle noise can all be cleaned by using spatial filtering techniques. Periodic noise requires the use of frequency domain filtering. This is because the other forms of noise can be modelled as local degradations, but periodic noise is a global effect.

7.8 Image Feature Extraction Methods

Feature extraction and matching are an essential component of many image processing applications. The first kind of features is specific locations in the images, such as corners, patches or peaks. These kinds of localised features are called interest or key point features and are described by the appearance of patches of pixels surrounding the point location. Another class of important features are edges. These features are matched based on their orientation and local appearance and are a good representation of object boundaries. There are two main approaches to finding features and their correspondences. The first is to find features in an image that can be accurately tracked using a local search technique, such as correlation or least squares. The second is to independently detect features in all the images and then match features based on their local appearances. The former approach is suitable when images are taken from nearby viewpoints or in rapid succession, while the latter is more suitable when large motion or appearance change occurs in object recognition applications. The feature detection and matching can be split into four separate stages. In the feature detection stage, each image is searched for locations that are likely to match in other images. In the feature description stage, each region around detected key point locations is converted into a more compact and stable invariant descriptor that can be matched against other descriptors. The feature matching stage searches for likely matching candidates in other images. The feature tracking stage is an alternative to the third stage that searches a small neighbourhood around each detected feature. An example of all these stages is found in David

Lowe's paper [12] which describes the development of scale invariant feature transform (SIFT). The different stages of this algorithm are: (i) construction of scale space, (ii) calculation of difference of Gaussians (DoG), (iii) location of DoG extrema, (iv) sub-pixel location of potential feature points, (v) filtering of edge and low contrast responses, (vi) assigning orientations for the key points and (vii) building of key point descriptors. The simplest possible matching criterion for comparing two image patches is their weighted summed square difference given by

$$E_{WSSD}(u,v) = \sum_i w(x_i, y_i)\left[f_1(x_i + u, y_i + v) - f_0(x_i, y_i)\right]^2 \qquad (7.59)$$

where f_0 and f_1 are the two images being compared, (u, v) is the displacement vector, $w(x, y)$ is a spatially varying weighting or window function and the summation i is over all the pixels in the patch. To compute how stable the metric is with respect to small variations in position $(\Delta u, \Delta v)$, an image patch is compared against itself, which is known as the autocorrelation function or surface given by

$$E_{AC}(\Delta u, \Delta v) = \sum_i w(x_i, y_i)\left[f_0(x_i + \Delta u, y_i + \Delta v) - f_0(x_i, y_i)\right]^2 \qquad (7.60)$$

Using a Taylor series expansion of the image function, the autocorrelation surface can be approximated as

$$E_{AC}(\Delta u, \Delta v) = \sum_i w(x_i, y_i)\left[f_0(x_i + \Delta u, y_i + \Delta v) - f_0(x_i, y_i)\right]^2$$

$$\approx \sum_i w(x_i, y_i)\left[f_0(x_i, y_i) + \nabla f_0(x_i, y_i) \cdot (\Delta u, \Delta v) - f_0(x_i, y_i)\right]^2$$

$$= \sum_i w(x_i, y_i)\left[\nabla f_0(x_i, y_i) \cdot (\Delta u, \Delta v)\right]^2$$

$$= (\Delta u, \Delta v)^T A(\Delta u, \Delta v) \qquad (7.61)$$

where $\nabla f_0(x_i, y_i) = ((\partial f_0/\partial x), (\partial f_0/\partial y))(x_i, y_i)$ is the image gradient at (x_i, y_i). The classic Harris detector [13] uses a $[-2\ -1\ 0\ 1\ 2]$ filter, but other variants convolve the image with horizontal and vertical derivatives of a Gaussian function. The autocorrelation matrix

$$A = w * \begin{bmatrix} f_x^2 & f_x f_y \\ f_x f_y & f_y^2 \end{bmatrix}$$

where the weighted summations are replaced with discrete convolutions with the weighting kernel w. This matrix can be interpreted as a tensor or multi-band image, where the outer products of the gradients ∇f are convolved with a weighting function to provide a per pixel estimate of the local quadratic shape of the autocorrelation function. The inverse of the matrix A provides a lower bound on the uncertainty in the location of a matching patch and provides an indicator of which patches can be reliably matched. An eigen value analysis is performed on A, and since the larger uncertainty depends on the smaller eigen value, maxima in the smaller eigen value is found to locate good features to track. A similar quantity proposed by Harris and Stephen is

$$\det(A) - \alpha \text{ trace } (A)^2 = \lambda_0 \lambda_1 - \alpha(\lambda_0 + \lambda_1)^2 \tag{7.62}$$

with $\alpha = 0.06$. Unlike eigen value analysis, this does not require the use of square roots and yet is still rotationally invariant and also down weights edge-like features where $\lambda_1 \gg \lambda_0$. Triggs [14] suggests using the quantity $\lambda_0 - \alpha\lambda_1$ with $\alpha = 0.05$, which reduces the response at one-dimensional (1D) edges. The 2×2 Hessian can be extended to detect points that are also accurately localisable in scale and rotation. Another quantity is the harmonic mean given by [15]

$$\frac{\det A}{tr A} = \frac{\lambda_0 \lambda_1}{\lambda_0 + \lambda_1} \tag{7.63}$$

which is a smoother function in the region where $\lambda_0 \approx \lambda_1$. An outline of a basic feature detection algorithm is given as: (i) horizontal and vertical derivatives of the image f_x and f_y are computed by convolving the original image with derivates of the Gaussians, (ii) the three images corresponding to the outer products of these gradients are computed, (iii) these images are convolved with a larger Gaussian, (iv) a scalar interest measure using one of the formulas is computed and (v) local maxima above a certain threshold is found and reported as detected feature point locations. While most feature detectors look for local maxima in the interest function, this leads to an uneven distribution of feature points across the image. The points will be denser in regions of high contrast. To mitigate this problem, features that are both local maxima and whose response is significantly greater than that of all of its neighbours within a radius r are detected. Detecting features at the finest scale may not be appropriate in many applications. When matching images with little high-frequency details, fine scale features may not exist. The solution is to extract features at a variety of scales by performing the same operations at multiple resolutions in a pyramid and then matching features at the same level. The scale of the object in the image is unknown for many applications. So, instead of extracting features at many different scales and matching all of them, features that are stable in both location and scale are extracted. Lindeberg [16] proposed using extrema in the LoG function as interest point locations. Lowe [12] proposed computing a set of sub-octave difference of Gaussian (DoG) filters, looking for a space and scale three-dimensional (3D) maxima using a quadratic fit. Three sub-octave levels are considered, which corresponds to a quarter octave pyramid. As with Harris operator, pixels where there is strong asymmetry in the local curvature of the DoG are rejected. This is evaluated by computing the local Hessian of the difference image D as

$$H = \begin{bmatrix} D_{xx} & D_{xy} \\ D_{xy} & D_{yy} \end{bmatrix}$$

and then rejecting key points for which $(tr(H)^2/\det(H)) > 10$. To add a scale detection mechanism to the Harris corner detector, the LoG function is evaluated at each detected point and only those points for which the Laplacian is larger or smaller than both its coarser and finer level values are retained.

To deal with image rotation, the descriptors have to be rotationally invariant. But such descriptors have poor discriminability as they map different looking patches to the same descriptor. A better method is to estimate a dominant orientation at each detected key point by extracting a scaled and oriented patch around the detected point and using it to form a feature descriptor. The simplest is the average gradient within a region around the key point with a Gaussian weighting function. However, the averaged signed gradient can be small and unreliable. A more reliable technique is to compute the histogram of orientations around the key point. In SIFT, a 36-bin histogram of edge orientations weighted by both gradient magnitude and Gaussian distance to the centre is computed and peaks within 80% of the global maximum are found. A more accurate orientation estimate is then computed using a 3-bin parabolic fit.

To introduce affine invariance, an ellipse is fitted to the autocorrelation or Hessian matrix and then the principal axes and ratios of this fit is used as the affine coordinate frame. Another important affine invariant region detector is the maximally stable extremal region (MSER) detector. To detect MSERs, binary regions are computed by thresholding the image at all possible grey levels. As the threshold is changed, regions whose rate of change of area with respect to the threshold is minimal are defined as maximally stable.

Local appearance of image patches vary from image to image and so, image descriptors should be made more invariant to such changes while preserving discriminability between different patches. For tasks that do not exhibit large amounts of foreshortening, simple normalised intensity patches perform reasonably well and are simple to implement. These multi-scale oriented patches are sampled at a spacing of 5 pixels relative to the detection scale, using a coarser level of the image pyramid to avoid aliasing.

SIFT features are formed by computing the gradient at each pixel in a 16×16 window around the detected key point, using the appropriate level of the Gaussian pyramid at which the key point was detected. The gradient magnitudes are down weighted by a Gaussian fall off function to reduce the influence of gradients far from the centre. In each 4×4 quadrant, a gradient orientation histogram is formed by adding the weighted gradient value to one of the eight orientation histogram bins. To reduce the effects of location and dominant orientation misestimation, each of the original 256 weighted gradient magnitudes is softly added to $2 \times 2 \times 2$ histogram bins using trilinear interpolation. The resulting 128 non-negative values form a raw version of the SIFT descriptor vector. To reduce the effects of contrast or gain, the 128-dimensional vector is normalised to unit length. To further make the descriptor robust to other photometric variations, values are clipped to 0.2 and the resulting vector is once again renormalised to unit length. A simpler way to compute descriptors inspired by SIFT is to compute the x and y derivatives over a 39×39 patch and then reduce the resulting 3042-dimensional vector to 36 using principal component analysis (PCA). Another popular variant of SIFT is SURF [17], which uses box filters to approximate the derivatives and integrals used in SIFT. A gradient location-orientation histogram (GLOH) descriptor [18] is a variant on SIFT that uses a log-polar binning structure. The spatial bins are of radius 6, 11 and 15, with eight angular bins except for the central region, for a total of 17 spatial bins and 16 orientation bins. The 272-dimensional histogram is then projected onto a 128-dimensional descriptor using PCA trained on a large database. GLOH has the best performance overall, and outperforms SIFT by a small margin.

The gauge SURF (G-SURF) descriptor [19] is based on second-order multi-scale gauge derivatives. While the standard derivatives used to build a SURF descriptor are all relative to a single chosen orientation, gauge derivatives are evaluated relative to the gradient direction at every pixel. Like standard SURF descriptors, G-SURF descriptors are fast to compute due to the use of integral images, but have extra matching robustness due to the extra invariance offered by gauge derivatives. With gauge coordinates, every pixel in the image is described in such a way that for the same 2D local structures, the description of the structure is always the same, even if the image is rotated. This is possible since multi-scale gauge derivatives are rotation and translation invariant. Instead of local first-order spatial derivatives, G-SURF descriptors measure per pixel information about image blurring and edge or detail enhancing, resulting in more discriminative descriptors. G-SURF descriptors outperform or approximate state of the art methods in accuracy while exhibiting low computational demands making it suitable for real-time applications. This family of descriptors comprises several descriptors of different dimensions based on second-order multi-scale gauge derivatives. The phase space-based SURF (P-SURF) descriptor [20] uses phase space to capture more structure information of local image patterns and improve the performance of SURF. This method is more efficient than SIFT since it neither computes the gradient orientation nor applies any interpolation to the feature representation while preserving considerable distinctiveness. Depending on the application some descriptors may be preferred instead of others. For example, for real-time applications a low-dimensional descriptor should be preferred instead of a high-dimensional one, whereas for image matching applications considering severe image transformations one can expect a higher recall by using high-dimensional descriptors.

The highly influential SIFT [12] features have been widely used in applications from mobile robotics to object recognition, but are relatively expensive to compute and are not suitable for some applications with real-time demands. Inspired by SIFT, the SURF features define both a detector and a descriptor. SURF features exhibit better results than previous schemes with respect to repeatability, distinctiveness and robustness, but at the same time can be computed much faster due to use of integral images [21]. Recently, Agrawal et al. [22] proposed some modifications of SURF in both the detection and description steps. They introduced Centre Surround Extrema (CenSurE) features and showed that they outperform previous detectors and have better computational characteristics for real-time applications. Their variant of the SURF descriptor, modified SURF (M-SURF), efficiently handles the descriptor boundary problem and uses a more intelligent two-stage Gaussian weighting scheme in contrast to the original implementation which uses a single Gaussian weighting step. All the mentioned approaches rely on the use of the Gaussian scale space framework to extract features at different scales. An original image is blurred by convolution with Gaussian kernels of successively large standard deviation to identify features at increasingly large scales. The main drawback of the Gaussian kernel and its set of partial derivatives is that both interesting details and noise are blurred away to the same degree. It seems to be more appropriate in feature description to make blurring locally adaptive to the image data so that noise will be blurred, while at the same time details or edges will remain unaffected. In this way, distinctiveness is increased when describing an image region at different scale levels. I-SURF [23] modifies the SURF descriptor by considering the boundary effect of the adjacent sub-regions, and introduces index vector to speed up matching. The descriptor used in CenSurE is based on Upright SURF descriptor which is a scale invariant only version of SIFT, and is called modified upright SURF (MU-SURF). The main difference between U-SURF and MU-SURF is that in MU-SURF each two adjacent sub-regions have an overlap of 2 pixels. The main difference is the larger size of MU-SURF's descriptor window and the sub-regions it uses to compute Haar wavelet responses. For each sub-region, the Haar wavelet responses are weighted with a pre-computed Gaussian centred on the sub-region centre. To reduce the matching process time, the features are indexed based on their signs, since CenSurE features are signed based on them being bright or dark blobs. Speeded up surround extrema (SUSurE) [24] exploit the notion of sparse sampling for both the detection and description stages as dictated by the filter responses from the image.

In spirit, non-linear diffusion shares some similarities with the geometric blur proposed by Berg and Malik [25], in which the amount of Gaussian blurring is proportional to the distance from the point of interest. From their definition, gauge derivatives are local invariants. Schmid and Mohr [26] used the family of local invariants known as local jet for image matching applications. Their descriptor vector contained eight invariants up to third order for every point of interest in the image.

When the performance of the local jet is compared against other descriptors such as steerable filters, image moments or SIFT, local jet exhibits poor performance compared to SIFT due to the fixed settings such as a fixed image patch size and a fixed Gaussian derivative scale. In addition, invariants of higher order are more sensitive to geometric and photometric distortions than first-order methods. Brown et al. [27] proposed a framework for learning discriminative local dense image descriptors from training data. They describe a set of building blocks for building discriminative local descriptors that can be combined together and jointly optimised to minimise the error of a nearest-neighbour classifier. Every pixel in the image is fixed separately in its own local coordinate frame defined by the local structure itself and consisting of the gradient vector $\bar{w}$ and its perpendicular direction $\bar{v}$

$$\bar{w} = \left(\frac{\partial L}{\partial x}, \frac{\partial L}{\partial y} \right) = \frac{1}{\sqrt{L_x^2 + L_y^2}} \cdot (L_x, L_y) \tag{7.64}$$

$$\bar{v} = \left(\frac{\partial L}{\partial y}, \frac{-\partial L}{\partial x} \right) = \frac{1}{\sqrt{L_x^2 + L_y^2}} \cdot (L_y, -L_x) \tag{7.65}$$

Here, L denotes the convolution of the image $f(x, y)$ with a 2D Gaussian kernel $g(x, y, \sigma)$, where σ is the kernel's standard deviation or scale parameter

$$L(x, y, \sigma) = f(x, y) * g(x, y, \sigma) \tag{7.66}$$

Derivatives can be taken up to any order and at multiple scales for detecting features of different sizes. To obtain gauge derivatives, directional derivatives with respect to a fixed gradient direction (L_x, L_y) are needed. The $\bar{v}$ direction is tangent to the lines of constant intensity, whereas $\bar{w}$ points in the direction of the gradient, thus $L_v = 0$; $L_w = \sqrt{L_x^2 + L_y^2}$. Every derivative expressed in gauge coordinates is an orthogonal invariant. The first-order derivative is the derivative in the gradient direction, and the gradient is an invariant itself. There is no change in the luminance if moved tangentially to the constant intensity lines. By using gauge coordinates, a set of invariant derivatives are obtained up to any order and scale that can be used efficiently for image description and matching. Of special interest, are the second-order gauge derivatives

$$
\begin{aligned}
L_{ww} &= \frac{L_x^2 L_{xx} + 2L_x L_{xy} L_y + L_y^2 L_{yy}}{L_x^2 + L_y^2} \\
L_{vv} &= \frac{L_y^2 L_{xx} - 2L_x L_{xy} L_y + L_x^2 L_{yy}}{L_x^2 + L_y^2}
\end{aligned}
\tag{7.67}
$$

These two gauge derivatives can be obtained as the product of gradients in $\vec{w}$ and $\vec{v}$ directions and the 2×2 second-order derivatives or Hessian matrix

$$
\begin{aligned}
L_{ww} &= \frac{1}{L_x^2 + L_y^2} (L_x \quad L_y) \begin{pmatrix} L_{xx} & L_{xy} \\ L_{yx} & L_{yy} \end{pmatrix} \begin{pmatrix} L_x \\ L_y \end{pmatrix} \\
L_{vv} &= \frac{1}{L_x^2 + L_y^2} (L_y \quad -L_x) \begin{pmatrix} L_{xx} & L_{xy} \\ L_{yx} & L_{yy} \end{pmatrix} \begin{pmatrix} L_y \\ -L_x \end{pmatrix}
\end{aligned}
\tag{7.68}
$$

Here L_{vv} is used as a ridge detector. Ridges are elongated regions of approximately constant width and intensity, and at these points the curvature of the isophotes is high. L_{ww} gives information about gradient changes in the gradient direction. The Gaussian kernel and its set of partial derivatives provide the unique set of operators for the construction of linear scale space under certain conditions. Some examples of algorithms that rely on the Gaussian scale space framework are SIFT and SURF invariant features. However, details are blurred in Gaussian scale space during evolution. The advantage of blurring is the removal of noise, but relevant image structures like edges are blurred and drift away from their original locations during evolution. In general, a good solution should be to make the blurring locally adaptive to the image yielding the blurring of noise, while retaining details or edges. In this family of descriptors, the first-order local derivatives L_x and L_y are replaced with the gauge derivatives L_{vv} and L_{ww} and do not perform any image evolution through a non-linear scale space. These descriptors measure information about blurring (L_{ww}) and edge enhancing (L_{vv}) for different scale levels. Another difference between first-order local derivatives and gauge ones is that gauge derivatives are intrinsically weighted with the strength of the gradient L_w. That is, the weighting is intrinsically related to the image structure itself, and no artificial weighting such as Gaussian weighting is needed. This is an important advantage over other descriptors, such as, for example, SURF, where different Gaussian weighting schemes have been proposed to improve the performance of the original descriptor.

7.9 Image Transformation and Filtering Approaches

The low-level image transformation operations can be categorised as follows: (i) if the output intensity level at a certain pixel is strictly dependent on only the input intensity level at that point, such an operation is known as type 0 or a point operation. Point operations are quite frequently used in image segmentation, pixel classification, image summing, differencing and so on; (ii) if the output intensity level at a pixel depends on the input intensity levels of the neighbouring pixels as well, then such operations are termed type 1 or local operations. Examples of local operations are edge detection (ED), image filtering and so on; and (iii) if the operations are such that the output level at a point is dependent on some geometrical transformation, these operations are termed type 2 or geometrical operations. The simplest of image transforms are point operators, where each output pixel's value depends on only the corresponding input pixel value. Examples of such operators include brightness adjustment, contrast adjustments, colour correction and transformations. In the image processing literature, such operations are also known as point processes. A general image processing operator is a function that takes one or more input images and produces an output image. In the continuous domain, this can be denoted as

$$g(x, y) = h(f(x, y)) \tag{7.69}$$

where the functions f and g operate over some range, which can either be scalar or vector valued as for colour images and video. For discrete sampled images, the domain consists of a finite number of pixel locations (x, y). So, an image can be represented either by its colour, as a grid of numbers, or as a 2D function. Two commonly used point processes are multiplication and addition with a constant as below

$$g(x, y) = af(x, y) + b \tag{7.70}$$

The parameters a (>0) and b are called the gain and bias parameters and they control contrast and brightness, respectively. The bias and gain parameters can also be spatially varying

$$g(x, y) = a(x, y)f(x, y) + b(x, y) \tag{7.71}$$

This is used when simulating the graded density filter used by photographers to selectively darken the sky or when modelling vignetting in an optical system. Multiplicative gain is a linear operation, since it obeys the superposition principle

$$h(f_0(x,y) + f_1(x,y)) = h(f_0(x,y)) + h(f_1(x,y)) \qquad (7.72)$$

Operators such as image squaring are not linear. Another commonly used two input or dyadic operator is the linear blend operator

$$g(x,y) = (1 - \alpha)f_0(x,y) + \alpha f_1(x,y) \qquad (7.73)$$

By varying α from 0 to 1, this operator can be used to perform a temporal cross dissolve between two images as seen in slide shows and film production, or as a component of image morphing algorithms. One highly used non-linear transform that is often applied to images before further processing is gamma correction, which is used to remove the non-linear mapping between input radiance and quantised pixel values. To invert the gamma mapping applied by the sensor, the following transformation is used:

$$g(x,y) = [f(x,y)]^{1/\gamma} \qquad (7.74)$$

where a gamma value of $\gamma = 2.2$ is a reasonable fit for most digital cameras.

7.9.1 Linear Filtering

This is a spatial averaging operation. It produces an output image, which is a smooth version of the original image, devoid of the high spatial frequency components that may be present in the image. In particular, this operation is useful in removing visual noise, which generally appears as sharp bright points in the image. Such high spatial frequencies associated with these spikes are attenuated by the low-pass (LP) filter. High-pass filtering (HPF) of an image, on the other hand, produces an output image in which the low spatial frequency components are attenuated. The cut-off frequency at which lower frequencies are attenuated is varied by the selection of filter coefficients. HPF is used for edge enhancement. Since the sharpness of an image is related to the content of high-frequency components, LP filtering leads to blurring, while HPF is used for de-blurring. Subtracting a blurred version of the image from the original image leads to sharpening of the image. As the name suggests the unsharp masking technique is used for crisping the edges. Such a technique is used in the printing industry. A signal proportional to the unsharp or LP filtered version of the original noisy image is subtracted from the image, such that the resulting image is a crisp high contrast image. A gradient or a high-pass (HP) signal may be added to the original image, which may result in a better high contrast image. From this view point, the unsharp masking operation can be represented by

$$g(x,y) = f(x,y) + \gamma h(x,y) \qquad (7.75)$$

where $\gamma > 0$ and $h(x, y)$ is a suitably defined gradient at (x, y). This is also referred to as high emphasis filter, where the high-frequency components are emphasised while retaining the low-frequency components of the image. LP filters always result in blurring the image and quite often the crisp edges are blurred by averaging. To minimise this effect, a directional averaging filter can be used. Spatial averages are calculated in several directions as

$$g(x, y; \theta) = \frac{1}{N_0} f(x - k, y - l),$$

$$(k, l) \in W_0$$

(7.76)

where W_0 is the neighbourhood selected in the direction θ. The key to the implementation of effective directional smoothing is to identify a specific direction θ^* for which $|f(x, y) - g(x, y; \theta^*)|$ is minimum. Such a θ^* for which the above objective function is minimum, yields the desired result. The directional smoothing operation often prevents the edges from getting blurred resulting from the smoothing operation.

7.9.2 Median Filtering

In this case, the input pixel is replaced by the median of the pixels contained in the neighbourhood. The algorithm for median filtering requires arranging the pixel grey values in the neighbourhood in increasing or decreasing order and picking up the value at the centre of the array. Generally the size of the neighbourhood is chosen as odd number so that a well-defined centre value exists. If, however, the size of the neighbourhood is even the median is taken as the arithmetic mean of the two values at the centre.

7.9.3 2D Transforms

The 2D transforms of an image are extremely important areas of study in image processing [28,29]. The image output in the transformed space may be analysed, interpreted and further processed for implementing diverse image processing tasks. These transformations are widely used, since by using these transformations, it is possible to express an image as a combination of a set of basic signals, known as the basis functions. In case of a Fourier transform of an image these basis signals are sinusoidal signals with different periods which describe the spatial frequencies in an image. This implies that an image is decomposed into its constituent sinusoids, using the Fourier transform, and the amplitudes of various frequencies constitute the frequency spectrum of the image. The process of inverse Fourier transform operation involves synthesising the image by adding up its constituent frequencies. The notion of frequency, more specifically spatial frequency, is not a mere mathematical abstraction. On the other hand, interestingly the human vision system, which is a biological system, essentially performs the frequency analysis of the image incident on the retina of our eyes. Thus such transforms, such as the Fourier transform, reveal spectral structures embedded in the image that may be used to characterise the image. The 2D discrete Fourier transform (DFT) of a 2D signal $f(x, y)$ of dimension $M \times N$ with integer indices x and y running from 0 to $M - 1$ and $N - 1$, is represented by

$$F(u,v) = \sum_{x=0}^{M-1} \sum_{y=0}^{N-1} f(x,y) \exp\left[-j2\pi\left(\frac{ux}{M} + \frac{vy}{N}\right)\right] \tag{7.77}$$

The equivalent 2D inverse DFT is

$$f(x,y) = \frac{1}{MN} \sum_{u=0}^{M-1} \sum_{v=0}^{N-1} F(u,v) \exp\left[j2\pi\left(\frac{ux}{M} + \frac{vy}{N}\right)\right] \tag{7.78}$$

DFT is normally implemented using fast Fourier transform (FFT) algorithms. The number of complex multiplications and additions to compute DFT is $O(N^2)$. By employing a divide and conquer approach based on successive division in FFT the computational complexity is reduced to $O(N \log_2 N)$.

Discrete cosine transform (DCT) is the basis for many image and video compression standards like JPEG and MPEG. The 2D forward DCT of a block of $M \times N$ pixels of a 2D image $f(x,y)$ is formulated as follows:

$$F(u,v) = \frac{2}{\sqrt{MN}} C(u)C(v) \sum_{x=0}^{M-1} \sum_{y=0}^{N-1} f(x,y) \cos\left[\frac{\pi(2x+1)u}{2N}\right] \cos\left[\frac{\pi(2y+1)v}{2M}\right]$$

for $u = 0,1,\ldots,N-1$ and $v = 0,1,\ldots, M-1$ where

$$C(k) = \begin{cases} \dfrac{1}{\sqrt{2}}, & \text{for } k = 0 \\ 1, & \text{otherwise} \end{cases} \tag{7.79}$$

The 2D inverse DCT is computed as

$$f(x,y) = \frac{2}{\sqrt{MN}} \sum_{u=0}^{N-1} \sum_{v=0}^{M-1} C(u)C(v)F(u,v) \cos\left[\frac{\pi(2x+1)u}{2N}\right] \cos\left[\frac{\pi(2y+1)v}{2M}\right] \tag{7.80}$$

The discrete Walsh Hadamard transform (WHT) of a function $f(x)$ is denoted by

$$W(u) = \frac{1}{N} \sum_{x=0}^{N-1} f(x)g(x,u)$$

$$g(x,u) = \frac{1}{N}\left[(-1)^{b_i(x)b_{n-1-i}(u)}\right]$$

(7.81)

$$n = \log_2 N$$

where $b_i(z)$ is the ith bit in binary representation of z. The recursive relation to generate a WHT kernel is represented as

$$H_N = \frac{1}{\sqrt{2}}\begin{bmatrix} H_{N/2} & H_{N/2} \\ H_{N/2} & -H_{N/2} \end{bmatrix}$$

(7.82)

The advantage of using the WHT is the simplicity in its computation due to the binary nature of the transform kernel.

7.9.4 Wavelet Transform

In Chapter 6, we have discussed independent component analysis and PCA for general signals and images. We also discussed wavelet transforms. Here, we discuss certain further aspects of these transforms. The PCA forms the basis of the Karhunen Loeve (KL) transform for compact representation of data. One of the major problems in image processing is dimensionality reduction. In practical problems, quite often the features that are chosen are correlated with each other and a number of them are useless so far as their discriminability is concerned. If the number of features can be reduced, that is, reduce the dimensionality of the feature space, then better accuracy will be achieved with lesser storage and computational complexity. The Fourier transform is an analysis of global frequency content in the signal. There are applications in image processing that require the analysis to be localised in the spatial domain. This can be handled by introducing spatial frequency into Fourier analysis. The classical way of doing this is through what is called the Windowed Fourier Transform. The central idea of windowing is reflected in the Short-time Fourier transform (STFT). The STFT [30] conveys the localised frequency component present in the signal during the short window of time. Fourier analysis expands an arbitrary signal in terms of an infinite number of sinusoidal functions of its harmonics. Fourier representation of signals is known to be very effective in analysis of time invariant stationary periodic signals. In contrast to a sinusoidal function, a wavelet is a small wave whose energy is concentrated in time. Wavelets [31,32] allow both time and frequency analysis of signals simultaneously because of the fact that the energy of wavelets is concentrated in time and still possesses the wave-like periodic characteristics. As a result, wavelet representation provides a versatile mathematical tool to analyse transient, time variant non-stationary signals that are not statistically predictable especially at the region of discontinuities, a feature that is typical of images having discontinuities at the edges. Wavelets are functions generated from one single basis function called the mother wavelet by dilations (scaling) and translations (shifts) in time (frequency) domain. If the mother wavelet is denoted by $\psi(t)$, the other wavelets $\psi_{a,b}(t)$ can be represented as

$$\psi_{a,b}(t) = \frac{1}{\sqrt{a}} \psi\left(\frac{t-b}{a}\right) \tag{7.83}$$

where a and b are two arbitrary real numbers. The variables a and b represent the parameters for dilations and translations, respectively, in the time axis. The parameter a causes contraction of $\psi(t)$ in the time axis when $a < 1$ and expansion or stretching when $a > 1$. That is why the parameter a is called the dilation (scaling) parameter. For $a < 0$, the function results in time reversal with dilation. The function is shifted to the right along the time axis by an amount b when $b > 0$ whereas it is a shift to the left along the time axis by an amount b when $b < 0$. That is why the variable b represents the translation in time (shift in frequency) domain. To define the discrete wavelet transform (DWT), it is essential to define the wavelets in terms of discrete values of the dilation and translation parameters a and b instead of being continuous. The most popular approach of discretising a and b is

$$a = a_0^m, \quad b = nb_0 a_0^m \tag{7.84}$$

where m and n are integers. The discrete wavelets can be represented by

$$\psi_{a,b}(t) = a_0^{-m/2} \psi(a_0^{-m} t - nb_0) \tag{7.85}$$

There are many choices to select the values of a_0 and b_0. The most common choice is $a_0 = 2$ and $b_0 = 1$. Hence, $a = 2^m$, $b = n2^m$. This corresponds to sampling of a and b in such a way that the consecutive discrete values of a and b as well as the sampling intervals differ by a factor of two. This way of sampling is popularly known as dyadic sampling and the corresponding decomposition of the signals is called the dyadic decomposition. Using these values, the discrete wavelets are represented as

$$\psi_{m,n}(t) = 2^{-m/2} \psi(2^{-m} t - n) \tag{7.86}$$

which constitutes a family of orthonormal basis functions. The wavelet coefficients can be derived as

$$c_{m,n}(f) = 2^{-m/2} \int f(t)\psi(2^{-m}t - n)dt \qquad (7.87)$$

This allows us to reconstruct the signal $f(t)$ from discrete wavelet coefficients as

$$f(t) = \sum_{m=-\infty}^{+\infty} \sum_{n=-\infty}^{+\infty} c_{m,n}(f)\psi_{m,n}(t) \qquad (7.88)$$

When the input function $f(t)$ and the wavelet parameters a and b are represented in discrete form, the transformation is the DWT of signal $f(t)$. The DWT became a very versatile signal processing tool after Mallat [33] proposed the MR representation of signals based on wavelet decomposition. The advantage of the DWT over Fourier transformation is that it performs MR analysis of signals with localisation both in time and frequency, popularly known as time frequency localisation. As a result, the DWT decomposes a digital signal into different sub-bands so that the lower frequency sub-bands have finer frequency resolution and coarser time resolution compared to the higher frequency sub-bands. The DWT is being increasingly used for image compression due to the fact that the DWT supports features like progressive image transmission (by quality, by resolution), ease of compressed image manipulation, region of interest coding, and so on. DWT is the basis of the new JPEG2000 image compression standard [34,35]. The Gabor filter is an example of wavelet filters widely used in many image processing applications such as texture analysis, segmentation, classification and so on [36]. In all such applications, it is necessary to analyse the spatial frequency components of an image in a localised fashion. For localised frequency analysis it is desirable to have a Gaussian envelope whose width adjusts with the frequency of the complex sinusoids. Gabor wavelets form class of self-similar functions which yield better localisation in space. The 2D Gabor filters optimally achieve joint resolution/localisation in space and spatial frequency domains. Gabor elementary functions are Gaussians modulated by complex sinusoids. The 2D Gabor functions are complex sinusoid gratings modulated by 2D Gaussian functions in the space domain, and shifted Gaussians in the spatial frequency domain which means that they are complex valued functions. Regardless of the region of frequencies passed, the 2D Gabor functions uniquely minimise the 2D space-frequency uncertainty principle for complex valued functions. Hence Gabor functions can be interpreted as the product of a modulating amplitude envelope with a complex carrier function whose argument is a modulating phase envelope, both of which can be computed and analysed separately. In the spatial domain, the Gabor function is a complex exponential modulated by a Gaussian function. The Gabor function forms a complete and non-orthogonal basis set and its impulse response in the 2D plane has the following general form:

$$G_x(x,y) = \frac{1}{2\pi\sigma_x\sigma_y}\exp\left[-\frac{1}{2}\left(\frac{x^2}{\sigma_x^2} + \frac{y^2}{\sigma_y^2}\right)\right]\exp\left(j2\pi u_0 x\right) \qquad (7.89)$$

where u_0 is the radial frequency of the Gabor function. The space constants σ_x and σ_y define the Gaussian envelope along the x- and y-axis. In a similar fashion, the Gabor function obtained by modulation of complex exponential function in y-direction by Gaussian function is defined. Each of the complex Gabor filters has the real and imaginary parts that are conveniently implemented as the spatial mask of $M \times M$ sizes. The Gabor filter bank has the parameters: (a) frequency half-peak bandwidth, (b) orientation half-peak bandwidth, (c) centre frequency and (d) orientation. The Gabor filters can be configured to have various shapes, bandwidths, centre frequencies and orientations by the adjustments of these parameters. To represent an image with fewer and fewer pixels at successive levels of approximation, the wavelet coefficients are considered as additional detail information is needed to go from a coarser to a finer information. Hence, in each level of decomposition the signal can be decomposed into two parts, one is the coarse approximation of the signal in lower resolution and the other is the detailed information that was lost because of the approximation. The wavelet coefficients describe the information or detail lost when going from an approximation of the signal at resolution 2^{m-1} to the coarser approximation at resolution 2^m.

7.9.5 Multi-Scale Image Decomposition

The concept of scale, or resolution of an image, is very intuitive. A person observing a scene perceives the objects in that scene at a certain level of resolution that depends on the distance to these objects. For instance, walking towards a distant building, he/she would first perceive a rough outline of the building. The main entrance becomes visible only in relative proximity to the building. Finally, the doorbell is visible only in the entrance area. As this example illustrates, the notions of resolution and scale loosely correspond to the size of the details that can be perceived by the observer. It is of course possible to formalise these intuitive concepts, and indeed signal processing theory gives them a more precise meaning. These concepts are particularly useful in image and video processing and in computer vision. A variety of DIP algorithms decompose the image being analysed into several components, each of which captures information present at a given scale. Let the images to be analysed be rectangular with $N \times M$ pixels. While several types of multi-scale image decompositions exist, some main methods are considered: In a Gaussian pyramid representation of an image, the original image appears at the bottom of a pyramidal stack of images. This image is then LP filtered and sub-sampled by a factor of 2 in each coordinate. The resulting $N/2 \times M/2$ image appears at the second level of the pyramid. This procedure can be iterated several times. Here resolution can be measured by the size of the image at any given level of the pyramid. In the original application of this method to computer vision, the LP filter used was a Gaussian filter, hence the terminology Gaussian pyramid. Another possible terminology in that case is simply LP pyramid. The total number of pixels in a pyramid representation is $NM + NM/4 + NM/16 + \cdots \approx (4/3)NM$. This is said to be an overcomplete representation of the original image, caused by an increase in the number of pixels.

i. The Laplacian pyramid representation of the image is closely related to the Gaussian pyramid, but here the difference between approximations at two successive scales is computed and displayed for different scales. An equivalent way to obtain the image at a given scale is to apply the difference between two Gaussian filters to the original image. This is analogous to filtering the image by using a Laplacian filter. Laplacian filters are band pass, hence the name Laplacian pyramid, also termed band pass pyramid.

ii. In wavelet decomposition, the image is decomposed into a set of sub-images or sub-bands that represent details at different scales. Unlike pyramid representations, the sub-images also represent details with different spatial orientations such as edges with horizontal, vertical and diagonal orientations. The number of pixels in a wavelet decomposition is only NM.

iii. Undecimated wavelet transform: The wavelet transform is not invariant to shifts of the input image, in the sense that an image and its translate will in general produce different wavelet coefficients. This is a disadvantage in applications such as ED, pattern matching and image recognition in general. The lack of translation invariance can be avoided if the outputs of the filter banks are not decimated. The undecimated wavelet transform then produces a set of band-pass images that have the same size as the original data set $N \times M$.

iv. Wavelet packets: Although the wavelet transform often provides a sparse representation of images, the spatial frequency characteristics of some images may not be best suited for a wavelet representation. Such is the case of fingerprint images, as ridge patterns constitute relatively narrow band pass components of the image. An even sparser representation of such images can be obtained by recursively splitting the appropriate sub-bands instead of systematically splitting the low-frequency band as in wavelet decomposition. This scheme is simply termed sub-band decomposition. An ingenious algorithm for finding the sub-band decomposition that gives the sparsest representation of the input image in a certain sense is proposed in Reference 37.

v. Geometric wavelets: One of the main strengths of 1D wavelet is their ability to represent abrupt transitions in a signal. This does not extend to higher dimensions. The extension of wavelets to two dimensions has limited ability to represent arbitrary patterns at arbitrary orientations and image edges. To represent a simple, straight edge, one needs many wavelets. To remedy this problem, ridgelet transform [38,39] was introduced, which decomposes images as a superposition of ridgelets. A ridgelet is parameterised by three parameters: resolution, angle and location. Ridgelets are also known as geometric wavelets, a growing family which includes exotically named functions such as curvelets [40], bandelets [41] and contourlets [42,43].

The above-mentioned hierarchical image representations are useful in many applications. In particular, they lend themselves to effective designs of reduced complexity algorithms for texture analysis and segmentation, ED, image analysis, motion analysis and image understanding in computer vision. Moreover, the Laplacian pyramid and wavelet image representations are sparse in the sense that most detail images contain few significant pixels. This sparsity property is very useful in image compression, as bits are allocated only to the few significant pixels; in image recognition, because the search for significant image features is facilitated, and in the restoration of images corrupted by noise, as images and noise possess rather distinct properties in the wavelet domain.

7.10 Image Fusion Mathematics

With the development of new imaging sensors arises the need of a meaningful combination of all employed imaging sources. The actual fusion process can take place at different levels of information representation; a generic categorisation is to consider the different levels as, sorted in ascending order of abstraction: signal, pixel, feature and symbolic level. These fusion levels and its corresponding applications are depicted in Figure 7.4.

7.10.1 Pixel-Level Fusion

To date, the result of pixel-level image fusion is considered primarily to be presented to the human observer. A possible application is the fusion of forward looking infrared (FLIR) and low light visible images (LLTV) obtained by an airborne sensor platform to aid pilots navigate in poor weather conditions or darkness. Most of the pixel-level fusion methods have been

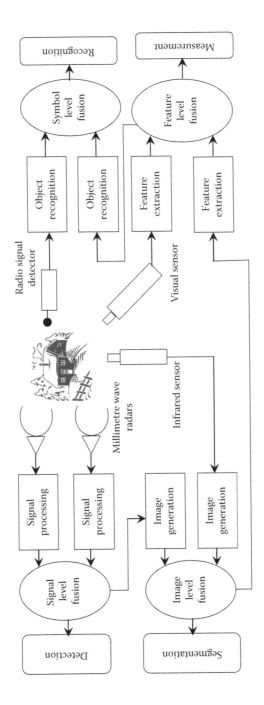

FIGURE 7.4
Different levels of image fusion and applications.

developed for the fusion of stationary input images which are spatially registered. In spatial domain techniques, the input images are fused in the spatial
domain, that is, using localised spatial features. Assuming that $g(\cdot)$ represents the fusion rule, that is, the method that combines features from the
input images, the spatial domain techniques can be summarised, as follows:

$$f_f(x,y) = g(f_1(x,y), f_2(x,y), \ldots, f_L(x,y)) \tag{7.90}$$

The main motivation behind moving to a transform domain is to work in a
framework, where the image's salient features are more clearly depicted than
in the spatial domain. It is important to understand the underlying image
structure for fusion rather than fusing image pixels independently. Most
transformations used in image processing are decomposing the images into
important local components, that is, unlocking the basic image structure.
Hence, the choice of the transformation is very important. Let $T\{\cdot\}$ represent a
transform operator and $g(\cdot)$ the applied fusion rule. Transform domain fusion
techniques can then be outlined, as follows:

$$f_f(x,y) = T^{-1}\{g(T\{f_1(x,y)\}, T\{f_2(x,y)\}, \ldots, T\{f_L(x,y)\})\} \tag{7.91}$$

The fusion operator $g(\cdot)$ describes the merging of information from the different input images. Many fusion rules have been proposed in the literature
[32–34]. These rules can be categorised, as follows:

- Pixel-based rules: The information fusion is performed in a pixel-by-
  pixel basis either in the transform or spatial domain. Each pixel (x, y)
  of the L input images is combined with various rules to form the corresponding pixel (x, y) in the fused image f_f. Several basic transform
  domain schemes were proposed [35], such as:
  - Mean rule – Fusion by averaging: fuse by averaging the corre-
    sponding coefficients in each image

$$T\{f_f(x,y)\} = \frac{1}{L} \sum_{i=1}^{L} T\{f_i(x,y)\} \tag{7.92}$$

    This rule gives the best performance especially for wave-
    let coefficients, as it seems to balance the high detail with the
    low-detail information. However, the fused image looks more
    blurred as the fusion rule has over smoothed the image details.
  - Maximum absolute value rule – Fusion by absolute maximum:
    fuse by selecting the greatest in absolute value of the correspond-
    ing coefficients in each image.

$$T\{f_f(x,y)\} = \text{sgn}(T\{f_i(x,y)\})\max_i | T\{f_i(x,y)\} | \qquad (7.93)$$

This rule seems to give very low performance as this scheme seems to highlight the important features of the images; however, it tends to lose some constant background information.

- Weighted combination rule – uses weighted combination of the transform coefficients

$$T\{f_f(t)\} = \sum_{i=1}^{L} w_i(t)T\{f_i(t)\} \qquad (7.94)$$

There are several parameters that can be employed in the estimation of the contribution $w_i(t)$ of each image to the fused image. In Reference 44, Piella proposed several activity measures. As each image is processed in $n \times n$ patches, the mean absolute value of each patch in the transform domain is used as an activity indicator in each patch. The weights $w_i(t)$ should emphasise sources that feature more intense activity, as represented by $E_i(t)$. Consequently, the weights $w_i(t)$ for each patch t can be estimated by the contribution of the ith source image over the total contribution of all the L source images at patch t in terms of activity. Hence

$$w_i(t) = \frac{E_i(t)}{\displaystyle\sum_{i=1}^{L} E_i(t)}, \quad E_i(t) = \| u_i(t) \|_1 \qquad (7.95)$$

The weighted combination rule seems to balance the pros and cons of the two previous approaches but the fused images seem sharper with correct constant background information.

- Sparse code shrinkage – Fusion by de-noising (hard/soft thresholding): simultaneous fusion and de-noising is performed by thresholding the transform coefficients.
- High/low fusion – combining the high-frequency parts of some images with the low-frequency parts of some other images.

The most straightforward way to obtain a fused image of several input images is performing the fusion as a weighted superposition of all input images. The optimal weights, with respect to information content and redundancy removal, can be determined by a PCA of all input image pixel values. By performing PCA of the covariance matrix of input images, the weights for each input image are obtained from the eigenvector corresponding to the largest eigen value. Another simple approach to image fusion is to build the fused image by the application of a simple non-linear operator such as maximum, minimum or median. If the bright objects of the input images are of interest, the fused image can be obtained by a pixel-by-pixel application of the maximum operator. An extension to this approach follows by the introduction of morphological operators such as opening or closing. One application is the use of conditional morphological operators by the definition of highly reliable core features present in input images and a set of potential features present only in one source, where the actual fusion process is performed by the application of conditional erosion and dilation operators.

Another approach to image fusion is that the fusion is expressed as a Bayesian optimisation problem. Using the multi-sensor image data and a priori model of the fusion result, the goal is to find the fused image which maximises the a posteriori probability. Due to the fact that this problem cannot be solved in general, some simplifications are introduced: All input images are modelled as Markov random fields to define an energy function which describes the fusion goal. Due to the equivalence of Gibbs random fields and Markov random fields, this energy function can be expressed as a sum of clique potentials, where only pixels in a pre-defined neighbourhood affect the actual pixel. The fusion task then consists of a maximisation of the energy function. Since this energy function is non-convex, stochastic optimisation procedures such as simulated annealing or iterated conditional modes are used.

Inspired by the fusion of different sensor signals in biological systems, many researchers have employed artificial neural networks in the process of pixel-level image fusion. The most popular example for the fusion of different imaging sensors in biological systems [45], is that rattlesnakes and the general family of pit vipers possess pit organs which are sensitive to thermal radiation through a dense network of nerve fibres. The output of these pit organs is fed to the optical tectum, where it is combined with the nerve signals obtained from the eyes. Newman and Hartline distinguished six different types of bimodal neurons merging the two signals based on a sophisticated combination of suppression and enhancement.

Image pyramids have been initially described for MR image analysis and as a model for binocular fusion in human vision. An image pyramid is a sequence of images where each image is constructed by LP filtering and subsampling from its predecessor. Due to sampling, the image size is halved in both spatial directions at each level of the decomposition process, thus leading to a MR signal representation. The difference between the input image and the filtered image is necessary to allow an exact reconstruction from the pyramidal representation. The image pyramid approach thus leads to a signal representation with two pyramids: smoothing pyramid containing the averaged pixel values, and difference pyramid containing the pixel differences, that is, the edges. So the difference pyramid can be viewed as a MR edge representation of the input image. The actual fusion process can be described by a generic MR fusion scheme which is applicable to both image pyramids and the wavelet approach.

There are several modifications of this generic pyramid construction method described above. Some authors propose the computation of non-linear pyramids, such as the ratio and contrast pyramid, where the multi-scale edge representation is computed by a pixel-by-pixel division of neighbouring resolutions. A further modification is to substitute the linear filters by morphological non-linear filters, resulting in the morphological pyramid. The gradient pyramid results, if the input image is decomposed into its directional edge representation using directional derivative filters.

A signal analysis method similar to image pyramids is the DWT. The main difference is that while image pyramids lead to an over complete set of transform coefficients, the wavelet transform results in a non-redundant image representation. The discrete 2D wavelet transform is computed by the recursive application of LP and HP filters in each direction of the input image (i.e. rows and columns) followed by sub-sampling. One major drawback of the wavelet transform when applied to image fusion is its well-known shift dependency, that is, a simple shift of the input signal may lead to completely different transform coefficients. To overcome the shift dependency of the wavelet fusion scheme, the input images must be decomposed into a shift invariant representation. There are several ways to achieve this.

The straightforward way is to compute the wavelet transform for all possible circular shifts of the input signal. In this case, not all shifts are necessary and it is possible to develop an efficient computation scheme for the resulting wavelet representation. Another simple approach is to drop the sub-sampling in the decomposition process and modify the filters at each decomposition level, resulting in a highly redundant signal representation. The actual fusion process can be described by a generic MR fusion scheme which is applicable both to image pyramids and the wavelet approach.

The basic idea of the generic MR fusion scheme is motivated by the fact that the human visual system is primarily sensitive to local contrast changes, that is, edges. Motivated from this insight, and in mind that both image pyramids and the wavelet transform result in a MR edge representation, it is straightforward to build the fused image as a fused multi-scale edge representation. The fusion process is summarised in the following: In the first step, the input images are decomposed into their multi-scale edge representation, using either any image pyramid or any wavelet transform. The actual fusion process takes place in the wavelet domain, where the fused multi-scale representation is built by a pixel-by-pixel selection of the coefficients with maximum magnitude. Finally the fused image is computed by an application of the appropriate reconstruction scheme. This is depicted in Figure 7.5.

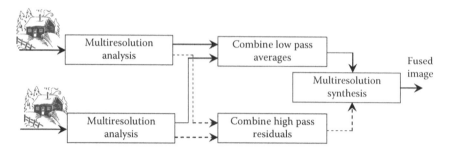

FIGURE 7.5
MR image fusion.

7.10.2 Feature Level Fusion

The basic idea is to perform a MR transform on each source image and, following some specific fusion rules, construct a composite MR representation from these inputs. The fused image is obtained by applying the inverse transform on this composite MR representation. This process is illustrated in Figure 7.6 for the case of two input source images. The combination algorithm consists of four modules: the activity and match measures extract information from the MR decompositions of the input images, which is then used by the decision and combination map to compute the MR decomposition of the fused image. MR analysis block computes a MR decomposition of the input source images $f_A(x, y)$ and $f_B(x, y)$. For every input image its MR representation y_A and y_B are obtained. The degree to which each coefficient in y_A and y_B is salient is expressed by the activity measure. The activity function block associates to every band image an activity, which reflects the local activity of the image. The match measure quantifies the degree of similarity between the source images. The decision map block is the core of the combination algorithm. For each level k, orientation band p and location n, the decision process assigns a value which is then used for the computation of the composite representation. The combination map yields the composite coefficient. Finally, the fused image is

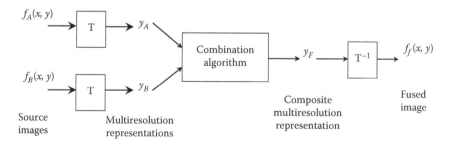

FIGURE 7.6
MR image fusion algorithm.

obtained by applying the inverse transformation on the composite MR decomposition $f_f(x, y) = T^{-1}\{y_F\}$ where T^{-1} is the inverse MR transform.

The MR representation comprises information at different scales. High levels contain coarse scale information while low levels contain finer details. Such a representation is suitable for image fusion, not only because it enables one to consider and fuse image features separately at different scales, but also produces large coefficients near edges, thus revealing salient information [38]. To perform a consistent fusion of objects at arbitrary scales, the decomposition over a large number of scales may appear necessary. However, using more levels does not necessarily produce better results; it may produce low-resolution bands where neighbouring features overlap. This gives rise to discontinuities in the composite representation and, thus, introduces distortions, such as blocking effects or ringing artefacts, into the fused image. The required analysis depth is primarily related to the spatial extent of the relevant objects in the source images. In general, it is not possible to compute the optimal analysis depth, but as a rule of thumb, the larger the objects of interest are, the higher the number of decomposition levels should be.

The computation of the activity depends on the nature of the source images as well as on the particular fusion application. Generally, based on the fact that the human vision system (HVS) is primarily sensitive to local contrast changes as in edges most fusion algorithms compute the activity based on energy. In the simplest case, the activity is just the absolute value of the coefficient. Alternatively, the contrast of the component with its neighbours in a 3×3 or 5×5 window, or some other linear or non-linear criteria can provide that measure. The match value or similarity measure can be defined as a normalised correlation averaged over a neighbourhood of the samples. Assuming that every composite coefficient is assembled from the source coefficients at the corresponding level, band and position, a simple choice of combining is a linear mapping where the composite coefficients are obtained by an additive or weighted combination. For the particular case where only one of the sources contributes to the composite, the combination is termed as selective combination or combination by selection. The decision map actually determines the combination of the various MR decompositions y_A and y_B and, hence, the construction of the composite y_F.

For a weighted combination scheme, the decision map controls the values of the weights to be assigned to each of the source coefficients. An approach is to assign to each coefficient a weight that depends increasingly on the activity. The resulting weighted average introduces the problem of contrast reduction in case of opposite contrast in different source images. This can be avoided by using a selective rule where the one with largest activity is chosen for the composite. This selective combination is known as maximum selection rule. It works well under the assumption that at each image location, only one of the source images provides the most useful information. At sample locations where the source images are distinctly different, the combination process selects the most salient component, while at sample locations where they are similar, the process averages the source components. In this manner, averaging reduces noise and provides stability where source images contain similar information, whereas selection retains salient information and reduces artefacts due to opposite contrast.

In general, the decision is made globally for a group of samples: for all bands p, all samples in the same level k and location n are assigned the same decision. Another method [46] consists in applying a majority filter to a preliminary decision map to remove possible wrong selection decisions caused by impulsive noise and the filtered decision map determines the combination of the images.

Because of their different physical meaning, the approximation and detailed images are treated by the combination algorithm through different procedures. Detail coefficients having large absolute values correspond to sharp intensity changes and hence to salient features in the image such as edges, lines and region boundaries. The approximation image is a coarse representation of the original image and has inherited some of its properties such as the mean intensity or texture information. In this case, an activity measure based on entropy, variance or texture, is a better alternative than based on energy. In many approaches, the composite approximation coefficients of the highest decomposition level, representing the mean intensity, are taken to be a weighted average of the approximation of the sources. A popular way to construct the composite is to use weighted average for the approximation coefficients and the selective combination based on the rule that picks the most salient component for the detail coefficients.

7.10.3 Region-Based Image Fusion

This combines aspects of feature and pixel-level fusion. The basic idea is to make a segmentation based on all different source images and to use this segmentation to guide the combination process. A major difference with other existing region-based approaches [40,41] is that the segmentation performed is: (i) multi-source, in the sense that a single segmentation is obtained from all the input images, and (ii) MR, in the sense that it is computed in a MR fashion and is not a segmentation of a sequence of images at different resolutions. For instance, in Reference 38, the regions are obtained by segmenting independently each of the approximation images and by exploiting the family relations, every detail coefficient is assigned to a region.

MR decompositions are used to represent the input images at different scales and, additionally, a multiresolution/multi-source (MR/MS) segmentation is introduced to partition the image domain at these scales. The activity and match measures are computed for every region in the decomposed input images. These measures may correspond to low level as well as intermediate level structures. Furthermore, the MR segmentation allows imposing data-dependent consistency constraints based on spatial as well as inter- and intra-scale dependencies. All this information, that is, the measures and the consistency constraints, is integrated to yield a decision map which governs the combination of the coefficients of the transformed sources. This combination results in a MR decomposition.

The linked pyramid structure was first described by Burt et al. [47]. It consists of a MR decomposition of an image with the bottom level containing the full resolution image and each successive higher level being a filtered/subsampled version derived from the level below it. The various levels of the pyramid are linked by means of child–parent relations between their pixels; such child–parent links are established during an iterative processing procedure. At first, an approximation pyramid is produced by LP filtering and sampling. Then, child–parent relations are established by linking each pixel in a level called child to one of the pixels in the next higher level called parent that is closest in grey value or in some other pixel attribute. The attribute values of the parents are then updated using the values of their children. The process of linking and updating is repeated until convergence. Finally, some pixels are labelled as roots. In the simplest case, only the pixels in the top level of the pyramid are roots. Every root and the pixels which are connected to it induce a tree in the pyramid. The leaves of each tree correspond to pixels in the full resolution image which define a segment or region. Thus, the linked pyramid provides a framework for an iterative process of image segmentation.

In order to exploit the image structure more efficiently, these schemes group image pixels to form contiguous regions and impose different fusion rules to each image region. In Reference 46, Li et al. created a binary decision map to choose between the coefficients using a majority filter, measuring activity in small patches around each pixel. In Reference 44, Piella proposed several activity level measures, such as the absolute value, the median or the contrast to neighbours. Consequently, she proposed a region-based scheme using a local correlation measurement to perform fusion of each region. In Reference 48, Lewis et al. produced a joint segmentation map out of the input images. To perform fusion, they measured priority using energy, variance or entropy of the wavelet coefficients to impose weighting on each region in the fusion process along with other heuristic rules.

Alternatively, the image is segmented into two regions: (i) active regions containing details and (ii) non-active regions containing background information. The threshold that will be used to characterise a region as active or non-active can be set heuristically to $2 \times \text{mean}_t\{E_i(t)\}$. Since the aim here is to create the most accurate edge detector, some tolerance is allowed around the real edges of the image. As a result, the following segmentation map $m_i(t)$ is formed from each input image:

$$m_i(t) = \begin{cases} 1, & \text{if } E_i(t) > 2 \times \text{mean}_t\{E_i(t)\} \\ 0, & \text{otherwise} \end{cases} \tag{7.96}$$

The segmentation map of each input image is combined to form a single segmentation map, using the logical OR operator. It is important to ensure that the segmentation map contains most of the strong edge information. Once the image has been segmented into active and non-active regions, these regions are fused using different pixel-based fusion schemes. For the active region, a fusion scheme that preserves the edges is used, that is, the max-abs scheme or the weighted combination scheme and for the non-active region, a scheme that preserves the background information is used, that is, the mean or median scheme. Consequently, this could form a more accurate fusion scheme that looks into the actual structure of the image itself, rather than fuse information generically. The region-based scheme manages to capture most of the salient areas of the input images. It performs reasonably well as an edge detector; however, it produces thicker edges, as the objective is to identify areas around the edges, not the edges themselves. The region-based fusion scheme produces similar results to the weighted fusion scheme. However, it seems to produce better visual quality in constant background areas, as the mean rule is more suitable for the non-active regions.

7.11 Image Fusion Algorithms

The first MR image fusion used a Laplacian pyramid and a sample-based maximum selection rule. Toet [49] presented a similar algorithm but using the ratio-of-LP pyramid. The fusion technique selects the highest local luminance contrast which is likely to provide better details to a human observer. Another variation of this scheme is obtained by replacing the linear filters by morphological ones. Burt and Kolczynski [50] proposed to use the gradient pyramid together with a combination algorithm that is based on local energy activity and a weighted average match measure. One of the first wavelet-based fusion systems considers the maximum absolute value within a window as the activity measure associated with the sample centred in the window. For each position in the transform domain, the maximum selection rule is used to determine which of the inputs is likely to contain the most useful information. This scheme performs better than the Laplacian pyramid-based fusion due to the compactness, directional selectivity and orthogonality of the wavelet transform. Wilson et al. [51] used a DWT fusion method and a perceptual-based weighting based on the frequency response of the HVS. The activity measure is computed as a weighted sum of the Fourier transform coefficients of the wavelet decomposition, with the weights determined by the contrast sensitivity. The perceptual distance between the sources is used together with the activity to determine the weights of the wavelet coefficients from each source. This perceptual distance is directly related to the matching measure: the smaller the perceptual distance, the higher the matching measure. The fused images obtained with this method are visually better than the ones obtained by fusion techniques based on the gradient pyramid or the ratio-of-LP pyramid.

Koren et al. [52] used a steerable wavelet transform for the MR decomposition. This is preferred because of its shift invariance and no aliasing properties. For each frequency band, the activity is a local oriented energy. Only the components corresponding to the frequency band whose activity is the largest are included for reconstruction. The detail coefficients are combined by a maximum selection rule, while the coarse approximation coefficients are merged by averaging. Pu and Ni [53] proposed a contrast-based image fusion method using DWT. The activity is measured as the absolute value called directive contrast and a maximum selection rule is used as the combination method of the wavelet coefficients. An alternative approach is proposed where the combination process is performed on the directive contrast itself. In another MR technique the fusion consists of retaining the modulus maxima of the wavelet coefficients from the different bands and combining them. Noise reduction can be applied during the fusion process by removing noise-related modulus maxima.

The intensity additive wavelet (AWL) method [54] is one of the existing MR wavelet-based image fusion techniques. It was originally designed for a three band (RGB) multispectral image. In this method, the spectral signature is preserved because the high-resolution panchromatic structure is integrated into the luminance L band of the original low-resolution multispectral image. Therefore, this method is only defined for three bands. This was subsequently extended to n bands. This maintains the spectral signature of an n-band image in the same way as AWL does with RGB images. This generalised method is called proportional AWL (AWLP). This method produces better results than standard wavelet algorithms, but the spatial improvement is in most cases still not acceptable.

The multiplicative method is derived from the four component technique of Reference 55. Of the four possible arithmetic methods only the multiplication is unlikely to distort the colours by transforming an intensity image into a panchromatic image. Therefore, this algorithm is a simple multiplication of each multispectral band with the panchromatic image. The advantage of the algorithm is that it is straightforward and simple. By multiplying the same information into all bands, however, it creates spectral bands of a higher correlation which means that it does alter the spectral characteristics of the original image data. The Brovey transformation was developed to avoid the disadvantages of the multiplicative method. It is a combination of arithmetic operations and normalises the spectral bands before they are multiplied with the panchromatic image. The spectral properties, however, are usually not well preserved. The colour normalisation (CN) spectral sharpening is an extension of the Brovey algorithm and groups the input image bands into spectral segments defined by the spectral range of the panchromatic image. The corresponding band segments are processed together in the following manner: Each input band is multiplied by the sharpening band and then normalised by dividing it by the sum of the input bands in the segment. This method works well for data from one sensor, but if the spectral range of the panchromatic image does not match the spectral range of the multispectral images, no spatial improvement is visible.

To fuse the images with the intensity–hue–saturation (IHS) fusion, three bands of a multispectral image are transformed from the RGB domain into the IHS colour space. The panchromatic component is matched to the intensity of the IHS image and replaces the intensity component. Modified IHS fusion was developed for a better fit of the fused multispectral bands to the original data. After the matching, the panchromatic image replaces the intensity in the original IHS image and the fused image is transformed back into the RGB colour space. This method works also well with data from one sensor, but for multi-temporal or multi-sensory (MS) fusion the results are in most cases not acceptable.

The Ehlers fusion is based on an IHS transform coupled with a Fourier domain filtering. This technique is extended to include more than three bands by using multiple IHS transforms until the number of bands is exhausted. A subsequent Fourier transform of the intensity component and the panchromatic image allows an adaptive filter design in the frequency domain. Using FFT techniques, the spatial components to be enhanced or suppressed can be directly accessed. The intensity spectrum is filtered with a LP filter whereas the panchromatic spectrum is filtered with an inverse HP filter. After filtering, the images are transformed back into the spatial domain with an inverse FFT and added together to form a fused intensity component with the low-frequency information from the low-resolution multispectral image and the high-frequency information from the high-resolution image. This new intensity component and the original hue and saturation components of the multispectral image form a new IHS image. As the last step, an inverse IHS transformation produces a fused RGB image. These steps can be repeated with successive three-band selections until all bands are fused with the panchromatic image. The Ehlers fusion shows the best spectral preservation but also has the highest computation time.

The principal component (PC) transform is a statistical technique that transforms a multi-variate data set of correlated variables into a data set of uncorrelated linear combinations of the original variables. For images, it creates an uncorrelated feature space that can be used for further analysis instead of the original multispectral feature space. The PC is applied to the multispectral bands. The panchromatic image is histogram matched to the first principal component (PC1) and sometimes to the second. It then replaces the selected component and an inverse PC transform takes the fused data set back into the original multispectral feature space. The advantage of the PC fusion is that the number of bands is not restricted. It is, however, a statistical procedure which means that it is sensitive to the area to be sharpened. The fusion results may vary depending on the selected image subsets.

The Gram–Schmidt fusion simulates a panchromatic band from the lower spatial resolution spectral bands. In general, this is achieved by averaging the multispectral bands. As the next step, a Gram–Schmidt transformation is performed for the simulated panchromatic band and the multispectral bands with the simulated panchromatic band employed as the first band. Then the high spatial resolution panchromatic band replaces the first Gram–Schmidt band. Finally, an inverse Gram–Schmidt transform is applied to create the pan sharpened multispectral bands. This method usually produces good results for fusion images from one sensor, but it is also a statistical procedure like the PC, so that the fusion results may vary depending on the selected data sets.

For the HPF fusion, first the ratio between the spatial resolution of the panchromatic and the multispectral image is calculated. A HP convolution filter kernel is created and used to filter the high-resolution input data with the size of the kernel based on the ratio. The HPF image is added to each multispectral band. Before the summation, the HPF image is weighted relative to the global standard deviation of the multispectral bands with the weight factors again calculated from the ratio. As a final step, a linear stretch is applied to the new multispectral image to match the mean and standard deviation values of the original input multispectral image. It shows acceptable results also for MS and multi-temporal data. Sometimes the edges are emphasised too much.

To apply the UNB fusion algorithm, histogram standardisation is calculated for the multispectral and panchromatic bands of the input images. The multispectral bands in the spectral range of the panchromatic image are selected and a regression analysis is calculated using a least-square algorithm. The results are used as weights for the multispectral bands. Via multiplication with the corresponding bands and a following addition, a new synthesised image is produced. To create the fused image, each standardised multispectral image is multiplied with the standardised panchromatic image and divided by the synthesised image. This method was designed for single sensor, single-date images and does not produce acceptable results for multi-sensor and/or multi-temporal fusion. It is used as the standard method for Quick-bird pan sharpening.

The Laplacian pyramid implements a pattern selective approach to image fusion, so that the fused image is constructed not pixel by pixel, but by different features. The idea is to perform a pyramid decomposition on each source image, then integrate the decompositions to obtain a composite representation, and finally reconstruct the fused image by performing an inverse pyramid transform.

Image pyramids have been initially described for a MR image analysis and as a model for the binocular fusion in human vision. An image pyramid can be described as collection of low or band pass copies of an original image in which both the band limit and sample density are reduced in regular steps. A MR pyramid transformation decomposes an image into multiple resolutions at different scales. A pyramid is a sequence of images in which each level is a filtered and sub-sampled copy of the predecessor. The lowest level of the pyramid has the same scale as the original image and contains the highest resolution information. Higher levels of the pyramid are reduced resolution and increased scale versions of the original image.

The first step in Laplacian pyramid transform is to LP filter the original image g_0 to obtain image g_1, which is a reduced version of g_0. In similar way g_2 is formed as a reduced version of g_1, and so on. The level-to-level averaging process is performed for levels $0 < l < N$ nodes $i, j, 0 \leq i < C_l, 0 \leq j < R_l$ following the equation:

$$g_l(i, j) = \sum_{m=-2}^{2} \sum_{n=-2}^{2} w(m, n) g_{l-1}(2i + m, 2j + n)$$ (7.97)

where N refers to the number of levels in the pyramid and C_l and R_l are the number of rows and columns of lth level images. Averaging is performed by convolution with one of a family of local, symmetric weighting functions. A window of size 5×5 is selected as it provides adequate filtering at low computation cost. This weighting kernel is chosen following certain constraints: separable for simplicity $w(m, n) = \hat{w}(m)\hat{w}(n)$, normalised so that sum of coefficients is 1 ($\sum_{m=-2}^{2} \hat{w}(m) = 1$), symmetric $\hat{w}(m) = \hat{w}(-m), m = 0, 1, 2$ and equal contribution so that all nodes at a given level contribute the same total weight to the nodes at the next higher level

$$\left(\begin{array}{c} \hat{w}(0) = a, \hat{w}(1) = \hat{w}(-1) = b, \hat{w}(2) = \hat{w}(-2) = c, a + 2c = 2b \\ \Rightarrow \hat{w}(0) = a, \hat{w}(1) = \hat{w}(-1) = 1/4, \hat{w}(2) = \hat{w}(-2) = 1/4 - a/2 \end{array} \right)$$

The reverse operation which is expansion is obtained by interpolating new node values between the given values

$$g_{l,n}(i, j) = 4 \sum_{m=-2}^{2} \sum_{n=-2}^{2} w(m, n) g_{l,n-1}\left(\frac{i - m}{2}, \frac{j - n}{2} \right)$$ (7.98)

for levels $0 < l < N$ and $0 \leq n$ and nodes $i, j, 0 \leq i < C_{l-n}, 0 \leq j < R_{l-n}$.

If Gaussian distribution is used as the window kernel, this pyramid is named the Gaussian pyramid. The Laplacian pyramid is a sequence of error images $L_0, L_1 \ldots, L_N$ and each is the difference between two levels of the Gaussian pyramid $L_l = g_l - \text{Expand}(g_{l+1}) = g_l - g_{l+1,1}$. There is no image g_{N+1} to serve as the prediction for g_N, $L_N = g_N$. The original image can be recovered without any loss by expanding, then summing all the levels of the Laplacian pyramid $g_0 = \sum_{l=0}^{N} L_l$. An efficient procedure is to expand L_N once and add it to L_{N-1}, then expand the resulting image once and adding it to L_{N-2}, and so on until the original image is recovered.

After the construction of the Laplacian pyramid, fusion is then implemented for each level using a feature selection decision mechanism. It can be used with several modes of combination, such as selection or averaging. In the first one, the combination process selects the most salient component pattern from the source and copies it to the composite pyramid, while discarding the less salient pattern. In the second one, the process averages the source patterns. This averaging reduces noise and provides stability where source images contain the same information. The former is used in locations where the source images are distinctly different, and the latter is used in locations where the source images are similar. One other possible approach is to select the maximum valued component and apply a consistency filter to eliminate the isolated points. For level N, it is the average of the source images.

An alternative to fusion using pyramid-based MR representations is fusion in the wavelet transform domain. The wavelet transform decomposes the image into low–high, high–low, high–high spatial frequency bands at different scales and the low–low band at the coarsest scale. The L–L band contains the average image information whereas the other bands contain directional information due to spatial orientation. Higher absolute values of wavelet coefficients in the high bands correspond to salient features such as edges or lines. With these premises, Li et al. [46] propose a selection-based rule to perform image fusion in the wavelet transform domain. Since larger absolute transform coefficients correspond to sharper brightness changes, a good integration rule is to select, at every point in the transform domain, the coefficients whose absolute values are higher.

Computationally efficient pixel-level image fusion (CEMIF) system proposed by Petrovic and Xydeas [56] is based on an adaptive, MR approach with a reduced number of levels. The goal of this technique is to reduce the computational complexity of MR systems, such as the Laplacian pyramid and Wavelet transform, while preserving the robustness and high image quality of MR fusion. Actually, the spectral decomposition employed in this system represents a simplified version of the conventional Gaussian–Laplacian pyramid approach. Multi-scale structure is simplified into two levels of scale only, the background and the foreground levels. The former contains the DC component and the surrounding baseband and represents large-scale features, on the other hand, the latter contains the high-frequency information, which means small-scale features. Signal fusion is performed at both levels independently. Background signals, obtained as the direct product of the average filtering, are combined using an arithmetic fusion approach. Foreground signals produced as the difference between the original and background signals are fused using a simple pixel-level feature selection technique. Finally, the resulting, fused, foreground and background signals are summed to produce the fused image.

A 2D averaging filter with templates of adaptively varying size is used to decompose the original signal into two sub-band signals. This kind of filter does not possess really good spectral characteristics, but they are used since they require only a fraction of the computational effort needed for templates with better spectral characteristics. With this filtering process the background signals are obtained. These background signals are then subtracted from the original image signals to obtain foreground signals. The foreground and background signals remain of the same resolution and size as the original input signals, no sub-sampling is done. There are two different arithmetic fusion approaches for background signals. They both give their optimal results for complimentary sets of statistical conditions. In cases where one background image dominates a direct elimination approach is employed. It is said that one background image dominates when it contains significantly more background information. For example, in night vision, an IR sensor may contain more information than a visible spectrum one. In this case, the dominant background image becomes the fused background image while the other background image is ignored. In cases where the energies of both input images have similar values, the fused signal is constructed as the sum of the non-DC input signals, and the average of the input image mean values. Foreground fusion is implemented using a simple feature selection fusion mechanism. Important information is easier to localise than in the background signals and a feature selection mechanism can be implemented on pixel level. Each pixel in the fused foreground is chosen to be the corresponding pixel with the highest absolute value from the input foreground image.

A pixel-level image fusion algorithm based on the spatial frequency was developed by Shutao et al. [57]. This method is computationally simple and can be used in real-time applications. Spatial frequency measures the overall activity level in an image. For an $M \times N$ image f, with the grey value at pixel position (x, y) denoted by $f(x, y)$, its spatial frequency is defined as $SF = \sqrt{RF^2 + CF^2}$ where RF and CF are the row and column frequency given by

$$RF = \sqrt{\frac{1}{MN} \sum_{x=1}^{M} \sum_{y=2}^{N} \left(f(x,y) - f(x,y-1)\right)^2} \tag{7.99}$$

and

$$CF = \sqrt{\frac{1}{MN} \sum_{x=2}^{M} \sum_{y=1}^{N} (f(x,y) - f(x-1,y))^2} \tag{7.100}$$

The spatial frequencies of two corresponding blocks are compared and the pixel value corresponding to maximum frequency block is retained. If frequencies are similar, then average of the pixel values is used. A saliency check is performed to remove isolated points.

The IHS technique is a standard procedure in image fusion, with the major limitation that only three bands are involved. Originally, it was based on the RGB true colour space. It offers the advantage that the separate channels outline certain colour properties, namely intensity (I), hue (H) and saturation (S). This specific colour space is often chosen because the visual cognitive system of human beings tends to treat these three components as roughly orthogonal perceptual axes. However, in remote sensing, arbitrary bands are usually assigned to the RGB channels to produce false colour composites for display purposes only. The IHS technique usually comprises four steps: (i) transform the RGB channels (corresponding to three multispectral bands) to IHS components; (ii) match the histogram of the panchromatic image with the intensity component; (iii) replace the intensity component with the stretched panchromatic image and (iv) inverse-transform IHS channels to RGB channels. The resultant colour composite will then have a higher spatial resolution in terms of topographic texture information. The original transformation matrix of the IHS transformation is orthogonal.

The Brovey transform is based on the chromaticity transform. It is a simple method for combining data from different sensors, with the limitation that only three bands are involved. Its purpose is to normalise the three multispectral bands used for RGB display and to multiply the result by any other desired data to add the intensity or brightness component to the image.

The principle of HPF is to add the high-frequency information from the high-resolution panchromatic image (HRPI) to the low-resolution multispectral image (LRMI) to get the high-resolution multispectral image (HRMI). The high-frequency information is computed by filtering the HRPI with a HP filter or taking the original HRPI and subtracting the LRPI, which is the LP filtered HRPI. This method preserves a high percentage of the spectral characteristics, since the spatial information is associated with the high-frequency information of the HRMIs, which is from the HRPI, and the spectral information is associated with the low-frequency information of the HRMIs, which is from the LRMIs. When boxcar filters are used, the filter length is crucial and must match the resolution ratio of the HRPI and LRMIs. A 3×3 boxcar filter is suitable for 1:2 fusion only, since the frequency response should have -6 dB cut-off at quarter the normalised frequency, where the frequency is normalised to the sampling frequency. For 1:4 fusion, a 5×5 boxcar filter with cut-off frequency at roughly 0.125 must be used. The frequency response has a smooth transition band accompanied by a large ripple outside the pass band. The principle of HP modulation is to transfer the high-frequency information of the HRMI to the LRMIs, with modulation coefficients, which equal the ratio between the LRMIs and the LRPI. The LRPI is obtained by LP filtering the HRPI.

The PCA method is similar to the IHS method, with the main advantage that an arbitrary number of bands can be used. The input LRMIs are first transformed into the same number of uncorrelated PCs. The PC1 image contains the information that is common to all the bands used as input to PCA, while the spectral information that is unique to any of the bands is mapped to the other components. Then, similar to the IHS method, the PC1 is replaced by the HRPI, which is first stretched to have the same mean and variance as PC1. As a last step, the HRMIs are determined by performing the inverse PCA transform. The transformation matrix contains the eigenvectors, ordered with respect to their eigen values. It is orthogonal and determined either from the covariance matrix or the correlation matrix of the input LRMIs. PCA performed using the covariance matrix is referred to as unstandardised PCA, while PCA performed using the correlation matrix is referred to as standardised PCA.

The term 'à trous' was originally introduced in Reference 58. It is based on the undecimated dyadic wavelet transform and is particularly suitable for signal processing since it is isotropic and shift invariant and does not create artefacts when used in image processing. The wavelet planes are computed as the differences between two consecutive approximations. To construct the sequence, the à trous algorithm performs successive convolutions with a filter obtained from the scaling function. The use of a B3 cubic spline yields a dyadic LP scaling function such as in one dimension. There are several possible ways of using the à trous algorithm in image fusion. MR analysis-based intensity modulation (MRAIM) was proposed by Wang [59]. It follows the GIF method, with the major advantage that it can be used for the fusion case in which the ratio is an arbitrary integer, with a very simple scheme.

7.12 Performance Evaluation

In many applications, the ultimate user or interpreter of the fused image is a human. Consequently, the human perception of the fused image is of paramount importance and therefore, fusion results are mostly evaluated by subjective criteria [83–85]. This involves human observers to judge the quality of the resulting fused images. Since the human quality measure depends highly on psycho-visual factors, these subjective tests are difficult to reproduce and verify, as well as time consuming and expensive. Hence, although it cannot be denied that subjective tests are important in characterising fusion performance, objective performance metrics appear as a valuable complementary method. To quantify a subjective impression such as image quality, quality is associated with the deviation of the experimental fused image from the ideal fused image. Then, another problem arises, namely, how to define the ideal fused image. A less usual approach is to design performance measures which, without assuming knowledge of a ground truth, can be used for quality assessment of the fused image. These performances measures quantify the degree to which the fused image is related to the input sources.

An example is that out of focus image fusion is evaluated by comparison of the fused image with an ideal composite created by a manual cut and paste process. Indeed, various fusion algorithms presented in literature have been evaluated by constructing some kind of ideal fused image and using it as a reference for comparing with the experimental fused results. Mean squared error-based metrics are widely used for these comparisons, and despite their well-known limitations, they can be helpful if used carefully. An example of such metrics is the root mean square error (RMSE).

Information theory-related metrics such as mutual information have also been proposed for fusion evaluation. Given two images x_F and x_R their mutual information is defined as the normalised grey level histograms of x_R, x_F, respectively, $h_{R;F}$ is the joint grey level histogram of x_R and x_F, and L is the number of bins. Let x_R, x_F correspond to the reference and fused images, respectively, and this indicates how much information the fused image x_F conveys about the reference x_R. Thus, the higher the mutual information between x_F and x_R, the more likely x_F resembles the ideal x_R. An example of an objective performance metric which does not assume the knowledge of ground truth is given in Reference [56]. In their approach [56], important visual information is associated with edge information measured for each pixel. Thus, they measure the fusion performance by evaluating the relative amount of edge information that is transferred from the input images to the fused image. Another non-reference objective performance metric is proposed by Qu et al. [60]. They evaluate image fusion by adding the mutual information between the fused image and each of the input images.

From the current literature, it can be concluded that objective performance assessment is an open problem which has received little attention. Most existing performance assessment methods are low level, that is, they act on the pixel level. High-level methods, that is, acting on region or even object level are non-existent. More research is required to provide valuable objective evaluation methods for image fusion, in particular, where if concerns region or object-based methods. To get an impression of the potential of the region-based versus the pixel-based scheme, a quality is used to evaluate the fused images where ground truth is not available. In all cases, region-based method gives a higher MI quality than the pixel-based method. A preliminary conclusion therefore is that the region-based scheme outperforms the pixel-based scheme. Moreover, use of region information for the combination of the approximation images as well as for the detail images improves the fused image considerably.

If the ground truth image $f_{gt}(x, y)$ is available, then explicit numerical evaluation of the fusion schemes can be performed. Assuming that the input images $f_i(x, y)$ are processed by the fusion schemes to create the fused image $f_f(x, y)$, to evaluate the scheme's performance, the following signal-to-noise ratio (SNR) expression can be used to compare the ground truth image with the fused image

$$\mathrm{SNR}_{\mathrm{dB}} = 10 \log_{10} \frac{\sum_x \sum_y f_{gt}(x,y)^2}{\sum_x \sum_y (f_{gt}(x,y) - f_f(x,y))^2} \tag{7.101}$$

As traditionally employed by the fusion community, the image quality index Q_0 can also be used as a performance measure. Assuming that m_f represents the mean of the image $f(x, y)$ and all images are of size $M \times N$, $-1 \le Q_0 \le 1$, the value of Q_0 that is closer to 1, indicates better fusion performance

$$Q_0 = \frac{4\sigma_{gtff} m_{fgt} m_{ff}}{(m_{fgt}^2 + m_{ff}^2)(\sigma_{fgt}^2 + \sigma_{ff}^2)},$$

$$\sigma_{ff}^2 = \frac{1}{MN-1} \sum_{x=1}^{M} \sum_{y=1}^{N} (f(x,y) - m_f^2), \tag{7.102}$$

$$\sigma_{fg} = \frac{1}{MN-1} \sum_{x=1}^{M} \sum_{y=1}^{N} (f(x,y) - m_f)(g(x,y) - m_g)$$

Recently, a universal image quality index (UIQI) [61] has been used to measure the similarity between two images. The UIQI is designed by modelling any image distortion as a combination of three factors: loss of correlation, radiometric distortion and contrast distortion. It is defined as follows:

$$Q = \frac{\sigma_{AB}}{\sigma_A \sigma_B} \cdot \frac{2\mu_A \mu_B}{\mu_A^2 + \mu_B^2} \cdot \frac{2\sigma_A \sigma_B}{\sigma_A^2 + \sigma_B^2} \tag{7.103}$$

The first component is the correlation coefficient for A and B. The second component measures how close the mean grey levels of A and B is and corresponds to an average luminance distortion with a dynamic range of [0, 1]. The third component measures a contrast distortion which is the similarity between the contrasts of A and B with a dynamic range of [0, 1]. The dynamic range of Q is [−1, 1]. If two images are identical, the similarity is maximal and equals 1.

Since image signals are generally non-stationary, it is appropriate to measure the number Q over local regions and then combine the different results into a single measure. A sliding window approach can also be used starting from the top-left corner of the two images A and B, a sliding window of fixed size moves pixel by pixel over the entire image until the bottom-right corner is reached. For each window w, the local quality index $Q(A, B|w)$ is computed for the values $A(x, y)$ and $B(x, y)$ where pixels (x, y) lie in the sliding window. Finally, the overall image quality index Q is computed by averaging all local quality indices:

$$Q(A, B) = 1/|W| \text{ sum } Q(A, B/w) \tag{7.104}$$

where W is the family of all windows and $|W|$ is the cardinality of W.

Piella [44] introduces three fusion quality measures based on Q_0. One measure is given by

$$Q(A, B, F) = \frac{1}{|W|} \sum_{w \in W} (\lambda_A(w) Q_0(A, F \mid w) + \lambda_B(w) Q_0(B, F \mid w)) \tag{7.105}$$

where $\lambda_i(w) = s(i|w)/(s(i|w) + s(j|w))$. $s(i|w)$ denotes some saliency of image i in window w. This should reflect the local relevance of image i within the window w, and it may depend on contrast, variance or entropy. $\lambda_i(w)$ is a local weight between 0 and 1, indicating the relative importance of image i compared to image j. Thus, in regions where image A has a large saliency compared to B, the quality measure Q is mainly determined by the similarity of F and input image A. On the other hand, in regions where the saliency of B is much larger than that of A, the measure Q is mostly determined by the similarity of F and input image B.

At this point, this model has produced a quality measure which gives an indication of how much of the salient information contained in each of the input images has been transferred into the composite image. However, the different quality measures obtained within each window have been treated equally. This is in contrast with the human visual system which is known to give higher importance to visually salient regions in an image. Another variant of the fusion quality measure is proposed by giving more weight to those windows where the saliency of the input images is higher. These correspond to areas which are likely to be perceptually important parts of the underlying scene. Therefore the quality of the composite image in those areas is of more importance when determining the overall quality. The overall saliency of a window is defined as $C(w) = \max(s(a|w), s(b|w))$. The weighted fusion quality measure is then defined as

$$Q_W(A, B, F) = \frac{1}{|W|} \sum_{w \in W} c(w)(\lambda_A(w) Q_0(A, F|w) + \lambda_B(w) Q_0(B, F|w)) \tag{7.106}$$

where

$$c(w) = C(w) \Big/ \Big(\sum_{w' \in W} C(w') \Big) \tag{7.107}$$

Another measure proposed takes into account some aspect of HVS, namely the importance of edge information. For this, Q_W is evaluated using edge images (norm of the gradient) instead of the original grey scale images. The edge dependent fusion quality index is given by

$$Q_E(A,B,F) = Q_W(A,B,F)^{1-\alpha} Q_W(A',B',F')^{\alpha} \tag{7.108}$$

where A' is the edge image corresponding with A. All the three measures have a dynamic range of $[-1, 1]$. The closer the value to 1, the higher the quality of the composite image.

If the ground truth image is not available, two image fusion performance indexes will be used: one proposed by Piella [44] and one proposed by Petrovic and Xydeas [56]. Both indexes are widely used by the image fusion community to benchmark the performance of fusion algorithms. They both attempt at quantifying the amount of interesting information (edge information) that has been conveyed from the input images to the fused image. In addition, as Piella's index employs the image quality index Q_0 to quantify the quality of information transfer between each of the input images and the fused image, it is bounded between -1 and 1.

The evaluation procedures are based on the verification of the preservation of spectral characteristics and the improvement of the spatial resolution. First, the fused images are visually compared. The visual appearance may be subjective and depends on the human interpreter, but the power of the visual cognition as a final backdrop cannot be underestimated. Second, a number of statistical evaluation methods are used to measure the colour preservation. These methods have to be objective, reproducible and of quantitative nature.

The correlation coefficient between the original multispectral bands and the equivalent fused bands. This value ranges from -1 to 1. The best correspondence between fused and original image data shows the highest correlation values. For a per-pixel deviation (PD), it is necessary to degrade the fused image to the spatial resolution of the original image. This image is then subtracted from the original image on a per-pixel basis. As final step, the average deviation per pixel is measured as digital number (DN) which is based on an 8-bit or 16-bit range. Here, zero is the best value.

The structure similarity index (SSIM) is a method that combines a comparison of luminance, contrast and structure and is applied locally in an 8×8 square window. This window is moved pixel-by-pixel over the entire image. At each step, the local statistics and the SSIM index are calculated within the window. The values vary between 0 and 1. Values close to 1 show the highest correspondence with the original images. The objective is to find the fused image with the optimal combination of spectral characteristics preservation and spatial improvement. Consequently as a third step two different quantitative methods are chosen to quantitatively measure the quality of the spatial improvement.

High-pass correlation (HCC): Correlation between the original panchromatic band and the fused bands after HP filtering. The HP filter is applied to the panchromatic image and each band of the fused image. Then the correlation coefficients between the HP filtered bands and the HP filtered panchromatic image are calculated.

ED in the panchromatic image and the fused multispectral bands: For this, a Sobel filter is used and a visual analysis is performed of the correspondence of edges detected in the panchromatic and the fused multispectral images. This was done independently for each band. The value is given in percent and varies between 0 and 100. A 100% means that all the edges in the panchromatic image were detected in the fused image.

However, as a reference image is not available to compare with, neither RMSE nor the Q can be computed. Instead, a measure based on mutual information is defined. This is obtained by adding the mutual information between the composite image and each of the inputs, and dividing it by the sum of the entropies of the inputs, that is

$$MI(A,B,F) = \frac{I(A,F) + I(B,F)}{H(A) + H(B)} \tag{7.109}$$

where $I(A,F)$ is the mutual information between A and F, and $H(A)$ is the entropy of A. In this way, the measure is normalised to the range $[0, 1]$.

Important visual information present in the output fused image can be associated with the edge information that is present in each pixel of the image. This visual to edge information association is supported by HVS studies. Furthermore, by evaluating the amount of edge information that is transferred from input images to the fused image, a measure of fusion performance can be obtained. Consider two input images f_A and f_B, and a resulting fused image f_F. A Sobel edge operator is applied to yield the edge strength $g(x, y)$ and orientation $\alpha(x, y)$ information for each pixel $f(x, y)$. Thus for an input image f_A

$$g_A(x,y) = \sqrt{e_A^x(x,y)^2 + e_A^y(x,y)^2} \tag{7.110}$$

$$\alpha_A(x,y) = \tan^{-1}\left(\frac{e_A^y(x,y)}{e_A^x(x,y)}\right) \tag{7.111}$$

where $e_A^x(x,y)$ and $e_A^y(x,y)$ are the output of the horizontal and vertical Sobel operators centered on pixel $f_A(x, y)$ and convolved with the corresponding pixels of image f_A. The relative strength and orientation values of $G_{AF}(x, y)$ and $A_{AF}(x, y)$ of an input image f_A with respect to f_F are formed as

$$G_{AF}(x,y) = \begin{cases} \dfrac{g_F(x,y)}{g_A(x,y)}, & \text{if } g_A(x,y) > g_F(x,y) \\[2ex] \dfrac{g_A(x,y)}{g_F(x,y)}, & \text{otherwise} \end{cases} \tag{7.112}$$

$$A_{AF}(x,y) = \frac{\left| \left| \alpha_A(x,y) - \alpha_F(x,y) \right| - \pi/2 \right|}{\pi/2} \tag{7.113}$$

The edge strength and orientation preservation values are given by

$$Q_{AF}^g(x,y) = \frac{\Gamma_g}{1 + e^{\kappa_g(G_{AF}(x,y)-\sigma_g)}} \tag{7.114}$$

$$Q_{AF}^\alpha(x,y) = \frac{\Gamma_\alpha}{1 + e^{\kappa_\alpha(A_{AF}(x,y)-\sigma_\alpha)}} \tag{7.115}$$

where $Q_{AF}^g(x,y)$ and $Q_{AF}^\alpha(x,y)$ model perceptual loss of information in f_F, in terms of how well the strength and orientation values of a pixel $f(x, y)$ in f_A are represented in the fused image. The constants $\Gamma_g, \Gamma_\alpha, \kappa_g, \kappa_\alpha, \sigma_g, \sigma_\alpha$ determine the exact shape of the sigmoid functions used to form the edge strength and orientation preservation values. Edge information preservation values are then defined as

$$Q_{AF}(x,y) = Q_{AF}^g(x,y)Q_{AF}^\alpha(x,y) \tag{7.116}$$

with $0 \le Q_{AF}(x, y) \le 1$. A value of 0 corresponds to the complete loss of edge information at location (x, y) as transferred from f_A to f_F. A value of 1 indicates fusion from f_A to f_F with no loss of information. A normalised weighted performance metric of a given fusion process F that operates on images f_A and f_B, and produces f_F is given by

$$Q_{AB/F} = \frac{\sum_{x=1}^{M}\sum_{y=1}^{N} Q_{AF}(x,y)w_A(x,y) + Q_{BF}(x,y)w_B(x,y)}{\sum_{x=1}^{M}\sum_{y=1}^{N}(w_A(x,y) + w_B(x,y))} \tag{7.117}$$

The edge preservation values $Q_{AF}(x, y)$ and $Q_{BF}(x, y)$ are weighted by $w_A(x, y)$ and $w_B(x, y)$, respectively. Edge preservation values which correspond to pixels with high edge strength influence $Q_{AB/F}$ more than those of relatively low edge strength. Thus, $w_A(x, y) = [g_A(x, y)]^L$ and $w_B(x, y) = [g_B(x, y)]^L$ where L is a constant and $0 \le Q_{AB/F} \le 1$.

7.13 Multimodal Biometric Systems and Fusion: Illustrative Examples

Unimodal or single biometric-based recognition systems do not fulfil the requirements of demanding secure applications in terms of universality, uniqueness, permanence, collectability, performance, acceptability and circumvention. This has motivated the current interest in multimodal biometrics in which these traits are simultaneously used for identification decision. Fusion in a multimodal system takes place in four levels: (i) sensor level, (ii) feature level, (iii) match score level and (iv) decision level. Levels of fusion are broadly classified into two categories [62]: (a) fusion before matching, called pre-classification – this approach includes sensor level and feature level fusion and (b) fusion after matching, known as post-classification – this includes match score level and decision level fusion.

In sensor level fusion different sensors are used to extract the same biometric trait. The multi-sensor data pertaining to a single trait increases the cost of the system but it helps in the segmentation and registration process. Information obtained from multiple sensors is complementary to each other. However, large memory space is required in sensor level fusion. In feature level fusion [63] different feature vectors of either the same biometric trait or from different biometric traits are extracted and combined. In a recogniser when feature vectors to be integrated are homogeneous, a single feature vector can be computed as a weighted average of the individual feature vectors. When feature vectors are non-homogeneous they are concatenated to form a single feature vector. This combination strategy is usually done by concatenation of the feature vectors extracted by each feature extraction module. This increases the size of the feature vector. Match score is a measure of the similarity between the input and template biometric feature vector. In match score level fusion scores are generated by multiple classifiers pertaining to different biometric traits and combined. In order to map scores of different classifiers into a single domain, normalisation technique is applied to the output of classifier before score fusion. Fusion at decision level [64] consists of combining the final output of different classifiers. In this fusion strategy, a separate authentication decision is made for each biometric trait. These decisions are then combined into a final vote. This approach overcomes the scores normalisation that would have been mandatory for the score level fusion.

7.13.1 Multimodal Biometric System Based on Feature Vector Fusion

Three biometric traits: fingerprint, face and signature are acquired separately and salient features of the traits are stored as templates in a database after pre-processing. Features from two or three biometric traits are concatenated to form a feature vector. Single matcher is used to evaluate the performance of the multimodal system. The features are sub-band coefficients of dual tree complex WT (DT-CWT) [65]. 2D DT-CWT does not only preserve wavelet transform's MR decomposition and perfect reconstruction but also adds new merits such as being insensitive to small image shifts and having multiple directional selectivity. DT-CWT provides shift invariance by introducing WT a'trous algorithm. Some of the feature vector fusion techniques considered are: (i) feature vector concatenation – the dimension of the resultant feature vector is thrice that of the dimension of a single feature vector as three traits are considered, (ii) feature vector averaging – the dimension of the resultant feature vector is same as the dimension of the individual feature vector as the average of the three vectors is considered as the feature vector, (iii) PCA features from concatenated feature vector – 2D PCA is applied on the concatenated feature vector to reduce the dimension by considering the PCs, (iv) PCA features before concatenation – 2D PCA is applied on the individual feature vectors and are concatenated to form the resultant feature vector and (v) feature concatenation of significant coefficients – required number of maximum magnitude coefficients are extracted from each feature vector and are concatenated to form the resultant feature vector.

The performance of the different feature vector fusion techniques mentioned above is evaluated on ECMSRIT (the college) database which has the three biometric traits of 100 users. The traits were recorded during two sessions in an interval of 1 month. A total of 10 samples of right thumb finger impressions are collected (from male and female staff and students of the department of electronics and communications engineering, MS Ramaiah Institute of Technology) using Nitgen 500 dpi fingerprint scanner. Each person is also requested to put his/her signature on an A4 size paper five times in each session which was scanned using a 300 dpi scanner. Frontal views of the face were also collected using LifeCam *Nx*-6000 digital camera which has a 2 MP sensor. Figure 7.7 shows the ECMSRIT database samples and the multimodal biometric system based on feature vector fusion. The size of the multimodal database considered is 10 samples, 3 biometric traits and 100 users equalling to 3000. One of the very popular classification algorithms support vector machine (SVM) is used for identification purpose and the performance is measured in terms of genuine acceptance rate (%GAR). GAR is defined as the fraction of genuine scores that exceed the given threshold. In the training phase, out of 10 samples of each user's biometric trait, 6 samples are considered for training and 4 samples are considered for testing the performance of the multimodal system. The database of each biometric trait is randomly split 30 times in the ratio 6:4 and the average GAR is considered for comparison. The feature vectors obtained from the training phase are stored in a database separately with fusion for multimodal biometrics and without fusion for unimodal biometrics. The test biometric trait is compared with the templates in the database using nearest-neighbour method employing Euclidean distance.

Table 7.1 gives the %GAR of different feature vector fusion algorithms obtained by combining two and three traits. The results are compared with

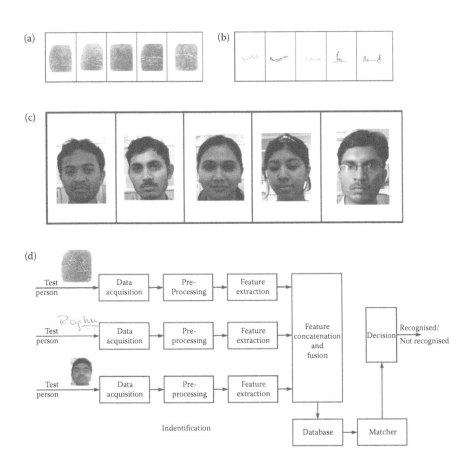

FIGURE 7.7
ECMSRIT databases and multimodal biometric system. (a) Sample of ECMSRIT fingerprint database, (b) samples of ECMSRIT signature database, (c) samples of ECMSRIT face database, (d) multimodal biometric system based on feature vector fusion.

the unimodal biometric trait performance. Results prove that combining three traits give better acceptance rate than combining two traits or unimodal systems. It also proves that feature vector fusion gives better performance than without fusion. Out of the fusion algorithms considered, PCA after concatenation gives maximum GAR% even with smaller feature dimension. This shows that PCA is a good technique for dimension reduction. Here, PCA is used to reduce the large feature vector dimension obtained by concatenating features of the three biometric traits. Also, combining three biometric traits gives significant improvement in GAR% (about 3%–5%) when compared to combining two biometric traits which in turn gives an improvement of about 3%–5% compared to single biometric trait. This proves the effectiveness of fusion-based algorithms.

TABLE 7.1

Comparison of GAR% of Different Feature Vector Fusion Algorithms

Fusion Method	Biometric Trait	Feature Dimension	GAR%[a]
PCA	Fingerprint	480	90.3
	Signature	480	92.2
	Face	480	96.3
Concatenation	Fingerprint + signature	1024	95.5
	Fingerprint + face	1024	96.7
	Face + signature	1024	94.1
	Fingerprint + face + signature	1536	97.6
Averaging	Fingerprint + signature	512	96.9
	Fingerprint + face	512	96.7
	Face + signature	512	93.7
	Fingerprint + face + signature	512	98.0
PCA after concatenation	Fingerprint + signature	512	95.8
	Fingerprint + face	512	96.7
	Face + signature	512	96.5
	Fingerprint + face + signature	480	98.8
PCA before concatenation	Fingerprint + signature	512	93.6
	Fingerprint + face	512	94.1
	Face + signature	512	93.8
	Fingerprint + face + signature	480	97.6
Concatenation of significant coefficients	Fingerprint + signature	512	90.4
	Fingerprint + face	512	89.0
	Face + signature	512	89.1
	Fingerprint + face + signature	512	93.0

[a] These results seem highly optimistic; however, these are included for illustrative purposes only, a reader/user should use some more and varied fusion methods and a few (more or other) biometric traits to generate results for their own data sets/samples. Only when more and similar studies are carried out for several types of data sets, biometric traits, fusion methods and classifiers, can one ascertain whether we can get such optimistic results. This is an open research topic in multimodal analysis and fusion.

7.13.2 Character Recognition System Based on Score Fusion

Fusion is one of the powerful methods for improving recognition rates of different character recognition algorithms. In fusion, to a certain extent the errors produced by one algorithm are taken care of by other algorithms and it emphasises the strengths and avoids weaknesses of individual algorithms. Fusion of data/information can be carried out on three levels of abstraction closely connected with the flow of the classification process: data level fusion, feature level fusion and classifier fusion [66]. Data level fusion in character recognition involves fusion of multiple samples of the same character which increases the computational complexity. Feature level fusion consists of combining different feature vectors extracted from the same character using different algorithms. When various features are combined as a single vector some large magnitude features may dominate the similarity measure and others may not have same impact on the classifier. A combination of individual classifiers' output is classifier fusion or score fusion in which individual classifier output overcome deficiencies of features and trainability of single classifiers. In this method if one algorithm fails to identify a class, another algorithm may support in identifying the class which improves the performance and the possibility of wrong identification by both the algorithms is reduced. Different classifiers potentially offer complementary information about the character to be classified which could be harnessed to improve the performance of the classifier. Therefore instead of relying on a single decision making, combination of classifiers can be used.

Three K-nearest-neighbour classifiers are used with $K = 3$, which are trained on 64 dimensional complex wavelet transform coefficients, 255 dimensional curvelet transform coefficients and 64 dimensional wavelet transform coefficients. The matching or distance scores associated with each classifier are combined using a fusion algorithm and then a decision is made based on the fused score. The distance scores are converted to similarity scores and normalised by min–max normalisation to map the scores to the interval [0, 1]. The fused score is a weighted combination of the individual scores. This fused score is compared with a threshold and the character is termed as recognised if the fused score is higher than the threshold. Otherwise, the character is termed as not recognised. The class to which the input character belongs is based on the class obtained by the majority voting scheme.

The database considered consists of 150 Kannada (an official Indian language of the state of Karnataka) text samples collected from 150 different writers. Writers from age of 10 years to age of 45 years were asked to write a page each in Kannada text taken from Kannada newspapers, magazines and textbooks. This document database was digitised at 300 dpi by a flatbed scanner. Characters were extracted by a fast joint segmentation and classification algorithm and this forms the character database of about 30,000 characters. This was partitioned into training and test data sets.

The Kannada character set was divided into: (i) main characters consisting of 15 vowels and 15 composite characters of 34 consonants which exist in top zone, and (ii) subscript characters consisting of 34 consonant conjuncts and 2 special symbols which exist in the bottom zone. Thus, the number of classes due to top zone characters is $15 + 34 \times 15 = 525$ and due to bottom zone characters is $34 + 2 = 36$, giving a total of $15 + 34 \times 15 + 36 = 561$ classes. Thus, the complete Kannada character set is represented by 561 characters or classes. The main characters are grouped into two subsets: subset 1 and subset 2 based on the width of the character in the top zone. The width of the characters less than or equal to 45 pixels are grouped as subset 1. The width of the characters greater than 45 pixels are grouped as subset 2. The consonant conjuncts and special symbols which exist in the bottom zone are identified by their position in a word and are grouped as subset 3. This results in 106 classes in subset 1, 180 classes in subset 2 and 36 classes in subset 3. Figure 7.8 shows the Kannada character databases and the character recognition system based on score fusion.

The training data set consisted of 5300 samples of subset 1 corresponding to 106 classes and each category had 50 patterns, 9000 samples of subset 2 corresponding to 180 classes and each category had 50 patterns and 360 samples of conjuncts corresponding to 36 classes and each category had 10 patterns. Test data set consisted of 2120 samples of subset 1 corresponding to 106 classes and each category had 20 patterns, 3600 samples of subset

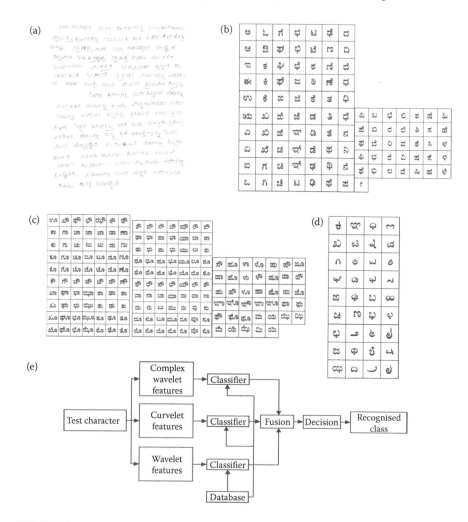

FIGURE 7.8

The language databases and the recognition system. a) Sample of handwritten Kannada characters, b) characters in subset 1, c) characters in subset 2, d) characters in subset 3, e) block diagram of the character recognition system based on score fusion.

TABLE 7.2

Results of Individual and Combined Classifiers for Kannada Character Database

Features	Feature Vector Dimension	Recognition Rate		
		Subset 1	Subset 2	Subset 3
Wavelet	64	84.9	76.9	54.5
Curvelet	255	84.9	79.9	59.2
Complex wavelet	64	89.4	89.3	63.2
Curvelet + wavelet + complex wavelet	–	94.4	94.7	88.6

2 corresponding to 180 classes and each category had 20 patterns and 360 samples of subset 3 corresponding to 36 classes and each category had 10 patterns. Segmented characters in the train and test data set of the character database were normalised to a size of 32×32 by using bilinear interpolation. Different features were extracted from the segmented characters in the train and test set. The train samples were used as a prototype for K-NN classifier and test samples were recognised based on the minimum Euclidean distance between train set and a sample in the test set and majority vote of its neighbours. The recognition rate of the classifier was defined as the ratio of number of correctly recognised characters to the total number of test characters. In this example, the weights used for fusion are all equal with value 0.33 and the threshold used for decision making is 0.5. Recognition accuracy of the individual classifier and combined classifier for the subsets are given in Table 7.2, from which it is clear that there is significant improvement in the recognition rate of the subsets with the decision of three classifier combination when compared with the recognition rate of the individual classifiers. This is because the test character is classified based on score level fusion and the decision based on the majority voting of the three classifiers.

EXERCISES

7.1 Draw the contours of constant distance pixels with a distance less than or equal to 2 from the centre pixel (x, y) for the following distance measures: (i) Euclidean distance, (ii) city block distance, (iii) chess board distance.

7.2 Find the output of the 3×3 averaging filter and four-neighbourhood Laplacian filter for the example image given. Consider only the given pixels without assuming any pixel values.

25	27	32	20	21
23	27	19	25	22
27	26	24	26	22
23	25	26	30	27
19	22	28	32	28

7.3 Derive and draw the input–output mapping functions of grey scale images for the following point operations. (i) adding 128, (ii) subtracting 128, (iii) negative, (iv) solarisation: complementing of pixels that are 128 or less and leaving other pixels unchanged, (v) complementing of pixels that are 128 or greater and leaving other pixels unchanged.

7.4 The following image has grey values coded using 3 bits per pixel. Compute the grey level histogram and the mapping that will equalise this histogram. Produce a 4×4 histogram equalised image. Draw the histogram before equalisation, mapping function and histogram after equalisation.

0	1	3	4
1	2	2	3
1	3	4	4
3	2	5	2

7.5 For the following example image, prove that a 3×3 median filtering removes the salt noise pixel added to the centre pixel. Do not assume any pixel values and consider only the given pixel values.

51	55	59	59	69
48	53	60	61	72
43	52	253	65	70
39	50	55	59	68
35	49	58	48	57

7.6 For the following examples of different edge profiles, find the first-order and second-order derivatives using the mask $[-1 \quad 0 \quad 1]$. Comment on the results.

  i. Upward step edge:

12	12	12	12	12	24	24	24	24	24

  ii. Downward step edge:

24	24	24	24	24	12	12	12	12	12

  iii. Ramp edge:

12	12	12	12	15	18	21	24	24	24

  iv. Roof edge:

12	12	12	12	24	12	12	12	12	12

7.7 For the example linear array $x = [10\ 25\ 40\ 55\ 70\ 85\ 100\ 115]$, compare the DFT and DCT reconstruction. Comment on the information packing capability of these transforms by retaining 50% of the coefficients. Repeat the example for 75%, 25% and 12.5% of the coefficients. Plot the reconstruction and compare the mean squared error in each case.

7.8 Apply a three-stage wavelet transform using Haar wavelet for the 1D example $x = [71\ 67\ 24\ 26\ 36\ 32\ 14\ 18]$. Consider all the wavelet coefficients obtained and reconstruct the original sample values. Reconstruct the sample values with a threshold of zero and consider only the coefficients greater than the threshold and prove that this reconstruction is very close to the original. Repeat the example with a threshold of value 9. Plot the reconstruction and calculate the mean squared error in each case.

References

1. Richardson, J. M. and Marsh, K. A. Fusion of multisensor data. *The International Journal of Robotics Research*, 7(6), 78–96, 1988.
2. Abidi, M. and Gonzalez, R. *Data Fusion in Robotics and Machine Intelligence*. Academic Press, Boston, 1992.
3. Canga, E. F. Image fusion. *Project Report for the Degree of MEng. in Electrical and Electronic Engineering, Signal and Image Processing Group*, University of Bath, June 2002.
4. Sternberg, S. Overview of image algebra and related issues. In *Integrated Technology for Parallel Image Processing* (Ed. Levialdi, S.). Academic Press, London, 1985.
5. Ritter, G. X. *Image Algebra*, University of Florida, Gainesville, available via ftp from ftp://ftp.cise.ufl.edu/pub/src/ia/documents, 1992.
6. Ravi, K. S. Probabilistic model-based multisensor image fusion. PhD thesis, Oregon Graduate Institute of Science and Technology, Portland, OR, 1999.
7. Bovik, A. C. *Handbook of Image and Video Processing*. Academic Press, San Diego, CA, 2000.
8. Kopf, J., Uyttendaele, M., Deussen, O. and Cohen, M. F. Capturing and viewing gigapixel images. *ACM Transactions on Graphics, Proceedings of SIGGRAPH*, 26(3), 93, 2007.
9. Gonzalez, R. C. and Woods, R. E. *Digital Image Processing*, 3rd Edition. Prentice Hall, USA, 2007.
10. Pizer, S. M., Amburn, E. P., Austin, J. D., Cromartie, R., Geselowitz, A., Greer, T., Romeny, B., Zimmerman, J. B. and Zuiderveld, K. Adaptive histogram equalization and its variations. *Computer Vision, Graphics and Image Processing*, 39, 355–368, 1987.
11. Canny, J. A computational approach to edge detection. *IEEE Transactions on Pattern Analysis and Machine Intelligence*, 8(6), 679–698, 1986.
12. Lowe, D. Distinctive image features from scale invariant keypoints. *International Journal of Computer Vision*, 60(2), 91–110, 2004.

13. Harris, C. and Stephens, M. J. A combined corner and edge detector. *Alvey Vision Conference*, 147–152, Manchester, UK, 1988.
14. Triggs, B. Detecting key points with stable position, orientation, and scale under illumination changes. *European Conference on Computer Vision, ECCV04 IV*, 100–113, Prague, Czech Republic, 2004.
15. Brown, M., Szeliski, R. and Winder, S. Multi-image matching using multi-scale oriented patches. *IEEE Computer Society Conference on Computer Vision and Pattern Recognition, CVPR*, San Diego, CA, 2005.
16. Lindeberg, T. Feature detection with automatic scale selection. *International Journal of Computer Vision*, 30(2), 79–116, 1998.
17. Bay, H., Tuytelaars, T. and Van Gool, L. SURF: Speeded up robust features. *Proceedings of European Conference on Computer Vision*, 110, 407–417, Graz, Austria, 2006.
18. Mikolajczyk, K. and Schmid, C. A performance evaluation of local descriptors. *IEEE Transactions on Pattern Analysis and Machine Intelligence*, 27(10), 1615–1630, 2004.
19. Alcantarilla, P. F., Bergasa, L. M. and Davison, A. J. Gauge SURF descriptors. *Journal of Image and Vision Computing*, 21(1), 103–116, 2013.
20. Liu, C., Yang, J. and Huang, H. P-SURF: A robust local image descriptor. *Journal of Information Science and Engineering*, 27, 2001–2015, 2011.
21. Viola, P. and Jones, M. J. Robust real time face detection. *International Journal of Computer Vision*, 57(2), 137–154, 2004.
22. Agrawal, M., Konolige, K. and Blas, M. R. CenSurE: Center surround extremas for real time feature detection and matching. *European Conference on Computer Vision, ECCV*, Marseille, France, 2008.
23. Huang, H., Lu, L., Yan, B. and Chen, J. A new scale invariant feature detector and modified SURF descriptor. *6th International Conference Natural Computation, ICNC*, 7, 3734–3738, Yantai, China, 2010.
24. Ebrahimi, M. and Mayol-Cuevas, W. SUSurE: Speeded up surround extrema feature detector and descriptor for real time applications. *Workshop on Feature Detectors and Descriptors: The State of The Art and Beyond, IEEE Computer Society Conference on Computer Vision and Pattern Recognition*, Miami, FL, 2009.
25. Berg, A. C. and Malik, J. Geometric blur for template matching. *IEEE Conference on Computer Vision and Pattern Recognition (CVPR)*, 607–614, Hawaii, USA, 2001.
26. Schmid, C. and Mohr, R. Local gray value invariants for image retrieval. *IEEE Transactions on Pattern Analaysis and Machine Intelligence*, 19(5), 530–535, 1997.
27. Brown, M., Gang, H. and Winder, S. Discriminative learning of local image descriptors. *IEEE Transactions on Pattern Analysis and Machine Intelligence*, 33(1), 43–57, 2011.
28. Acharya, T. and Ray, A. K. *Image Processing: Principles and Applications*. John Wiley and Sons, Inc., Hoboken, NJ, 2005.
29. Jain, A. *Fundamentals of Digital Image Processing*. Prentice Hall, Englewood Cliffs, NJ, 1989.
30. Allen, J. B. and Rabiner, L. R. A unified approach to short time Fourier analysis and synthesis, *Proceedings of IEEE*, 65, 1558–1564, 1977.
31. Mallat, S. *A Wavelet Tour of Signal Processing*. Academic Press, New York, 1999.
32. Meyer, Y. Wavelets: Their past and their future. *Progress in Wavelet Analysis and its Applications*, SIAM, Philadelphia, 1993.

33. Mallat, S. G. A theory of multiresolution signal decomposition: The wavelet representation. *IEEE Transactions on Pattern Analysis and Machine Intelligence*, 11(7), 674–693, 1989.

34. Taubman, D. S. and Marcellin, M. W. *JPEG 2000 – Image Compression Fundamentals, Standards and Practice*. Kluwer Academic Publishers, Boston, 2000.

35. Rabbani, M. and Joshi, R. An overview of the JPEG 2000 Still image compression standard. *Signal Processing: Image Communication*, 17(1), 3–48, 2002.

36. Gabor, D. Theory of communication. *Journal of IEE*, 98, 429–457, 1946.

37. Coifman, R. R. and Wickerhauser, M. V. Entropy based algorithms for best basis selection. *IEEE Transactions on Information Theory*, 38(2), 713–718, 1992.

38. Candès, E. J. and Donoho, D. L. Ridgelets: A key to higher dimensional intermittency? *Philosophical Transactions of the Royal Society A*, 357(1760), 2495–2509, 1999.

39. Do, M. N. and Vetterli, M. The finite ridgelet transform for image representation. *IEEE Transactions on Image Processing*, 12(1), 16–28, 2003.

40. Candès, E. J. and Donoho, D. L. Curvelets – A surprisingly effective non-adaptive representation for objects with edges. In Curves and Surfaces (Eds. Rabut, C., Cohen, A. and Schumaker, L. L.). Vanderbilt University Press, Saint-Malo, 105–120, 2000.

41. Pennec, E. L. and Mallat, S. Sparse geometric image representations with bandelets. *IEEE Transactions on Image Processing*, 14(4), 423–438, 2005.

42. Do, M. N. and Vetterli, M. Contourlets. In *Beyond Wavelets* (Ed. Welland, G. V.). Academic Press, New York, 2003.

43. Do, M. N. and Vetterli, M. The contourlet transform: An efficient directional multiresolution image representation. *IEEE Transactions on Image Processing*, 14(12), 2091–2106, 2005.

44. Piella, G. A general framework for multiresolution image fusion: From pixels to regions. *Information Fusion*, 4, 259–280, 2003.

45. Newman, E. A., Gruberg, E. R. and Hartline, P. H. The infrared trigemino – Tectal pathway in the rattlesnake and in the python. *Journal of Comparative Neurology*, 191, 465–477, 1980.

46. Li, H., Manjunath, B.S. and Mitra, S.K. Multisensor image fusion using the wavelet transform. *Graphical Models and Image Processing*, 57(3), 235–245, 1995.

47. Burt, P. and Adelson, E. Laplacian pyramid as a compact image code. *IEEE Transactions on Communication*, 31(4), 115–123, 1983.

48. Lewis, J. J., O'Callaghan, R. J., Nikolov, S. G., Bull, D. R. and Canagarajah, C.N. Region based image fusion using complex wavelets. *Proceedings of 7th International Conference on Information Fusion*, 555–562, Stockholm, Sweden, 2004.

49. Toet, A. Image fusion by a ratio of lowpass pyramid. *Pattern Recognition*, 9, 245–253, 1989.

50. Burt, P. J. and Kolczynski, R. J. Enhanced image capture through fusion. *Proceedings of the 4th International Conference on Computer Vision*, 173–182, Berlin, Germany, May 1993.

51. Wilson, T. A., Rogers, S. K. and Meyers, L. R. Perceptual based hyperspectral image fusion using multiresolution analysis. *Optical Engineering*, 34(11), 3154–3164, 1995.

52. Koren, I., Laine, A. and Taylor, F. Image fusion using steerable dyadic wavelet transforms. *Proceedings of the IEEE International Conference on Image Processing*, 232–235, Washington, DC, 1995.

53. Pu, T. and Ni, G. Contrast based image fusion using the discrete wavelet transform. *Optical Engineering*, 39(8), 2075–2082, 2000.
54. Nunez, E., Otazu, X., Fors, O., Prades, A., Palà, V. and Arbiol, R. Multiresolution based image fusion with adaptive wavelet decomposition. *IEEE Transactions on Geoscience and Remote Sensing*, 37(3), 1204–1211, 1999.
55. Crippen, R. E. A simple spatial filtering routine for the cosmetic removal of scanline noise from LANDSAT TM P-Tape imagery. *Photogrammetric Engineering and Remote Sensing*, 55(3), 327–331, 1987.
56. Petrović, V. and Xydeas, C. Computationally efficient pixel-level image fusion. *Proceedings of Eurofusion99*, Stratford-upon-Avon, 177–184, 1999.
57. Shutao, L., James, T. K. and Yaonan, W. Combination of images with diverse focuses using the spatial frequency. *Information Fusion*, 2, 169–176, 2001.
58. Shensa, M. J. The discrete wavelet transform: Wedding the à trous and Mallat algorithms. *IEEE Transactions on Signal Processing*, 40(10), 2464–2482, 1992.
59. Wang, Z., Ziou, D., Costas, A., Li, D. and Li, Q. A comparative analysis of image fusion methods. *IEEE Transactions on Geoscience and Remote Sensing*, 43(6), 1391–1402, 2005.
60. Qu, G., Zhang, D. and Yan, P. Information measure for performance of image fusion. *Electronics Letters*, 38(7), 313–315, 2002.
61. Wang, Z. and Bovik, A. C. A universal image quality index. *IEEE Signal Processing Letters*, 9(3), 81–84, 2002.
62. Nandakumar, K. Multibiometric system: Fusion strategies and template security. PhD thesis, Department of Computer Science and Engineering, Michigan State University, 2008.
63. Ross, A. and Govindarajan, R. Feature level fusion using hand and face biometrics. *Proceedings of SPIE Conference on Biometric Technology for Human Identification*, 5779, 196–204, Orlando, Florida, March 2005.
64. Chatzis, V., Bors, A. G. and Pitas, I. Multimodal decision level fusion for person authentication. *IEEE Transactions on Systems, Man and Cybernetics, Part A*, 29(6), 674–680, 1999.
65. Kingsbury, N. Complex wavelets for shift invariant analysis and filtering of signals. *Journal of Applied and Computational Harmonic Analysis*, 10(3), 234–253, 2001.
66. Gader, P. D., Mohamed, M. A. and Keller, J. M. Fusion of handwritten word classifiers. *Pattern Recognition Letters*, 8, 31–40, 1995.

8

Decision Theory and Fusion

8.1 Introduction

Decision theory (DT) and making appropriate decisions now play a very important role in the area of decision fusion (DeF), mainly because of the complex situations encountered in multi-sensory (MS) data fusion (DF) processes. Also, as seen in Chapters 4 and 5, the problem of statistical estimation and associated decision making (which is often a concealed process) constitutes a complete problem and solution of target tracking. Selecting an appropriate cost/loss/criterion function for the derivation of optimal estimation/ filtering algorithms is in itself a decision making process, and is often taken for granted, but it is one of the most important aspects in the area of control and estimation technology, and now more so in DF technology.

DT is mainly about how to make decisions and what decisions to take, on/ about the status of an object, scenario and/or event, based on some objective analysis (sometimes subjective analysis) carried out from the given data/ information sets [1–3]. Sometimes, decisions are made based on available data from sensors and many other sources. Often these data/information sets are processed and then some appropriate decision/s are taken or made. At times decisions are made based on derived information from various competing sensors/sources/knowledge bases. The ultimate idea of such (intermediate) decision making is to aid humans in making higher level, sophisticated decisions/commitments/firm actions. We routinely see that in our lives most actions initiated by us involve making decisions. However, DT focuses on some aspects of human activities. In many situations, there are often a few options to choose from, and we tend to choose these options in some non-random and definite way, since random decisions are seldom good. In almost all cases, our choices should be goal-directed activities and actions, and as such DT is concerned with goal-directed behaviour in the presence of several or a few options, and sometimes these options might be of a conflicting nature. In our daily activities, we do decide (upon something) continually (in discrete manner), though not continuously (in continuous manner). Modern DT has contributions from several disciplines, and there is a large overlap, and the subject of DT has benefited from various methods/concepts/approaches pursued by researchers of different back-grounds [1–3].

We would like to emphasise here that, in a kind of decision process (DP) that is called normative DT (NDT), one discusses about how decisions should be made suggesting the methods of making decisions, whereas in the other kind of DP called the descriptive DT (DDT) one studies about how decisions are actually or really made [1,2]. It is usual to recognise that NDT is about how decisions should be made (rather than how actually they are made) – the methodology of this decision approach suggests rational decision mak-ing in order to be rational, this latter aspect being one part in the definition of AI (artificial intelligence), that is, it is then related to the AI definition in respect of the humans who think rationally. In the DDT the AI part that is operative is how 'humans act rationally'. This aspect also includes the aspect of coordination of the decisions in social decision procedures, so, we see how the DP (NDT/DDT) further suggests an important part of AI. In fact AI has several facets and sub-topics such as (i) perception, (ii) planning, (iii) logical decision making, (iv) actions and (v) learning, a few amongst many such aspects of AI.

The DP can be further divided into various active sub-components [1–3]: (i) figuring out the situation and/or the identification of the actual problem, (ii) basic data-gathering, collecting necessary information about the prob-lem, (iii) arriving at feasible solutions – by some intermediate processing steps, (iv) evaluation of these solutions – intermediate analysis (intermediate or final decision/s), (v) performance evaluation of the decision/s made and (vi) implementation of the decision-making process, that is, taking actions to reach the goal, for example, a mobile vehicle heading towards a destination point. Some steps can be carried out in concurrent manner, whereas some in sequential manner. When we make decisions (i.e. make a choice between various options), we try to obtain/select/use as good an outcome (or result) as feasible/possible. We need to use some measure for this selection – this is done according to some standard criterion (many such criteria are used in estimation theory/practices) that can be used to judge what is good or bad. In DT such a standard criterion is known or specified, and then one proceeds to express this criterion in a precise and useful way making it depend on cer-tain variable of the problem at hand. Then by an optimisation procedure the cost function/criterion is minimised/maximised depending upon the situ-ation and the type of the criterion defined. In a general sense this standard criterion is related to the cost or utility function.

Any way of making a certain decision would be incomplete and not an assuring and confident way unless it is supported by some mathematical expressions/formulae and expressed by some specific criterion. What we really need is statistical decision theory (SDT) which may sound merely a description, but when expressed and written in mathematical terms turns out to be a formal way of making decisions [1–3]. However, the SDT is a valuable aid to scientific management, sciences and the engineering decision-making processes. Here, the term 'statistical' denotes our reliance on quantitative methods and criteria. This means that the DT process should be based on some quantitative criterion, or loss/cost function, and the minimisation of this function or the maximisation of the utility function would give the optimal solution to the decision-making process. This loss function is also called cost function or simply function. Thus, SDT helps us select the best option from many competing and alternative ones. However, this best decision would depend on the available information used in the DP. In the future, if some more data and information are available and incorporated in the process of making decisions, then we need to revisit and revise the decision process again. This is in the line with the process of scientific study and investigation. As we have seen earlier for DP, the SDT has also the same and several ingredients/elements of decision making [1–3]: (i) definition of the problem (about the object of investigation, test scenario, or a situation), (ii) specification and choice of an appropriate criterion – this is the kernel of the SDT, (iii) accurate determination of the environmental situation and its parameters if possible, since the criterion is made dependent on these parameters, (iv) description of all the alternative/intermediate actions – scientific/managerial/engineering ones, (v) development and enunciation of the DP or procedure, if the procedure is an involved one then it is laid down by an appropriate flow diagram for quick understanding – and in case we are going to use fuzzy logic (FL) then the decision process is described in the form of If … Then … rule, (vi) arriving at an optimal or an adequate solution of the problem and (vii) making the final decision.

Most and perhaps all decision-making problems are characterised by one or more of the following aspects [2]: (i) the aim to attain an established or specified goal, (ii) the availability of many alternative, possible and feasible actions and (iii) certain environment that exists with respect to these actions, for example, risk, certainty/uncertainty, conflict and ignorance.

Often, in many science and engineering problems and closely related DF systems, we encounter a situation wherein we have several similar options and many results of the analysis of the available data/information about one or more (but nearly similar) scenarios/objects of classification [3,4]. Then we need to make a decision as to which one option or result we should chose and accept so that a desired goal is met with reasonable prediction accuracy (PA). The point here is that in any DF task based on the sensors' data/measurements/information, the intrinsic/inherent accuracy of the source/sensor does not necessarily increase or improve. And hence, with the help of the DF based on the information from several sensors/sources what we attain is the overall enhancement of the PA, or enhanced total information and the reduction of uncertainty. It is important to keep this point in mind that by sensor DF we do not increase the intrinsic accuracy of any sensor/data, but obtain the overall enhanced information and reduced uncertainty of prediction and hence the enhanced PA.

Often this decision need not be optimal. In this process of decision making we need to use the available knowledge and information and use some logical, statistical or heuristic method to arrive at an appropriate decision. The application of artificial neural networks (ANNs)/FL/genetic algorithm (GA) and closely related methods are termed as heuristic approaches for DP. In the discipline of computer science and AI these paradigms are termed as soft computing methods. Although, these heuristic approaches are different from the statistical approaches, they can be augmented thus. This means that one can also combine these heuristic approaches with some (formal and proper) statistical approaches or concepts leading to more powerful theories of DT. Then DT decision making is generally and perhaps always coupled with the knowledge acquisition, assimilation and representation (aspects from AI), more so with the latter. This is derived from an inference engine, a process operating over the measured data and/or over the collected raw information. Thus, we see that this approach closely resembles the one with the new paradigm called the FL-based approach. Of course, the same DP classically can be handled with the help of so-called crisp logic. The FL logic adds an extra parameterisation into the DP, and can handle complex decision-making processes.

In Chapter 3, we have studied the concepts of fuzzy logic type 1 (T1FL) and interval type 2 fuzzy logic (IT2FL) that can be very useful for decision making and sensor DF, wherein we use fuzzy inference system (FIS) with the If ... Then ... rules. Also, we have studied the application of FL in conjunction with Kalman filtering (KF) algorithms [4] to obtain better performance of the fusion process. As seen in Chapters 1 to 5, the low-level DF is mainly concerned with the problem of mathematical modelling and combining information in some logical and appropriate way using statistical/probabilistic methods. In almost all the control and systems engineering domains the decision making is concerned with: (i) obtaining an estimate of the true state of the dynamic system from the measured data (Chapters 4, 5), because the true state either is not known or is not measureable, (ii) performing a control action and/or (iii) deciding to acquire more information to enhance the process of estimation and decision making [3]. In mid-level DF situations, for example, in feature/image processing and fusion (Chapters 6 and 7), significant and more sophisticated decision-making processes are involved. In this process all the information available is utilised, and the DP then generates an action that corresponds to the intended decision, that is, the final fused image. In this DP, a loss function or a utility function provides a means of evaluating and comparing different actions, DF processes, allowing for direct comparison of alternative decisions, that is, fusion rules/formulae. Based on a specified loss function it is possible to model the decision-making process. This further helps evaluate different DF methods including DeF approaches [3,4], which are also called symbol level fusion.

8.2 Loss and Utility Functions

The decision making starts with an unknown state of nature x (in/of a set X), for example, in engineering problems and systems. The point is that we are interested in determining the state, at the current time (and many times also for the future, say one-time step ahead) of a dynamic system from the available and mostly noisy measurements. In this situation of gathering information, an experiment is conducted that generates an outcome or set of measurements z (in/of a set Z), and we need to make sure that these measurements that we have made (or obtained) are in some sense related to the state we want to determine, the point is that the state of the system is directly not known or not measurable to us [3]. It is assumed that the system is observable (and also controllable i.e. the conditions of observability and controllability are satisfied in the sense specified by the control and system theory). In fact this requirement is specifiable for the mathematical model of the system that the model represents. On the basis of these measurements, an action 'a' (in/of a set A) must be computed, this is done specifically to make an estimate of the true state, since the true state is not known as noted above, in which case 'a' belongs to the set X [3]. In general, it is to choose one of a specified set of possible actions in A, and a decision rule 'δ' is defined as a function taking measurements into actions; that is, $\delta: Z \rightarrow A$. Thus, for every possible measurement z, 'δ' defines what action 'a' must be taken; that is, $a = \delta(z)$. This action is taken based on the measurements taken from the sensors that actually observe the system of interest.

A loss function $L(.,.)$, (or $f(.,.)$ – called a cost function) or correspondingly, utility function $U(.,.)$, is defined as a mapping from pairs of states and actions to the real line (on x–y plane) by the expressions: $f: X \times A \rightarrow R$ and $U: X \times A \rightarrow R$ [3]. A loss function $f(x, a)$ signifies that 'f' is the loss (or cost) incurred in taking the action a (when the true state of the system x is of course not known), since we are using the estimate of the true state and our action is based on this estimate. Similarly, the utility function $U(x, a)$ signifies that it is the gain obtained in taking the action a when the true state of the system is x. For a fixed state of the system both the utility and loss/cost (function) will help in inducing an ordering on the set of actions A. The loss and utility functions are supposed to obey certain rules called the utility or 'rationality axioms' which assure a preference and consistent pattern. For fixed x (of X) [3]: (i) given any a_1, a_2 (in A), either $U(x, a_1) < U(x, a_2)$, $U(x, a_1) = U(x, a_2)$ or $U(x, a_1) > U(x, a_2)$, that is, given any two actions we can assign real numbers which indicate our preferred alternatives – less than, equal or more than; (ii) if $U(x, a_1) < U(x, a_2)$ and $U(x, a_2) < U(x, a_3)$ then obviously $U(x, a_1) < U(x, a_3)$, that is, if we prefer action a_2 to a_1 and action a_3 to a_2, then it is imperative that we (must) prefer action a_3 compared to action a_1, here, the preference/preferred ordering is transitive in nature and (iii) if $U(x, a_1) < U(x, a_2)$, then $\alpha U(x, a_1) + (1 - \alpha)U(x, a_3) < \alpha U(x, a_2) + (1 - \alpha)U(x, a_3)$, for any $0 < \alpha < 1$. We interpret this as: if the action a_2 is preferred to the action a_1 then, in a choice between two random situations that both a_1 and a_2 occur with probability α, the situation with action a_2 will be preferred.

It is very important that the loss (or utility) function faithfully and accurately represents the value/weight and hence, the importance placed on different decisions. However, many of the decision rules are insensitive to the exact form of the loss function, for example, symmetric ones [3], and in general a loss function in the form of $f(x, a)$ is not very useful because the true state of nature of 'x' will not be known with precision and the true loss incurred in taking an action will not be known, however, we do obtain some good help by choosing appropriate and good cost/loss/utility function to help us make decisions form various alternatives. In general there will be probability distribution function (PDF), $P(x)$, and/or probability density function (pdf), $p(x)$, summarising all the (probabilistic) information about the state at the time of decision making. Then one natural method of defining loss is as an expected loss (Bayesian expected loss) which for continuous-valued state variables is simply given by [3]

$$\beta(a) \equiv E\{L(x,a)\} = \int_{-\infty}^{\infty} L(x,a)p(x)dx \tag{8.1}$$

and for discrete-valued state variables is given by

$$\rho(a) \equiv E\{L(x,a)\} = \sum_{x \in X} L(x,a)p(x) \tag{8.2}$$

Bayesian expected loss weighs the loss incurred by the probability of occurrence (an average loss).

8.3 Bayesian DT

As we have seen in Chapters 4 and 5, the probabilistic information concerning x is obtained after taking a number of measurements in the form of a posterior distribution $p(x|Z_k)$. An *expected utility* (or loss function) β following an action a can be defined with respect to a specified posterior distribution as [3]:

$$\beta(p(x|Z_k),a) \equiv E\{U(x,a)\} = \int_x U(x,a)\,p(x|Z_k)dx \tag{8.3}$$

The Bayesian action a is defined as the strategy which maximises the posterior expected utility function

$$\hat{a} = \arg \max_{a} \beta(p(x|Z_k), a) \tag{8.4}$$

This is equivalent to maximising

$$\int_{x} U(x,a)p(Z_k|x)p(x)dx \tag{8.5}$$

In the estimation problems, the action set is made equal to the set of possible states $(A = X)$, and the MMSE (minimum mean square error) estimate is defined by

$$\hat{x} = \arg \min_{a \in X} \int_{x} (x - a)^T (x - a)p(x|Z_k)dx \tag{8.6}$$

This is clearly a Bayesian action with respect to a squared error loss function defined by $L(x, \hat{x}) = (x - \hat{x})^T (x - \hat{x})$.

8.4 Decision Making with Multiple Information Sources

As we have seen earlier the decision-making process with several sensor/sources of data is much more complex than that for the single sensor/source. The reasons are [3]: (i) how can we compare the utilities of two different decision makers unless there is a common measure of value, this being known as the problem of inter-personal utility comparison; and (ii) what should the decision makers aim for: (a) should they maximise a utility which expresses only local preferences? or (b) should they evaluate its actions with respect to some common or group utility function? However, for specific decision-making problems in which concepts of utility may be simplified and made very precise, we can arrive at consistent/useful solutions to the utility comparison and the group utility problem.

8.4.1 Super Bayesian

Suppose we have a system consisting of 'm' information sources/sensors and a single overall decision maker. The decision maker has to combine probabilistic information from all these sources and make decisions based on the global posterior pdf. If this global pdf is $p(x|Z_k)$, then the Bayesian group action is given by [3]

$$\hat{a} = \arg\max_a \beta(p(x|Z_k), a)$$
$$= \arg\max_a E(U(x|a), Z_k) \qquad (8.7)$$

In Equation 8.7, $U(x, a)$ is a group utility function and the solution is well defined in terms of classical Bayesian analysis. This approach is called 'super Bayesian'.

8.4.2 Multiple Bayesians

We now consider a system of 'm' Bayesians. Each of the systems is able to get its own probabilistic information and shares with all the other Bayesian information before computing a posterior pdf [3]. This posterior pdf obtained by each Bayesian i is given by $p(x_i|Z_{ki})$. Each Bayesian is now called upon to compute an optimal action which is (supposed to be) consistent with those of the other Bayesians. Thus, a fully decentralised situation is described and it is a difficult challenge for which the universal solution cannot be found. This is because a generally applicable criterion for determining rationality/optimality is not available. Also, there is a dilemma whether one should follow a group or an individual optimality. The simplest case is wherein the optimal action a_i at each Bayesian 'i' would turn out to be the same for all the Bayesians yielding a group action. However, in general the local decisions are not the same. Then a solution is achieved as described next.

Each Bayesian i computes a permissible (feasibly acceptable) set of actions $A_i \subseteq A$ and if the set $A = \cap_j A_j$ is non-empty, then from the set A, the group action is selected. Then an acceptable class of actions is obtained by maximising [3]

$$\beta(p(x|Z_k), a) = \sum_j w_j E\{U_j(x, a)|Z_k\} \qquad (8.8)$$

with $0 \le w_j \le 1$ and $\Sigma_i w_j = 1$. From the LL principle, Equation 8.8 can be written as a maximisation of the following expression:

$$\sum_j w_j \int U_j(x, a) p(Z_{kj}|x) p(x) dx \qquad (8.9)$$

It is clear from Equations 8.8 and 8.9 that acceptable actions are represented. Yet, it is difficult to choose a single optimal action a from this set, because the individual utilities U_i cannot be compared. This is further so, because these may correspond to completely different scales of value. In case each local utility function were described on a common scale, then one can get the optimal group action based on Equation 8.8 or 8.9. With a few additional restrictions on the form of U_i, such comparisons can be made to obtain sensible and acceptable decisions. Where the comparison between local utilities is justified, the optimal group decision is obtained from [3]

$$\hat{a} = \arg\max_a \left\{ \sum_j w_j [E\{U_j(x,a)\} - c(j)]^\gamma \right\}^{1/\gamma} \tag{8.10}$$

Here, $c(j)$ is a decision maker j's security level. It plays the role of 'safeguarding j's interests'. The weight w_j is as given in Equation 8.8 and we have $-\infty \le \gamma \le \infty$. For $\gamma = 1$, we obtain a solution that minimises Equation 8.8. For $\gamma = -\infty$ and $\gamma = \infty$ we obtain the Bayesian max–min and min–max solutions, respectively. For $\gamma = 1$ and $\gamma = \infty$ we get a solution for an individual decision-maker with a net loss of expected utility.

8.4.3 Product Solution

The Nash-product solution (NPS) that consists of the product of the individual utilities can be used when the direct comparison of the utilities is not justifiable, and is obtained from the following criterion [3]:

$$\hat{a} = \arg\max_a \left\{ \sum_j [E\{U_j(x,a)\} - c(j)] \right\} \tag{8.11}$$

In Equation 8.11, some level of individual optimality can be maintained by using proper values of $c(j)$.

8.5 Fuzzy Modelling Approach for Decision Analysis/Fusion

We often encounter a situation wherein the sensors are geographically/physically separated and they observe a common object. These sensors are linked to a global processor, or to a DF centre. In a usual CP (Chapters 4 and 5) system, these sensors transmit their measurements without any processing at local nodes. Then the global processor incorporates these measurements in an optimal manner and obtains the global objective [5]. However, these communication channels might have finite capacities. Also, in several such practical cases, the mathematical model used to represent the physical phenomenon in the decentralised detection and/or estimation (DCDE) might not be fully known, and there would certainly be some measurement inaccuracies associated with the sensors and data. Often these uncertainties might be of a non-random nature that might be associated with the local sensors and also with the decisions, and these can be represented by means of fuzzy sets and fuzzy logic (FLS). This non-random uncertainty can be specified in the form of a suitably defined membership function, which in turn defines a gradual transition from full membership to zero membership. Also, all the parameters of the mathematical models would not be known, or known with some inaccuracies. The FL-based modelling approach could be of benefit in this regard also.

In the fuzzy modelling approach, each sensor receives a set of measurements and then derives a local decision regarding the current situation, and since these local sensor decision-rules are given, the aim is to combine these local decisions. In this case, the central fusion centre (CFC) would receive the decisions from the local sensor nodes with varying degrees of accuracy. The local sensor error probabilities might not remain constant and hence, can be modelled using FL. This way we can incorporate the local sensor decision inaccuracies in the global decision function. Then the aim is to arrive at the optimal DeF rule.

A particular sensor $S_k(k = 1, ..., M)$ derives a binary decision u_k that is conveyed to the global decision maker (GDM). Here, it is combined with the local decisions received from the other sensors to produce the global decision u_0. Also, it is important to assume that the local decisions are statistically independent across the sensors, and also it is assumed that the prior probabilities of occurrence of hypotheses, H_1, and H_0, denoted by p_1 and p_0, are known [5].

The Bayesian cost assignment used is $C_{ij} \rightarrow$ when the GDM decides $u_0 = i$, when the hypothesis H_j is true, for $i; j = 0; 1$; and obviously in this assignment $C_{10} > C_{00}$ and $C_{01} > C_{11}$ are true. The optimum fusion rule that minimises the global risk for the particular cost assignment is obtained. The Bayesian risk function connected with the global decision u_0 is written as [5]

$$R = \sum_{i=0}^{1} \sum_{j=0}^{1} C_{ij} P(u_0 = i \, H_j) \tag{8.12}$$

Introducing the local decision vector $U = (u_1, u_2, ..., u_M)$ and rearranging the above equation we get the following expression:

$$R^* = \sum_{u} P(u_0 = 1|U)\{(C_{10} - C_{00})P(U|H_0)p_0 - (C_{01} - C_{11})P(U|H_1)p_1\} \tag{8.13}$$

Here, $R = R^* +$ the remainder, and for a given local decision vector U, the aim is to design the global decision u_0 such that R^* and hence R are minimised.

Let $C_f = p_0(C_{10} - C_{00})$ and $C_d = p_1(C_{01} - C_{11})$ then we have the following decision rule: decide, $P(u_0 = 1|U) = 1$, if the local decision vector U is such that $C_f P(U|H_0) \leq C_d P(U|H_1)$, then decide u_0 otherwise. Since, $P(U|H_j)$, $j = 0,1$ are fuzzy quantities, finally the crisp decision is taken based on the approach defined in Reference 5.

Thus, in this FL-based (fuzzy modelling) DeF, the GDM, receives the fuzzy local sensor decisions, and then employs the Bayesian decision criterion to derive the global decision. The fuzziness of the input data yields a fuzzy global Bayes risk. Due to the requirement of generating a clear cut (crisp) global decision regarding the hypothesis present, the global Bayes risk has to be defuzzified by mapping the FL quantities (in the global Bayes risk) onto the real line. The total distance criterion (TDC) can be used in doing this defuzzificaion [5]. Hence, the decision process involves obtaining the fusion rule that would result in the smallest equivalent real number for the fuzzy Bayes risk in the sense of the TDC. This decision rule consists of a ratio of the fuzzy ranks of the local decision vector (under hypotheses H_1 and H_0, respectively) which is then compared against a fixed threshold.

8.6 Fuzzy-Evolutive Integral Approach

A theoretical presentation of the standard and improved fuzzy integral algorithms for DeF is given in Reference 6. The fuzzy-evolutive integral (FEI) is a hybrid method that optimises the mixing modes of the outputs that are assigned to more neural classifiers. It uses the standard fuzzy integral (Sugeno's fuzzy integral which is a non-linear functional) to realise the suitable output combination of some distinctive neural networks (NNW). The development of the fuzzy integral is given in Reference 7. The NNWs are selected based on the weights (weighting factors) assigned to them by a GA. The chromosomes in the GA will encode the (real) fuzzy densities g_i^j as a vector $C_j = (g_1^j, g_2^j, ..., \lambda_j)$, and the fitness function for C_j is given by the following formula [6]:

$$E(C_j) = \sum_{A \in B(X)} g_{\lambda_e}(A) - \frac{1}{\lambda_j}\left[\prod_{x_i \in A}(1 + \lambda_j g_i^j) - 1\right] \tag{8.14}$$

Here, $g(A)$ of the first terms are the initially assigned values for the fuzzy measures. These are computed using g_i^j and λ_j. In the proposed method, the FL provides the representation of high-level knowledge (since it is specified by human experts), the NNW provides the basic structure of the algorithm and GA provides the information for the development. The detailed development of this method is given in Reference 7. The authors [6] have used the

FEI technique to increase the target recognition performances. This is based on high-resolution radar and video imagery using a real input database.

8.7 Decision Making Based on Voting

It is possible to arrive at decisions based on the concept of voting that is used in hardware/software (HW/SW) fault detection. The voting schemes and associated terminology in dependable computing have been originally derived from certain concepts in socio-political systems, for example, elections. The main idea in voting is that with inputs drawn from a small set of integers, we seek whether the similarity between the two domains is strong. For example, we use radar image analysis to classify approaching aircraft type as (a) civilian (0), (b) fighter (1) and (c) bomber (2); with the numbers in the parenthesis indicating that type of aircraft. If the three independent decision makers' units arrive at the conclusions $\langle 1, 1, 2 \rangle$, then the presence of a fighter plane is suggested. In this case, we see that the candidate '1' wins a majority of the vote. When we have a large or infinite input domain/volume of data, then the voting can have a new meaning.

A voter receives inputs X_1, X_2, ..., X_N from M-of-N (i.e. M out of N) cluster set and generates a representative output [8,9]. This simplest voter that could have been HW implemented (or SW implemented) does a bit-by-bit comparison of the outputs and checks if a majority of the N inputs are identical. If this is so, then it outputs the majority, say, M. This scheme works well if we can assure and guarantee that every module generates an output that matches the output of every other functional module bit-by-bit. This would not be a problem if the modules: (i) are identical (HW) processors, (ii) use identical inputs, (iii) use identical algorithms/software and (iv) have mutually synchronised clocks. However, if the modules are different processors and/or are running different SW for the same problem, then it is possible for two correct outputs to diverge slightly in the lower significant bits (LSBs). In that case, we can declare two outputs x and y as practically and very nearly identical if $|x-y| < delta$ for some specified delta. This property being 'practically identical' is not a transitive one [8]. For such an approximate agreement we can do what is known as plurality voting, which looks for a set of at least 'k' practically identical outputs. This is the k-plurality voting scheme. It is a set in which each member is practically identical or very near to all other members, and picks any of them (or the median) as the representative one. In this case, we have assumed that each output has an equal chance (probability) of being faulty, in case of decision making in fault-detection systems, however, this may not be true in some cases. The HW/SW producing one output may have a different failure probability than does the HW/SW producing another output. In this case, each output is assigned a weight (i.e. weightage/weighting) that is related to its probability of being correct. It is possible to use fuzzy membership functions to decide this weight. It would be a good idea to use FL in conjunction with the voting schemes. The voter then does weighted voting and produces an output that is associated with over half the sum of all the weights.

8.7.1 A General Framework for Voting

It is feasible that virtually all the voting schemes of practical interest and use can be formulated in terms of the so-called generalised weighted voting model. We are given n input data objects $x_1, x_2, ..., x_n$ and with their associated non-negative real votes $v_1, v_2, ..., v_n$ with $Sv_i = V$, then compute output y and its vote w such that y is supported by a set of input objects with votes totalling w. Here w satisfies a condition associated with the voting sub-scheme, and the possible voting sub-schemes are [9]: (i) unanimity, that is, $w = V$; (ii) majority, that is, $w > V/2$; (iii) supermajority if $w \geq 2V/3$ and (iv) threshold if $w >$ a preset lower bound.

In *M-of-N* voting in which an output choice is accepted if at least 'M' votes agree with the decisions of 'N' sensors. We can use a variant, the so-called *T-out-of-v* voting in which the output is accepted if $H = \Sigma_i w_i v_i > T$. Here 'w' are the user-defined weights, the 'v' are decisions of the algorithm, and T is a user-defined threshold [9]. For example, in studying the infrared (IR) spectrum related to the occurrence of fire or not, the decision parameter v can take binary values 0 and 1, corresponding to normal case and the existence of, say fire, respectively. The decision parameter v_1 is 1 if the pixel (from IR spectrum) is a moving pixel and 0 if it is stationary; and the decision parameter v_2 is taken as 1 if the pixel is coloured and 0 otherwise.

8.8 DeF Using FL for Aviation Scenarios

As we have seen in Chapter 3, FL is one form of logic that can be used in design, decision making and operation of many intelligent control systems. We need to use available knowledge and information along with some logic and/or statistical methods to take an appropriate decision from among many competing possibilities. In case of aviation examples, one of the objectives of DeF is to take one final decision and action in an entire surveillance volume. For example, in an environment of a flying aircraft, at any instant of time using outputs from different levels of DF: level 1 – object refinement (L1OR) and level 2 – situation refinement (L2SR) [4]. The L1OR forms object assessment by combining the location, parametric and identity information to obtain refined representation of individual objects such as emitters, platforms and weapons, and in L2SR an attempt is made to find a description of the relationship between objects and observed events. In this section, we consider formation flight and flying along an air lane and apply FL-based DeF. This FL-decision system is realised using MATLAB/ SIMULINK®, and the results are generated using numerically simulated data [4,10].

8.8.1 Decision Level Fusion, FL and FIS

DeF (symbol level fusion is synonymously used) represents the high-level, higher than both the kinematic level and image level DF, information wherein the symbol represents a decision. DeF mainly depends upon the external knowledge and/or on inference derived from it. The results obtained/fused from the DeF system are also useful to classify images, detect changes and detect and recognise targets. The L2SR for DF signifies SA by relating the objects to the current SA or by relating object assessments mutually, and the SR helps in developing a description of current relationships among objects and events in the context of such an environment. The SA aids in decision making: the various decisions/actions that a pilot can take are: (i) avoid any collision with any closely flying aircraft, (ii) decipher the intentions of the enemy, if suspected so and (iii) communicate with other friendly aircraft if need be.

The application of FL at the higher levels of DF is a good choice for precise decision making: if one output of SA is 'aircraft is non-friendly and targeting tank' then it can be interpreted as a decision that 'aircraft is non-friendly and targeting tank'. The fusion comes into picture if there is more than one such decision for the same object seen by multiple sensors of different types/ accuracies. These different accuracy levels of each sensor dictate different confidence levels while making the decision. In order to arrive at an accurate decision it is necessary to fuse the decisions, the outputs form SA using FL approach. In case of symbolic data/information the inference methods from AI can be used, for example, FL, since the fusion of symbolic information would require reasoning and inference in the presence of modelling and/ or data uncertainty (this uncertainty is modelled by FL). At the core level in FL-based DeF systems, the (fuzzy) If … Then … rules via fuzzy implication functions are processed in the so-called FIS. Here, the approach used is based on heuristic knowledge acquired from domain experts, and it is believed to be more realistic in treating the SA. The knowledge is captured in the If … Then … rule base.

As seen above the core of FL-based system is FIS in which the fuzzy 'If … Then' rules are processed using fuzzy implication functions (FIF) to finally obtain output as fuzzy sets. A FIF plays a very significant role in a successful design of the FIS/FLS. As a result it becomes necessary to select an appropriate FIF from the existing FIFs/fuzzy implication methods. These

FIFs should satisfy some of the intuitive criteria of generalised modus ponens (GMP) and generalised modus tolens (GMT) of forward and reverse chains of logic, respectively [4,10] for: (a) GMP, we have premise 1: u is A'; premise 2: IF u is A THEN v is B; then consequence: v is B'; and (b) GMT, we have premise 1: v is B'; premise 2: IF u is A THEN v is B; then consequence: u is A'. If we have a new FIF, then it should also satisfy some of the intuitive criteria of GMP/GMT. Fuzzy implication function/method provides mapping between input and output fuzzy sets so that fuzzified inputs can be mapped to desirable output fuzzy sets. Basically, a fuzzy If … Then … rule, provided by domain expert, is interpreted as a fuzzy implication. Then fuzzified output of inference engine (FIE/FIS) obtained using sup (supremum-star) composition is given by $B = RoA$ where 'o' is a compositional operator and 'R' is a fuzzy relation in the product space $U \times V$, and in the form of the FMF it is given by $\mu_B(v) = \mu_R(u, v)o\mu_A(u)$. A fuzzy implication, $\mu_{A \to B}(u, v)$, is also a kind of relation that provides I/O mapping, then the above equation can be rewritten as $\mu_B(v) = \mu_{A \to B}(u, v)o\mu_A(u)$.

There are many standard ways and interpretations of the fuzzy If … Then … rule to define the fuzzy implication process [4], by different combinations of t-norms and s-norms (Chapter 3). We can find ways to interpret the fuzzy If … Then … rules, that is, some common fuzzy implications which satisfy one or more of the intuitive criteria are [4]: (i) mini-operation rule of fuzzy implication (MORFI–Mamdani), that is, the standard intersection (SI) operator of T-norms in the expression $R_{MORFI} = \mu_{A \to B}(u, v) = \min(\mu_A(u), \mu_B(v))$; (ii) product-operation rule of fuzzy implication (PORFI–Larsen), the algebraic product (AP) operator of T-norms in expression $R_{PORFI} = \mu_{A \to B}(u, v) = \mu_A(u)\mu_B(v)$ and (iii) the arithmetic rule of fuzzy implication (ARFI–Zadeh/Lukasiewicz), bounded sum (BS) operator of S-norms and the complement operator.

8.8.2 Performance Evaluation: Illustrative Examples

In this section, we consider two applications of a decision system using FL: (i) formation flight and (ii) flying along an air lane. By properly defining the If … Then … rules and combinations of FMFs and FIFs the DeF can be appropriately extended to more than two objects.

8.8.2.1 DeF 1: Formation Flight

In this case, we use an FL-based decision software/system (FLDS) residing in (its) own platform to decide whether two enemy fighter aircraft have formation flight or not [4,10]. The simulated data are generated for two aircraft of the same class and identity in pitch plane only with [4,10]: (i) initial state of aircraft 1: $X_1 = [x \quad \dot{x} \quad z \quad \dot{z}] = [0\,m \quad 166\,m/s \quad 1000\,m \quad 0\,m/s]$; (ii) initial state of aircraft 2: $X_2 = [x \quad \dot{x} \quad z \quad \dot{z}] = [0\,m \quad 166\,m/s \quad 990\,m \quad 0\,m/s]$; (iii) sampling rate as 1 Hz;

(iv) simulation time as 30 s; (v) aircraft motion with constant velocity and (vi) kinematic model as $X_i(k+1) = FX_i(k) + Gw_i(k)$ with k as scan/index number, i (= 1,2) as the aircraft number, F as the state transition matrix, G as process noise gain matrix, and w as white Gaussian process noise with covariance matrix

$$Q = 0.1 * \text{eye}(4,4), \quad F = \begin{bmatrix} 1 & T & 0 & 0 \\ 0 & 1 & 0 & 0 \\ 0 & 0 & 1 & T \\ 0 & 0 & 0 & 1 \end{bmatrix} \text{ and } G = \begin{bmatrix} T^2/2 & 0 & 0 & 0 \\ 0 & T & 0 & 0 \\ 0 & 0 & T^2/2 & 0 \\ 0 & 0 & 0 & T \end{bmatrix}$$

The two aircraft maintain a formation flight from $t = 0$ to 5 s, then split apart at $t = 5$ s and remain in that mode for up to $t = 10$ s From 10th second to 15th second, they fly with constant separation and start approaching each other from 15th second onwards. Later on they form a pair from 20th second, and stay in formation flight for another 10 s.

The performance of the FL-decision fusion (FLDF) depends on proper selection of FMFs, fuzzy rules, FIFs, aggregation methods and defuzzification method. The inputs to FMFs/FLDF are the numerical differences of the aircraft's bearing, elevation, separation distance along z-axis, speed, identity and class. For each I/O there is a FMF which fuzzifies the data between 0 and 1. The trapezoidal membership functions are chosen. The fuzzy rules used to decide whether two aircraft form a pair or not are: (i) rule 1: if two aircraft have the same bearing, elevation and speed then they have same kinematics and (ii) rule 2: if two aircraft have the same kinematics, identity, class and are at a short distance from each other then they form a pair. These rules are processed by FIF and PORFI/Larsen implication is used. We use BS operator of T-conorm/S-norm for the aggregation process. The aggregated output is then defuzzified using centre of area (COA) method. The FLDF is implemented in MATLAB/SIMULINK environment, Figure 8.1 [4,10]. There are certain other inputs such as speed, bearing, aircraft identity and class to decide whether both the aircraft form pair or not. From Table 8.1 [10], we infer that the FLDF is able to correctly detect the aircraft pair and split periods.

8.8.2.2 DeF 2: Air Lane

Now, we describe the FLDF to decide whether a particular aircraft is flying along the air lane or not. The data for this case are obtained with: (i) initial state of aircraft: $X = [x \quad \dot{x} \quad y \quad \dot{y}] = [2990$ m 0 m/s 0 m 332 m/s 2990 m 0 m/s 0 m 332 m/s]; (ii) air lane located along y-axis at $x = 3000$ m; (iii) sampling rate as 1 Hz; (iv) simulation time as 30 s; (v) aircraft motion with constant velocity and (vi) the model as $X(k+1) = FX(k) + Gw(k)$, F as state transition matrix (as in previous example), G as process noise gain and w as white Gaussian process noise with covariance Q: 1.0 eye (4,4). The inputs are fuzzified distance (absolute separation between aircraft and air lane along y-axis),

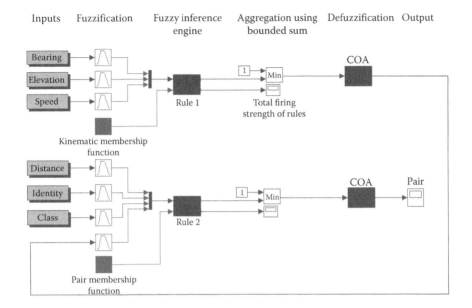

FIGURE 8.1
FLDF-1 system for formation flight. (From Raol, J. R. and Sudesh, K. K. *Journal of Aerospace Sciences and Technologies*, 65(3), 273–286, 2013. With permission.)

TABLE 8.1

Decision Results for FLDF-1

Time (s)	Output – After Defuzzification	Decision Whether Pair Is Formed or Not
1	3.799	Yes
3	3.799	Yes
5	3.799	Yes
7	0	No
9	0	No
11	0	No
13	0	No
15	0	No
17	0	No
19	0	No
21	3.799	Yes
23	3.799	Yes
25	3.799	Yes
27	3.799	Yes
29	3.799	Yes

Source: Adapted from Raol, J. R. and Sudesh, K. K. *Journal of Aerospace Sciences and Technologies*, 65(3), 273–286, 2013. With permission.

absolute value of bearing difference between them, and class of the aircraft. The FMFs are of trapezoidal shape. The fuzzy rules are: (a) rule 1: if aircraft has the same bearing as air lane AND if it is close to the air lane, then the aircraft is flying along the air lane and (b) rule 2: if aircraft is civilian then there is high possibility that aircraft is flying along the air lane. The PORFI is used for this case. The BS/algebraic sum (AS)/standard union (SU) operators of T-conorm/S-norm are also used one by one in the aggregation process. The aggregated output fuzzy set is defuzzified using COA method, and the FLDF system block diagram is shown in Figure 8.2 [10]. Table 8.2 [10] illustrates the comparison of final outputs obtained for different aggregation methods. We see that: (i) in case of the SU operator, for aggregation, constant output is observed irrespective of whether aircraft is along the air lane or not; (ii) for the AS operator, the smooth transition takes place between 0 and 1 meaning thereby that a hard decision is made by the FLDF system about the aircraft, whether it is along the lane or not; (iii) for BS operator, non-smooth transition is observed between 0 and 1, it gives confidence level while taking the decision about an aircraft flying along the air lane or not (the larger the final output, the more the confidence) and (iv) use of BS in the aggregation process provides better results than other methods studied here. The aim in the two examples considered here has been to illustrate the use and application of T1FL to two aviation scenarios in decision making/ fusion. The decision-making process has been incorporated in the rules: rule 2, 'if two aircraft have the same kinematics, identity, class and are at a short distance from each other then they form a pair' – (a decision here for *DeF 1*), and rule 2 'if aircraft is civilian then there is high possibility that aircraft is flying along the air lane' – (a decision here for *DeF 2*).

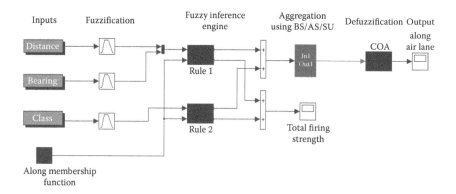

FIGURE 8.2
FLDF-2 system for air lane. (From Raol, J. R. and Sudesh, K. K. *Journal of Aerospace Sciences and Technologies*, 65(3), 273–286, 2013. With permission.)

TABLE 8.2

Decision Results for Output for Air Lane Decision (Using Standard Fuzzy
Operator/Implication Functions): FLDF-2

Time (s)	Output – After Defuzzification (SU)		Output – After Defuzzification (AS)		Output – After Defuzzification (BS)[a]	
1	1.0000	Yes	0	No	0.0001	No
3	1.0000	Yes	1.0000	Yes	0.7330	Yes
5	1.0000	Yes	1.0000	Yes	1.0001	Yes
7	1.0000	Yes	1.0000	Yes	1.0001	Yes
9	1.0000	Yes	0	No	0.0001	No
11	1.0000	Yes	1.0000	Yes	0.649	Yes
13	1.0000	Yes	1.0000	Yes	1.0001	Yes
15	1.0000	Yes	1.0000	Yes	0.2906	No
17	1.0000	Yes	1.0000	Yes	0.149	No
19	1.0000	Yes	0	No	0.0001	No
21	1.0000	Yes	0	No	0.0001	No
23	1.0000	Yes	0	No	0.0001	No
25	1.0000	Yes	1.0000	Yes	0.158	No
27	1.0000	Yes	1.0000	Yes	1.0001	Yes
29	1.0000	Yes	1.0000	Yes	0.852	Yes

Source: Adapted from Raol, J. R. and Sudesh, K. K. *Journal of Aerospace Science and Technologies,* 65(3), 273–286, 2013. With permission.

[a] Yes, for value greater than 0.5 and No, for less than 0.5.

8.9 DeF Strategies

In analysis and design of pattern recognition systems (PRS), we need to achieve the best possible classification performance for a given problem. As we already know from Chapter 1, there are three main fusion strategies: (i) information/DF (or called low-level fusion), (ii) feature fusion (FF) (called intermediate-level or mid-level fusion) and (iii) DeF (called high-level fusion). DeF can provide and operate on a set of classifiers in order to obtain a better and unbiased result [11]. These classifiers may be of the same or different type, and they can have the same or different feature sets. Various classifiers are: (i) support vector machine (SVM), (ii) ANN, (iii) k-NN and (iii) GMM. It is also true that a single classifier may not be well suited for a given application. In this case, a set of classifiers is used and the outputs of all the classifiers are combined by some methods to obtain the final output. For such a problem, the DeF methods can be used. It has been demonstrated that some of these combination methods consistently outperform a single best classifier. The methods of (FF) deal with the selection and combination of features of the objects under study with a view to removing redundant and irrelevant features. A feature is irrelevant if it does not correlate greatly with the class information. Then, the final set of features is combined and fused together to obtain a better feature set. Subsequently, this fused feature set (FFS) is presented to a classifier to obtain the final result. Thus, it is expected that the DeF and FF would give better classification accuracy. Areas of application of DeF and FF are [11]: (i) remote sensing, (ii) medical image processing, (iii) target and object refinement (OR), (iv) face recognition, (v) speech processing, (vi) video classification/retrieval and (vii) gene detection in DNA sequences.

The DeF aims at combining the beliefs (in the sense of probabilities or D-S beliefs) of the set of models used into a single (decision), consensus (-reaching agreement) belief. There are three very popular DeF approaches at the top level [12]: (i) the linear opinion pool (LOP), (ii) the logarithmic opinion pool (LgOP) and (iii) the voting or ranking (VR) approach. The LOP is a very simple technique (Section 2.6.3.1/Equation 2.48). The fusion output is the weighted sum of the probabilities coming from each senor/node model. It is possible that the combined DeF output might be multimodal [12]. The LgOP also yields the probability distribution, and the output would be unimodal. The LgOP is the weighted product of the individual model outputs as can be seen by Equation 2.49, Section 2.6.3.2. The main feature of the LgOP is that if any individual pdf is zero, then the combined posterior is also zero. Here, there is a strict 'veto' of that model, whereas in LOP the zero pdf is averaged out in the posterior pdf [12]. This can be easily ascertained by comparing Equations 2.48 and 2.49.

As we have mentioned in Section 8.7, a simple method for combining the results of multiple models and local decisions is to use a voting procedure, which for each model must generate a decision instead of a score. Several voting methods are in vogue for decision making and fusion: (i) maximum, (ii) minimum and (iii) median votes. Also, the ranking methods are appropriate for problems that involve numerous classes; they utilise the order of classes as estimated by the model(s). The ranking methods utilise the class set reduction (CSR) method to reduce the number of class candidates without losing the true class – by reducing this number and re-ordering the spare classes, the true class is expected to move to the top in the ranking. The Borda count (BC) is very popular rank-based method. Other computational methods for DeF are (i) D-S method and (ii) the fuzzy integral method. These techniques combine the beliefs of various individual models, and not just their respective decisions, into an overall consensus/agreed belief.

8.9.1 Classifier Systems

The reasons for preferring a multi-classifier process/system (MCS) over a single classifier are: (i) to improve the accuracy and efficiency of the classification system; (ii) often the volume of data to be analysed by a single classifier is too huge, training a classifier would not be practical, whereas in an

MCS the data are partitioned into subsets, trained with different classifiers for different subsets and the outputs combined; (iii) compared to a single classifier, MCS with a subset of features may provide a better performance and (iv) MCS might improve the generalisation performance, since, a single classifier may not perform well for a certain input when it is trained with a limited data set. Classifier combinations are based on the two types of study [11]: (i) the decision optimisation method makes and obtains an optimal combination of classifiers' decision (given a fixed set of highly specialised classifiers) and (ii) the coverage optimisation method generates a set of mutually complementary and generic classifiers that are then combined to achieve optimal accuracy (assuming a fixed combination rule). MCS can be achieved by [11]: (i) variation of initial parameters of the classifiers – a set of classifiers is created, then each classifier is trained with the same training data; (ii) training the same classifier with different training data sets and (iii) training different types of classifiers (SVM, ANN) with the same training data set. The two types of classifier combination strategies are: (i) classifier fusion – every classifier is provided with complete information and the outputs from the classifiers are combined, every classifier contributes to making of a final decision and (ii) classifier selection – every classifier is an expert in a specific domain of the feature space, the local expert alone decides the output of the ensemble. The MCS employing the DeF fusion are also known variously [12]: (i) the decision combination, (ii) the mixer of experts, (iii) classifier ensembles, (iv) classifier fusion, consensus aggregation, (v) the dynamic classifier selection and (vi) hybrid methods.

8.9.2 Classifier Fusion and Selection

The methods of combining the outputs of the L classifiers in an ensemble depend on the information obtained from the individual classifiers. The information fusion at the decision level is divided into the categories based on the type of classifier outputs [11]: (i) type 1 (the abstract level): each classifier D_i produces a class label $s_{i,i} = 1, \ldots, L$, that is, for any object $x \in \Re n^n$ to be classified, the L classifier outputs define a vector s $[s_1, \ldots, s_L]^T \in \Omega^L$ where $\Omega = \{\omega_1, \omega_2, \ldots, \omega_C\}$ is the set of class labels; (ii) type 2 (the rank level): the output of each D_i is a subset of Ω, with the alternatives ranked in order of plausibility of being the correct label and (iii) type 3 (the measurement level): each classifier D produces a c-dimensional vector $[d_{i,1}, \ldots, d_{i,c}]^T$, the value d_{ij} representing the support for the hypothesis that vector x used for the classification is from class ω_j. Further, the output of a classifier may be a crisp label or probabilistic confidences or ranking of classes in a decreasing order. The combination techniques/the outputs of classifiers used might be [11]: (i) crisp labels (abstract level), (ii) a subset of possible matches sorted in a decreasing order of confidences (rank level) and (iii) probabilistic confidence measures obtained over each class (measurement level).

8.9.2.1 Combining Class Labels: Crisp Outputs

It is assumed that the class labels are available from the classifier outputs. The decision of the ith classifier is defined as $d_{ij} \in 0, 1, i = 1, 2, ..., L$ and $j = 1, 2, ...,C$ (C is the number of classes). If the ith classifier chooses class ω_j, then $d_{ij} = 1$ and 0 otherwise.

8.9.2.1.1 Majority Voting

Here, the ensemble chooses a class when any of these situations are considered [11]: (i) all the classifiers agree on the specific class, the unanimous voting; (ii) predicted by at least one more than half the number of classifiers, the simple majority voting or (iii) it receives the highest number of votes. The ensemble decision for the majority voting is as follows: choose class ω_k, if

$$\sum_{i=1}^{L} d_{i,k} = \max_{j=1}^{c} \sum_{i=1}^{L} d_{i,j} \tag{8.15}$$

The majority voting is an optimal combination rule under certain assumptions [11]: (i) the number of classifiers, L, is odd, (ii) the probability for each classifier to give the correct class label is p for any instance x and (iii) the classifier outputs are independent. The plurality voting and simple majority are identical and the ensemble makes the correct decision if at least $(L/2) + 1$ classifiers choose the correct label. The probability of the success of the ensemble is given by the following:

$$P_{maj} = \sum_{m=\lfloor L/2 \rfloor+1}^{L} \begin{bmatrix} L \\ m \end{bmatrix} p^m (1 - p)^{L-m} \tag{8.16}$$

There is a possibility of the ensemble giving a better accuracy over the individual accuracy p, if p is greater than 0.5.

8.9.2.1.2 Weighted Majority Voting

In a situation wherein the classifiers (in the ensemble) do not show identical performance, then we can give greater weights/weightage/coefficient/ importance to more competent classifiers to make the final decision, then the label (crisp) outputs are represented as degrees of support for the classes

$$d_{ij} = \begin{bmatrix} 1, & \text{if } D_i \text{ labels } x \text{ in } w_j \\ 0, & \text{otherwise} \end{bmatrix} \tag{8.17}$$

In this case, the discriminant function for class j obtained via Equation 8.17 is given by

$$g_j(x) = \sum_{i=1}^{L} b_i d_{i,j} \tag{8.18}$$

In equation b_j is the weight of classifier D_i, thus, the value of the discriminant function will be the sum of the coefficients for these members whose output for input x is class j, and the classifiers whose decisions are combined via weighted majority voting would choose class w_k if the following condition hold true [11]:

$$\sum_{t=1}^{L} b_t d_{t,k} = \max_{j=1}^{c} \sum_{t=1}^{L} b_t d_{t,j} \tag{8.19}$$

If voting weights satisfy the following proportionality, then the accuracy of the ensemble P_{wmaj} would be maximised:

$$b_t \propto \log \frac{p_t}{1 - p_t} \tag{8.20}$$

In Equation 8.20, p_t is the accuracy of the classifier t.

8.9.2.1.3 Behaviour Knowledge Space

This method, behaviour knowledge space (BKS) uses a lookup table based on the classification of training data – here, the information is kept on how each label-combination is produced by the classifiers, this label vector s gives an index to a cell in a lookup table, Z. Each $z_j \in Z$ is placed in the cell indexed by s for that object, the numbers of the elements in each cell are tallied and the most representative class label is selected for that cell. Then, the highest score would correspond to the highest estimated posterior probability $P(k/s)$. If any ties occur, then these are resolved arbitrarily and the empty cells are labelled in an appropriate way.

8.9.2.1.4 Naive Bayesian Combination

In the naive Bayesian combination (NBC) method the classifiers are considered to be mutually independent, when a class label is given. If we have $P(s_j)$ as the probability that classifier D_j classifies a sample x in class s_j, then the conditional independence permits us to have the following representation [11]:

$$P(s|w_k) = P(s_1, s_2, \ldots, s_L |w_k) \prod_{i}^{L} P(s_i |w_k) \tag{8.21}$$

Then we have the posterior probability that is required to label the sample x given as follows:

$$P(w_k|s) = \frac{P(w_k)P(S|w_k)}{P(S)} \tag{8.22}$$

The support for the class is computed as follows:

$$\mu_k(x) \propto P(w_k)\prod_i^L P(s_i|w_k) \tag{8.23}$$

For each classifier D_i, a $C*C$ confusion matrix CM^i is computed by applying D_i to the training data set. The (k,s)th entry of this matrix $CM^i_{k,s}$ is the number of elements of the data set whose true class label was ω_s and was assigned by D_i to class ω_s support can also be represented as follows:

$$\mu_k(x) \propto \frac{1}{N_k^{L-1}}\prod_{i=1}^L CM^i_{k,s_i} \tag{8.24}$$

In Equation 8.24, an estimate of the probability, $P(S_i|w_k)$ is given by $CM^i_{k,s_i}/N_k$, where the N_s denotes the total number of a data set that belongs to the class w_s.

8.9.2.2 Class Ranking

We sort the possible matches in the decreasing order of our confidence and the output of each of such classifiers is a subset of these matches. The decisions by the classifiers are now presented as rankings of classes with a view to comparing them across different types of classifiers/situations of a problem. The two main approaches when fusion is applied in MCS are [11]: (a) CSR and (b) class set reordering (CSRe).

8.9.2.2.1 CSR Approaches

Our aim here is to consider a reduced set of classes by reducing the number of classes (as small as possible), but still assuring that the correct class is represented in the reduced set. A CSR method tries to find the trade-off between minimising the class set and maximising the probability including a true class. This can be taken care of by: (i) intersection of the neighbourhoods – firstly these are determined by the ranks of true classes for the worst case in the training data set, then, the lowest rank given by any classifier is taken as the threshold. Only the classes ranked above are used for further processing and (ii) union of neighbourhoods – the threshold for each classifier is computed as the maximum (worst) of the minimum (best) ranks of true classes over the training data set, then, the redundant classifier is determined easily, since its threshold would be equal to zero (i.e. its output is incorrect) [11].

8.9.2.2.2 *CSRe Approaches*

In a CSRe method, the idea is to improve the overall rank of the true class. There are three approaches [11] as discussed below:

i. The highest rank method: Each possible match is assigned the highest (minimum) rank as ascertained from different matchers for class re-ranking or class set re-ordering. We assume that for each input pattern, *m* classifiers are applied to rank (i.e. for ranking) a given set of classes. The minimum of these *m* ranks is allotted to each class. Then, the classes are sorted according to the new ranks with a view to obtaining a combined ranking for that input. To ascertain the individual class as a final decision-output, the class that is at the top of the reordering ranking is chosen. If there is any tie, it is broken arbitrarily to obtain a strict linear ordering. This approach is utilised if there are few classifiers and a large number of classes. The merit is that the strength of every individual classifier is utilised. A demerit is that of a possibility of occurrence of ties when combined ranking is done, the number of classes sharing the tie depending on the number of classifiers used. As a result the method is useful only for a small number of classifiers.

ii. The BC method: Here, the sum of the ranks is allocated by the individual matchers to calculate the combined ranks. This is specified as a mapping from a set of individual rankings to a combined ranking leading to the most relevant decision. The BC B_j for a particular class j is the sum of the number of classes ranked below class j by each classifier as follows:

$$B_j = \sum_{i=1}^{L} B_i(j) \qquad (8.25)$$

The BC method is based on the assumption of additive independence among the contributing classifiers, the demerit being that it treats all classifiers equally and the individual classifier capability is disregarded.

iii. The logistic regression method: Here the weighted sum of the individual ranks is computed and the weights are determined by regression. Here, we assign the weights to each classifier reflecting their importance in an MCS. We assume that the responses $(x_1, x_2, ..., x_m)$

from $m(m)$ classifiers are the highest for the classes ranked at the top of the ordered list. Then, the logistic response function is given by

$$\pi(x) = \frac{e^{(\alpha+\beta_1 x_1+\beta_2 x_2+\cdots+\beta_m x_m)}}{1 + e^{(\alpha+\beta_1 x_1+\beta_2 x_2+\cdots+\beta_m x_m)}} \tag{8.26}$$

The coefficients in the exponential terms are the constants. Then, a confidence measure is defined as follows:

$$L(x) = \log \frac{\pi(x)}{1 - \pi(x)} = \alpha + \beta_1 x_1 + \beta_2 x_2 + \cdots + \beta_m x_m \tag{8.27}$$

We can determine a threshold value so that the classes which have the confidence value below it are discarded.

8.9.2.3 Combining Soft Outputs

In this approach, the output supplied by a classifier for a given class is interpreted as the degree of support (DOS) given to that class. This DOS is accepted as an estimate of the posterior probability for that particular class. A decision profile matrix $DP(x)$ is used for decision combination. In a typical MCS, a feature vector (FV) x is classified into one of the C classes using L classifiers, and Ω is given as $\{\omega_1, \omega_2, \ldots, \omega_c\}$, the set of class labels; each classifier D_i in the ensemble gives output c degrees in the interval [0,1]. The classifier D_i provides $d_{ij}(x)$ as a support for the class label ω_j. The larger it is, the more the likelihood of the sample getting assigned to class label w_j, and the output of L classifiers for a particular input x is organised as a decision profile $DP(x)$ as given below

$$\begin{bmatrix} d_{1,1}(x)\ldots d_{1,j}(x)\ldots d_{1,c}(x) \\ d_{i,1}(x)\ldots d_{i,j}(x)\ldots d_{i,c}(x) \\ d_{L,1}(x)\ldots d_{L,j}(x)\ldots d_{L,c}(x) \end{bmatrix} \tag{8.28}$$

8.9.2.3.1 Class-Conscious Techniques

Given $DP(x)$, these Class-Conscious methods operate class wise on each column of $DP(x)$ and obtain $\mu(x)$. It can use several rules: (i) the sum rule computes the soft class label vectors as $\rightarrow \mu_j(x) = \sum_{i=1}^{L} d_{i,j}; j = 1, 2, \ldots, C$; (ii) the product rule computes these vectors as $\mu_j(x) = \prod_{i=1}^{L} d_{i,j}; j = 1, 2, \ldots, C$; (iii) the min rule computes $\rightarrow \mu_j(x) = \min_{i=1}^{L} d_{i,j}; j = 1,2,\ldots,C$ and (iv) the max rule computes $\rightarrow \mu_j(x) = \max_{i=1}^{L} d_{i,j}; j = 1,2,\ldots,C$.

8.9.2.3.2 Class-Indifferent Techniques

In the CIM, the entire $DP(x)$ is used for computation of $\mu(x)$, ignoring the classes and there are two such approaches: (i) decision templates (DeT) and (ii) D-S combination:

i. DeT: In the DeT, we remember the most typical decision profile for each class w_j called the decision template, DeT_j this is then compared with the decision profile using a similarity measure, the closest match will label x and the measures of similarity used are as follows:

- Squared Euclidean distance $DeT(E) \rightarrow$ the ensemble support for ω_j is computed by

$$\mu_j(x) = 1 - \frac{1}{L*C} \sum_{i=1}^{L} \sum_{k=1}^{C} \{DeT_j(i,k) - d_{i,k}(x)\}^2 \qquad (8.29)$$

with $DeT_j(i,k)$ as the (i,k)th entry in DeT_j.

- Symmetric difference $DeT(S) \rightarrow$ the support for ω_j is computed as follows:

$$\mu_j(x) = 1 - \frac{1}{L*C} \sum_{i=1}^{L} \sum_{k=1}^{C} \max[\min\{DeT_j(i,k),(1 - d_{i,k}(x))\}, \min\{(1 - d_{tj}(i,k)), d_{i,k}\}] \qquad (8.30)$$

ii. D-S: let DeT^i denote the ith row of DeT_j, and $D_i(x)$ the output of D_i, then the proximity between DeT_j^i and the output of classifier D_i for the input x is expressed as

$$\Phi_{j,i}(x) = \frac{\left(1 + \left\| DeT_j^i - D_i(x) \right\|^2 \right)^{-1}}{\sum_{k=1}^{c} \left(1 + \left\| DeT_j^i - D_i \right\|^2 \right)^{-1}} \qquad (8.31)$$

with L proximities for each DeT. Then, the D-S degree of belief is computed for every class $j = 1, \ldots, C$ and for every classifier $i = 1, 2, \ldots, L$, by the following expression:

$$b_j(D_i(x)) = \frac{\Phi_{j,i}(x) \prod_{k \neq j} (1 - \Phi_{k,i}(x))}{1 - \Phi_{j,i}(x)\{1 - \prod_{k \neq j} (1 - \Phi_{k,i}(x))\}} \qquad (8.32)$$

Subsequently, the final DOS is given, with K as the normalising coefficient, by

$$\mu_j(x) = K \prod_{i=1}^{L} b_j\{D_i(x)\}; \quad j = 1, 2, \ldots, c \quad (8.33)$$

8.9.3 Selection of the Classifiers

A single classifier is used to correctly classify the input pattern, exactly one classifier is chosen to make the decision. There are static and dynamic classifier selection methods (CSMs). In the static CSM, during the training phase, the selection regions are specified prior to classifying the unlabelled vector x, and in the operation phase, the region of x is first found; for example, R_{j}, and it is processed further by the respective classifier $D_i(j)$, responsible for this region. The training phase is carried out in either of the ways [11]: (a) specify the region and then assign a responsible classifier for each region or (b) given the set of classifiers, find a region where each classifier is the best one. In a dynamic CSM, the choice of the classifier to label x is made during the operation phase, the choice being based on the certainty of the current decision – the preference given to more certain classifiers. Then, the competency of each classifier is determined in the vicinity of x as the classifier's accuracy. The classifier with the highest competence is authorised to label x.

8.10 SA with FL and DeF for Aviation Scenarios: Illustrative Examples

With the advent of powerful long-range radars and beyond-visual range (BVR) missiles, pilots increasingly rely on board sensors of the aircraft to make tactical decisions. Real-time decisions require the pilot to assimilate and process large amounts of data from multiple sources in a short span of time. This aspect is further compounded when there are multiple unknown targets or, in the case of an extremely complex air scenario, involves multiple friendly and enemy aircraft. Hence, there is a need for an automated system which is comprehensive as well as provides an aid in the decision making activity. Towards this we present, here, the results of SA systems: (i) that use T1FL for decision making/fusion in four aviation scenarios, (ii) studies with modified SA models and (iii) studies with noisy inputs to the SA models [13–17]. The main aim is to replicate a pilot's mental thought process while in a given situation, and implement the same in the form of an automated system. These situations are simulated with multiple variations of the standard aviation scenarios. The results indicate that some existing FL implication functions work very well for the examples considered. This opens up newer possibilities of applications of FL-based DeF and SA technology to varieties of aerospace/aviation problems in decision making.

8.10.1 SA and Decision-Level Fusion

The ultimate aim is to obtain a total/final picture of the opponent's objective (in defence applications). The procedure based on T1FL described here is applicable to other civilian data analysis and fusion systems also, for example, aircraft landing in bad weather. Conventionally, the FL that is well known is called type 1 (T1FL). T1FL is used to incorporate the heuristic knowledge of the human expert via fuzzy If … Then … rules. Thus, here we use T1FL to help in decision making for the four aviation scenarios, wherein we use appropriate rules. The T1FL method has found applications in industrial control systems, home appliances, robotics and aerospace engineering, for it is one form of logic that can be used in design, decision and operation of (artificial) intelligent (AI) control systems. We very briefly describe the FIS, and the MATLAB-GUI-based tool for the evaluation of existing FIF for DeF. We consider four aviation scenarios that might arise during air flight and/ or air combat: (i) formation flight, (ii) attack, (iii) threat assessment and (iv) air lane flying. The decision process is realised using MATLAB/SIMULINK toolboxes, and the results are generated using numerically simulated data. The approach used here is based on heuristic knowledge acquired from the domain experts (pilots/aviation experts) and hence, it is more realistic in treating the SA problems, and this knowledge is captured and expressed in the If … Then … rule base. Hence, it is strongly felt that the FL-based DeF definitely would be a value addition to the performance of the DeF process and systems, especially because it incorporates the knowledge of the real experts. Currently, the aim is to apply T1FL and see how it performs for the aviation scenarios considered.

8.10.2 A MATLAB/GUI Tool for Evaluation of FIFs

All FIFs might not completely satisfy the intuitive criteria of forward/ reverse chain of logic. Hence, a tool that can be user friendly and interactive to evaluate these FIFs against these criteria was developed. It provides a methodology to determine if any of the existing FIFs satisfies a given set of intuitive criteria – it is a MATLAB and graphics based tool, and evaluates any FIF with respect to these criteria. All the GMP/GMT criteria and the FIFs are integrated in the tool, and can be selected and evaluated by specifying a particular combination using GUI. The tool helps in visualising the results analytically as well as graphically. The tool is of generic nature to evaluate any existing or newly developed FIF. In addition, it helps user to devise a new FIF by use of graphs and evaluate the new FIF for the satisfaction of the forward/reverse chain criteria. Whether any new FIF is really useful needs to be ascertained by its use in FIS for a given design/control problem. The inferences can be made if the criteria are satisfied or not. If the criteria are satisfied or not, the difference does not show up in the FIF, the FIF remains as is. It just says whether the criteria are satisfied by the particular FIF or not. It is up to the reader/user to further use this FIF or not. Using this tool and the procedure described above, all the existing FIFs were evaluated and it was found that this tool worked very well. This tool is also used to generate some results presented here. Since, there is a possibility of existence of more FIFs that can be obtained by using various unexplored combinations of fuzzy operators, an effort was made to derive a few new FIFs using material implication, propositional calculus and fuzzy operators. Then the consequences of these new FIFs were tested with GMP/GMT criteria using the same MATLAB/graphics-based tool described above. The evaluation approach gives a feasibility to arrive at new FIF which might be useful in certain applications of FL in analysis and design of control/AI systems. The tool is flexible and new avenues can be explored based on the intuitive experience of the user and designer and her/his special needs for the control design and MS DF-cum-AI processes. Hence, it is only ascertained here that the derivation of any new FIFs could be a good possibility in search of new FIFs and does not guarantee that these new FIFs would be superior to existing ones.

8.10.3 FL and DeF

Here, we consider four aviation scenarios and study the performance of some modified SA models. The SA process aids in decision making, as for example, for a pilot of the fighter aircraft the various decisions-based actions that a he/she can take are: (a) avoid any collision with any nearby flying object, (b) access the intentions of the enemy, if suspected to be an enemy aircraft and (c) communicate with other nearby aircraft if found to be friendly. An application of FL at the higher levels of the DF process could be a good choice for precise decision making. If one of the outputs of SA is 'aircraft is non-friendly and targeting tank' then it can be interpreted as the situation assessor taking the decision that 'aircraft is (really) non-friendly and (actually) targeting tank'. The fusion plays a role when there are more than one such decision for the same object of interest seen by multiple sensors of different types and accuracies and different accuracy levels of each sensor dictate different confidence levels while taking a decision. In order to have an accurate decision it is necessary to fuse the decisions (outputs form situation assessor) using the FL approach.

8.10.4 Performance of FL-Based Decision Systems

In this section, three applications of a decision system using T1FL are studied: (i) formation flight, (ii) attack and (iii) threat assessment. With proper selection of If … Then … rules and combinations of MFs and FIFs, the DeF based on FL can be extended to other aviation scenarios and to more than two objects.

8.10.4.1 Scenario 1: Formation Flight

An FLDS residing in (its) own ship (platform) to decide whether two enemy fighter aircraft have formation flight or not is studied. The model to assess the pair formation for two aircraft of unknown origin is shown in Figure 8.3. This SA model is modified by the addition of two new inputs, Figure 8.4, and subsequently has modified rules. The inputs: 'speed', 'elevation' and 'bearing' are computed/processed to determine if the two aircraft have the same kinematics. The introduction of the two new inputs: 'altitude' of the two aircraft and the 'aspect (angle)' between them is used with the existing model's remaining inputs of 'distance', 'ID' and 'speed' to determine whether the two aircraft are flying in a formation. Aspect angle (AA) is the number of degrees measured from the tail of an aircraft to the other aircraft, it has nothing to do with bearing and its value ranges from 0° to 180°. AA when used in addition to the distance between the two aircrafts gives an accurate view of the lateral displacement between the two vehicles and hence, gives a more accurate representation of whether the two aircraft form a pair. The altitudes of both the aircraft are checked to see if they are higher than the minimum required level and if not, then the aircraft are considered not in a formation flight.

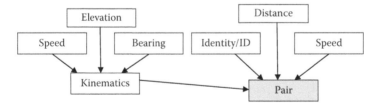

FIGURE 8.3
SA model for pair formation.

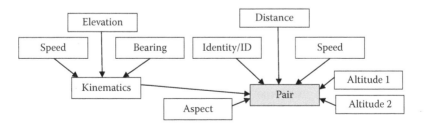

FIGURE 8.4
Modified SA model for pair formation.

The kinematics are generated for the aircraft of the same class and identity in pitch plane (with no motion in x–y plane) using the following initial conditions and state-space model:

(i) Initial state of aircraft 1: $X_1 = [x \quad \dot{x} \quad z \quad \dot{z}] = [0 \text{ m } 166 \text{ m/s } 1000 \text{ m } 0 \text{ m/s}]$;
(ii) initial state of aircraft 2: $X_2 = [x \quad \dot{x} \quad z \quad \dot{z}] = [0 \text{ m } 166 \text{ m/s } 990 \text{ m } 0 \text{ m/s}]$;
(iii) sensor-measurement update rate is 1 Hz; (iv) total simulation time is 50 s; (v) aircraft motion is with constant velocity and (vi) kinematic model is $X_i(k + 1) = FX_i(k) + Gw_i(k)$ with $k(=1, 2, \ldots, 50)$ as scan/index number, $i(=1,2;$ T is sampling interval) is the aircraft number, F is the state transition matrix, G is the process noise gain matrix and w is white Gaussian random process noise with covariance matrix $Q = 0.1*\text{eye}(4,4)$, and we have

$$F = \begin{bmatrix} 1 & T & 0 & 0 \\ 0 & 1 & 0 & 0 \\ 0 & 0 & 1 & T \\ 0 & 0 & 0 & 1 \end{bmatrix} \quad \text{and} \quad G = \begin{bmatrix} \dfrac{T^2}{2} & 0 & 0 & 0 \\ 0 & T & 0 & 0 \\ 0 & 0 & \dfrac{T^2}{2} & 0 \\ 0 & 0 & 0 & T \end{bmatrix}$$

Figures 8.5 and 8.6 depict the flight scenario in terms of the trajectories and elevation angles of the aircraft. In the first 5 s, the distance between the two aircraft is small and at the end of these 5 s, their flight paths begin to diverge. In the subsequent 5 s the two aircraft travel with a constant parallel

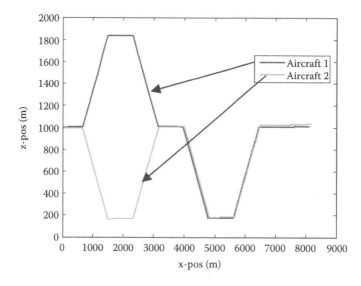

FIGURE 8.5
Simulated flight trajectories of the two aircraft.

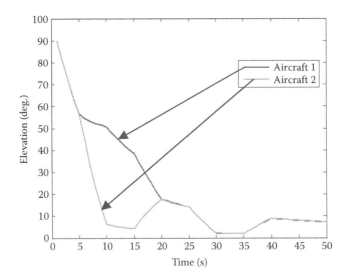

FIGURE 8.6
Elevation angles of the two aircraft.

separation. From the 15th second to the 20th second, their flight paths converge and they then fly at a constant altitude for the next 5 s. From the 25th
second, both aircraft begin descending until they reach 200 m at the 30th
second, remain at that altitude for 5 s and then start their ascent to 1000 m,
after which they fly at a constant altitude. As we can see from Figures 8.5 and
8.6, at the 29th second, and also at the 9th second in the case of the second
aircraft, the altitudes of the two aircraft fall below 460 m (~1400 ft), which is
the minimum altitude that the aircraft should possess for formation flight.
The AA between the two aircraft goes below the required range (30°) at the
44th second. Two important assumptions are made for this simulation, the
aircraft are: (i) friendly to each other and (ii) always within the vicinity of
the sensor/s. The inputs to the FL decision system are: the numerical differences of the aircrafts' bearing, elevation, separation distance along the z-axis,
speed, identity and class. The new input, aspect, is the AA between the two
aircraft. The altitudes of the two aircraft are input separately, Figures 8.7
and 8.8. Trapezoidal MFs are used to fuzzify each input and output data
between 0 and 1. It should be noted that the limits of these functions are
provided for the sake of concept proving, based on the designers'/authors'
intuition. In practice, these limits should be provided by an expert in the
relevant domain. The rules used to decide whether two aircraft form a pair
or not are as follows:

Rule 1: If two aircraft have the same bearing, elevation and speed, then
they have the same kinematics.

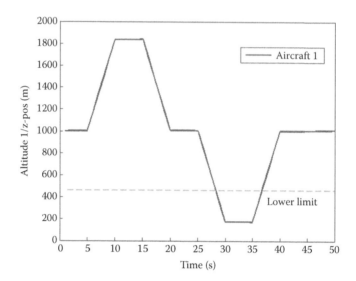

FIGURE 8.7
Altitude of aircraft 1.

> *Rule* 2: If two aircraft have the same kinematics, the same identity, the same class and are at a short distance from each other, then they form a pair.

> *Rule* 3: If the altitude of either aircraft is below 460 m or aspect lies outside of the 30–60° range, then they do not form a pair.

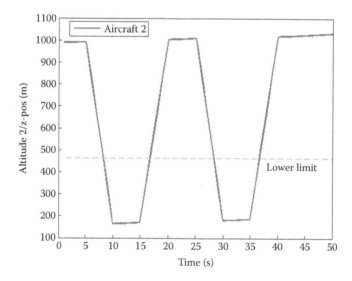

FIGURE 8.8
Altitude of aircraft 2.

Here, the PORFI-Larsen implication method is used and the BS operator of the T-conorm/S-norm is used in the aggregation process. The aggregated output fuzzy set is defuzzified using the COA method. The decision system is implemented in MATLAB/Simulink as shown in Figure 8.9. The outputs of the previous/original and the modified/new models are compared in Figure 8.10. We can see that the system is able to correctly detect the aircrafts' pair and split periods. Table 8.3 illustrates the numerical comparison of the final outputs from the two models, original and new, to determine pair formation.

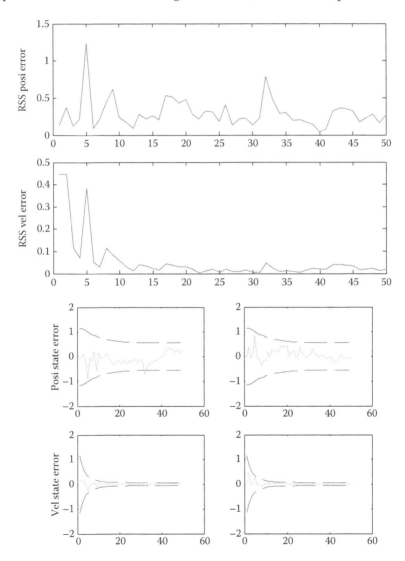

FIGURE 8.9
FL-based DeF system in Simulink for formation flight using the modified SA model.

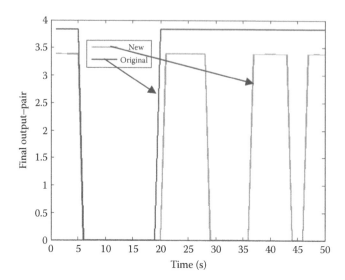

FIGURE 8.10
Outputs of the DeF system to determine formation flight for original SA model and the new SA model.

It is observed that there is a minor change in the value of the final defuzzified output representing the decision 'Yes'. We observe that the output of the new model is different from that of the original at the 29th and 44th second as a result of the two new inputs, altitude and AA, causing the third rule to be satisfied at those times.

8.10.4.2 Scenario II: Attack

Next, we consider the possibility of an attack from another aircraft. Figure 8.11 depicts the SA model proposed to determine such a possibility. The speed of the aircraft, aspect with respect to own ship and distance from own ship are checked to see if the aircraft is closing in on own ship. Then, the identity of the aircraft and its class are used to predict whether the closing aircraft is planning an attack on own ship. The inputs speed and distance have three MFs: small, medium and large. Aspect has three MFs: small, medium and high. The intermediate output closing has two MFs: Yes and No. The input ID represents the status of the IFF (identification friend or foe) and has three MFs: friend, foe or unknown, and the input class has four MFs: fighter, bomber, transport and missile. Finally, the output attack has two MFs: Yes and No. The rules used to decide the possibility of an attack by the other aircraft are

> *Rule* 1: If an aircraft has high speed, has a close distance to another aircraft and is heading towards it (high aspect), then the aircraft is trying to close in on the other.

TABLE 8.3

Numerical Comparison of the Results of
the Formation Flight for the Two Models

Time (s)	Old Model	New Model
1	3.7984	3.385
3	3.7984	3.385
5	3.7984	3.385
7	0	0
9	0	0
11	0	0
13	0	0
15	0	0
17	0	0
19	0	0
21	3.7984	3.385
23	3.7984	3.385
25	3.7984	3.385
27	3.7984	3.385
29	3.7984	0
31	3.7984	0
33	3.7984	0
35	3.7984	0
37	3.7984	3.385
39	3.7984	3.385
41	3.7984	3.385
43	3.7984	3.385
45	3.7984	0
47	3.7984	3.385
49	3.7984	3.385

Rule 2: If an aircraft is closing in on another aircraft, has a different ID
and is a fighter aircraft, then the aircraft has the intention of attacking the other.

In the FIS, MORFI (Mini [-mum] operation rule of fuzzy implication,
Mamdani) and SU are used as the implication and aggregation methods,
respectively, and COA is applied for defuzzification. Figure 8.12 depicts
the implementation of the decision system in MATLAB/Simulink using
the features of the FL toolbox to build the inference system. To check the
validity of the system, the data for simulation are generated in MATLAB
for 20 s. The state-space model similar to the scenario I is used. A fighter
enemy aircraft is assumed. Initially, the aircraft is at a large distance with
respect to own-ship and a medium AA. After 10 s, the aircraft has moved

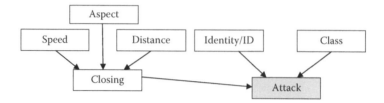

FIGURE 8.11
SA Model to predict attack.

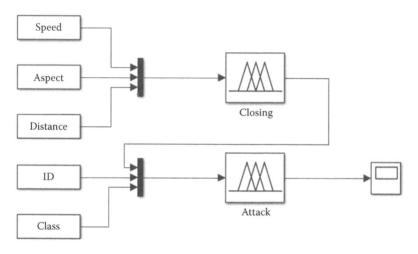

FIGURE 8.12
FL-based DeF system in Simulink to predict attack.

closer and the AA is now high. The velocity of the fighter is high for the entire period of the simulation. As seen from Figure 8.13, the system is able to correctly detect the action of the aircraft to be attacking or non-attacking. The modified SA model for attack, shown in Figure 8.14, takes into account the velocity of the own ship and that of the other aircraft whose intentions are unknown. These two velocities are compared in order to determine the speed advantage of the other aircraft over own ship. The RWR (radar warning receiver) sensor reading is also factored in. This sensor detects the radio emissions of radar systems. Its primary purpose is to issue a warning when radar signals that might be a threat, such as that from an enemy aircraft, are detected. This, when combined with the other information available, will help in assessing the intentions of the other aircraft. In the modified SA model for the attack, the inputs own velocity and closing velocity have three MFs: low, medium and high. Speed has four MFs: high advantage, medium advantage, low advantage and disadvantage. The input RWR has

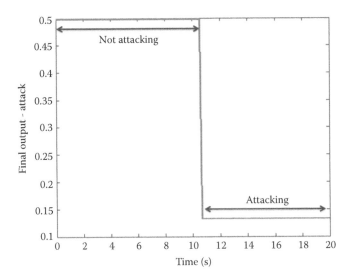

FIGURE 8.13
Output of the DeF system for attack.

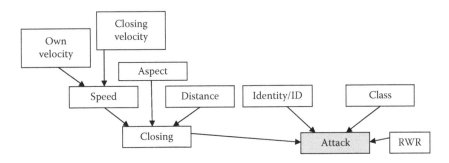

FIGURE 8.14
Modified SA model for attack.

two MFs: illuminating and non-illuminating. All other inputs and the output, attack, remain the same. The rules used to decide the possibility of attack by the other aircraft are

Rule 1: If an aircraft has high speed advantage, has a close distance to another aircraft and is heading towards it (high aspect), then the aircraft is trying to close in on the other.

Rule 2: If an aircraft is closing in on another, has a different ID, is a fighter aircraft and RWR is illuminated, then the aircraft is attacking the other.

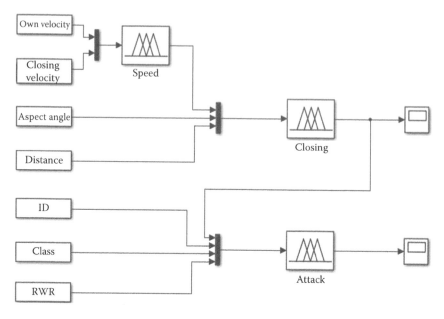

FIGURE 8.15
FL-based DeF system in Simulink to predict attack using modified SA model.

In the FIS to determine speed advantage, the implication method used is
PORFI and the aggregation method is BS, while MORFI and SU are used as
the implication and aggregation methods in closing and attack FISs. All three
FISs use COA for defuzzification. Figure 8.15 shows the implementation of
the system in MATLAB/Simulink using features of the FL toolbox. To check
the validity of the system, the data for simulation is generated in MATLAB
for 20 s. A fighter enemy aircraft is assumed. Initially, the aircraft is at a large
distance with respect to own-ship and a medium AA. After 10 s, the aircraft
has moved closer and the AA is now high. The velocity of own ship is low
for the entire period of the simulation whereas that of the fighter is high. The
RWR sensor reading is 'illuminated' for the first 15 s and 'non-illuminated'
for the reminder of the time. As seen from Figure 8.16, the system is able to
correctly detect the action of the aircraft to be attacking or non-attacking.

8.10.4.3 Scenario III: Threat Assessment

Threat assessment builds on the attack model to develop a comprehensive
system. This system, as shown in Figure 8.17, combines the conventional
inputs such as velocity, AA, deviation angle (angle off), elevation, RWR
sensor reading, class, ID and range in a systematic manner to compute
various intermediary parameters such as combat geometry, energy driven
positional geometry, sensor driven positional geometry and situational
geometry to determine the action of the unknown aircraft. The inputs own

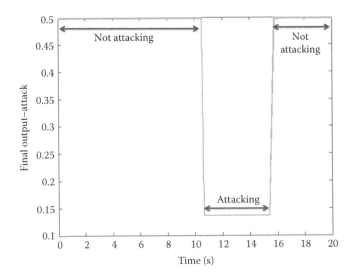

FIGURE 8.16
Output of the DeF system for modified SA attack model.

velocity and closing velocity have three MFs: low, medium and high. Speed has four MFs: high advantage, medium advantage, low advantage and disadvantage. AA and deviation angle, each with the MFs low, medium and high, are combined in order to assess the combat geometry of the two aircraft. Elevation represents the elevation of the other aircraft with respect to own ship and has three MFs: negative low, positive low and positive medium. The input RWR has two MFs: illuminating and non-illuminating.

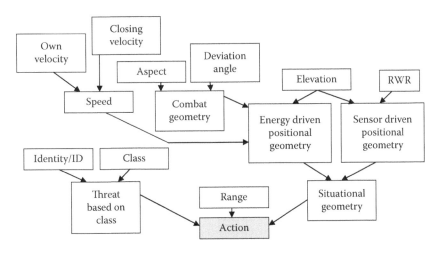

FIGURE 8.17
SA model for threat assessment.

Speed, combat geometry and elevation are used to compute the energy-driven positional geometry (EDPG). EDPG is a measure of the advantage possessed by an aircraft from a kinetic and potential energy point of view. Sensor-driven positional geometry (SDPG) is a measure of the sensor advantage that the aircraft's elevation and RWR puts it under. Situational geometry represents the overall situation based on 'EDPG' and 'SDPG'. An aircraft at a higher elevation might be at an energy advantage but at a sensor disadvantage. Situational geometry is a parameter which takes both into consideration. Combat geometry, EDPG, SDPG and Situational geometry have five MFs: high advantage, advantage, disadvantage, mutual disadvantage and neutral. Class represents the type of aircraft and has four MFs: fighter, bomber, missile and transport. ID has three MFs: friend, foe or unknown. Class and ID of the unknown aircraft are taken into consideration to compute the threat based on class. It has four MFs: high threat, medium threat, low threat and benign. Range represents the distance between own ship and the other aircraft and has three MFs: short, medium and long. Threat based on class, range and situational geometry is used to predict the unknown aircraft's actions. The output, action, has four MFs: offensive, evasive, defensive and passive. Appropriate rules are defined in each FIS, taking all possible scenarios into consideration. The implication method used is PORFI and BS is used for aggregation. COA method is used to defuzzify the output. Figure 8.18 shows the system's implementation in MATLAB/Simulink environment, using FL toolbox to model the inference systems. To verify the proper working of the system, a possible situation is simulated using MATLAB for 20 s. In this scenario, an enemy bomber is considered. Initially, the bomber is at a large distance from own ship, with a low elevation, speed and AA. During this time, the deviation angle is

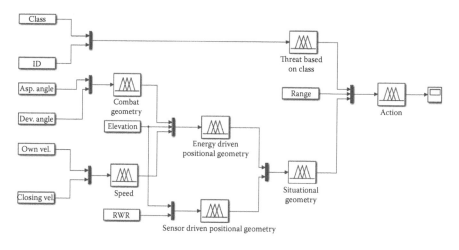

FIGURE 8.18
FL-based DeF system in Simulink to predict attack using modified SA model.

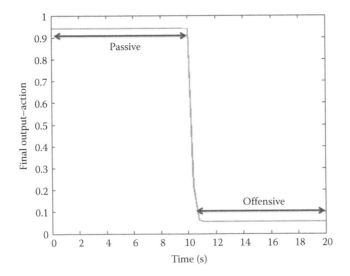

FIGURE 8.19
Output of the DeF system to predict attack using the modified SA model.

medium. After 10 s, the aircraft has moved closer and is now travelling at high speed. It has moved higher and turned such that it has a high AA and low deviation angle (facing own ship head-on). The velocity of own ship is a constant low for the entire simulation period and the RWR is always illuminated. Figure 8.19 shows the decision given by the system for this scenario. The system is able to correctly identify the threat posed by the enemy aircraft.

8.10.4.4 Study of Effect of Noise on FL-Based Decision Fusion Systems

In practical situations the inputs sourced from the sensors are often contaminated with noise and hence, it is necessary to test the DeF systems' performance in the presence of noise. In order to test the systems, most inputs to the system are exposed to noise and the performance evaluated. The amount of noise added to the inputs is varied by trial and error and the minimum signal-to-noise ratio (SNR) of the inputs for which the system produces the expected performance is determined.

8.10.4.4.1 Threat Assessment

Various inputs have been contaminated with random input signal, and these are shown in Figures 8.20 through 8.24. A comparison between the outputs of the system with and without the addition of noise is shown in Figure 8.25. From trial and error, it was found that the decision system could produce accurate results for or above SNR of 20 dB with reasonable tolerance levels.

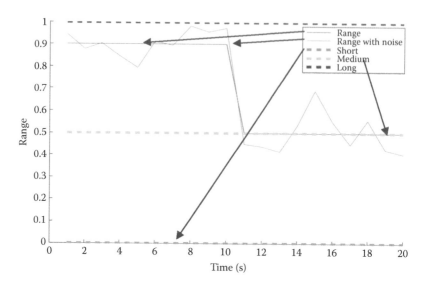

FIGURE 8.20

Range input with noise.

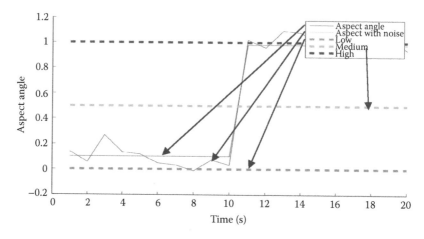

FIGURE 8.21

AA with noise.

8.10.4.4.2 *Flying along Air Lane*

Now, we describe a system that is used to decide if a particular aircraft is flying along the air lane. A sensitivity study is carried out to check the performance of the system in the presence of noise. The SA model for this system is shown in Figure 8.26, and the inputs to the system are distance (the absolute separation between the aircraft and the air lane along the *y*-axis),

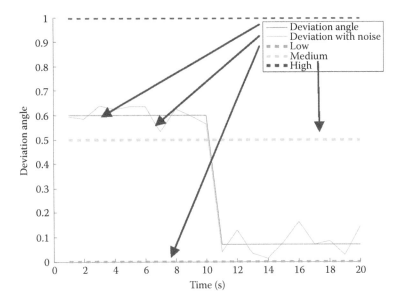

FIGURE 8.22
Deviation angle with noise.

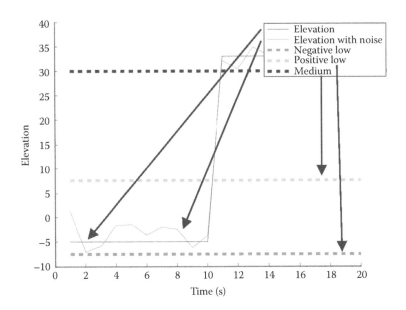

FIGURE 8.23
Elevation with noise.

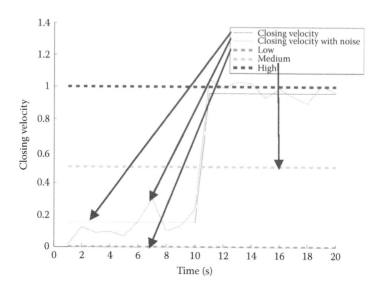

FIGURE 8.24
Closing velocity with noise.

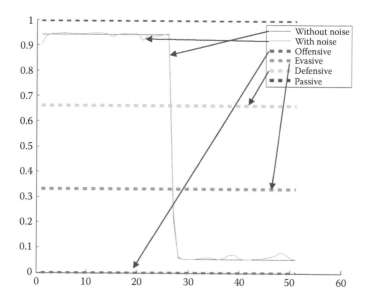

FIGURE 8.25
Output of the FL-based decision system for threat assessment with noisy inputs.

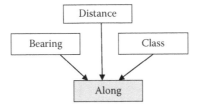

FIGURE 8.26
SA model for flying along air lane.

the absolute value of bearing difference between them, and the class of the aircraft. Trapezoidal MFs are used to fuzzify the inputs and outputs. The rules used to decide whether a particular aircraft flies along an air lane or not are as follows:

    Rule 1: If the aircraft has the same bearing as the air lane and if it is close to the air lane, then the aircraft is flying along the air lane.

    Rule 2: If the aircraft is civil, then there is a high possibility that the aircraft is flying along the air lane.

PORFI is used as the implication method and BS operator of the T-conorm/S-norm is used in the aggregation process. The aggregated output fuzzy set is defuzzified using COA method. The Simulink model to determine if an aircraft is following the air lane uses the FL toolbox for the inference system and is shown in Figure 8.27. The trajectory of the aircraft and its relative motion with respect to the air lane is simulated and the position of the air lane and aircraft are shown in Figure 8.28. The bearings of the aircraft as well as that of the air lane are also generated and shown in Figure 8.29. Noise is added to the system's inputs to simulate the presence of noise in various sensors that might be interfaced with the decision system. The two primary inputs processed to decide if the aircraft is following the air lane are distance and bearing. Upon the addition of noise to the distance between the aircraft and air lane, the input

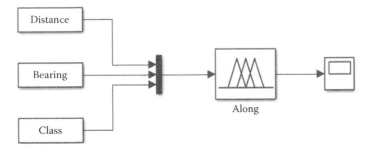

FIGURE 8.27
FL-based DeF system in Simulink for flying along an air lane.

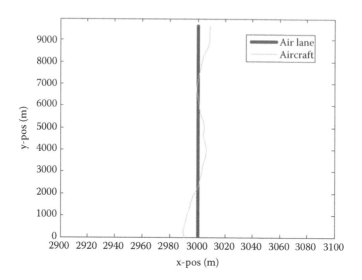

FIGURE 8.28
Position of air lane and aircraft.

becomes as shown in Figure 8.30. The bearing difference is also exposed to noise and the subsequent plot is shown in Figure 8.31. From trial and error, it was found that the system could produce accurate results for or above an SNR of 16 dB with reasonable tolerance levels. A comparison between the outputs of the system with and without the addition of noise is shown in Figure 8.32.

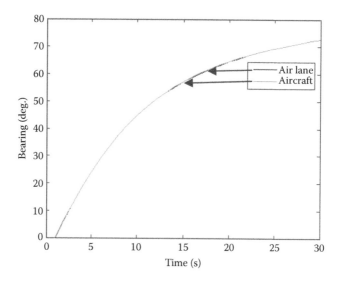

FIGURE 8.29
Bearings of aircraft and air lane.

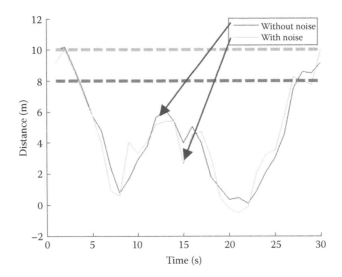

FIGURE 8.30
Distance between the aircraft and air lane.

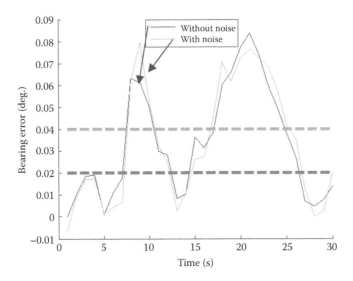

FIGURE 8.31
Bearing difference between aircraft and air lane.

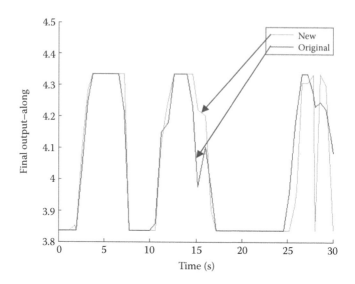

FIGURE 8.32
The outputs of the DeF system for air lane in the presence and absence of noise.

8.10.5 Discussion of the Results

The main aim of this work has been to illustrate the use and application of T1FL to the three aviation scenarios in decision making/fusion using SA models. The decision making has been incorporated in the rules: (i) if two aircraft have the same kinematics, identity, class and are at a short distance from each other then they form a pair', (ii) if the altitude of either aircraft is below 1400 ft (460 m) or aspect lies outside of the 30°–60° range, then they do not form a pair, (iii) if an aircraft is closing in on another, has a different ID and is a fighter aircraft, then the aircraft is attacking the other and (iv) if an aircraft is closing in on another, has a different ID, is a fighter aircraft and RWR is illuminated, then the aircraft is attacking the other. The results of this work are qualitative as well as quantitative and are presented in Table 8.3 and Figures 8.10, 8.13, 8.14 and 8.18. Also, these decision systems have been tested extensively in the presence of noise (in several inputs) to simulate the noise associated with the sensors in the actual environment. The systems proved to be stable with the presence of noise. It was found that the threat assessment system provided accurate outputs above a minimum SNR of 20 dB, and the along air lane system provided accurate results above a minimum of 16 dB SNR. Thus, the FIF-evaluation tools as applied to the existing FIFs have been proved to be working satisfactorily for the SA in the aviation scenarios presented in this example. The proposed methodology can be easily extended to more general aviation scenarios as well as to the problems of multiple dimensions. The results indicate that some existing fuzzy implication functions work well for the examples considered. While the results are very encouraging, some more studies can be made to evaluate these FIFs and their applicability in general control systems as well as in aerospace DF and DeF systems. This definitely opens up new possibilities of applications of DeF based on FL to varieties of aerospace and aviation problems in decision making. The next research direction is to extend the present study to application of IT2FL to DeF for SA.

EXERCISES

8.1 Where in a target-tracking problem is the decision making involved?

8.2 Take some numerical values and verify the inequalities: if $U(x, a_1) < U(x, a_2)$ and $U(x, a_2) < U(x, a_3)$ then obviously $U(x, a_1) < U(x, a_3)$.

8.3 Take some numerical values and verify the inequality: if $U(x, a_1) < U(x, a_2)$, then $\alpha U(x, a_1) + (1 - \alpha)U(x, a_3) < \alpha U(x, a_2) + (1 - \alpha)U(x, a_3)$, for any $0 < \alpha < 1$.

8.4 Why and how can FL be used in decision making?

8.5 How is decision making useful in low-level and mid-level DF process?

8.6 What is k-plurality voting in decision making?

8.7 Why and how can FLS be fundamentally considered as a decision making and DeF paradigm?

8.8 What is the role of FL in hypothesis testing?

8.9 Why in the derivation of the NPS, would the actual value of c not play any role:

$$\hat{a} = \arg \max_{a} \left\{ \sum_{j} [E\{U_j(x,a)\} - c(j)] \right\}_{?}$$

8.10 How can DeF be used for detection of a fault in a dynamic system?

References

1. Hansson, S. O. Decision theory: A brief introduction. Department of Philosophy and the History of Technology, Royal Institute of Technology (KTH), Stockholm, 1994. http://home.abe.kth.se/~soh/decisiontheory.pdf, accessed November 2011.

2. Duft, K. D. Statistical decision theory. Extension Marketing, Cooperative Extension, College of Agriculture and Home Economics, Washington State University, Pullman, WA, www.agribusiness-mgmt.wsu.edu/.../cash-asset/Stat_Dec_Theory.pdf, accessed November 2011.

3. Durrant-Whyte, H. Multi sensor data fusion. Australian Centre for Field Robotics, University of Sydney NSW 2006, Australia, hugh@acfr.usyd.edu.au, 22 January 2001.

4. Raol, J. R. *Multi-Sensor Data Fusion with MATLAB.* CRC Press, FL, 2010.

5. Samarasooriya, V. N. S. and Varshney, P. K. A fuzzy modelling approach to decision fusion under uncertainty. *Journal of Fuzzy Sets and Systems,* 114, 59–69, 2000.

6. Vizitiu, C. I., Serban, V., Molder, C. and Stanciu, M. Decision fusion method to improve the performances of multispectral ATR system. *Proceedings of the 1st WSEAS International Conference on Sensors and Signals,* SENSIG '08, Bucharest, Romania, November 7–9, 2008.

7. Vizitiu, I. C. *Neuro-Fuzzy-Genetic Architectures: Theory and Applications,* MTA Press, Bucharest, 2011.

8. Koren, I. and Krishna, C. M. *Fault Tolerant Systems.* Morgan Kaufmann Publishers Inc., San Francisco, CA, USA, 2007.

9. Behrooz, P. Fault-tolerant computing – software design methods, University of California, Santa Barbara, November 2007, USA. www.ece.ucsb.edu/~parhami/pres.../f33-ft-computing-lec18-agree.ppt, accessed January 2013.

10. Raol, J. R. and Sudesh, K. K. Decision fusion using fuzzy logic type I for two aviation scenarios. *Journal of Aerospace Sciences and Technologies (the Journal of Aeronautical Society of India),* 65(3), 273–286, 2013.

11. Mangai, U. G., Samanta, S., Das, S. and Chowdhury, P. R. A survey of decision fusion and feature fusion strategies for pattern classification. *Journal of IETE Technical Review,* 27(4), 293–307, 2010. http://tr.ietejournals.org/text.asp?2010/27/4/293/64604, accessed May 2013.

12. Sinha, A., Chen, H., Danu, D. G., Kirubarajan, T. and Farooq, M. Estimation and decision fusion: A survey. *Neurocomputing,* 71, 2650–2656, 2008. www.elsevier.com/locate/neucom.

13. Rao, N. P., Kashyap, S. K. and Girija, G. Situation assessment in air-combat: A fuzzy-Bayesian hybrid approach. *International Conference on Aerospace Science and Technology,* Bangalore, INCAST 2008-094, 26–28 June 2008.

14. Bonanni, P. *The Art of the Kill,* 1st Edition. Spectrum HoloByte, USA, 165, 1993, ISBN- 9780928784831.

15. Rao, N. P., Kashyap, S. K. Girija, G. and Debanjan, M. Situation and threat assessment in BVR Combat. *AIAA Guidance, Navigation, and Control Conference,* Portland, OR, 8–11 August 2011.

16. Shrinivasan, L., Prabhu, A., Manivannan, H., Sridhar, K. and Ahmed, S. Decision fusion using fuzzy logic type 1 for situation assessment in three aviation scenarios. *(accepted for) 2014 International Conference on Advances in Electronics, Computers and Communications (ICAECC),* Reva Institute of Technology and Management, Bangalore, 10–11 October 2014.

17. Prabhu, A., Manivannan, H., Sridhar, K., Ahmed, S. and Shrinivasan, L. Study of noise on situation assessment systems developed for aviation scenarios. *1st IEEE National Conference on Electrical and Electronics Engineering,* NCEEE-2014, HKBK College of Engineering, Bangalore, 5 June 2014.

9

Wireless Sensor Networks and Multimodal Data Fusion

9.1 Introduction

Sensor data fusion (DF) activity in wireless sensor networks (WSNs) assumes a greater importance in present day complex engineering scenarios, be they aerospace flight vehicles, or large-scale security network (NW) systems, or even the world wide web, the internet. In such situations and systems it is very important to sense and monitor environmental conditions cooperatively, and this is why one would constitute and design a WSN with distributed and autonomous devices/sensors. Further uses of the WSNs are in [1,2]: (i) environmental and health monitoring situations, (ii) prediction, detection and management of natural calamities, (iii) aerospace and large structural health monitoring systems, (iv) medical-situations-monitoring systems and (v) distance-based multi-robot/vehicles coordination. These sensor networks (SNWs) consist of small and yet a large number of autonomous sensor nodes which are supposed to be fairly inexpensive, and disposable/replaceable. Depending on situations and specific requirements these sensor nodes are distributed onto a wide area (and onto long-ranging/various heights) and often these devices are operated remotely. These nodes have several constraints though [1,2]: (a) there could be limited storage resources, (b) there is a requirement of large computational capability, (c) they need continuous power supply (at remote sensor nodes in the fields/forests, battery power to operate the devices is limited; because these devices in such areas might only be operating on battery power) and (d) they have a need of a large communication bandwidth (BW). These sensor/devices nodes might be in the form of one or more clusters. These nodes forward their acquired data/information to their respective cluster head/s. These heads then send the data to the specialised node that is called sink node. The sink node is also called base station (BS). This process of sensor-data communication is done via a multi-hop wireless communication procedure [1–4]. In the larger domain of WSNs, the smart (miniaturised) devices and sensors (SDSs) integrated into aerospace structures (e.g. large aircraft wing/large spacecraft structure), into large machinery in industrial plants or power generating stations, and in any related large environment, coupled with the very efficient delivery and dissemination of the sensed information data, certainly provide tremendous benefits to the technological base of any country or society [4]. There are other possibilities in WSNs such as: multiple BSs and/or mobile nodes. The full spectrum of WSNs requires a good study of several aspects such as [1–4]: (i) various communication functions, (ii) data/source information delivery models, (iii) NW control, routing, and monitoring, (iv) collaborative information and data processing, (v) data querying, (vi) (multi) tasking, (vii) statistical learning/reinforce learning, (viii) situation adaptation, (ix) evolution/development of the systems/nodes, (x) need to use FL for modelling, control and incorporating expert knowledge, (xi) neuro-computing for learning strategies and adaptation, (xii) swarm intelligence and (xiii) artificial immune systems. Despite wide knowledge, many WSNs face certain challenges [1–4]: (a) wireless ad hoc nature, (b) effect of mobility and topology changes, (c) handling energy/power limitations, (d) (large area-based) physical distribution of the nodes and their management and coordination, (e) design/deployment aspects and issues, (f) localisation of data/information and control if required, (g) data aggregation (DAG) and sensor fusion aspects, (h) energy aware (-ness) routing for efficient utilisation of resources, and clustering, (i) task/power scheduling (in huge electrical grid NWs), (j) security of various systems and (k) assurance of quality service in time and location.

Despite these limitations we can have several benefits of a WSN [4]: (i) fewer catastrophic failures of the systems that use such WSNs in their functioning – because of redundancy provided by the NW of SDSs, (ii) conservation of natural resources – by quick information sharing about the disaster and other environmental failures, (iii) highly improved manufacturing productivity in an industrial application/power system generation and distribution – due to distributed NWs and (iv) improved emergency response of the systems (e.g. security system, tsunami/flood alert systems). The most important aspect of the WSNs is freedom from the hundreds of (physically) connecting wires, and the huge maintenance cost of such wires and the maintenance of these complicated wired-connections. An ideal WSN should (a) be (highly) scalable, (b) consume (much) less power/energy, (c) be smart employing smart sensors and micro-electro-mechanical systems/sensors (MEMS), (d) be easily SW-programmable/restructurable, (e) be capable of fast data/information acquisition, (f) be highly reliable/redundant and accurate over very long-term usage, (g) not cost (too) much to purchase, install and maintain and (h) not require much maintenance.

9.2 Communication Networks and Their Topologies in WSNs

There are a number of topologies for radio communications NWs. We here discuss some that are very suitable for WSNs [4].

9.2.1 Star Network

In a star NW (SNW), single BS sends and/or receives a message to a number of remote SD nodes. In SNW the remote nodes can only send or receive a message from the single BS. These nodes are not allowed to send messages to each other in SNW, this means that the individual SDSs are not directly connected – they are connected through only the single BS. This type of WSN is simple and it keeps the power consumption of the remote node to a minimum. The SNW allows a low latency (delay) communications between the BS and the remote node. The demerit of the SNW is that the BS should be within the RF transmission range of all the nodes. The SNW is not very robust for it depends only on one node to manage the NW.

9.2.2 Mesh Network

The mesh NW (MNW) allows communications from any node to any other node within its RF range, thus, multi-hop communications is feasible [4]. If a node wants to send a message to another node (out of its range) it uses an intermediate node to forward the message to the desired node. The MNW has the merit of redundancy and scalability. The MNW's range is not limited by the range in between single nodes, since more nodes can be added to the system. The demerit is in higher-power consumption for the nodes that implement the multi-hop communications. Also, as the number of communication hops increases, the time of delivery of the message also increases.

9.2.3 Hybrid Star: Mesh Network

The hybrid star – MNW (HSMNW) provides robust and versatile communications. It keeps the power consumption of the wireless sensor nodes to a minimum. In HSMNW, the lowest power SDSs are not enabled with the ability to forward messages, thus allowing for minimal power consumption, while other nodes have multi-hop capability.

9.3 Sensor/Wireless Sensor Networks

A sensor network/wireless sensor network (sensor NW/WSN) consists of a large number of SDSs that have a capability of sensing, computation and communication of the data/results of computations [2]. These SDSs can be termed as somewhat 'smart' devices/sensors, since they have their own capacity of computation/data processing. These SDSs perform the data gathering activity collaboratively. These sensor NWs were originally developed for use by military applications. The WSN encompasses the areas of wireless communication, computer NWs, MEMS, system and control aspects and algorithm development (computer science). The SDSs nodes (called motes) have the sensing, computing and communication capability and these are the energy source units. The SDSs detect a signal and via other nodes transmit the data to the outer world such as (i) satellite and (ii) remote local monitoring station. These data are then received by a remote monitoring facility. These nodes are separated from each other by, say, a distance of 250–300 m. These data could also be transmitted to other alternative needs: (a) mobile robot, (b) surveillance aircraft and (c) any BS that requires this information. The WSN could be self-forming and adaptive, modular and flexible. It should be possible to add or delete the nodes as and when required.

9.3.1 Need in WSNs

The WSN is a wireless communication media with a large number of (mostly) densely deployed motes [2]. It could have a constraint of size, cost, computational power, communication BW and energy. Any WSN is prone to failure, like as any other system of NWs. This could be due to any obstruction, and hence loss of motes. There could be other failures in HW system/components, computer processing units, algorithm divergence (and undetected-SW bugs), power, and so on. The WSNs are preferred to be self-organising and distributed in nature and topology. The WSN hence would have modularity, flexibility and some local/self-computing/data processing capability with SDS nodes. The major use and application of the WSNs in the distributed and monitoring modes [2] are in: (i) the military for reconnaissance and detection, (ii) the environment for fire/flood detection, bi-complexity mapping, (iii) industry and business for process control and inventory management, (iv) civilian home management. In the target-tracking mode, the use and application of WSNs are: (a) the military for surveillance and targeting, (b) public traffic control, (c) healthcare and rescue in tracking the movement of elderly persons and in drug administration and (d) business for human tracking. Other very important usages of WSNs are [2]: (i) monitoring in remote sensing areas with harsh environmental conditions and (ii) gathering of environmental data to help understand the processes that connect climate changes and rock fall in selected areas of monitoring.

9.3.2 WSN Challenges

Any WSN being a communication NW, has its own fundamental challenge of communication aspects [2]: (i) architecture and protocol task at the NW level and (ii) topological – (a) position of the SDSs, for example random or regular/uniform layout, (b) homogeneous or heterogeneous distributed architectures of these SDSs, (c) should it be a dynamic or a static NW? and (d) what type of clustering is to be employed? Some more and further challenges lie in the area of sensor management and control (latter if required) [2]: (i) efficient resources (SDSs) allocation of the tasks to the sensors, (ii) security of the SDSs and (iii) fault tolerance of the SDSs and related HW/SW issues – it could be a link/connection or a node failure. The usual HW challenge can be met with an adequate design of HW platform by realising low cost and tiny SDS nodes. The latter is done by using MEMS/NEMS (nano) technologies, and is often known as MINA technology.

As usual it is very important to evaluate the performance of any WSN from the point-of-view of its technical ability to adequately perform the tasks handed over to the NW. Hence, it is important to measure the performance of the WSN of its important sub-parts/sub-systems or as a whole system quantitatively. This should be done in terms of accuracy, latency (inherent delay), scalability, stability and fault tolerance. Another very important challenge and aspect of the WSN is the task of proper sensing/collection of data and subsequent DF – how to fuse these data coming from various and many sensors using local communication for further decision making at the local or global level.

9.4 Wireless Sensor Networks and Architectures

The DF architectures that can be used in WSNs are the same as the classical or conventional ones used for any DF process [2,3] (discussed in Chapters 1, 4 and 5):

i. Centralised – This is the simplest of all the three types of the architectures possible, in which a central processor/central fusion station (CP/CFS) fuses the RRs collected by all other SDS nodes. The data are relayed to a central sink/node or processor. The advantage is that an erroneous report can be easily detected. The demerit is that it is inflexible to sensor changes and the workload is concentrated at one point. Further this architecture is not scalable, it suffers the problem of data congestion and reliability.

ii. Decentralised – The DF process occurs at each SDS node on the basis of local measurements and information gathered from its neighbouring nodes. There is no CP/central node. This architecture, as we have seen in Chapters 1 and 5, is scalable and fault tolerant. It is also tolerant to the addition or loss of some SDS nodes or any dynamic changes in the WSN.

iii. Hierarchical – In this architecture, the SDS nodes are segregated into hierarchical levels. The sensing nodes are at the level 0 and the BS one at the highest level, then the reports move from the lower levels to the higher ones. The merit is that the workload gets balanced among several SDS nodes/motes.

The distributed data fusion (DDF/Chapter 5) is particularly suitable for the WSNs, whereby the distributed algorithms are used for processing the sensor data. In this DDF architecture, the data are used to compute local estimates. Then, the (reports/results) RRs are forwarded to nearby nodes. The receiving nodes fuse the data/results and update the local estimates. The design objectives of such a DDFA [2] are: (i) scalability which is deployed and used in very large WSNs, (ii) higher efficiency, which means less transmission and less computations, (iii) robustness and reliability of the WSN, which means there is no centralised weak point, and that NW's delayed information can be handled and (iv) autonomy, meaning self-adaptability.

9.4.1 Distributed Kalman Filtering

This distributed KF-DKF algorithm is applicable in large-scale WSNs with limited capabilities that have local communication and routing [2]. In Chapter 5, however, of course the information filter (IF) estimation scheme is advocated for the DDF architecture. The performance of the DKF algorithm is analysable in terms of the properties of the NW. It has very good robustness features regarding various NW imperfections, including delays, loss of link, fragmentation of the NW and its asynchronous operation. The DKF assumes the identical sensing models (measurement models) across WSN. The idea is to decompose the conventional KF into n collaborative micro-KFs (MKFs) with local communication. The estimation then for each MKF involves two dynamic consensus systems using two consensus KFs (CF): (i) low pass CF $\rightarrow$ fusion (average) of measurements and (ii) band pass CF $\rightarrow$ fusion (average) of inverse covariance matrices. Certain aspects of distributed sensing/monitoring, distributed estimation and distributed sensor fusion have been discussed in Chapter 5.

9.5 Sensor Data Fusion in WSN

We now know that the WSN comprises of the sensor nodes and a BS at the least. Each SDS node/mote is powered by a battery and is equipped with: (i) integrated sensors, (ii) local data processing capabilities and (iii) short-range radio communications [3]. The motes perform in-network (iNW) DF due to their limited power and shorter communication range. In the DF process, the DF node collects the results from multiple nodes. It fuses the results with its own results based on a decision criterion (Chapter 8). Then, the node sends the fused data to another node and/or BS. The merits are: (i) reduced traffic load and (ii) conservation of energy of the sensors. This is mainly due to the fact that the loads of original data need not be transmitted, but only the results of the data processing need to be communicated. This reduces the communication overheads on the NW. The very important aspects in the

WSN-DF are: (i) what is the distance at which the mote report a sensed event? (ii) how does a mote fuse the multiple reports into one piece of information? and (iii) what DF architecture needs to be used?

9.5.1 Reporting

There are three aspects in reporting: (i) periodical reporting in which DS nodes periodically send the reports to the BS, (ii) in case of the BS inquiry response reporting, the BS queries sensors in specific regions for current sensed information and (iii) in the case of event triggered reports, the occurrence of a certain event triggers reports from sensors of that particular region.

9.5.2 Decision of Fusion

The voting (also, used in fault tolerant systems, Chapter 8) is the oldest method and is widely used in DeF method, and can be easily employed in WSNs. The mote arrives at a consensus/agreement by a voting scheme like: (a) majority voting, (b) complete agreement and (c) weighted voting. The voting approach is popular due to it being very simple and accurate. There are other decision (fusion) methods that one can use that are based on the probability-based Bayesian model (Chapter 8) and on stack generalisation.

9.5.3 Cluster-Based Data Fusion

The motes need to perform efficient DF to extend the lifetime of the WSN due to their energy limitations. It is also important to note that the lifetime of a sensor NW is the number of rounds of DF that it can perform before the first mote fully wears or drains out [3]. In the WSN-DF this aspect is called the maximum lifetime data aggregation (MLDAG). Then the main aim is, given the location and energy of each mote and the BS, to determine an efficient manner to gather and aggregate RRs from the sensors to the BS. For this a cluster-based heuristic approach (CMLDA) discussed next can be utilised.

A system model is first defined with n motes $(1, …, n)$ and $n + 1$ BSs $(1, …, n + 1)$. The data packet size is fixed to k. Initial energy of a sensor i is given as e_i. The received energy is given as $RX_i = k*e_{ele}$. Then the transmitted energy is given as $TX_{i,j} = k*(e_{ele} + e_{amp}*d^2_{i,j})$. Based on these preliminaries, an algorithm that can be used in the two phases is given below [3]:

a. Phase 1

Step 1: The motes are grouped into clusters named as super-sensors – we call these as SDSs.

Step 2: Each SDS consists of a minimum number of DSs.

Step 3: The energy of a SDS is the sum of the energy of all the DSs within its cluster/zone.

Step 4: The distance between the two SDS is the maximum distance between two DSs where each resides in a different SDS.

Step 5: Then apply the MLDA algorithm.

b. The MLDA Algorithm

The ILP is utilised to find a near-optimal permissible flow NW. The objective is to maximise the lifetime (T) of NW under the given energy constraints. Then generate the schedule(s) from the permissible flow NW.

c. Phase 2

Step 1: Initialise the aggregation schedule = 0.

Step 2: Lifetime T is set to 0.

Step 3: From Phase 1 choose a scheduler

Step 4: Initialise aggregation tree, A with the BS.

Step 5: Visit each super cluster and add the nodes to the tree such that the residual energy at each edge is maximum.

Step 6: Add A to the aggregation scheduler.

Step 7: Increment the lifetime T to $T + 1$.

Step 8: Repeat steps 3–7 until a node drains out.

This algorithm provides a set of DF schedules that maximise the lifetime of the NW, and the clustering of the motes reduces the time needed to solve the ILP.

9.5.4 Synchronisation among Nodes

Another important aspect of the DF in the WSN is that of the synchronisation among the nodes [3]. In the DF process, the internal motes at each level wait for a certain period of time before these motes fuse the received RRs. In such a case if the motes at each level wait for the same period of time then an internal node may timeout before receiving RRs from all of its constituent nodes. With such insufficient RRs, the credibility of a sensed event can be questionable. To handle this aspect an efficient DF protocol with the following features is described next. These features are: (i) it synchronises the motes at different levels, (ii) the motes at higher levels wait longer before fusing the data, (iii) a fixed time duration/period is assigned from the sensing of an event to the time it is received by the BS and (iv) it provides a balance between NW latency and accuracy [3]. The multi-level fusion synchronisation (MLFS) protocol with the following parameters can be described next: (a) MAX – time the BS waits before fusing the received RRs, (b) at consecutive levels the time difference dt in waiting and (c) K – the distance in hops from the sink. The algorithmic steps are given below [3]:

Step 1: A leaf node reports to its parent node upon the detection of an event.

Step 2: As a result of this action the timer of the parent node is triggered.

Step 3: Then a 'start' message is sent by the parent node to trigger the timer of its neighbouring nodes.

Step 4: The timer at a mote expires after $(MAX - K^*dt)$ s.

The parameters are set based on the following aspects: (i) if the BS knows the depth of the fusion tree, then it computes the values of MAX and dt, (ii) in a learning phase the BS queries the motes with different values of MAX and dt and (iii) and they adjust the values based on the RRs credibility and application requirements. The merits of this algorithm are that: (a) it synchronises motes at different levels and (b) the MAX and dt can be tuned. The demerits are: (a) the RRs arriving after timeout are discarded and (b) a collision can occur if the 'start' messages cause a latency value larger than the MAX.

9.5.5 Resistance against Attacks

We have so far assumed that the motes performing the DF process are secured. However, a wrong-malicious DF node is likely to send some fictitious (bogus) RRs to the BS, and the BS is incapable of detecting this fictitious information since the motes do not directly send the RRs to the BS. In such a situation, one can use the witness-based scheme to ensure that the BS accepts only valid DF results. The main idea is that to ascertain the validity of a RR, the fusion node should provide some proof from several witnesses which are the nodes that also perform DF but do not send their RRs to the BS. The merit of the scheme is that it ensures that only the valid RRs are accepted by the BS. The demerit is that the multiple copies of similar RRs are fused by the witnesses, and the scheme is not energy efficient. The algorithmic steps [3] for this procedure are given below:

Step 1: Let there be m witnesses and one DF node.

Step 2: Each witness w_i shares a unique key with the BS k_i.

Step 3: On receiving RRs from the motes, each witness performs DF and obtains the result r_i.

Step 4: It then sends a message authentication code (MAC) to the DF node as $MAC_i = MAC(r_i, w_i, k_i)$.

Step 5: The DF node computes its result and sends its MAC key with its witnesses to the BS.

Step 6: The BS exercises a voting mechanism to determine the validity of the report.

Step 7: If the report is corrupted, the BS discards it. It then polls one of the witness nodes for the correct report.

The voting schemes used are described next. The BS can utilise two voting schemes to determine the validity of the fused report [3]: (i) $m + 1$ out of $m + 1 \rightarrow$ the result is valid if supported by all the witnesses and (ii) n out of $m + 1 \rightarrow (1 = <n< = m + 1)$, the result is valid if supported by at least n witnesses.

   a. Scheme 1: The $m + 1$ out of $m + 1$ voting scheme [3]

      Step 1: After receiving all the $MACs$ from the witness nodes, the DF node computes

$$MAC_F = MAC\ (S_F, F, MAC_F, w_1, MAC_1, \ldots w_m, MAC_m).$$

      Step 2: F sends $(S_F, F, w_1, \ldots, w_m, MAC_F)$ to the BS.

      Step 3: The BS computes the $MAC_i = MAC\ (S_F, w_i, k_i)$ for each w.

      Step 4: Finally $MAC'_F = MAC\ (S_F, F, K_F, MAC_1\ xor \ldots xor\ MAC_m)$ is computed.

      Step 5: If $(MAC_F = MAC'_F)$ then accept the report.

      In the above $m + 1$ out of $m + 1$ scheme a corrupt witness node can always send an invalid MAC and achieve denial of service attack. The second algorithm prevents that from happening.

   b. Scheme 2: The n out of $m + 1$ voting scheme [3]

      Step 1: F should not merge all the $MAC_i's$ but instead forward them all

$$R = (S_F, F, MAC_F, w_1, MAC_1, \ldots w_m, MAC_m).$$

      Step 2: If at least n out of $m + 1$ $MAC's$ match, then the result S_F is accepted, otherwise the result is dropped.

9.6 Multimodality Sensor Fusion

Due to the availability of inexpensive motes, it has become feasible to configure a large (sensor) NW of these motes, economically and technically, for collecting distributed information [5]. In such a WSN, the motes of different principles and/or modalities generate multimodal signal outputs/data (MMSD) irrespective of spatial/temporal overlapping/coverage. As a result it is important to study the aspects involved in building the motes, configuring these SNWs, and developing mathematical models for managing and interpreting the RRs received from these motes of different modalities. These areas are: (i) MMSD management, (ii) MMSD analysis and interpretation, (iii) sensor data interaction and (iv) real-world system development (HW/SW/ algorithms) and deployment.

9.6.1 Multimodal Sensor Data Management

At the current sensor technology level there are various types of sensing modalities: (i) optical, (ii) chemical, (iii) mechanical, (iv) thermal, (v) electrical, (vi) chromatographic, (vii) magnetic, (viii) biological, (ix) fluidic, (x) ultrasonic and (xi) mass sensing [5]. It is quite possible that several or a few types of these sensors might have been in use in any WSN. Such a variety of sensors would give signals/data with various types of features/modalities. What is important to note is that the signals generated from these multimodal motes need to be sampled, filtered, compressed (if needed), transmitted, fused and stored. These data-management (DM) aspects should be performed in a very efficient manner in order to conserve resources/power. Some situations in the MMSD analysis and management are: (a) fusion from multi-camera data, (b) infrared-, thermal- or acoustic-guided video surveillance systems, (c) operational results/RR of video-guided remotely operated medical–surgical instruments, (d) dynamic configuration of large-scale mobile-camera NWs, (e) inventory control using sensor-RF integrated systems and (f) habitat monitoring using multi-sensory data/information [5].

9.6.2 Multimodal Sensory Data Interpretation

The collected sensory data and RRs should be analysed for interpretation and for ensuing subsequent actions for control and prediction of phenomena, such as (vehicle) path finding and motion control of a mobile vehicle/robot (more useful for multi-robot/swarm of UAVs coordination). This is called mapping of the data/information to the semantics of interest (MDSI). Such MDSI can be performed by using statistical analysis methods, and these techniques should work in the hostile environment if needed, since some of the motes might be situated in such environments itself. Also, these techniques should be capable of working with limited resources and in the presence of missing and noisy data. Some of the special features that these data-processing algorithms should possess are: (i) one-pass processing ability (or often the recursive-data processing ability is required, since the off-line batch-iterative mode would not be suitable), since we might not be able to store the data due to their huge volume – this calls for the real-time processing/performance/analysis, (ii) due to restricted power, the computational resources are constrained, for example in terms of CPU cycles, RAM, disk space, NW BW and (iii) the algorithms should be fault tolerant, also, since the data might be incomplete, and uncalibrated, we need to use recursive robust filtering/estimation algorithms. Some of the important aspects in this direction are: (a) intelligent use of the data/RRs, (b) use of multimodal decision theory, (c) fault detection, identification (FDI), and management – FDI of false alarms, (d) methods of multimodal sequence data learning, (e) multimodal signal sampling and filtering and (f) state estimation and pattern recognition under constraints of the resources [5].

9.6.3 Human–Sensor Data Interaction

It is also important to build models and methods for human–sensor/human–computer interaction (HSCI) process, especially if we increasingly use the smart sensors/MEMS/smart processing devices (motes) in WSNs. The idea here is that these interactions systems/SW/algorithms sense human intentions, for example (in/for a) large-security systems. These techniques provide the tools to visualise the RRs/sensors' states, help to diagnose and debug the remote sensors, and configure and manage the SNWs.

9.6.4 Real-World System Development and Deployment

The development and deployment of WSNWs (HW/SW/algorithms) for real-world applications is very crucial for the benefit of society: (i) biomedical health monitoring, diagnostic and therapeutic systems, (ii) sensor systems that are used for crisis management, (iii) large-scale surveillance systems, (iv) tracking/monitoring of mobile units – endangered species/inventory control/transportation and (v) sensor assessment – reliability, verification, validation [5]. All these activities need multimodal sensory data sampling, acquisition, transmission, processing, analysis and interpretation for which very efficient techniques need to be developed.

9.6.5 Multimodal Fusion Methodology

Multimodal fusion (MMF) also provides certain benefits to multimedia analysis situations. In one sense the integration of multimedia, their distinct features, and even intermediate decisions is referred to as MMF in Reference 6 (which is an excellent review work on MMF). In yet another closely related sense, the MMF also covers the aspect of fusion of human signatures, finger prints, face-images in the area of person identification for security reasons. Also, one can use the MMF in case of biometric analysis/fusion using finger prints, face-images and the voice of a person. In such cases one can use FL for decision making. The multimedia data are sensory such as audio, video and RFID and/or web-net resources and databases. Then the idea in MMF is to enhance the total information of the event, system or an environment using such data with differing features and types which then help in providing an accurate overall decision.

Very important features of such an endeavour are: (i) different media are gathered at different frame rates (and formats) adding to the complexity of handling this asynchrony, (ii) the processing times are also different, (iii) the various modalities could be either independent or correlated, which could be appropriately used, (iv) these modalities might have differing confidence levels and (v) cost in units of time or other measures [6]. These characteristics of the MMF pose several difficulties in decision making: (a) deciding the level of fusion that is very suitable for the situation at hand, that is to use feature or decision level fusion, (b) how to fuse these modalities, (c) how to synchronise various data from these modalities and when to fuse – this is due to the fact that these modalities operate at differing rates and (d) what to fuse.

9.6.5.1 Data Fusion Levels

In feature level for MMF, the features are first combined and then sent to a single analysis unit. The feature fusion (FF) combines, say skin colour and motion cues, into a bigger feature vector (FV). This FV becomes an input to the face detector. The most commonly used features are: (i) visual features (colour histogram, blobs), (ii) text features (from automatic speech recogniser, OCR), (iii) audio features (based on STFT/FFT, linear predictive coding), (iv) motion parameters (pixel variation, direction/magnitude histogram) and (v) other supplementary information (names, place of origin, time-instants) [6]. FF utilises the correlations between multiple features extracted from different modalities. The time synchronisation is difficult because the closely coupled modalities might be captured at different time-instants. In decision level fusion, the local decisions are made based on the extracted individual features. These local decisions are fused in decision level fusion. Often hybrid multimodal fusion (HMMF) is utilised. Table 9.1 shows the rule-based DF approaches for multimedia/modality fusion [6].

9.6.5.2 Techniques for Multimodal Fusion

Several techniques of MMF are briefly discussed in this section. In linear weighted fusion, the information from various modalities is combined in a linear fashion. The normalised weights are obtained by several methods: min-max, decimal scaling, sore, *tanh* estimators and sigmoid functions. The extracted features' vectors from various modalities are combined using either sum or product rule (here I signifies features from some images or other feature vectors)

$$I_f = \sum_{i=1}^{n} w_i I_i \tag{9.1}$$

$$I_f = \prod_{i=1}^{n} I_i^{w_i} \tag{9.2}$$

Another method is based on majority voting that we have discussed in Chapter 8. A production rule-based decision level fusion is also used, wherein the weighting factor and a condition–action part are used. The methods so far described can be called rule-based techniques and are domain specific. Hence, we need to have adequate knowledge of the domain.

TABLE 9.1

MMF Methodology Based on Fusion Rule

Fusion Method	Fusion Level	Modalities	Multimedia Task
Linear weighted rule	Feature/s	Video – trajectory coordinates (TC)	Human tracking (HT)
		Video – colour, motion, texture (CMT)	HT
		Video – CMT	Face detection, monologue detection, traffic monitoring
	Decision	Audio (phonemes) and visual	Speaker recognition (SR)
		Audio, video	Spoken word recognition
		Audio, video, synchrony score	Monologue detection, semantic concept detection, annotation in video
		Image (several features)	Image retrieval
		Text, audio, video, motion	Video retrieval (VR)
		Text, video	VR
		Audio, video index	Person identification, from audio–video sources
Majority voting	Decision	Raw speech (patterns)	Speaker identification from audio sources
Custom defined rule	Decision	Visual, text	Semantic sports video indexing
		Speech, 2D gesture	HCI
		Speech, 3D pointing gesture	MM interaction with robot
		Speech, pen gesture	MM dialog system

Source: Adapted and modified from Springer Science+Business Media: *Multimodal Fusion for Multimedia Analysis: A Survey. Multimedia Systems,* 16, 2010, 345–379, Atrey, P. K. et al.

The other MMF methods are based on classification of the MM observations/measurements. The idea is that these observations are classified/categorised/cast into pre-specified classes/patterns. These MMF techniques are: (i) support vector machine (SVM), (ii) Bayesian inference (BI) (Chapter 2), (iii) Dempster–Shafer approach (DS) (Chapter 2), (iv) dynamic Bayesian networks (DBNWs), (v) artificial neural networks (ANNs) (Chapter 10), (vi) and maximum entropy method (MEM) (Chapter 2).

The SVMs are very popular and are being used for feature/text categorisation, concept classification, face detection and MMF. The SVM is a supervised learning method and is considered as an optimal binary linear classifier. In MMF it is used for pattern classification with inputs as the scores from the individual classifiers. As variants the gradient-descent-optimisation linear fusion (GDOF) is used for fusing the kernel-matrices and the super-kernel non-linear fusion (SNLF) is used for the combination of the individual classifier models.

In BI-based MMF, the MM information is combined based on probability theory (Chapter 2). For the BI MMF we have the following expression:

$$p(H|d_1, d_2, ..., d_n) = (1/N) \prod_{k=1}^{n} p(d_k|H)^{w_k}$$

(9.3)

In Equation 9.3 ds are the decisions or the feature vectors from n modalities, then we get p as the jt. pdf of an hypothesis H based on the fused feature vectors/decisions, and the ws are the weights of the individual modalities. The hypothesis with the maximum probability is computed using the following expression:

$$\hat{H} = \arg \max_{H \in E} p(H|d_1, d_2, ..., d_n)$$

(9.4)

The MAP estimation method is used to determine the optimum hypothesis from all the feasible hypotheses E.

IN D-S approach [7,8] the MMs are combined using the following expression:

$$(m_i \oplus m_j)(H) = \frac{\sum_{I_i \cap I_j = H} m_i(I_i)m_j(I_j)}{1 - \sum_{I_i \cap I_j = \varnothing} m_i(I_i)m_j(I_j)}$$

(9.5)

In Equation 9.5, the ms are the belief mass functions like the probabilities (used in BI theory) but are not quite the probabilities, and D-S gives the mass of a hypothesis H based on the two modalities, where the I are the features or the local decisions, as the case may be.

In a DBNW, that is based on the same principle as BI, the nodes represent random variables like audio/video as measurements/states and the edges denote the their probabilistic dependencies.

ANNs can also be used for MMF at features level to fuse the low-level features to recognise the images, or at the decision level to fuse the decisions from multiple trained NN-classifiers.

The MEM can be used as a classifier. It provides a probability of a measurement that belongs to a particular class. This is done on the basis of the information content that a particular measurement has. Let us have two input measurements as u_i and u_j of different types. The probability of their belonging to a class C can be given as follows:

$$P(X|u_i, u_j) = \frac{1}{N(u_i, u_j)} e^{f}$$

(9.6)

In Equation 9.6, we have N as the normalising factor and f is the combined feature or decision vector: $f = f(u_i, u_j)$.

Next, we have estimation (Chapters 4 and 5) based MM fusion methods: (i) Kalman filter and (ii) particle filter-based methods. These methods are useful for the estimation of the states of a moving object using MM data. The MM data could be audio signals, and/or multiple features (spatial position, shape and colour information). The features from several cameras are fused in KF/ EKF to obtain the states of the object. The particle filters (PFs) are a set of simulation based approaches. They obtain the estimates of the state-distribution of the non-Gaussian, non-linear state-space model. They are also known as sequential Monte Carlo methods. The particles are the random samples of the state variable. Each particle is characterised by an associated weight. The time-propagation part of the PF advances each particle as per its dynamic (system-model) and the data-update part of the PF updates the particle based on the current sensor information/data, and the measurement model.

In addition to MMF, we need to evaluate the correlation among different modalities, since this can give additional information that can be advantageously used for the data fusion process (DFP) itself. For this we can compute the correlation coefficients and mutual information (Chapter 2) from differing modalities. If the correlation between certain modalities is high, then, we can use these modalities for the DF, otherwise we do not use these modalities, and similarly for the MI between the modalities, and depending upon the applications, and type of modalities to be combined, the value of the threshold correlation coefficient could be chosen from 0.8 to 1.

9.7 Decision Fusion Rules in WSN

In a WSN a number of sensors collect and (pre-) process the REUD measurements, make local decisions, transmit these decisions over channels that might be fading (as well as might be noisy) to a CFS [9]. At this CFS, these local decisions are processed jointly, and then a final decision is made.

9.7.1 System Model

A three-layer system model can be used: (i) local sensor layer, (ii) fading channel layer and (iii) fusion centre [9]. At the local sensor layer all the M sensors collect the measurements under a specific hypothesis. Then each sensor makes a binary decision: $u_i = 1$ is sent if H_1 is decided and $u_i = -1$ is sent otherwise, with $i = 1, 2, \ldots, M$. The detection performance is characterised by its probability of false alarm and detection, and are specified as P_{fa_i} and P_{d_i} respectively. In the fading channel layer, the local sensor-decisions (u_i) are transmitted over the parallel channels which might fade independently. For the fusion rules the magnitude of the fading channel is considered as constant during the transmission. Then we have the following model for the output of the channel:

$$z_i = x_i + v_i \tag{9.7}$$

$$z_i = h_i u_i + v_i \tag{9.8}$$

We have h (measurement model) as the attenuation of the fading channel, and v is a zero mean Gaussian random variable. Then, the CFS decides which hypothesis is more likely to be true based on the data received z.

9.7.2 Fusion Rules

Optimal likelihood ratio fusion rule is given by the following formula [9]:

$$\Lambda = \sum_{i=1}^{M} \log \left[\frac{P_{d_i} \exp\{-(z_i - h_i)^2/2\sigma^2\} + (1 - P_{d_i})\exp\{-(z_i + h_i)^2/2\sigma^2\}}{P_{fa_i} \exp\{-(z_i - h_i)^2/2\sigma^2\} + (1 - P_{fa_i})\exp\{-(z_i + h_i)^2/2\sigma^2\}} \right] \tag{9.9}$$

In Equation 9.9, σ^2 is the variance of the additive white Gaussian noise for all the channels and we need local sensor performance indices and as well as the channel knowledge.

The expression for the Chair–Varshney fusion rule is given as [9]

$$\Lambda = \sum_{sign(z_i)=1} \log \frac{P_{d_i}}{P_{fa_i}} + \sum_{sign(z_i)=-1} \log \frac{1 - Pd_i}{1 - Pfa_i} \tag{9.10}$$

The rule of Equation 9.10 is mathematically equivalent to the rule of Equation 9.9 for large signal-to-noise ratios (SNRs), and we see that the channel knowledge (h) is not required [9]. The fusion rule using a maximum ratio combining statistic is given as [9]

$$\Lambda = 1/M \left\{ \sum_{i=1}^{M} h_i z_i \right\} \tag{9.11}$$

It only requires the knowledge of the channel gain, and does not need probabilities, though it is assumed that the detection and false alarm probabilities are the same. The fusion rule using an equal gain combining statistic is given as [9]

$$\Lambda = 1/M \left\{ \sum_{i=1}^{M} z_i \right\} \tag{9.12}$$

The rule of Equation 9.12 requires a minimum amount of the information and outperforms the rule of Equations 9.10 and 9.11 in terms of its detection performance [9]. A decision fusion rule that requires only the knowledge of the statistics of the fading channel is fully developed in Reference 9.

9.8 Data Aggregation in WSN

In a WSN, the motes coordinate among themselves to form a communication NW like a single multi-hop or a hierarchical NW organisation which can have several clusters and cluster heads [10]. In such a NW, the data generated from neighbouring sensors are in many cases redundant and highly correlated. Also, these data would be very huge in volume for the BS to process them very effectively. Thus, there is a requirement to combine some of these data into a very high quality of information at the motes or the intermediate (smart) sensor-nodes. This is to be done to reduce the packet transmission so that the energy and the BW are conserved. This process of combining the data in WSN is called DAG. Thus, DAG is the process of gathering and aggregating the data from several sensors in order to eliminate redundant transmission and subsequently provide the fused information to the BS. Also, the DAG might involve primarily the fusion of (certain) data from these intermediate SDSs and then transmission of the aggregated data to the BS. There are several features that one should keep in mind while studying the WSNs and especially the DAG methods [10]: (i) energy efficiency (EE), (ii) network lifetime (NWLT), (iii) data accuracy (DAc) and (iv) latency.

The DAG scheme is considered to be energy efficient, if it maximises the functionality of the NW [10], that is, to minimise the energy consumption of each mote. This is specifically encompassed by the NWLT. The NWLT is formally defined as the number of DAG rounds till the $m\%$ of the SDSs die. The m is to be specified by the WSN designer. The idea is to continue performing the DAG task such that there is uniform drainage of the energy in the NW. The energy efficiency and NWLT are synonymous aspects for improving EE and enhancing the NWLT. In a target localisation situation the DAc is determined by the estimate of the target location at the BS. The delay in data transmission, routing and DAG determines the latency. Thus, the latency is measured as the time delay between the received data packets at the BS and the data generated at the source nodes.

Since, the architecture of the SNWs plays a crucial role in the performance of different DAG protocols, it is important to have a general idea of several DA protocols. These protocols have been specifically designed for different NW architectures. Several such DAG protocols [9] are given in Table 9.2. The comparison of DAG in hierarchical NWs and flat NWs is given in Table 2

TABLE 9.2

Hierarchical Data Aggregation Protocols

Name of the Protocol	Type of the Organisation	Objective	Features
LEACH	Cluster	NWLT to see the number of the nodes that are functional, latency	Randomised cluster head rotation, it has non-uniform drainage of energy across different DSs
HEED	Cluster	NWLT: number of rounds until the first node dies out	Assumption of multiple power levels. Cluster heads are well distributed, achieves better performance than LEACH
PEGASIS	Chain	NWLT: average energy expended by a node	We need the global knowledge of the NW, huge energy saving compared to LEACH
Hierarchical chain based	Chain	Energy × delay	Binary chain-based scheme is eight times better than LEACH and the three level scheme is five times better than PEGASIS
EADAT	Tree	NWLT: number of alive sensors at the end of simulation time	The approach is based on the sink initiated broadcasting
PEDAP-PA	Tree	NWLT: time until the last node dies	The approach is based on minimum spanning tree. Achieves performance improvement two times better than LEACH, PEGASIS

Source: Adapted and modified from Rajagopalan, R. and Varshney, P.K. *Data Aggregation Techniques in Sensor Networks: A Survey*. Department of Electrical Engineering and Computer Science, Syracuse University, NY. http://surface.syr.edu/eecs, accessed April 2013.

of Reference 10. Table 3 in Reference 10 gives the summary of NW flow-based DA algorithms and Table 4 of Reference 10 gives summary of different approaches that characterise the tradeoffs involved in the DAG.

9.9 Hybrid Data and Decision Fusion in WSN

Often in WSNs there is a strong spatial and temporal correlation of measured and transmitted data. Such correlations are characterised by relatively simple physical models: a set of partial differential equations (PDEs) [11]. As we know such PDEs are used to describe the diffusion of a gas in air or of a fluid in water. In a distributed WSN it is worthwhile to have some of the motes processing measurements from the neighbouring motes for robust estimation. Often the model parameters are estimated from such nodes using their own data. Then these resultant estimates are exchanged. In general, the DF outperforms decision fusion. Here, we study these aspects in the context of the WSN.

In this WSN, we consider a finite set of sensor nodes $S = \{s_i\}$ which are deployed over the one dimensional (1D) sensor field $[0, L]$. The motes measure samples of the scalar field $x(\xi,t)$ for $t > 0$, and space $x < \xi < L$. The time evolution can be modelled by the following PDE equation [11]:

$$x_t(\xi,t) = ax_{\xi\xi}(\xi,t) + bx_\xi(\xi,t) + cx(\xi,t) \tag{9.13}$$

We have $\xi = 0, \xi = L$ are the boundary points. We assume that the node measurements are affected by zero mean white Gaussian noise. Our idea is to identify the parameters: a, b and c in Equation 9.13. These estimates are useful for NW prediction of certain phenomena.

9.9.1 Identification of Parameters of the Partial Differential Equation

The estimation of these parameters is carried out at some cluster heads in DF mode. The discrete equivalent model of the PDE is used for parameter estimation using EKF (Chapters 4, 5). The discretisation is done in space as well as time coordinates. Subsequently, the lumped model is used

$$x(k + 1) = \phi(\theta)x(k) + B(\theta)u(k) \tag{9.14}$$

$$z_j(k) = H_jx(k) + v(k) \tag{9.15}$$

In Equation 9.14, the *theta* are the parameters to be estimated and j is the cluster. The cluster heads have the same state-space model given by Equation 9.14, but could have a different measurement model. The unknown parameters are treated as the additional states and a joint/augmented state vector is estimated by using the EKF equations (Section 4.2.6).

9.9.2 Hybrid Data/Decision Fusion Approach: Illustrative Example

The distributed algorithm based on the EKF achieves the so-called parameter estimation and is actually the process of the measurement level DF (Section 4.2.7). This is done by passing their own measurements to the cluster head which achieves the parameter estimation. So, this is the DF process achieved at the cluster head. The transmission of the gathered time series from the member nodes could be quite expensive in terms of energy consumption. This is because a large amount of data samples might be transmitted. In that case one can allow each node in a cluster to process its own measurements and then pass on the parameter estimates to the cluster head. This latter act may be termed as the DeF for the purpose of this section. In this case, the cluster head averages the received estimates by following equation [11]:

$$\hat{\theta}_j = \frac{1}{|S_j|} \sum_{s_l \in S_j} \hat{\theta}_{sl} \tag{9.16}$$

The denominator in Equation 9.16 is the number of member node s_l in the cluster S_j. Often the weighted average DF rule instead of just averaging could be used. Naturally in this case since only the measurements are available from one node, the estimates might not be very accurate. In a hybrid approach some preliminary iterations are carried out within each cluster using the DF strategy. Then, the algorithm switches to the DeF mode. This preserves energy, since for a few initial iterations the data (time series) are transmitted. This needs the criteria for switching mechanism: (i) a function based on the diagonal terms of the covariance matrix or (ii) the number of samples being sent; thus, the algorithm can switch to the decision fusion if either

$$f(p_i(\theta)) < t_1 \tag{9.17}$$

$$n_{it} > t_2 \tag{9.18}$$

The ts are the thresholds and n is the number of samples to the fusion. The condition of Equation 9.18 can be used to restrict excessive battery consumption.

In Reference 11, the authors present the result of the application of the hybrid approach of DF for the case studies: (i) DF for three sensor nodes, (ii) only one sensor node processes its measurements, (iii) DF with three nodes, and then switched after 20 iterations to only one node and (iv) the effect of noise and the location of nodes that sample the field. They also, study the performance with three nodes and compare the results of the DF with the DeF obtained with the average and the weighted rules. The performance metrics are given in Table 9.3. The results with the weighted rules seem to be

TABLE 9.3

Data/Decision Fusion Results – % Errors

Performance Metric	Data Fusion	Decision Fusion	Weighted Decision Fusion
Mean error	0.44	1.85	1.78
STD	3.57	7.98	2.61

Source: Adapted and modified from Rossi, L. A., Krishnamachari, B. and Jay Kuo, C. C. *Vehicular Technology Conference*, 7, 4616–4620, 26–29 September, 2004. www.ceng.usc.edu/~bkrishna/research/papers/VTC2004FallLR.pdf, accessed May 2013.

more acceptable. The weights for the weighted DF rule were taken as proportional to the distance of the nodes to the boundaries.

9.10 Optimal Decision Fusion in WSN

A decision fusion architecture is considered that has a fusion centre (CP), and M distributed sensors. These are the local decision makers. We then say that mth sensor device measures a feature vector x_m [12]. Then, x_m will be assigned to a class label within the set of N feasible labels denoted as $C = \{C_1, C_2, ..., C_N\}$ according to a local decision rule. Thus, we have the following expression for the local decision rule:

$$l_m(x_m) = d_m \in C \tag{9.19}$$

So, we have d_m as the decision, and if $x_m \in C_n$, then $d_m = C_n$. Then, the global feature vector is given as the following composite vector:

$$x = \{x_1^T \; x_2^T \; ... x_N^T\}^T \tag{9.20}$$

As we have seen the local decisions are made using Equation 9.19. The idea is that a decision rule maps a given (local feature vector) into a particular class label. Also, it can be said that the decision rule partitions the feature vector space N into disjoined domains. Then, the feature vectors in each domain are assigned to the same label. Since, there are M sensors, the local decisions are forwarded to CP, where a decision fusion rule is used to compute a final decision.

9.10.1 Complementary Optimal Decision Fusion

The decision fusion is based on M-dim feature vector d. Due the fact there being N classifiers, there will be N^M different decision vectors. As said earlier, these different decision vectors will partition the feature space into the same number of disjoined domains. Additionally, these domains would be assigned to a specific class label by a DeF algorithm. The resultant decision fusion method would be optimal, if the probability of the correct decision assignment is maximised. This can happen if only the decision vector is used and only for the said purpose. In the complementary optimal decision fusion (CDF) method, when the ODF does not work, the non-ODF is used. So, the hybrid approach is used and it is called CDF method. We discuss several CDF rules.

a. Non-weighted threshold voting

The simple method is to have the non-weighted voting as follows:

$$\sum_{i=1}^{M} w_i d_i(x) \quad \begin{bmatrix} l = C_1 \\ > \quad th \\ < \\ l = C_2 \end{bmatrix} \tag{9.21}$$

Here $d_i(x) = 1$ if $x \in C_1$ and $= 0$ if $x \in C_2$. Fusing decisions from the M nodes, $M - 1$ threshold (th) possibilities exist. An optimal threshold is computed to minimise the error

$$th = \arg \min_{0 \le m \le M-1} e(m + 1/2) \tag{9.22}$$

In Equation 9.21, since all the nodes are equally weighted, the performance is likely to be poor.

b. Weighted LS threshold

The weights are determined by using a LS estimation method. Here, we have the following quantities defined:

$$D = [d(x_1), d(x_2), ..., d(x_j)]^T \tag{9.23}$$

Equation 9.23 is the matrix of the decision vectors. We also have

$$d(x) = [d_1(x), d_2(x), ..., d_M(x)] \tag{9.24}$$

as the decision vector for the sample x, and w is the weight vector, so that the following LS estimator is feasible:

$$w = D^{-1} \times l \tag{9.25}$$

In Equation 9.25, l is the label vector and we emphasise that it might be necessary to use the pseudo-inverse. The error is defined as follows:

$$e = l - D \times w \tag{9.26}$$

The error of Equation 9.26 is minimised in the sense of the LS, the solution given in Equation 9.25 is then very obvious as the LS estimates of the weighing coefficients.

c. Optimal linear threshold
   For this, we use the steepest descent method as follows to obtain the
   updates of the weights:

$$w(k + 1) = w(k) - \eta\Delta\{e(k)\} \tag{9.27}$$

The gradient in Equation 9.27 is defined as follows with the error
vector given by Equation 9.26:

$$\Delta\{e(k)\} = \left\{ \frac{\partial e(k)}{\partial w_1(k)}, \frac{\partial e(k)}{\partial w_2(k)}, \cdots, \frac{\partial e(k)}{\partial w_M(k)} \right\}^T \tag{9.28}$$

d. Local classifier accuracy weighting
   The weights are chosen and assigned to the different decisions proportional to their accuracy level. If the classifier is more likely to be correct
   it is assigned a larger weight. These weights are normalised as follows:

$$w_i = \frac{r_i}{\sum_{i=1}^{M} r_i} \tag{9.29}$$

In Equation 9.29, r is the classification rate. If the accuracy of the
classifier remains fairly accurate and constant, then the method
would give acceptable decisions.

e. Follow-the leader-rule
   This method assigns the label assigned by the classifier as most
   likely to be correct as the decision result, as follows:

$$w_i = \begin{bmatrix} 1 & \text{if } i = \arg\max_{1 \le i \le M} r_i \\ 0 & \text{otherwise} \end{bmatrix} \tag{9.30}$$

For this rule also, the behaviour of the classifiers is required to
remain constant among the different sets of samples.

EXERCISES

9.1 Why is multimodal DF in a WSN a generally difficult and com-
     plex task?

9.2 Give one example of a static WSN.

9.3 Give one example of a dynamic WSN.

9.4 Give some examples of WSNs from real-life applications.

9.5   What are some of the activities for which WSNs are used?

9.6   Give an outline or the block diagram of a smart sensor/device that can be used in a WSN.

9.7   Why in some (or many) WSNs do we talk of battery power, limited energy and so on?

9.8   In a WSN, if we want the minimal power loss $(P_t - P_r)$, deduce an approximate relationship between the distance d (of the receiver from the source) and the transmission frequency f_t.

9.9   Why feedback (FB) control is used in a WSN in the context of DA?

9.10   Discuss the role of DAG in DF and DF in DAG in WSNs.

References

1. Kulkarni, R. V. and Venayagamoorthy, G. K. Computational intelligent in wireless sensor networks: A survey. *IEEE Communications Surveys and Tutorials*, 13(1), 68–96, First Quarter 2011.

2. Khaleghi, B. Distributed data fusion in sensor networks. PAMI Research Group, ECE Department, University of Waterloo, accessed August 2012. https://www.google.co.in/webhp?sourceid=chrome-instant&ion=1&espv=2&ie=UTF-8#q=Khaleghi+Distributed+data+fusion+in+sensor+networks+PAMI.

3. Khan, A. Data fusion in sensor networks. University at Buffalo, the State University of New York, accessed August 2012. https://www.google.co.in/webhp?sourceid=chrome-instant&ion=1&espv=2&ie=UTF-8#q=Khan%2C+A.+Data+fusion+in+sensor+networks.

4. Wilson, J. S. Wireless sensor networks: Principles and applications. In *Sensor Technology Handbook*, Chapter 22 (Eds. Townsend, C. and Arms, S.). MicroStrain, Elsevier Inc., Burlington, MA, USA, 2005. http://www.globalspec.com/reference/46556/203279/chapter-22-wireless-sensor-networks-principles-and-applications.

5. Chang E. Y. *Foundations of Large-scale Multimedia Information Management and Retrieval: Mathematics of Perception*. Springer-Verlag, Berlin, Heidelberg, 2011. http:// www.springer.com/gp/ book/ 9783642204289.

6. Atrey, P. K., Hossain, M. A., El Saddik, A. and Kankanhalli M. S. *Multimodal Fusion for Multimedia Analysis: A survey*. Springer Multimedia Systems Journal, Springer-Verlag, NY, USA, 16, 345–379, 2010. www.comp.nus.edu.sg/~mohan/papers/fusion_survey.pdf.

7. Challa, S. and Koks, D. Bayesian and Dempster–Shafer fusion. *Sadhana (Engineering Journal of the Indian Academy of Sciences,* Bangalore, India), *Multisource Multisensor Information Fusion*, Sp. Edition (Guest Ed. Raol, J. R.), 29(2), 145–176, April 2004.

8. Raol, J. R. *Multisensor Data Fusion with MATLAB*. CRC Press, FL, 2010.

9. Niu, R. Chen B. and Varshney, P. K. Decision fusion rules in wireless sensor networks using fading channel statistics. *Conference on Information Sciences and Systems*, The Johns Hopkins University, 12–14 March, 2003.

10. Rajagopalan, R. and Varshney, P. K. *Data Aggregation Techniques in Sensor Networks: A Survey*. Department of Electrical Engineering and Computer Science, Syracuse University, NY. http://surface.syr.edu/eecs, accessed April 2013.

11. Rossi, L. A., Krishnamachari, B. and Jay Kuo, C. C. Hybrid data and decision fusion techniques for model based data gathering in wireless sensor networks. *Vehicular Technology Conference*, 7, 4616–4620, 26–29 September, 2004. www.ceng.usc.edu/~bkrishna/research/papers/VTC2004FallLR.pdf, accessed May 2013.

12. Duarte, M. F. and Hu, Y. H. Optimal decision fusion with application to target detection in wireless ad hoc sensor networks. *IEEE International Conference on Multimedia and Expo* (ICME), 27–30 June 2004, Taipei, Taiwan, 1803–1806, 2004.

10

Soft Computing Approaches to Data Fusion

10.1 Introduction

Soft computing (SC), in fact, greatly differs from the conventional (we can call it hard computing, or better firm computing! for a software [SW] or an algorithm/program resides in the HW) because it is tolerant of (i) imprecision, (ii) uncertainty, (iii) partial truth and (iv) approximation, and the role model for SC is as such the human mind, since the human mind (the humind) does not always take hard decisions [1], it takes soft or often vague decisions. The directive principle of SC is: exploit the tolerance for imprecision (of definitions/expressions of the views), uncertainty in partial decisions/data/ information, (available) partial truth and approximation (of computing new decisions/estimations) to achieve tractability/solvability, robustness and low cost of solutions [1]. The principal paradigms of SC are fuzzy logic (FL), neural computing (artificial neural networks [ANNs], NC), evolutionary computation (EC → GAs), machine learning (ML) and probabilistic reasoning (PR). Our biological neural networks (BNNs) and our many decision-making abilities work as if on the principles of imprecision and uncertainty. The belief networks (NWs) (D-S theory of decision making), chaos theory and parts of learning theory are also the supportive disciplines of some of these components that do or carry out the SC. Hence, the SC is a sharing partnership in which each one of the components contributes a distinct method with its own features and the characteristic way of addressing a problem, in totality, in its domain by using the available information/data that are partially imprecise, uncertain and incomplete. Our BNNs and our (humans) decision-making approaches also always try to make sense out of such data and in such environments, and in such cases we hardly use formal mathematics, be it learning, adaptation and/or data fusion (DF). Thus, SC paradigms in fact then, model our behaviours and activities and also provide computational tools so that we can formally and legitimately use these techniques for solving difficult problems and even develop and build artificial intelligent (AI) systems that work like humans. Such AI-based systems HW/SW/algorithms/procedures) can also work in hazardous situations and areas, and can also perform tasks of data/image processing, decision making and DF.

Hence, the SC paradigms in fact are complementary rather than competitive to handle very complex problems that require the application and building of SC-based AI systems. Also, SC is viewed as a foundation component for the conceptual and computational (artificial) intelligence/AI.

Hence, interestingly enough, the paradigms of ANNs, FL and genetic algorithms (GAs), including the support vector machines (SVM) are collectively called SC approaches to solve many and complex numerical problems in many science, engineering and business-related areas. The fundamental research on these methods and their applications has advanced at a rapid pace during last four/five decades. Also, the techniques of ANN, FL and GA are being considered increasingly for multi-sensory data fusion (MSDF).

The ANNs possesses good ability of learning, using a suitable learning algorithm, patterns/features/inherent characteristics from the experimental data presented to ANN. ANNs also have fault tolerance due to their inherent topology that is a massive one, they have massive parallelism. Of the various types of ANNs, the feed forward neural networks (FFNNs) and recurrent neural networks (RNNs) have found successful applications in References 2–7: (i) pattern recognition, (ii) non-linear curve fitting/mapping, (iii) aero-flight mechanics data analysis, (iv) aircraft modelling for design and performance optimisation, (v) adaptive control, (vi) system identification/parameter and (vii) robotics/mobile vehicles-path planning and motion control.

FL is a multi-valued logic (Chapter 3) used for modelling vagueness in data, this vagueness is quite different from the usual uncertainty that we deal with in the probability models (Chapter 2). We can also incorporate our knowledge of the environment/scenario using fuzzy If … Then … rules into the fuzzy inference system (FIS), that is used in the fuzzy inference engine (FIE) to analyse/design the FL-based control systems. Thus, the FL with its rule-based system and FIS/E (Chapter 3) is a powerful tool to design such systems based on AI. With fuzzy logic type 2 (FLT2/IT2FLS), one can expand the scope of modelling and also representing vagueness in the rules. Because of the inherent structures of ANNs and FL/S, their modelling abilities of linear as well as non-linear dynamic systems and adaptive learning abilities (from experimental/empirical data e.g. adaptively determining the fuzzy rules), these techniques are becoming very important in the development of multi-sensor DF strategies [6].

GAs are optimisation paradigms based on direct-search methods inspired by nature's evolutionary systems, like population reproduction and survival [8]. GAs can obtain (most of the times) global (and robust) parametric solutions to many problems in science and engineering which require optimisation based on some pre-defined cost function. The applications of GAs can be easily extended to the problems of economics, psychology and biology. As we know, the natural systems such as the biological species have evolved over last several million years. Because these biological systems have gone through thousands and thousands of iterations over many generations, their processes and systems have become very robust by way of crossovers, mutation, adaptation and learning. As a result of this strong evolutionary 'experience', nature offers good and robust solutions even if these solutions are not optimal, in the sense of classical term 'optimal'. We find that most or at least several such biological systems are generally more robust, efficient and flexible compared to the most sophisticated man-made and optimal systems. Nature invariably seeks good working solutions rather than optimal ones, and such paradigms like ANNs, FLs and GAs can provide such solutions and systems that are more robust rather than more optimal.

Thus, the appropriate use of each individual system/approach, ANNs, FL/S or GA and/or an appropriate combination/conjunction/aggregation of two or more of these paradigms can be utilised for building efficient and robust MSDF strategies and/or systems, and even parallel computers (based on RNNs/ANNs). In this chapter, we describe these methods with sufficient mathematics and study the possibilities of their use for MSDF.

10.2 Artificial Neural Networks

The ANNs mimic certain simple features of the BNN system, the latter being massively parallel (chemical–electrical) circuits and both consist of very simple processing elements [2,3]. This is not to say that the BNNs are overly simple systems, only the superficial functional levels are simple and similar. These elements are called artificial neurons or neuronal nodes. Important features of a BNN (e.g. in a human nervous system) and ANN are shown in Figure 10.1 [5]. The neurons are connected by the so-called synapses weights. The ANN in its entirety conducts a parallel distributed computing

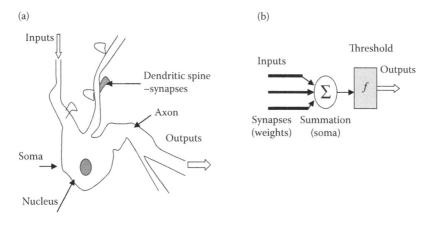

(a)

(b)

FIGURE 10.1
Biological and artificial neural NWs – a simple comparison. (a) Biological neuron-model (BNM) and (b) artificial neuron-model (ANM).

task. The ANNs are useful for modelling of systems from their I/O (input to output) data. The ANNs' weights are trained (obtained) by some suitable training algorithm and this is called learning from data – supervised learning. As such, the ANNs can adequately approximate the system's behaviour in an overall sense. In this chapter, we discuss FFNNs, RRNs and RBNN (radial basis neural network) and see how some of these can be used for DF.

10.2.1 Feed-Forward Neural Networks

The FFNNs are signal-data processing mechanisms of a very large number of simple processing elements as seen in Figure 10.2. They are non-linear black-box modelling topologies, the weights/coefficients of which are determined by any conventional estimation method. The FFNNs have non-cyclic and layered topology – the information flows from say, left to right in the forward direction. They are very suitable for mapping of non-linear systems, can be used for linear systems too, avoiding the non-linear activation function. Basically the FFNN is trained using the training data set. After adequate training is accomplished, it is used for prediction of the outputs from a different input data set (that belongs to the same class of data and was not used in first place for training the FFNN). This validates the estimated model based on FFNN.

10.2.1.1 Back Propagation Algorithm for Training

The weights/coefficients/synapses of the FFNN are estimated using the conventional back propagation-steepest descent (BPSD) gradient optimisation method. Because of the layered arrangement of the weights, the estimation requires propagation of the estimation error (the output error) from the output layer towards the hidden/input layers in the backward direction (i.e. right to left). The following variables/symbols in a typical FFNN are defined: u_0 as the

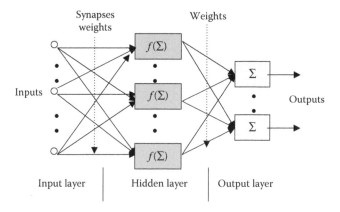

FIGURE 10.2
FFNN – left to right topology with one hidden layer.

actual input signal/data (used for training) to the FFNN, n_i as number of input neurons, n_h as number of the hidden layer neurons, n_o as number of output neurons, W_1 as $n_h \times n_i$ weight matrix connecting input and hidden layers, W_{10} as $n_h \times 1$ bias vector, W_2 as $n_o \times n_h$ weight matrix connecting the hidden and output layers, W_{20} as $n_o \times 1$ output layer bias vector and μ as a learning rate parameter. In the forward pass, from the left to the right, the data processing/computations are carried out using the equations as follows [4] (since the input signal u_0 is known and the initial guesstimates of the weights are already made).

First, the intermediate output from the initial input is computed as follows:

$$y_1 = W_1 u_0 + W_{10} \tag{10.1}$$

Then, we have the intermediate input as the non-linear function of the previous output

$$u_1 = f(y_1) \tag{10.2}$$

Here, y_1 is a vector of intermediate values from the input layer and u_1 is the input to the hidden layer. The $f(y_1)$ is a non-linear sigmoid function represented by

$$f(y_i) = \frac{1 - e^{-\lambda y_i}}{1 + e^{-\lambda y_i}} \tag{10.3}$$

The signal between the hidden and output layers is given as

$$y_2 = W_2 u_1 + W_{20} \tag{10.4}$$

$$u_2 = f(y_2) \tag{10.5}$$

Since, the output ($u_2 = f(y_2)$) can be computed, and the desired output of the dynamic/algebraic system (to be modelled) is known, a quadratic function is defined as follows:

$$J = \frac{1}{2}(z - u_2)(z - u_2)^T \tag{10.6}$$

This function is optimised with respect to the weights to determine the weights of the output layer. Using the steepest descent (SD) method for optimisation, we obtain the following differential equation:

$$\frac{dW}{dt} = -\mu(t) \frac{\partial J(W)}{\partial W} \tag{10.7}$$

for any general weight vector. Since, the output error is defined as $e = z - u_2$, the expression for the gradient of the cost function is given as

$$\frac{\partial J}{\partial W_2} = -f'(y_2)(z - u_2)u_1^T \tag{10.8}$$

Here, u_1 is the gradient of y_2 with respect to W_2 and the derivative f' of the node activation function f is obtained as

$$f'(y_i) = \frac{2\lambda_1 e^{-\lambda_1 y_i}}{(1 + e^{-\lambda y_i})^2} \tag{10.9}$$

Then, the modified error of the output layer is expressed as

$$e_{2b} = f'(y_2)(z - u_2) \tag{10.10}$$

Then, the recursive weight update expression for the output layer is given as

$$W_2(i + 1) = W_2(i) + \mu_2 e_{2b} u_1^T + \Omega[W_2(i) - W_2(i - 1)] \tag{10.11}$$

Equation 10.11 is the discrete-time update rule and the additional constant Ω which is a momentum constant is used to smooth out the (very large) weight changes and to accelerate the convergence of the algorithm. Then, the BP (back propagation) of the output error to the input layer and the weight update rule for the weights W_1 of the input layer are given as

$$e_{1b} = f'(y_1)W_2^T e_{2b} \tag{10.12}$$

$$W_1(i + 1) = W_1(i) + \mu_1 e_{1b} u_0^T + \Omega[W_1(i) - W_1(i - 1)] \tag{10.13}$$

The training data are inputted to the FFNN in a sequential/recursive manner, and then again and again presented, but with (new) initial weights as outputs from the previous cycle. The process is stopped after the convergence is attained, hence the training process is recursive iterative. The values of μ in Equations 10.11 and 10.13, and Ω need not necessarily be the same.

10.2.1.2 Recursive Least Squares Filtering Algorithms

Next, we briefly describe two recursive training algorithms for ANNs: (i) for non-linear output and (ii) linear output, that are essentially based on the KF theory (Chapter 4).

10.2.1.2.1 BP Recursive Least Squares Filtering Algorithm: With Non-Linear Output Layer

In this training algorithm during the forward pass, the signals y and u are computed for each layer as was done in the BP algorithm. The gains K_1 and K_2 are computed for both the layers and the forgetting factors f_1 and f_2 are properly chosen. The usual scalar data processing scheme is given below.

For layer 1, the updates for filter gain K_1 and covariance matrix P_1 are given as [4]

$$K_1 = P_1 u_0 (f_1 + u_0 P_1 u_0)^{-1} \tag{10.14}$$

$$P_1 = (P_1 - K_1 u_0 P_1)/f_1 \tag{10.15}$$

For layer 2, the updates for filter gain K_2 and covariance matrix P_2 are given as [4]

$$K_2 = P_2 u_1 (f_2 + u_1 P_2 u_1)^{-1} \tag{10.16}$$

$$P_2 = (P_2 - K_2 u_1 P_2)/f_2 \tag{10.17}$$

Then the modified output error is given as

$$e_{2b} = f'(y_2)(z - u_2) \tag{10.18}$$

The BP of the output error to inner/hidden layer gives inner layer error as

$$e_{1b} = f'(y_1) W_2^T e_{2b} \tag{10.19}$$

Finally, the weight update rule for the output layer is given as

$$W_2(i + 1) = W_2(i) + (d - y_2) K_2^T \tag{10.20}$$

In the above d is given by

$$d_i = \frac{1}{\lambda} \ln \left[\frac{1 + z_i}{1 - z_i} \right] \quad \text{and } z_i \neq 1 \tag{10.21}$$

For the hidden layer, the weight update rule is

$$W_1(i + 1) = W_1(i) + \mu e_{1b} K_1^T \tag{10.22}$$

In this algorithm, the computations of (Kalman) gains is required, otherwise the procedure for training is similar to the BP algorithm. When weight update rule of Equation 10.22 is used, the range of values of μ is not the same as that when rule of Equation 10.13 is applied.

10.2.1.2.2 BP Recursive Least Squares Filtering Algorithm: With Linear Output Layer

In this linear case, the output layer does not have any non-linearity, but only the inner layer has, and hence, the linear KF concept is directly applicable. Since, the output layer is linear, the output is computed as

$$u_2 = y_2 \tag{10.23}$$

The KF gain computations are the same as discussed in the case above. Since, the output has no non-linearity, the error for output layer is

$$e_{2b} = e_2 = (z - y_2) \tag{10.24}$$

The BP of the output error gives

$$e_{1b} = f(y_1)W_2^T e_{2b} \tag{10.25}$$

The weight update rules are

$$W_2(i + 1) = W_2(i) + e_{2b}K_2^T \tag{10.26}$$

$$W_1(i + 1) = W_1(i) + \mu e_{1b}K_1^T \tag{10.27}$$

The recursive-iterative process of data processing for this learning algorithm is again the same as that of the BP algorithm. The converged/computed output is compared with the desired/available output (used for training/learning) in order to judge the NW's ability for prediction. The same criterion as mean square error (MSE) (Chapters 2 and 4) can be used for this purpose.

10.3 Radial Basis Function Neural Network

The radial basis function (RBF) NW is a part of two-layer ANN. In the ANN then, each hidden unit implements a radial activated function. Then, the output is a weighted sum of outputs from the hidden units. The input (unit) is non-linear whereas the output is linear [9]. The weights are found by a suitable training algorithm using a MSE between the training set output and the NW-predicted output. The RBF NWs have been used for [9]: (i) modelling of chaotic time series, (ii) system identification, (iii) control engineering, (iv) modelling of electronic devices, (v) (communications) channel equalisation, (vi) speech recognition, (vii) image registration, (viii) shape from shading identification, (ix) three-dimensional (3D) object modelling, (x) motion estimation and moving object segmentation and (xi) sensor data fusion (SDF).

10.3.1 NW Structure

Figure 10.3 depicts the RBF NW topology. The RBFs are wired into the two-layer FFNN, with a set of I/Os. It has one hidden layer with certain units and each one of these units is a RBF. Various functions are used as activation functions [9]: (i) thin-plate spline, (ii) Gaussian function and (iii) mixture of Gaussian functions. The Gaussian activation function is given by

$$\phi_1(u) = \exp\left\{-(u - m_i)^T P_i^{-1}(u - m_i)\right\} \tag{10.28}$$

Here, we have u as the input signal-data/features, the hidden layer units as $i = 1,2,\ldots,n_h$.

The m and P are the mean and covariance matrix of the ith Gaussian function, respectively. The RBF geometrically represents a bump in the multi-dimensional space, then the mean vector is the location and P is the shape

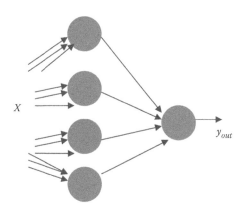

FIGURE 10.3
Radial basis neural network.

of the activation function, and in effect the activation function models a pdf with m and P as the first- and second-order statistics. Thus, the output layer results in the following expression:

$$U_j(u) = \sum_{i=1}^{n_h} W_{ij}\phi_i(u) \tag{10.29}$$

The *W*s are the contribution of the hidden units to the output units, and we have as the number of output units $j = 1,2,\ldots,n_0$.

In pattern classification, the RBF used is a sigmoid function as given below [7]

$$\phi_j(u) = \frac{1}{1 + \exp\{-U_j(u)\}} \tag{10.30}$$

10.3.2 RBF Properties

RBF NW is specified by its localisation and an activation hyper surface, and if it is Gaussian function, then these are characterised by the two parameters: mean and covariance (P). If this matrix P is diagonal, with all its elements equal, then the hyper surface is a hyper sphere, otherwise it is hyper ellipsoid. In the latter case, the influence of the activation function decreases according to the Mahalanobis distance (MD) from the centre of the ellipsoid. The normalised MD is given by the following expression:

$$d(u,v) = (u - v)P^{-1}(u - v)^T \tag{10.31}$$

Here, P is the covariance of the input data. What this simply means is that the samples located at a larger MD from the centre of the RBF will not be able to activate the basis function effectively. Hence, the maximum activation is achieved when the data coincide with the mean vector. The Gaussian basis functions are quasi-orthogonal. Also, the product of the two bases functions is almost zero if their centres are far away from each other. The RBFs can be considered as the potential functions. The hidden units with the output weights *W*s having the same sign for a certain output unit, will have their activation fields joined together, in the same way as the electrical charges of the same sign form the electrical fields, and the weights with the different signs will form the activation fields like the electrical fields of the opposite signs [9]. The RBFs are found to be suitable for interpolation, functional modelling, for modelling the pdfs and even for implementing the Bayesian rule [9].

10.3.3 RBF Training Algorithm

For RBF training the cost function is defined as [9]

$$J = \sum_{n=1}^{N} (U_j(u_n) - V(u_n))^T (U_j(u_n) - V(u_n)) \tag{10.32}$$

Here N is the total numbers of the vectors from the training-set data (TSD). The U is the RBF output vector, and the V is the vector associated with the data samples, u_n, the TSD. The cost function J is minimised with the RBF NW parameters. Several techniques can be used to obtain the optimal training algorithm: (i) orthogonal least squares (LS) method with Gram–Schmidt algorithm, (ii) SD algorithm, (iii) expectation maximisation (EM) algorithm and (iv) learning vector quantisation. The centres of the RBF are initialised randomly. The Euclidean distance is used for adaptation to the closest centre. The centre is updated as follows:

$$\hat{m}_i = \hat{m}_1 + \eta(u_n - \tilde{m}_i) \tag{10.33}$$

The training rate in Equation 10.33 can be taken as the reciprocal of the total number of the data samples. Then, the centre is the classical first-order statistical estimation. Then, the output weights are estimated by using the LS method [4].

10.4 Recurrent Neural Networks

The RNNs have a special structure: the FFNNs with feedback of some or all the output states [4], Figure 10.4. It is shown in Reference 4 that RNNs can be

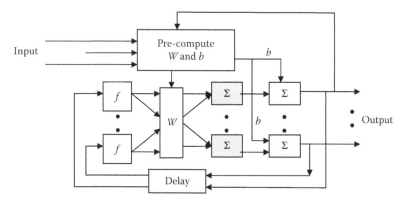

FIGURE 10.4
Recurrent neural network.

used very advantageously to estimate the parameters of any linear (-non-linear) dynamic system from its I/O data. The RNNs are dynamic neural NWs, and hence amenable to explicit parameter estimation in state-space models. The RNNs are very amenable to be incorporated into the equation error formulation and hence, for parameter estimation of dynamic systems. These are batch iterative procedures wherein a set of data is processed to compute the gradient of a cost function and estimation error. The estimates are then refined using an iterative procedure based on the improved estimates of error and its gradients. We study certain variants of RNNs from the point of view of explicit parameter estimation. We see that these variants are related to each other by an affine or linear transformation of their states. These variants are categorised in the way in which the sigmoid non-linearity operates on [4]: (i) the states, (ii) the weighted states, (iii) the residual of the NW signal and (iv) forcing input.

10.4.1 RNN-S/Hopfield Neural Network

This NW variant is the Hopfield neural network (HNN) which has a number of mutually interconnected information processing units called neurons. The outputs of the NW are non-linear function of the states (S) of the NW. The dynamic model of the NW is given as (Figure 10.5a) [4]

$$\dot{x}_i(t) = -x_i(t)R^{-1} + \sum_{j=1}^{n} W_{ij}\beta_j(t) + b_i; \quad j = 1,...,n \tag{10.34}$$

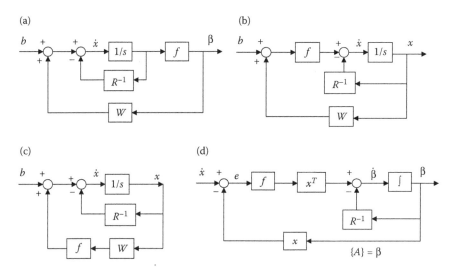

FIGURE 10.5
Various structures of RNN. (a) RNN-S structure, (b) RNN-FI structure, (c) RNN-WS structure, (d) RNN-E structure.

We have x as the internal states, β as the output state, $B_j(t) = f(x_j(t))$, w_{ij} as the neuron weights, b as the bias input to the neurons and f as the sigmoid non-linearity. R (like a resistance in an electrical circuit) is the neuron impedance and n is the dimension of the neuronal state. Equation 10.34 is also written in compact form as

$$\dot{x}(t) = -x(t)R^{-1} + W\{f(x(t))\} + b \tag{10.35}$$

Equation 10.35 represents the 'classical' neuro-dynamics. It obtains a simple system retaining essential features: (i) neuron as a transducer of input to output, (ii) a smooth sigmoidal response up to a maximum level of output and (iii) feedback nature of the connections. Thus, the model retains two aspects: dynamics and non-linearity.

10.4.2 RNN-Forcing Input

In RNN-forcing input (RNN-FI) the non-linearity operates on the forcing input, FI = weighted states + input to the NW → modified input = $f(Wx + b)$. The NW dynamics are given as (Figure 10.5b)

$$\dot{x}_i(t) = -x_i(t)R^{-1} + f\left(\sum_{j=1}^{n} w_{ij}x_j(t) + b_i\right) \tag{10.36}$$

with $f(\cdot) = f(FI)$. This NW is related to RNN-S by an affine transformation as is shown below. We use $x_H(t) = Wx + bR$ in Equation 10.35 to obtain the following [3]:

$$
\begin{aligned}
W\dot{x} &= -(Wx + bR)R^{-1} + Wf(Wx + bR) + b \\
W\dot{x} &= -WxR^{-1} - b + Wf(Wx + bR) + b \\
\dot{x} &= -xR^{-1} + f(Wx + bR) \\
\dot{x} &= -xR^{-1} + f(FI)
\end{aligned} \tag{10.37}
$$

We have *FI* as the modified input vector, due to *bR* term. We assume that W is invertible. We see that Equation 10.37 has exactly the same form as Equation 10.36, the RNN-FI.

10.4.3 RNN-Weighted States

In RNN-weighted states (RNN-WS), the non-linearity operates on the weighted states. The NW dynamics are described by (Figure 10.5c)

$$\dot{x}_i(t) = -x_i(t)R^{-1} + f(s_i) + b_i \tag{10.38}$$

Here, $s_i = \sum_{j=1}^{n} w_{ij}x_j$. It is seen that this NW is related to RNN-S by linear transformation. Substitute $x_H(t) = Wx$ in Equation 10.35 to get

$$W\dot{x} = -(Wx)R^{-1} + Wf(Wx) + b$$
$$\dot{x} = -xR^{-1} + f(s) + W^{-1}b$$

(10.39)

With the modified input vector, a matrix W should be invertible.

10.4.4 RNN-Error

In the RNN-error (RNN-E), the non-linearity directly operates on the residual error or equation error. The function f or its derivative f' does not enter into the dynamic equation. However, it does affect the residuals by way of quantising them and thereby reducing the effect of any measurement outliers, and hence, this NW structure can be used for deriving robust estimators that can handle the output outliers. The NW dynamics are given by (Figure 10.5d)

$$\dot{x}_i(t) = -x_i(t)R^{-1} + \sum_{j=1}^{n} w_{ij}x_j(t) + b_i$$

(10.40)

Here, we say that the internal state x_i is β_i, the parameters of the general dynamic system.

10.5 FL and Systems as SC Paradigm

In Chapter 3, we have discussed the concepts of FL type 1 and 2 (T1FL and IT2FL). However, we further elaborate several (related) aspects of FL here (as a part of the SC paradigms), based on Reference 10. Hence, the material of this section is complementary to that of Chapter 3.

10.5.1 Traditional Logic

In this traditional logic (TL), one main aim is to provide rules that are used to determine whether a particular argument is true of false. Interestingly enough, the language of logic is based on mathematics and the reasoning process is precise and unambiguous (this is more so for the crisp logic that is based on the hard decisions, yes or no, see Chapter 3). The logical arguments consist of certain statements and a statement is a sentence that unambiguously is true or false. Many facts in our daily life are easily expressible by such simple factual statements. We use predicates like: 10 is an even number; we write it as follows:

$$10 \in \{x \mid x \text{ is an even number}\} \text{ or } 10 \in \{x \mid P(x)\} \tag{10.41}$$

In Equation 10.41, the P stands for the property of x. It is known as a predicate. The predicate is not a statement until some particular value is specified. Then its truth or falsity can be verified or worked out. We use the universe of discourse (UOD) to assure that a statement is evaluated for relevant values. For the example of Equation 10.41, the UOD is the set of natural numbers. For example, we have for all x and y that $x^2 - y^2$ is the same as $(x + y)(x - y)$ which can be written in a mathematical form as follows:

$$\forall x, \ y((x, y \in R) \wedge (x^2 - y^2) = (x + y)(x - y)) \tag{10.42}$$

Here, $\wedge$ stands for the logical operation AND and R is UOD (of real numbers). Thus, the predicate of Equation 10.42 is true only for real numbers, and is all quantifier. Thus, the concept of the UOD is very useful in avoiding any logical paradoxes. We also have the existential quantifier interpreted as 'there exists' or 'for some'. It describes a statement as being true for at least one member of the set: $(\exists x)((river\ (x) \wedge name\ (Kaveri)))$. Also, there are a number of connectives whose purpose is to allow us to join predicates or statements in order to form more complicated ones: AND $\wedge$, OR $\vee$ and NOT $\sim$. Actually the NOT is not a connective, for it only applies to a single predicate or statement. In traditional logic the main tools of logical reasoning are tautologies, such as the modus ponens: $(A \wedge (\rightarrow B)) \rightarrow B$, here $\rightarrow$ means 'it implies'. There are associated truth tables based on these connectives. The Aristotelian concepts have been useful but they have the following problems: (i) they cannot express ambiguity, (ii) they lack quantifiers and (iii) cannot handle exceptions. If we consider a temperature as a variable and say 'he has a high temperature' as a predicate, then the ambiguity is what is 'high' to some, is not 'high' to others. The 'most of the people came before noon', has a lack of quantifier, since the 'most' cannot be expressed in the terms of UOD and/or existential quantifiers. One more limitation of the traditional predicate is expressing things that are sometimes true but not always.

10.5.2 Fuzzy Sets and Uncertainty

FL is conceptually based on the fact that a variable high temperature may mean different degree of 'high' to different people, and also it might depend on the application problem. Thus, FL is based on the notion of 'fuzzy sets' (FSs). In traditional logic, the boundaries are very clear and crisp, whereas in FL/S partial membership is permitted. FL/S-based analysis and design involves: fuzzification, evaluation of rules (If ... Then ... rules), and defuzzification. In case of FL/S, the generalised modus ponens is utilised that allows *A* and *B* to be characterised by FSs. The FL permits to violate the laws of non-contradiction, because the element could be a member of more than one set. Also, the intersection of a set with its complement does not necessarily result into an empty set → the excluded middle is not applicable. Every crisp set is a FS, better said as crisp set is a special case of FL/S. A FS *A* is a subset of a UOD *U*, where *A* is characterised by a membership function $\mu_A(x)$. This membership function gives the grade of membership in/for FS *A*. Several FL/S operations have been described in Chapter 3.

SC tolerates uncertainty. Hence, it is in order here to discuss the aspects of uncertainty. That the probability of raining today is 0.9 is a stochastic uncertainty. We will probably have a successful financial year is lexical uncertainty. The uncertainty could be in the form of ambiguity, incompleteness and/or noisy data. The classical-sense uncertainty is dealt with probability (of course very successfully), Bayesian rules, Shannon theory (entropy/lack of information) and D-S belief theory.

10.5.3 Production Rules

As we have seen in Chapter 3, the FIS needs the: (i) fuzzification of the linguistic variables and assigning of a membership value in a FS, (ii) production rules – If ... A, Then ... B, and so on and (iii) defuzzification. The production rules represent the knowledge that a human expert has of the system under analysis/design. We also know that the concept of knowledge is highly ambiguous, and it is our attempt to capture the knowledge in the form of these rules as much as is possible, so that this knowledge is usable in the design of control systems. We have seen in Chapter 3 that the uncertainty in the fuzzy rule base (FRB) cannot be handled by FL type 1, but it can be so done by using T2FL/IT2FL. By defining various (knowledge) production rules the knowledge is captured in some sense so that it can be integrated into the FIS. We can say that the production If ... Then ... rules are modelling the knowledge for easy understanding and proper use in building FL/S-based expert/AI control systems. This process of building FIS and FL/S-based systems is very much a part of the SC philosophy as against hard computing, because now the SC allows the mathematical representation of uncertainty, ambiguity and vagueness in the computational procedures. The knowledge could be based on [10]: rules, facts, truths, reasons, defaults, heuristics and the knowledge engineer utilises some techniques for capturing the knowledge in respect of the problem at hand.

In the If ... A, Then ... B rule the left-hand side 'If' represents the antecedent or conditional part. The right-hand side of the rule 'Then' signifies the conclusion or action (consequence) part. The rules are used to encode the empirical associations between the incoming data-patterns/feature vectors and the actions that are required to be performed by the system as a consequent part. These rules can be derived by using induction learning methods. In traditional production rule only one rule gets executed, even though more may apply. In FL/FIS all the applicable rules contribute to the output. Since the FIS embeds more knowledge and information, fewer rules might be required.

10.5.4 Linguistic Variables

FL/S and FIS use the linguistic variables [10]: (i) name (x), (ii) term set ($T(x)$), (iii) UOD $\rightarrow X$, (iv) syntactic rule $\rightarrow$ generates the names of the values of x and (v) semantic rule for associating its meaning with each value. If x is a temperature, the term set T (temperature) could be

$$T(temperature) \; = \; \{cold, \; cool, \; warm, \; hot\} \; over \; the \; UOD = \{0, 250\} \qquad (10.43)$$

10.5.5 Fuzzy Logic and Functions

Thus, in essence the FL is the collection of: (i) (fuzzy) MFs, (ii) specification of If … Then … rules, (iii) method of combining the FL outputs: FL operators, FIFs and aggregation methods $\rightarrow$ definition of FSs, operations, identities and transformations and (iv) the process of final defuzzification. The details of these have been discussed in Chapter 3. So, in essence the word fuzzy refers to the linguistic uncertainty $\rightarrow$ the word *tall* $\rightarrow$ means a range/variation is introduced $\rightarrow$ a grade is described. Thus, a FL rule uses MFs as variables.

10.5.6 Fuzzy Hedges

As we know cold, cool, warm and hot are the linguistic values of the linguistic variable temperature, see expression (10.43), in general a value of a linguistic variable is a composite term $x = x_1, x_2, …, x_n$, where each u is an atomic term. Then, from one atomic term we can create more terms by utilising the concept of *hedges* $\rightarrow$ like *very, most, rather, slightly, more or less* and so on. Hence, with the hedges we can create even a larger set than that in Equation 10.43. We can define these hedges by the process of normalisation, intensifier, concentration and dilation, for example, using concentration, we *very x* is defined as

$$very \; x \; = \; x^2 \quad and \; very \; very \; x = x^4 \qquad (10.44)$$

Assume, we have the variable defined as $x = 1/0 + 0.7/20 + 0.3/40 + 0/60 + 0/80 + 0/100$; then we have the following expressions for various hedges [10]:

a. *very slow* $= x^2 = 1.0/0 + 0.49/20 + 0.09/40 + 0.0/60 + 0.0/80 + 0.0/100$
b. *very very slow* $= x^4 = 1.0/0 + 0.24/0 + 0.008/40 + 0.0/60 + 0.0/80$
$\qquad + 0.0/100$
c. *more or less slow* $= x^{0.5} = 1.0/0 + 0.837/20 + 0.548/40 + 0.0/60 + 0.0/80$
$\qquad + 0.0/100$

$$(10.45)$$

The hedge *rather* is a linguistic modifier that moves each membership by an appropriate amount, and we can have (after setting that amount to unity)

d. *rather slow* $= 0.7/0 + 0.3/20 + 0.0/40 + 0.0/60 + 0.0/80$ $\qquad\qquad (10.46)$

The *slow but not very slow* is a modification that is using the connective *but*, which in turn is an intersection operator, and the membership functions in discrete forms are found as follows [10]:

e. *slow* $= 1.0/0 + 0.7/20 + 0.3/40 + 0.0/60 + 0.0/80 + 0.0/100$
f. *very slow* $= 1.0/0 + 0.49/20 + 0.09/40 + 0.0/60 + 0.0/80 + 0.0/100$
g. *not very slow* $= 0.0/0 + 0.51/20 + 0.91/40 + 1.0/60 + 1.0/80 + 1.0/100$
h. *slow but not very slow* $= min(slow, \ not \ very \ slow) = 0.0/0 + 0.51/20$
$\qquad + 0.3/40 + 0.0/60 + 0.0/80 + 0.0/100$

$$(10.47)$$

The *slightly* hedge is the FS operator for intersection acting on the FSs *plus slow* and 'not' (*very slow*); *slightly slow* = Int (norm(*plus slow* and NOT *very slow*)). Here, *plus slow* is *slow* to the power of 1.25, and is the intersection operator [10]:

i. *slow* $= 1.0/0 + 0.7/20 + 0.3/40 + 0.0/60 + 0.0/80 + 0.0/100$
j. *plus slow* $= 1.0/0 + 0.64/20 + 0.222/40 + 0.0/60 + 0.0/80 + 0.0/100$
k. *not very slow* $= 0.0/0 + 0.51/20 + 0.91/40 + 1.0/60 + 1.0/80 + 1.0/100$
l. *plus slow and not very slow* $= min(plus \ slow, \ not \ very \ slow) = 0.0/0$
$\qquad + 0.51/20 + 0.222/40 + 0.0/60 + 0.0/80 + 0.0/100$
m. *norm* (*plus slow and not very slow*) $=$ (*plus slow and not very slow*/*max*)
$\qquad = 0.0/0 + 1.0/20 + 0.435/40 + 0.0/60 + 0.0/80 + 0.0/100$
n. *slightly slow* $= int (norm) = 0.0/0 + 1.0/20 + 0.87/40 + 0.0/60$
$\qquad + 0.0/80 + 0.0/100$

$$(10.48)$$

With a study of Equations 10.44 through 10.48, we seem to be in a much better situation to appreciate the meaning of the syntactic and semantic aspects of the rules. A syntactic rule defines, in a recursive fashion, more term sets by using a *hedge* → *T(slow)* = {*slow, very slow, very very slow,...* }. The semantic rule defines the meaning of terms such as *very slow* which can be defined as *very slow = (slow)²*. We can obviously either generate new hedges or modify the meaning of existing ones.

10.5.7 FIS/FLC Developmental Process

The steps involved in the developmental process of analysis/design of FL-based inference system-cum-control system are given next [10]:

1. To establish the characteristics of the system, and to define the specific operating properties of the proposed fuzzy model:
   a. To establish step 1 one can use the traditional systems analysis and knowledge engineering techniques; the analyst-cum-designer should identify the relevant and appropriate inputs to the system including the basic transformations if required; also identify what outputs are expected from the system under evaluation/design process.
   b. The designer should decide if the FLCS is a subsystem of a global system; if so then define also where this sub-system fits into the global architecture; the FLCS could be the global system itself.
   c. The numerical ranges of I/Os should be specified.
2. To decompose each control variable (I/Os) into FS and to give unique names to them:
   d. The number of labels associated with a control variable should generally be odd and between five and nine.
   e. In order to obtain a smooth transition from a state to another each label should overlap somewhat with its neighbours; overlapping of 10%–50% is fine.
   f. The density of the FS should be highest around the optimal control point of the system and eventually thin out as the distance from that point increases.
3. To obtain the production rules that relate the input values the output values:
   g. Begin with an exhaustive and good list of If ... Then ... rules and deal with the redundant, impossible and implausible rules later; if the number of rules increases may be several rule bases might be constructed; each rule base to deal with a particular situation/ condition of the system to be modelled.

    h. Since each rule 'declares' a small quantum of knowledge, the order in the knowledge base (KB) is unimportant.

    i. To maintain the KB one should group the rules by their premise variable; how many rules is obviously dependent on the application; it is related to the number of control variables.

4. To decide on the way to be used in order to convert an output FS into a crisp solution variable:

    j. As seen in Chapter 3, there are many ways to perform the conversion but by and large, the control applications use the centroid method.

    k. Remaining steps are similar to any modelling exercise; at the end of the fuzzy system construction the process of simulation begins; the model is compared against known test cases and the results are used/refined for validation until the desired performance is achieved.

10.6 FL in Kalman Filter for Image-Centroid Tracking: A Type of Fusion

We consider in this section the FL-based target-image (TI) centroid detection and tracking algorithm (FL-CDTA) that combines both the object and motion recognition methods for practical target tracking situations from imaging sensors – an implicit way of (data–information–image) fusion [6].

    In TI tracking, an algorithm analyses the image sequences and outputs the location states and velocity of the moving target within the image sequence. We use FL in KF as a part of some adaptive process. Detection and tracking of the TI involves: (i) intra-scan (single image) level – identifying potential targets by image segmentation methods, calculation of centroid of the identified target and (ii) inter-scan (between images) level – tracking centroid using single or multiple target tracking techniques, separation of true and false targets by association based on both object and motion characteristics. We use centroid tracking algorithm using nearest neighbour (NN) (NNKF, Chapter 5) method, with the combination of FL and KF. The pixel intensity is discretised into several layers of grey levels and it is assumed that sufficient target pixel intensities are within the limits of certain target layers. The algorithm implementation involves the conversion of the data from the image into a binary image by applying upper and lower threshold limits for the target layers. The binary TI is then converted to clusters using NN criterion. For the known target size, the information is used to set limits for removing those clusters that differ sufficiently from the size of the target cluster to reduce computational burden. Then the centroid of the clusters is calculated and this information is used for tracking the target.

10.6.1 Segmentation/Centroid Detection Method

The two kinds of segmentation are: texture segmentation and particle segmentation (Chapter 7). In texture segmentation, an image is partitioned into different micro-regions and each region is defined by a set of feature characteristics. In particle segmentation, an image is partitioned into object regions and background regions, and the segmentation is a task of extracting object or particles of interest as precisely as possible from the image. In FL-CDTA algorithm, particle segmentation is used to separate the target, that is, the object of interest, from the background, when the target is not fully visible. The pixel intensities are discretised into 256 grey levels. The particle segmentation is done in two steps: (a) the grey level image is transformed into a binary image using lower and upper threshold limits of the target, the thresholds of target are determined using the pixel intensity histograms from the target and its surroundings and (b) the detected pixels are grouped into clusters with NN technique, the grey image $Im(i,j)$ is converted into a binary image with intensity $\beta(i,j)$ by a hard limit on the intensity [6]:

$$\beta(i, j) = \begin{cases} 1 & I_L \leq Im(i, j) \leq I_U \\ 0 & \text{otherwise} \end{cases} \tag{10.49}$$

Here, I_L and I_U are the lower and upper threshold limits of the target intensity. The detection probability (DP) of the pixel (i,j) is defined as [6]

$$\begin{aligned} P\{\beta(i, j) = 1\} &= p(i, j) \\ P\{\beta(i, j) = 0\} &= 1 - p(i, j) \end{aligned} \tag{10.50}$$

Here, $p(i, j) = (1/\sigma\sqrt{2\pi})\int_{I_L}^{I_U} e^{(-(x-\mu)^2)/2\sigma^2}\, dx$, considering the grey image $I(i,j)$ as having a Gaussian probability distribution/density (pdf) with mean μ and variance σ^2. A pixel is considered as belonging to the cluster only if the distance between this pixel and at least one other pixel of the cluster is less than the proximity distance d_p. The d_p is chosen as

$$\sqrt{\frac{1}{p_t}} < d_p < \sqrt{\frac{1}{p_v}} \tag{10.51}$$

Here, p_t and p_v are DPs of target and noise pixels, respectively. By choosing the proximity distance a fewer noise clusters are obtained. In practice, it is better to use d_p close to $\sqrt{1/p_t}$ to minimise the gaps in the TI. The centroid of the cluster is determined using non-convolution method as [6]

$$(x_c, y_c) = \frac{1}{\sum_{i=1}^{n}\sum_{j=1}^{m} I_{ij}} \left(\sum_{i=1}^{n}\sum_{j=1}^{m} iI(i,j), \sum_{i=1}^{n}\sum_{j=1}^{m} jI(i,j) \right) \tag{10.52}$$

Here, (x_c, y_c) is the centroid of the cluster, I_{ij} is the intensity of the (i,j)th pixel and n and m are the dimensions of the cluster. After centroid computation the location of the moving object is tracked using FLKF (described in Section 10.6.2) from frame to frame.

10.6.2 FL-Based KF

In this case study, we use FL concept at the measurement update level of the KF processing cycle for TI-centroid tracking. We call this process as fuzzy error mapping (FEM) and the filter as FL-based KF (FLKF). The modified measurement update equation is given by

$$\hat{X}(k+1|k+1) = \tilde{X}(k+1|k) + KC(k+1) \tag{10.53}$$

Here, $C(k+1)$ is FEM vector as an output from FIS and is a non-linear function of the innovation vector e and its time derivative (e-dot). We use FLKF in target tracking application and assume that only position (x–y-axis) measurements of TI are available. Hence, the FEM vector consists of modified innovation sequence for x- and y-axis, that is,

$$C(k+1) = [c_x(k+1) \quad c_y(k+1)] \tag{10.54}$$

To compute $C(k+1)$, the innovation vector e is first separated into its x and y components, μ_n and e_y. We assume that TI motion in each axis is independent and develop the FEM for the x direction and then generalise the result to include y direction. We generalise this application by using the (first-order) numerical derivative of the innovation sequence and hence the FEM vector consists of two inputs, that is, e_x and $\dot{e}_x$, and single output $c_x(k+1)$, where $\dot{e}_x$ is computed by forward finite difference method

$$\dot{e}_x = \frac{e_x(k+1) - e_x(k)}{T} \tag{10.55}$$

10.6.3 Realisation of Fuzzy Error Mapping Using FIS

The various input/output MFs are as per Chapter 7 of Reference 6. The main difference between the work of this section and Reference 6 is that we use FL in the KF wherein the tracking data are the TIs. The labels used in linguistic variables to define membership functions are: LN (large negative), MN (medium negative), SN (small negative), ZE (zero error), SP (small positive), MP (medium positive) and LP (large positive). The rules for the inference in FIS are based on the past experiences and intuitions of the analyst. For example, one such rule is

$$\text{IF} \quad e_x \text{ is } LP \quad \text{AND } e_y \text{ is } LP \quad \text{THEN } c_x \text{ is } LP \tag{10.56}$$

This particular rule is created based on the fact that having e_x and $\dot{e}_x$ with large positive values indicate an increase in innovation sequence at faster rate. The future value of e_x and hence of $\dot{e}_x$ can be reduced by increasing the present value of c_x ($\approx Z - H\tilde{X}$) with a large magnitude (in fact Table 9.15 of Reference 5 summarises the 49 rules needed to implement FEM, whereas Table 9.16 of the same reference gives just four important rules for the similar task). The output c_x, at any instant of time is computed using the inputs e_x and $\dot{e}_x$, input membership functions, rules mentioned in this table, aggregator, and defuzzification.

10.6.4 Simulation of Synthetic Image Data

The mathematical model of FLIR (forward looking infrared sensor) for the generation of a synthetic image is used. Consider two-dimensional (2D) array of pixels

$$m = m_\xi \times m_\eta \tag{10.57}$$

where each pixel is represented by a single index $i = 1,2,\ldots,m$ and the intensity I of pixel i is given by

$$I_i = s_i + n_i \tag{10.58}$$

Here, s_i is the target intensity and n_i is the noise intensity in pixel i, which is assumed to be Gaussian with zero mean and covariance σ^2. The total TI related intensity is given by

$$s = \sum_{i=1}^{m} s_i \tag{10.59}$$

If the number of pixels covered by the target is denoted by m_s, then the average target intensity over its extent is given by

$$\mu_s = \frac{s}{m_s} \tag{10.60}$$

The average pixel signal-to-noise ratio (SNR) (over the extent of the target) is

$$r' = \frac{\mu_s}{\sigma} \tag{10.61}$$

Using the above procedure the synthetic images in a frame is generated with the following inputs: (i) target pixel intensity (s_i): $N(\mu_t, \sigma_t^2)$, (ii) noise pixel intensity (n_i): $N(\mu_n, \sigma_n^2)$, (iii) TI – rectangle (with base NX and height NY) and (iv) position of the target in each scan (x-position and y-position). The motion of the TI in the frame is numerically simulated by using kinematic models of target motion. Constant velocity (CV) kinematic model is used for generation of the data which determines the position of the target in each scan.

10.6.4.1 State Model

The state model used to describe the CV target motion is given by

$$X(k+1) = \begin{bmatrix} 1 & T & 0 & 0 \\ 0 & 1 & 0 & 0 \\ 0 & 0 & 1 & T \\ 0 & 0 & 0 & 1 \end{bmatrix} X(k) + \begin{bmatrix} T^2/2 & 0 \\ T & 0 \\ 0 & T^2/2 \\ 0 & T \end{bmatrix} w(k) \tag{10.62}$$

Here $X(k) = [x \;\; \dot{x} \;\; y \;\; \dot{y}]^T$, T is sampling period and $w(k)$ as zero mean Gaussian noise with variance $Q = \begin{bmatrix} \sigma_w^2 & 0 \\ 0 & \sigma_w^2 \end{bmatrix}$

10.6.4.2 Measurement Model

The measurement model is given by

$$Z(k+1) = \begin{bmatrix} 1 & 0 & 0 & 0 \\ 0 & 0 & 1 & 0 \end{bmatrix} X(k+1) + v(k+1) \tag{10.63}$$

Here $v(k)$ is the centroid measurement noise with zero mean and covariance matrix as

$$R = \begin{bmatrix} \sigma_x^2 & 0 \\ 0 & \sigma_y^2 \end{bmatrix} \qquad (10.64)$$

Both the process noise and centroid measurement noise are assumed to be uncorrelated with each other.

10.6.5 Gating and NN Data Association

The NN method is applicable for data association in various applications such as image processing, sonar, wireless communications and radar. The steps are: (i) consider all entries as the new targets, (ii) find the corresponded entries to each target by considering gating intervals around the past position of each target (the entries inside the gating intervals are related to the corresponded targets), (iii) choose the nearest entry of each interval as the corresponding final observation of each target, (iv) run the tracking algorithm to estimate the paths based on their past position and the corresponded observations to each target and (v) all paths which are smaller than a pre-determined length should be removed. The gating method has been discussed in Chapters 4 and 5. The measurement falling within the gate is to be considered as a valid measurement for the data association whether it may or may not be related to the TI. In gating the problem is which measurement is to be taken for updation. This is the situation where data association comes into the picture. Let us consider measurement received from the sensor at scan k as given by

$$Z(k + 1) = HX(k + 1) + v(k + 1) \qquad (10.65)$$

At scan k, the FLKF tracking forms the prediction $\tilde{X}(k + 1 | k + 1)$ for the state vector $X(k)$ to be used at time $k + 1$ (Chapter 4) and is given by

$$\hat{X}(k + 1 | k + 1) = \tilde{X}(k + 1 | k) + KC(k + 1) \qquad (10.66)$$

Here, $C(k + 1) = [c_x(k + 1) \quad c_y(k + 1)]$
The associated residual covariance matrix S is given by

$$S(k + 1) = H\tilde{P}(k + 1 | k)H^T + R \qquad (10.67)$$

It is assumed that the measurement $z(k+1)$ at frame $k+1$ conditioned on z^k is Gaussian

$$p[z(k+1)|z^k] = N[z(k+1)\ \tilde{z}(k+1|k),\ S(k+1)] \qquad (10.68)$$

Here $N[.,.]$ is the Gaussian pdf with argument the vector-valued random variable $z(k+1)$, mean $\tilde{z}(k+1|k)$ and innovation covariance matrix $S(k+1)$. Then the true measurement will be in the region defined by the equation

$$V(k+1,\ \gamma) \cong \left\{ [z(k+1) - \tilde{z}(k+1|k)]^T S^{-1}(k+1)[z(k+1) - \tilde{z}(k+1|k)] \leq \gamma \right\} \qquad (10.69)$$

$$= \vartheta^T(k+1)S^{-1}(k+1)\vartheta(k+1) \qquad (10.70)$$

Here, γ is the gate threshold that determines the probability that the measurement $z(k+1)$ is inside the gate. The quantity $\vartheta^T(k+1)S^{-1}(k+1)\vartheta(k+1)$ is the sum of the squares of n_z independent Gaussian variables with zero mean and unit standard deviation, and hence the quantity is $\chi^2_{n_z}$ distributed. The gate threshold γ can be obtained from the tables of the χ^2 distribution.

10.6.6 Simulation of FL-CDTA: Illustrative Example

A 2D array of 64×64 pixels is considered for the background image, which is modelled as white Gaussian random field with mean μ_n and variance σ_n^2, that is, $N(\mu_n, \sigma_n^2)$. A 2D array of pixels, which is modelled as white Gaussian random field with a mean μ_t and variance σ_t^2 (i.e.) is used to generate a rectangular target of size (9×9). We use 49 rules to build up FEM for this case. The total number of scans is 50 and image frame rate is 1 frame/s. The initial state vector of the target in the image frame is $X = [x\ \dot{x}\ y\ \dot{y}]^T = [10\ 0\ 10\ 0]^T$. Various time-histories, state errors and innovations are shown in Figures 10.2 through 10.4 (generated by the MATLAB code **imtrackFLDFM.m**). Because the images are very noise, the estimates are a little bit noisy but still the FL-CDTA tracks the target. Table 10.1 shows the performance metrics. The normalised state error and the normalised innovations error are 1.4117 and 0.6684 respectively. From Figures 10.6 through 10.8 and Table 10.1 it can be observed that various performance parameters are within acceptable limits and hence show the consistency of the FL-CDTA for TI tracking.

TABLE 10.1

PFEs in Both x- and y-Position and the (%) RMS Position/Velocity Errors (FL-CDTA)

PFEx	PFEy	RMSPE	RMSVE
0.7079	0.5819	0.3512	0.01108

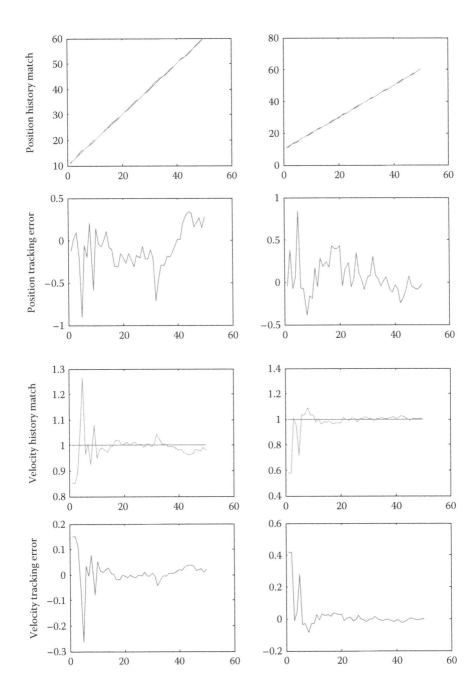

FIGURE 10.6
Position/velocity time history match/tracking error (FL-CDTA).

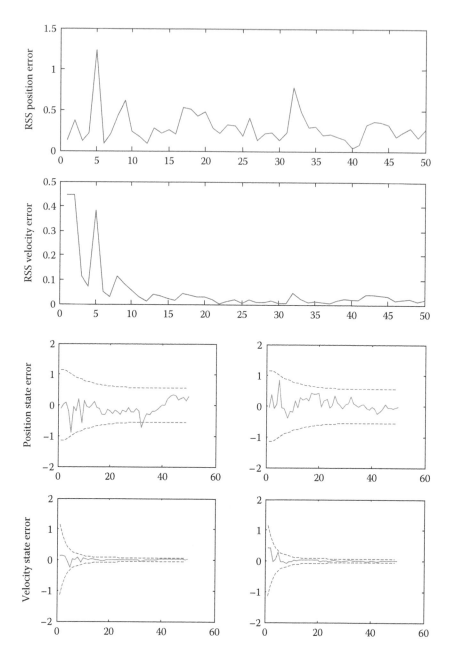

FIGURE 10.7
RSS errors and theoretical bounds (FL-CDTA).

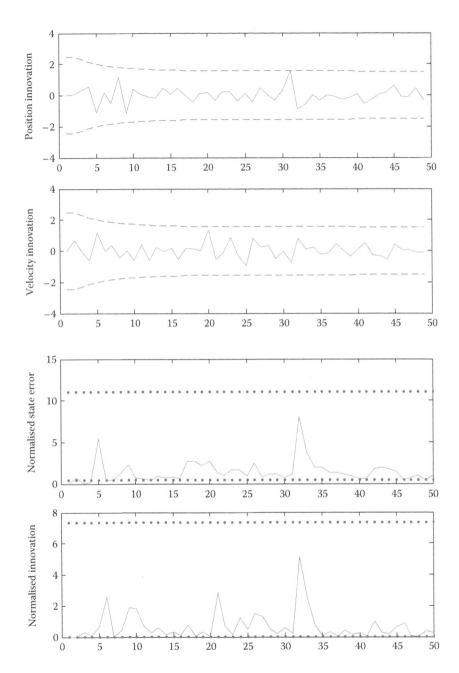

FIGURE 10.8
Innovations and state errors (FL-CDTA).

10.7 Genetic Algorithms

The GAs are part of the evolutionary computation (EC) that simulates evolution on a digital computer. From such a simulation emerges certain optimisation algorithms. These are based on certain simple rules. The goal is to obtain a feasible solution of the optimisation problem. The EC mimics the biological evolution and is a search procedure that probabilistically applies search operators in the search space to obtain the solution. Thus, the GAs are computational optimisation schemes with an approach that seems rather unconventional. These algorithms solve the problems imitating nature in the way that nature has been working for millions of years on the evolution of life forms/biological species. Based on biological systems, GAs adopt the rules of natural selection and genetics to obtain robust solutions [4,8]. In fact the solutions are supposed to be globally optimal (but not always) and robust. A simple procedure based on the premise of survival of the fittest (from nature's evolution mechanism), a population or samples of feasible solutions is combined in a manner similar to the combination of chromosomes in a natural genetic system. The fitter population members/samples pass on their structures as genes in a far greater measure than their less fit samples do. As the generations evolve, the net effect is evolution of the population towards an optimum. Thus, the GAs operate by combining the information present in different possible solutions such that a better solution is obtained in the next/future generations.

10.7.1 Biological Basis

The genetic information is contained in the nucleus of the biological-animal cell. The so-called chromosomes are part of the nucleus cell. These chromosomes store the genetic information. Each chromosome is built of DNA. These form pairs in humans and there are 23 such pairs. These chromosomes are divided into parts called genes. The possibilities of the genes for one property, is called 'allele' and every gene has a unique position on the chromosomes and is called 'locus'. The reproduction is the process of recombination or crossover and mutation. This process is repeated until a robust and stable solution is found that is judged based on the concept of fitness.

10.7.2 Components and Operations in GA

We next describe briefly the constituents of a typical GA.

10.7.2.1 Chromosomes

These represent encoding of information in a string of fixed (finite) length and each one consists of a string of bits (binary digit: 0 or 1); it could be a symbol from a set of more than two elements; generally for function optimisation the chromosomes are constructed from binary strings. These chromosomes consist of genes, and each gene represents a unit of information and it can take different values. Thus, the strings composed of features or detectors, assume values like 0 or 1 that are located at different positions in the string. The total package or system, that is, the combination is called the genotype or structure. The phenotype occurs when the interaction of genotype with environment takes place.

10.7.2.2 Population and Fitness

GA operates on population of possible samples/(candidate) solutions with chromosomes. The population members are known as individuals/samples. Each individual/sample is assigned a fitness-value based on the objective function. Better individuals (samples/solutions) have higher fitness values and weaker ones have lower fitness-values.

10.7.2.3 Initialisation and Reproduction

A population of possible initial solutions is created by randomly selecting information from the search space and encoding it. In reproduction process the individual strings are copied as per their fitness values. The strings with a greater fitness-value have higher probability of contributing one or more offsprings to the next generation. The ones with very low fitness values are neglected.

10.7.2.4 Crossover

In a crossover process, a site/location is selected randomly along the length of the chromosomes (i.e. onto its encoded length of bits), and each chromosome is split into two pieces at this crossover site. The new strings are formed by joining the top piece of one chromosome with the tailpiece of the other in the crossover process. Also, it is possible to use the same string of the chromosome and just exchange some consecutive bits from left to right and right to left.

10.7.2.5 Mutation

In this small operation a bit (only one bit) in a string is changed at a random location (0 is flipped to 1 or 1 is flipped to 0). The idea is to break the monotony and add a bit of novelty. The mutation helps gain information not available to the rest of the population and lends diversity to the population.

10.7.2.6 Generation

Each iteration in the GA-based optimisation process is called a generation, and in each generation pairs are chosen for crossover operation, the fitness is determined, and mutation is carried out during the crossover process. Then a new population evolves which is carried forward.

10.7.2.7 Survival of the Fittest

The individuals/samples/candidates may be fitter or weaker than some other population members. So, they must be ranked as per their fitness value. In each generation, the weaker ones are allowed to wither out and the ones with good fitness values take further part in the genetic operation. The final result is the evolution of the population towards the global optimum.

10.7.2.8 Cost Function, Decision Variables and Search Space

In many optimisation problems, the goal is to find optimal parameters to increase the production and/or to reduce the expenditure/loss. That is done to get maximum profit by reorganising the system/its parameters that affect the cost function. The parameters of the system that decide the cost are called the decision variables. The search space is a Euclidean space in which the parameters take different values and each point in the space is a probable solution, and is subject to evaluation using some norm in the Euclidean vector space.

10.7.3 Steps in Genetic Algorithm

A simple GA is described, which uses a binary coding.

First define domain of the variables, fitness function, the crossover probability and mutation probability. Then carry out the following steps:

Step 1: Create a population of N samples from a chosen search space denoting the decision variables. Generate this randomly, say: x_1, x_2, ..., x_N. Produce series of 0s and 1s to create chromosomes of fixed length, that is, encoding the decision variables.

Step 2: Calculate the cost function values and assign fitness (values) to each member: $f(x_1), f(x_2), ..., f(x_N)$.

Step 3: Sort the members accordingly to their respective fitness values; the parent chromosomes might be selected with a probability proportional to their fitness values.

Step 4: Carry out crossover operation taking two chromosomes at a time, and mutate the chromosomes with a chosen probability of mutation.

Step 5: Evaluate these new chromosomes based on the fitness function, and retain the best members of the population (with higher fitness values) and remove the weaker members with the lower fitness values.

Step 6: Replace the old generation by the just created new one (and with some new samples to retain the number of samples as the same, since the weaker samples are removed from the population) and repeat the process until convergence is obtained.

10.7.4 Other Aspects of GAs

We need to know when to stop the GA iterations. If we choose the population size as small and fixed, then more generations are needed for the convergence of a GA to an optimal solution. We can track the fitness value and stop the GA if the value further does not improve. We can also specify definite pre-determined number of generation/iterations. We can choose insignificant change in the norm of the estimated parameters. We can evaluate gradient of the cost function and use the conventional approaches for assessing the quality of the estimates and stop the GA. The GA becomes more involved in coding the chromosomes by zeros and ones for more complex problems. For the problems of science and engineering we can use real numbers. For the crossover one can use the average of the two samples. Once the best individual is selected a mutation is used – a very small amount of noise is added, or a small change in one digit is made. For example, after crossover, we obtain the new individual as $(\beta_1 + \beta_2)/2$; and for mutation one would obtain $\beta_3 = \beta_1 + \varepsilon^* v$, where ε is a constant and v is a number chosen randomly between –1 and 1. Thus, the feature of using the GA without coding (but, with real numbers) is very well suited for all engineering problems: parameter estimation, control, optimisation and signal processing.

The GA can be used for multimodal, multi-dimensional and multi-objective optimisation problems in science and engineering, and in business fields. As such the computations in GA operations are very simple, but they tend to become very complex as the number of iterations grows. So, the GA procedure can be parallelised and the power of the parallel computers can be advantageously utilised. The GA can work on population samples concurrently and hence their natural parallelism can be exploited to implement them on parallel computers [4].

10.8 SDF Approaches Using SC Methods: Illustrative Examples

We discuss here certain aspects of sensor/DF using ANNs and GAs. In these cases, we consider the DF as the fusion of two estimated parameter vectors (called parameter vector fusion, PVF), like the state vector fusion (SVF, Chapter 4). However, depending on the situation one can consider either data level fusion or (parameter-) estimation level fusion. If one uses the SIMO or

MIMO state-space model of the dynamic system for parameter estimation, then the RNNs (also including FFNNs), and the GA automatically become the (measurements-) data level fuser like the KF as shown in Section 4.2.7. However, currently we consider only the parameter vector fusion (PVF).

10.8.1 DF for Parameter Estimation Using RNN

First, we derive the procedure for parameter estimation of a linear dynamic system using RNN (HNN). We consider the following system:

$$\dot{x} = Ax + Bu; \quad x(0) = x_0 \tag{10.71}$$

For parameter estimation using RNN (HNN) we have $\beta = \{A,B\}$ as the parameter vector (of dimension $nx1$) to be estimated. We consider the dynamics being affected by the non-linear function f, that is, $\beta_i = f(x_i)$, and define the following cost function [4]:

$$E(\beta) = 1/2 \sum_{k=1}^{N} e^T(k)\, e(k) = 1/2 \sum_{k=1}^{N} (\dot{x} - Ax - Bu)^T (\dot{x} - Ax - Bu) \tag{10.72}$$

Then we have $e(k)$ as the equation error expression

$$e = \dot{x} - Ax - Bu \tag{10.73}$$

Using the SD method, we have

$$\frac{d\beta}{dt} = -\frac{\partial E(\beta)}{\partial \beta} = -\frac{1}{2} \frac{\partial \left\{ \sum_{k=1}^{N} e^T(k)\, e(k) \right\}}{\partial \beta} \tag{10.74}$$

Here, β contains the elements of A and B, hence, we obtain expressions $\partial E/\partial A$ and $\partial E/\partial B$ for A and B vectors, with $\Sigma(\cdot) = \sum_{k=1}^{N}(\cdot)$:

$$\frac{\partial E}{\partial A} = \Sigma(\dot{x} - Ax - Bu)(-x^T) = A\Sigma x\, x^T + B\, \Sigma u\, x^T - \Sigma \dot{x}\, x^T$$

$$\frac{\partial E}{\partial B} = \Sigma(\dot{x} - Ax - Bu)(-u) = A\Sigma x\, u + B\, \Sigma u^2 - \Sigma \dot{x}\, u \tag{10.75}$$

Further expanding the expression of Equation 10.75, we obtain, for $A(2,2)$ and $B(2,1)$

$$\begin{bmatrix} \dfrac{\partial E}{\partial a_{11}} & \dfrac{\partial E}{\partial a_{12}} \\ \dfrac{\partial E}{\partial a_{21}} & \dfrac{\partial E}{\partial a_{22}} \end{bmatrix} = \begin{bmatrix} a_{11} & a_{12} \\ a_{12} & a_{22} \end{bmatrix} \begin{bmatrix} \Sigma x_1^2 & \Sigma x_1 x_2 \\ \Sigma x_2 x_1 & \Sigma x_2^2 \end{bmatrix} + \begin{bmatrix} b_1 \\ b_2 \end{bmatrix} \begin{bmatrix} \Sigma u x_1 & \Sigma u x_2 \end{bmatrix}$$

$$- \begin{bmatrix} \Sigma \dot{x}_1 x_1 & \Sigma \dot{x}_1 x_2 \\ \Sigma \dot{x}_2 x_1 & \Sigma \dot{x}_2 x_2 \end{bmatrix} \tag{10.76}$$

For simplicity of explanation and clarity, we use 2 DOF (degree of freedom) state-model in Equation 10.76 and simplifying, we obtain the following expressions for the gradients [4]:

$$\frac{\partial E}{\partial a_{11}} = a_{11}\Sigma x_1^2 + a_{12}\Sigma x_2 x_1 + b_1 \Sigma x_1 u - \Sigma \dot{x}_1 x_1$$

$$\frac{\partial E}{\partial a_{12}} = a_{11}\Sigma x_1 x_2 + a_{12}\Sigma x_2^2 + b_1 \Sigma u x_2 - \Sigma \dot{x}_1 x_2$$

$$\frac{\partial E}{\partial a_{21}} = a_{21}\Sigma x_1^2 + a_{22}\Sigma x_2 x_1 + b_2 \Sigma u x_1 - \Sigma \dot{x}_2 x_1 \tag{10.77}$$

$$\frac{\partial E}{\partial a_{22}} = a_{21}\Sigma x_1 x_2 + a_{22}\Sigma x_2^2 + b_2 \Sigma u x_2 - \Sigma \dot{x}_2 x_2$$

Also, we have for parameter of B

$$\frac{\partial E}{\partial b_1} = a_{11}\Sigma x_1 u + a_{12}\Sigma x_2 u + b_1 \Sigma u^2 - \Sigma \dot{x}_1 u$$

$$\frac{\partial E}{\partial b_2} = a_{21}\Sigma x_1 u + a_{22}\Sigma x_2 u + b_2 \Sigma u^2 - \Sigma \dot{x}_2 u \tag{10.78}$$

We assume that the impedance R is very high, we have the following dynamics of RNN-S:

$$\dot{x}_i = \sum_{j=1}^{n} w_{ij}\beta_j + b_i \tag{10.79}$$

For RNN-S (HNN), we also have $E = -1/2 \Sigma_i \Sigma_j W_{ij}\beta_i\beta_j - \Sigma_i b_i\beta_i$ as the energy landscape, and we get

$$\frac{\partial E}{\partial \beta_i} = -\sum_{j=1}^{n} w_{ij}\beta_j - b_i; \quad \text{or} \quad \frac{\partial E}{\partial \beta_i} = -\left[\sum_{j=1}^{n} w_{ij}\beta_j + b_i\right] = -\dot{x}_i \quad \text{or} \quad \dot{x}_i = -\frac{\partial E}{\partial \beta_i} \tag{10.80}$$

Since, we have

$$\beta_i = f(x_i), \quad \dot{x}_i = (f^{-1})'\dot{\beta}_i \tag{10.81}$$

Thus, we have

$$(f^{-1})'\dot{\beta}_i = -\frac{\partial E}{\partial \beta_i} \tag{10.82}$$

Here ''' denotes the derivative w.r.t. β. Hence, we have parameter estimation equation as

$$\dot{\beta}_i = \frac{1}{(f^{-1})'(\beta_i)} \frac{\partial E}{\partial \beta_i} = \frac{1}{(f^{-1})'(\beta_i)} \left[\sum_{j=1}^{n} w_{ij}\beta_j + b_i \right] \tag{10.83}$$

Finally we get the expressions for the weight matrix W and the bias vector b as [4]

$$W = - \begin{bmatrix} \Sigma x_1^2 & \Sigma x_2 x_1 & 0 & 0 & \Sigma u x_1 & 0 \\ \Sigma x_1 x_2 & \Sigma x_2^2 & 0 & 0 & \Sigma u x_2 & 0 \\ 0 & 0 & \Sigma x_1^2 & \Sigma x_2 x_1 & 0 & \Sigma u x_1 \\ 0 & 0 & \Sigma x_1 x_2 & \Sigma x_2^2 & 0 & \Sigma u x_2 \\ \Sigma x_1 u & \Sigma x_2 u & 0 & 0 & \Sigma u^2 & 0 \\ 0 & 0 & \Sigma x_1 u & \Sigma x_2 u & 0 & \Sigma u^2 \end{bmatrix} \tag{10.84}$$

$$b = - \begin{bmatrix} \Sigma \dot{x}_1 x_1 \\ \Sigma \dot{x}_1 x_2 \\ \Sigma \dot{x}_2 x_1 \\ \Sigma \dot{x}_2 x_2 \\ \Sigma \dot{x}_1 u \\ \Sigma \dot{x}_2 u \end{bmatrix} \tag{10.85}$$

The parameter estimation algorithm based on equation error-RNN scheme for a dynamic system can be described as: (i) compute the weight matrix W, and the bias vector b, since the measurements of x, $\dot{x}$ and u are available for a certain time interval T; (ii) randomly choose the initial values of β_i and (iii) solve the following differential equation (10.86): as $\beta_i = f(x_i)$ and the sigmoid non-linearity is a known function, by differentiating and simplifying Equation 10.83, we get [4]

$$\frac{d\beta_i}{dt} = \frac{\lambda(\rho^2 - \beta_i^2)}{2\rho}\left[\sum_{j=1}^{n} w_{ij}\beta_j + b_i\right] \tag{10.86}$$

with

$$f(x_i) = \rho\left(\frac{1 - e^{-\lambda x_i}}{1 + e^{-\lambda x_i}}\right) \tag{10.87}$$

Thus, we see that the integration of Equation 10.86 yields the solution to the parameter estimation problem based on equation error-RNN the structure. Tuning is achieved by proper selection of λ and ρ. Often λ is chosen less than 1.0. The scheme is termed as non-recursive, since the computations of elements of W and b are performed by considering all the data. The discrete form of Equation 10.86 is given as

$$\beta_i(k+1) = \beta_i(k) + \frac{\lambda(\rho^2 - \beta_i^2(k))}{2\rho}\left[\sum_{j=1}^{n} w_{ij}\beta_j(k) + b_j\right] \tag{10.88}$$

The recursive scheme for parameter estimation using RNN-EE formulation is given in Reference 4.

10.8.1.1 DF: Parameter Vector Fusion

The DF for parameters is illustrated using the RNN parameter estimator of Equation 10.89. Consider the 2 DOF state-space model described by [4]

$$\dot{x} = \begin{bmatrix} -0.7531 & 1 \\ -1.3760 & -1.1183 \end{bmatrix} x + \begin{bmatrix} 0 \\ -2.49 \end{bmatrix} u \tag{10.89}$$

The 100 data samples are created using a doublet input, u, and initial state of the system with $x(0) = [0.1\ 0.01]$. The parameter estimation was carried out for various SNRs. The data are generated using one SNR and then PE carried out. Then the data are generated using another SNR, and the PE carried out (as per the pairs shown in Table 10.1). The tuning parameters λ and ρ were used as 0.1 and 100, respectively, for generating the results. Then the following fusion rules are used:

$$Betaf\ (fused) = 0.5\ (beta1 + beta2) \tag{10.90}$$

$$Betaf\ (fused) = (peen(2) * beta1 + peen(1) * beta2)/(peen(1) + peen(2)) \tag{10.91}$$

TABLE 10.2

RNN-Based PE and Fusion with Different Rules

Case	SNRs (Realisation R1 and R2)	PVF by Average Fusion Rule – Final Peen	PVF by Peen-Weighted Fusion Rule – Final Peen
1	5, 10 R1	8.7279	7.9467
	5, 10 R2	8.2477	7.4938
2	10, 15 R1	5.1721	4.9970
	10, 15 R2	4.8339	4.6580
3	5, 15 R1	7.7504	6.1186
	5, 15 R2	7.3189	5.7246
4	2, 5 R1	17.1664	15.1923
	2, 5 R2	16.2815	14.3961

The results of the fusion of parameter estimates are shown in Table 10.2. The final peen is computed using the true parameters and the fused parameters. (The peen is the parameter estimation error norm.) The parameter estimation results are generated using the MATLAB code **parestrnn1JRRDF.m** and related modules. The realisations (R1, R2) are generated using different seed numbers (for R1, the seed number = 1234; and for R2, the seed number = 4321) for the measurement noise processes (with different SNRs), simulating as if the data have arrived from two and yet similar-type sensors. We see that the results of the fusion of parameter estimates using the peen-weighted fusion rule are better than the fusion with the average fusion rule. The fusion attempted here is like the SVF, Chapter 4.

10.8.2 DF for Parameter Estimation Using GA

The application of GA to parameter estimation does not primarily need the utilisation of the gradient of the cost function. We define the data equation for parameter estimation as follows:

$$z = H\beta + v; \quad \hat{z} = H\hat{\beta} \tag{10.92}$$

The cost function is given as

$$E = \frac{1}{2}\Sigma(z - \hat{z})^T(z - \hat{z}) = \frac{1}{2}\Sigma(z - H\hat{\beta})^T(z - H\hat{\beta}) \tag{10.93}$$

In GA, we use Equation 10.93 or its equivalent as the fitness function [4]. Fitness value:

$$\left[\frac{1}{2}\sum_{k=1}^{N}(z(k) - \hat{z}(k))^T \hat{R}^{-1}(z(k) - \hat{z}(k)) + \frac{N}{2}\ln(|\hat{R}|)\right]^{-1} \tag{10.94}$$

where

$$\hat{R} = \frac{1}{N} \sum_{k=1}^{N} (z(k) - \hat{z}(k))(z(k) - \hat{z}(k))^T \qquad (10.95)$$

We consider 3 DOF state-space system described by [4]

$$\begin{bmatrix} \dot{x}_1 \\ \dot{x}_2 \\ \dot{x}_3 \end{bmatrix} = \begin{bmatrix} -2 & 0 & 1 \\ 1 & -2 & 0 \\ 1 & 1 & -1 \end{bmatrix} \begin{bmatrix} x_1 \\ x_2 \\ x_3 \end{bmatrix} + \begin{bmatrix} 1 \\ 0 \\ 1 \end{bmatrix} u \qquad (10.96)$$

The measurement equation/model is given by

$$z = [2 \quad 1 \quad -1] \begin{bmatrix} x_1 \\ x_2 \\ x_3 \end{bmatrix} \qquad (10.97)$$

The data for the dynamic system are generated with doublet input, u, and with total simulation time of 20 s ($\Delta t = 0.1$ s; no. of data samples = 200). The estimation of 15 parameters is done by using file **parestgaJRRDF.m** and related modules. Other constants are: the initial state of the system, $x(0)$ =[10 1 0.1]; POPSIZE = 20 (number of sets of parameters/population size); and MAXITER = 30 (number of GA iterations). The results of the fusion of parameter estimates (like SVF) using GA are shown in Table 10.3. The rules used are from Equations 10.90 and 10.91, and the data realisations were generated in the similar manner as for the example of estimation with RNN. We see that the parameters are estimated with very good accuracies, and hence except for the two realisations, there is no difference between the two fusion rules for the PVF.

TABLE 10.3

GA-Based PE and Fusion with Different Rules

Case	SNRs (Realisation R1 and R2)	PVF by Average Fusion Rule – Final Peen	PVF by Peen-Weighted Fusion Rule – Final Peen
1	2, 5 R1	0.4313	0.4304
	2, 5 R2	0.4649	0.4646
2	1, 2 R1	0.4294	0.4294
	1, 2 R2	0.4608	0.4608

10.8.3 Multiple Neural NWs Using DF Method for Non-Linear Process Modelling

Control and supervision of advanced industrial control and fusion systems require accurate mathematical/computational models. Often ANNs are employed for this modelling tasks [11]. Often, the combination of multiple ANNs is used to improve the generalisation performance of the ANNs. The candidate ANN in the aggregated NW structure would model the same relationship (plant model) but is developed using different data sets and/or different NW-training techniques/algorithms, or each ANN could have a different structure. The idea is not to select one best ANN structure, but all the working ANNs are combined, since the best one might not be the really best to the unseen data. In Reference 11, DF technique is used to combine these multiple ANNs. A proper model is selected at each sampling instant using Bayesian approach. The probability of a candidate ANN being a (nearly) true model is determined using the SSE (sum of squared error) and using a sliding window (SDW). The ANN with the highest probability is selected. In DF situation, we have three broad approaches [11]: (i) positional (kinematic) fusion, (ii) (object) identity fusion and (iii) ancillary support algorithm. In Reference 11, the decision level identity fusion (DLIF) based on the Bayesian approach is employed for combining ANNs, where for the joint identity declarations will be the final prediction from the aggregated ANNs – a joint system. As such there are several methods for decision fusion [11]: (i) classical inference, (ii) Bayesian inference (BI), (iii) D-S theory, (iv) generalised evidence and (v) heuristic method. In Reference 11, the second one is used to obtain a good DF process/system. From various sensors we get the individual identity declarations, and then all these identity declarations are combined using a Bayesian combination formula, which then drives the decision logic to output the final fused identity declaration.

10.8.3.1 Aggregation of ANNs Using DF Method: Illustrative Example

In the BI, we have $H_1, H_2, \ldots, H_i$ that represent hypotheses that explain an event E, say measurement (i.e. an observation of the object/process), then we can compute the posterior probability of the hypothesis H_i being true, $P(H_i|E)$, based on the evidence E

$$P(H_i|E) = \frac{P(E|H_i)P(H_i)}{\sum_i (P(E|H_i))P(H_i)} \tag{10.98}$$

The $P(H_i)$ is the a priori probability that the H_i is true. $P(E|H_i)$ is the probability of observing the evidence E such that H_i is true. The philosophy of Equation 10.98 can be extended to the number of ANNs for combination.

For the dynamic process modelling we can represent the individual NW models as follows:

$$y_t = y_t^k[y(t-1),...,y(t-n_0),u(t-1),...,u(t-m_i)] + e_t^k, \quad k = 1,2,...,n \qquad (10.99)$$

The combined NW model (output) is represented as follows:

$$y_t^a = w_t^1 y_t^1 + w_t^2 y_t^2 + \cdots + w_t^n y_t^n \qquad (10.100)$$

Here, the predicted output is given by the weighted fusion rule. In Reference 11, a NN method is used to estimate the likely errors that would have occurred at discrete time t. In the NN the data point of the given model is compared with the model input data in the training and testing sets. Then the NN of this data point is obtained. The NW prediction error for this data point is likely to be close to that of its NN. Thus, we use the error of NN as an estimate of the error in the NW. We can also use the average of the several NN errors. We can then compute the $P^k(t-1)$ using the estimated SSE of the individual NWs within a SDW. Then we can get the $P^k(t)$ as follows [11]:

$$p_t^k = \frac{(1/(2\pi\sigma)^{0.5})p_{t-1}^k e^{-\left(y_t - y_t^k/\sigma\right)^2}}{\sum_{m=1}^{n} (1/(2\pi\sigma)^{0.5})p_{t-1}^m e^{-\left(y_t - y_t^k/\sigma\right)^2}} \qquad (10.101)$$

The prior predictions are used as an individual declaration from the different sensors. These priors give some information about the previous history of the data as we can see from the following equations:

$$SSE^k(t) = \sum_{j=t-w}^{t-1} [e^k(j)]^2 \qquad (10.102)$$

$$p_{(t-1)}^k = \frac{1/(SSE^k(t))}{\sum_{j=1}^{n} 1/(SSE^j(t))} \qquad (10.103)$$

Let us have the data sets {X, Y} with X (px1) vector of the inputs and Y (qx1) vector of the outputs, with N as the number of data samples. In order to develop the aggregated NW-model (with n-models) one can re-sample the original data sets to form n (say, $n = 20$) replication of the original data set.

Subsequently, the BI is used to select the final value of the prediction output with the maximum value of the posterior probability as follows:

$$\hat{y}_t^a = \arg\ \max(p_t^1, p_t^2, ..., p_t^n) \tag{10.104}$$

If the computed probability for the kth NW is the largest, then the prediction from this NW is considered as the prediction from the combined NW, Equation 10.104. Some 20 NWs with fixed structures and 20 NWs with various structures were developed for modelling dynamic non-linear processes [11]. The data are rearranged in the following form:

$$X = [y(L-1) \quad ... \quad y(L-n_0) \quad u(L-1) \quad ... \quad u(L-m_i)$$
$$y(L) \quad ... \quad y(L-n_0+1) \quad u(L) \quad ... \quad u(L-m_i+1)$$
$$\cdot \qquad\qquad \cdot \qquad\qquad \cdot \qquad\qquad \cdot$$
$$\cdot \qquad\qquad \cdot \qquad\qquad \cdot \qquad\qquad \cdot \tag{10.105}$$
$$\cdot \qquad\qquad \cdot \qquad\qquad \cdot \qquad\qquad \cdot$$
$$y(N-1) \quad ... \quad y(N-n_0) \quad u(N-1) \quad ... \quad u(N-m_i)]$$

$$Y = [y(L) \quad y(L+1) \quad ... \quad y(N)]^T \tag{10.106}$$

In Equations 10.105 and 10.106, we have $L = \max(m_i, n_0) + 1$. Then every individual NW, single layer and FFNN, was trained by the Levenberg–Marquardt algorithm. The hidden layer used the sigmoid activation function whereas the output layer used the linear one. All the data were scaled to zero and STD equal to 1. The data were divided into (a) training set, (b) validation set and (c) unseen validation set. The NW with the lowest SSE on the testing set and the lowest SSE on the unseen data were used. The long-range predictions as discussed in Reference 11 were used to study and evaluate the performance of the process model building. Three DF methods were used as follows:

10.8.3.1.1 Average of All the NWs
This DF rule is given as

$$\hat{y} = \frac{1}{n}[\hat{y}_1(t) + \hat{y}_2(t) + \cdots + \hat{y}_n(t)] \tag{10.107}$$

All the NWs are given the same weight.

10.8.3.1.2 Bayesian Combination-Predictor Rule
$$\hat{y}(t) = p_t^1 \hat{y}(t) + p_t^2 \hat{y}_2(t) + \cdots + p_t^n \hat{y}_n(t) \tag{10.108}$$

For Equation 10.108, the weights are calculated using Equations 10.101 through 10.103.

10.8.3.1.3 DF of the NWs

For this case, Equation 10.104 $\hat{y}_t^a = \arg\max(p_t^1, p_t^2, \ldots, p_t^n)$ is used.

The three case studies discussed in Reference 11 are: (i) modelling of reactant concentration in an irreversible exothermic reaction process, (ii) modelling of pH in a neutralisation process and (iii) modelling of real-world data for water discharged in the Langat River. The various results indicated that [11]: (i) the long-range predictions were very accurate and (ii) overall, the DF-based combination scheme gave significantly improved generalisation capability of the combination NWs since they obtained much more accurate long-range predictions.

10.8.4 DF Using Neural NWs: Illustrative Example

In recent times, the ANN and FL-based method for MSDF/DF has gained much impetus [12]. In this approach the neurons are trained to represent the sensor data, and then via 'recall' the combinations can be activated in response to different sensor stimuli [12]. Thus, in Reference 12, a scheme of measurement fusion using a self-organising neural NW is studied. A single layer neuron with Hebbian-type learning/adaptation can be seen to evolve into a filter for the first PC (principal component, Chapter 6) for the input distribution [12]. The model output is a linear combination of its inputs and is given as follows:

$$y = \sum_{i=1}^{m} w_i x_i \tag{10.109}$$

In Equation 10.109, we have x as the inputs to the NW, and there are m weights. The learning rule is given as follows:

$$\omega_i(k+1) = \omega_i(k) + \eta \, y(k) x_i(k), \quad i = 1, 2, \ldots, m \tag{10.110}$$

The learning rule with the effect of normalisation [12] to curtail the unnecessary growth of the weights is given as

$$\omega_i(k+1) = \omega_i(k) + \eta \, y(k)[x_i(k) - y(k)\omega_i(k)], \quad i = 1, 2, \ldots, m \tag{10.111}$$

The rule of Equation 10.111 has two important terms [12]: (i) positive feedback for self-amplification for evolution of the weights according to its external input, and (ii) negative feedback via $-y(k)$ for controlling the unnecessary growth of the weights. In Reference 12, the results of the conventional and

TABLE 10.4

Comparison of the Performance of the DF Rules

Errors-Metrics (MSE) in (.) Axis	MSF1	MSF2	SVF1	SVF2	NNWF
x	0.1045	0.0671	0.3960	0.03498	0.000085
y	0.3009	0.2759	0.5943	0.05849	0.000399
z	0.2199	0.1655	0.5212	0.06621	0.000048

Source: Adapted from Yadiah, N., Singh, L., Bapi, R.S. et al. In *International Joint Conference on Neural Networks*, Sheraton Vancouver Wall Centre Hotel, Vancouver, BC, Canada, 16–21 July 2006.

ANN-based DF schemes for the practical data from four radars are given. For the conventional methods the radar measurement data in the polar coordinates were used. The estimates of the states x, y, z of the rocket were obtained using KF. Also, the ANN in the measurement/data level fusion was used for the same problem of the rocket tracking with Hebbian learning rule with the learning rate as 0.1. The conventional methods used are: (i) the measurement level fusion (MLF) method based on the combined observations (Chapter 4), (ii) the weighted-MLF2 (Appendix B.5), (iii) the conventional SVF (Chapter 4) and (iv) the modified SVF (Appendix B.6). These MSDF rules are discussed in Chapter 4 and Appendix B. Table 10.4 shows the comparison of the performances of the five DF methods for the rocket-target tracking using the real data [12]. The performance of neural NW (NNWF) seems to be very encouraging.

10.9 Machine Learning

The machine learning algorithms (MLA) has emerged as a field of statistical-computational theories of learning processes [13]. It is a very vast subject and is being treated as a separate discipline and subject, and hence here only a brief introduction is given to indicate a possible use in DF. The designed MLA are now routinely used in many systems: (i) speech processing, (ii) computer vision, (iii) data mining and (iv) many related commercial/business applications, especially the discipline of business data-analytics, or more sophisticatedly recently called system analytics. The system analytics is concerned with the application of mathematical model building for understanding business processes, and then forecasting and controlling these processes so that much greater benefits can be derived from this businesses. The MLA concerns itself with the question of how to build the computer systems/software/algorithms/procedures/processes that automatically improve with the experience [13]. It also seeks to answer the question of what the fundamental laws/rules that govern all the learning processes are.

These questions for the field for MLA cover various tasks as how: (a) to design autonomous mobile robots that learn to navigate from their own experience, that is, find their own path and reach the goal/destination; (b) to undertake data mining of the historical medical records to learn which future patients will respond best to what kind of treatment and (c) to build search engines that automatically customise to their users' interest. These learning tasks are known as data mining, autonomous data/pattern discovery, database updating and programming by example. We learn many processes from nature, the environment, and from our past/gathered experiences to solve the problems that we face today, and we also predict the course of actions to solve the problems that are most likely to be faced in future. Machine learning is also considered as an area of AI and is concerned with the study and use of computer algorithms that improve automatically with experience [14].

10.9.1 Taxonomy of the Machine Learning

The ML(A) taxonomy can be outlined next [14]. First aspect is to build or postulate a full or partial probabilistic model for the situation at hand. The nonprobabilistic model then pertains to finding a discriminant or the so-called regression function that can be further used for developing the proper algorithms and generate the results of analysis. Then one makes reasoning from the observed training cases to the general rules. This reasoning is then applied to the test cases. This process is called the induction. The transduction process is then reasoning from the observed specific training cases to the specific test cases. So basically we start from a priori knowledge and assumptions about the problem at hand. Then we estimate the model, by iteration over the initially postulated model by using some estimation method, for example, regression/LS, for which we could use the TSD called induction. From this estimated model we deduce the predicted output. Going from training data to the predicted output is the process of transduction. There are mainly three types of ML: (i) supervised learning – the examples from the training data are used to train the algorithm, which learns a function, the learner then should be able to generalise from the presented data to unseen examples, (ii) unsupervised learning – the algorithm only received the data from the input vector space, then a mathematical model is fitted to these observations and (iii) reinforce learning – here, an agent is used to explore the environment, then it receives an award (positive or negative), the agent only knows this, but how is not told, and maybe at the end of the process/game the agent knows. The processing of the training data could be in a batch or online learning mode. The task of the ML could be data classification or regression. The desirable features of the ML algorithm/method are: (a) it should be simple and simple solutions are favoured, (b) it should be a powerful technique to learn the solutions of a given problem, (c) the algorithm should be stable in iterations, and the error should not grow, (d) it should converge in a finite time and (e) it should be scalable to then number of training examples, input features and test examples [14].

10.9.2 Machine Learning Diversity Metric for DF

Interestingly the supervised machine learning (SML), especially for the classification task, has some parallels with the information retrieval (IR). In both the domains the documents (items) are to be classified into discrete classes. So, there is similarity between the classifier ensembles and the IR DF task. In case of classifier the evidences are combined from multiple classifiers [15]. Here, the DF is viewed as the process of combining the ranked lists of documents obtained from the multiple IR systems. The process is supposed to enhance the performance. In machine learning classification the problem of selecting the correct class for a data point from a discrete set of class labels is studied and solved. One of the important metrics in this domain studies the relationship between the classifier outputs at certain level. Here, at each data point a classifier scores 1 if it correctly classifies and 0 if it does not. These scores are aggregated across the entire training-points data set. This aggregation should give the measure of the diversity. The metric based on the concept of entropy (Chapter 2) is given as [15]

$$ E = \frac{1}{N} \sum_{j=1}^{N} \frac{1}{(L - L/2 - 1)} \min\left\{ \sum_{i=1}^{L} y_{j,i}; L - \sum_{i=1}^{L} y_{j,i} \right\} \tag{10.112} $$

In Equation 10.112, L is the number of classifiers in the ensemble and N is the total number of data points (in the training set). $y_{j,i}$ is the value (1 or 0) that the ith classifier has received on the jth data point. The value of E would vary between 0 and 1 (0 → no difference, and 1 → the highest possible diversity).

10.9.2.1 Preliminary Results Using E-Measure: Illustrative Example

What is very important is to map the metrics between the domains of SML and IR. The ANNs can be considered as a part of the SML, where the ANNs can be used as data/pattern classifiers. In the context of MLA, a metric operates at the level, that is, the output of each classifier is either correct or incorrect, the classifiers may agree/disagree and be right/wrong. Similarly in the context of IR, a document may be relevant/non-relevant and IR systems may agree/disagree on this aspect. If an IR system returns a document then it is considered as evidence that it considers it to be relevant. The absence of a document is an affirmation that the system viewed it to be non-relevant. The metrics operate on unordered sets of outputs, whereas the ranked list returned by IR system imposes an ordering between the documents. The major emphasis is on (i) the ability of a metric to focus and capture diversity between document (IR) result sets and (ii) whether it is possible to postulate a relationship between diversity, accuracy and combined performance. One needs to make a number of decisions with respect to the parameters and scope of the experiment to investigate these aspects. In one experiment [15], the inputs from the

TREC 2004 Web Track were used. Some 51 queries were chosen and there were between 8 and 147 relevant documents. Several queries only had one relevant document associated with them. In that case, all the systems would frequently return that particular relevant document somewhere in the thousands of documents they returned. This resulted in no diversity between them. In order to minimise the impact of differing-length result sets, some 35 inputs were used that tended to return 1000 documents (the TREC maximum). Then the teams of five systems were fused using the sliding window data fusion (SWD/F) [16]. The diversity was quantised using the E-measure of Equation 10.112, which was calculated on a per-query basis. These statistics were averaged across all queries that resulted in a single value for the entropy of the fused system. Then the average MAP score for the inputs and the combined MAP score (of the fused output) were similarly combined across the query set. The first 250,000 combinations of the 35-candidate systems into ensembles of size 5 were fused using the SWD/F rule. The results were not promising [15]. To investigate this aspect further, the results were sorted out with respect to the average MAP score and the top 5000 highest performing/most accurate combinations were set aside. The correlation between each set of the variables was measured using Pearson's correlation coefficient, r. Here, again, no clear relationship between either the average and combined MAP, ($r = 0.19$), (ACMAP) or between entropy and average MAP ($r = -0.12$) (EAMAP) emerged [14]. However, there appeared a pattern with respect to entropy and the combined MAP score ($r = 0.80$) (ECMAP) as shown in Figure 10.9 [15], and it is clear that the CMAP scores are higher for systems with a larger, hence more diverse, E-value. This preliminary study

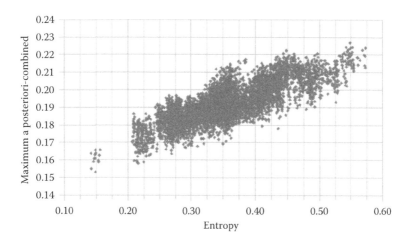

FIGURE 10.9
Entropy versus combined MAP (ECMAP) for the top 5000 systems. (From Leonard, D., Lillis, D. and Zhang, L. Applying machine learning diversity to data fusion in information retrieval. http://lill.is/pubs/Leonard2011.pdf, accessed April 2013. With permission.)

shows an evidence of a relationship between the combined performance of fused result sets, the average performance of the input result sets and the diversity between the relevant documents (returned in the component result sets as measured using the *E*-metric) [15].

10.9.3 Machine Learning Algorithm and DF for Estimating Energy Expenditure

It is a well-known fact that the regular and accurate self-monitoring of physiological parameters and energy expenditure can provide important information for the enhanced self-awareness of an individual's health. A physiological monitoring device/system should [17]: (i) provide the estimates of energy expenditure, (ii) be accurate, (iii) provide continuous feedback to the user, (iv) be easy to use and (v) be cost effective. The effective use of MLA methods and a combination of basic sensors can be advantageously used. MLA can face several challenges [17]: (i) need of high-quality data, (ii) variability present in the data, (iii) sometimes violation of the assumption of MLA that the training and testing data should be the same, (iv) the mathematical models used should satisfy multiple objectives, (v) the algorithms need to be upgraded when the HW systems improve periodically, (vi) often we need the results in real time which puts an additional requirement in terms of computational time and accuracy (the finite word-length implementation of the recursive data processing algorithms, one can use the SRIF algorithm, Chapters 4 and 5) and (vii) choice of efficient methods and algorithms in terms of space/time complexity. The modelling process would generally involve: (a) data collection and editing (preliminary filleting/noise removal → data pre-processing), (b) feature generation, (c) development of regression models and (d) internal and external validation of these models.

The features can be derived using standard MLA feature generation techniques such as principal component analysis (PCA) and independent component analysis (ICA) (Chapters 6 and 7); some new approaches can be developed based on intuition and visual observation with experience. Several regression models can be built that would provide energy expenditure estimates. These models can be for a specific activity. Then, these regressions are combined according to the probabilities output from the activity classifiers. It is possible to use many AI-based regression techniques: (i) robust regression and (ii) locally weighted regression for fitting the data. Feature selection/training for the regression models can be performed using *k*-fold by cross-validation method. Often the regressions are tuned to predict relative measures that are subsequently adjusted for the given test-subject (person or test object as the case may be). For the validation of each algorithm release cycle, certain data sets are not used for the entire development period. Then, the performance of the model/s is evaluated on these validation data sets.

The models are approved and released if they pass pre-defined criteria on the validation data sets and the training sets.

With healthcare expenditure spiraling each year, people can benefit from an effective, inexpensive, easily wearable and accurate physiological monitoring device. The BodyMedia armbands (BMA) [17] attempt to provide such a solution in a conventional manner by using SDF and state-of-the-art MLA and AI approaches. The modelling process for estimating physiological parameters, especially energy expenditure, has been presented in Reference 17. Their results demonstrate the capability of the armband sensors and models to provide accurate results for various activities for a large range of users. These users come both from labs and free-living situations. There is a scope of the integration of new sensors and the ongoing development of data models for extracting new physiological features. Many of the health-care and monitoring systems and the related smart sensors and devices can effectively use a combination of certain MLA techniques and the SDF approaches to derive very accurate and reliable energy expenditure estimates and other parameters. These applications are extremely data-driven approaches and can derive benefits from the ability of the MLA methods and the enhanced accuracy of prediction due to the SDF.

10.10 Neural-Fuzzy-Genetic Algorithm Fusion

It is often felt that one can combine the results of ANNs to improve the overall results of data/image analysis, including classification, and recognition rates in multimodal biometric applications. In this situation, one needs to optimise the choice of the ANNs for the combination as well as to decide an approach to do this combination [18]: (i) modular and (ii) ensemble. In the modular approach the main problem is partitioned into sub-tasks. In the second approach, a set of ANNs is trained on the same task, thereby obtaining more reliable (ensemble-) output. We discuss here, the three methods of SC, ANNs, FL and GA for combining the ANNs.

10.10.1 ANN-Based Approach

ANN is considered as a mapping (operator) from input to the output of the NW, that is, $y = f(x)$, where y is from the set of the output, and x is the set from the input. Also, since the usual classification problem is a mapping case from the presented feature vectors to the output class, an ANN can be called a classifier. Thus, we consider an ANN with a two-layered NW with T neurons (number of input features) in the input layer, H neurons in the hidden layer and c neurons (the number of classifiers) the output-classifier layer. Then, the functioning of the classifier NW (CNW) is a non-linear decision-making process. Thus, we obtain the following for expression for the CNW output [18]:

$$P(\omega_i|X) \approx y_i = f\left\{\sum_{k=1}^{H} \omega_{ik}^{om} f\left(\sum_{j=1}^{T} \omega_{kj}^{mi} x_j\right)\right\} \tag{10.113}$$

In Equation 10.113, we have $X = (x_1, x_2, \ldots, x_T)$ and $\Omega = (\omega_1, \omega_2, \ldots, \omega_c)$, the unknown input and the class set, respectively, and each output neuron produces the y_i that belongs (to the class) by the quantity given by Equation 10.113. The w_{kl}^{mi} is the weight between the jth and the kth neurons, and w_{ik}^{om} is the weight between kth and ith neurons, and f is the usual sigmoid function. The highly weighted neuron is selected as the corresponding class.

We have mainly two methods of combining the ANN classifiers. In the method based on the fusion technique the classification is based on a set of real-valued measurements [18]:

$$P(\omega_i|X), \quad 1 \le i \le c \tag{10.114}$$

Equation 10.114 represents the probability that X comes from each of the classes c under the condition X itself. Then, in the NW-combined scheme, each NW k estimates a set approximation of those true values as given by

$$P_k(\omega_i|X), \ 1 \le i \le c; \ \ 1 \le k \le n. \tag{10.115}$$

One can use the approach of averaging as a NW-classifier-fusion (NWCF) rule as follows:

$$P(\omega_i|X) = \frac{1}{n}\sum_{k=1}^{n} P_k(\omega_i|X), \quad 1 \le i \le c \tag{10.116}$$

One can also use the following NWCF rule that takes into account the reliability (Section 2.5) of the NWs:

$$P(\omega_i|X) = \frac{1}{n}\sum_{k=1}^{n} R_k^i P_k(\omega_i|X), \quad 1 \le i \le c \tag{10.117}$$

With the sum of the reliability weighting factors/coefficients, as usual is 1.

Another method of combining the CNWs is based on voting techniques (Chapter 8). One can use any of the prevalent voting approaches [18]:

(i) unanimity, (ii) majority, (iii) plurality and (iv) Borda count (BC). We discuss here the method based on the BC which is the sum of the number of classes ranked below i by each NW $\rightarrow B_k(i)$ being such a number. The BC is given by the following expression:

$$P(\omega_i|X) = \sum_{k=1}^{n} B_k(i), \quad 1 \le i \le c \tag{10.118}$$

Equation 10.118 can be easily extended to incorporate the reliability coefficients as follows:

$$P(\omega_t|X) = \frac{1}{n}\sum_{k=1}^{n} R_k^i B_k(i), \quad 1 \le i \le c \tag{10.119}$$

The resulting decision is made by selecting the class label for which the BC is the largest.

10.10.2 Fuzzy Logic-Based Approach

This approach utilises the fuzzy integral [6,18] for combining the ANNs for final decision making. FI is a non-linear functional defined with respect to a fuzzy measure (FM). The FM is given [18] as
A set $g: 2^Y \rightarrow [0, 1]$ is an FM with the following conditions:

a. $g(\phi) = 0; g(Y) = 1$
b. $g(A) \le g(B)$ if $A \subset B$ $\qquad$ (10.120)
c. *if* $[A_i]_{i=1}^{\infty}$ is an increasing sequence of measureable sets,

Then we have the sub-condition as

$$\lim_{i \to \infty} g(A_i) = g(\lim_{i \to \infty} A_i) \tag{10.121}$$

Next, FI is defined; Y a finite set and $h: Y \rightarrow \{0, 1\}$ is a fuzzy subset of the set Y. The FI over Y of the function h is defined as follows [6,18]:

$$h(y) \circ g(.) = \max_{E \subseteq X}\{\min(\min_{x \in E} h(y), g(E))\} \tag{10.122}$$

$$= \sup_{\alpha \in [0,1]} \{\min(\alpha, g(h_\alpha))\} \tag{10.123}$$

Here, $h_\alpha = \{y | h(y) \geq \alpha\}$. We interpret this as: (i) $h(y)$ is the degree to which the h is satisfied by y, (ii) min(y) measures the degree to which the h is satisfied by all the elements in E, (iii) $g(E)$ is a measure of the degree to which the subset of the objects E satisfies the concept that is measured by g. The min signifies the degree to which E satisfies the measure g and min $h(y)$. The biggest of these terms is taken by the max operation. FI signifies the maximal grade of agreement between the objective evidence and the expectation. FI can be used to arrive at the consensus in ANN-classification problem. We specifically note here that the FM is not necessarily an additive measure. Sugeno introduced the so-called g_λ-FM [18] that satisfies the following additional property:

$$g(A \cup B) = g(A) + g(B) + \lambda g(A)g(B) \quad \text{for some } \lambda > -1 \quad (10.124)$$

In Equation 10.124, we have $A, B \subset Y$ and $A \cap B = \varphi$. It also means that the union measure can be computed by using the component measures.

Now, we can use the FI for combining NWs. Let Y be given as $Y = \{y_1, y_2, \ldots, y_n\}$, and let $h{:}Y \rightarrow [0,1]$ be a function, with $h(y_1) \geq h(y_2) \geq, \ldots, h(y_n)$. Then, the FI, e (with respect to FM g) is computed by the following expression:

$$e = \max_{i=1}^{n}\{\min(h(y_i), g(A_i))\} \quad (10.125)$$

with $A_i = \{y_1, y_2, \ldots, y_i\}$. Also, when g is a g_λ-FM, then the values of $g(A_i)$ can be computed recursively as follows:

$$g(A_1) = g(\{y_1\}) = g^1 \quad (10.126)$$

$$g(A_i) = g^i + g(A_{i-1}) + \lambda g^i g(A_{i-1}) \quad \text{for } 1 < i \leq n \quad (10.127)$$

In Equation 10.127, λ is obtained from the following equation:

$$\lambda + 1 = \prod_{i=1}^{n}(1 + \lambda g^i) \quad (10.128)$$

In Equation 10.128, $\lambda \in (-1, +\infty)$; $\lambda \neq 0$. The computation of the FI requires the knowledge of g^i, which can be interpreted as the degree of importance of the source y_i towards the final (decision) evaluation.

Let us have $\Omega = \{\omega_1, \omega_2, \ldots, \omega_c\}$ as a set of classes of interest, $Y = \{y_1, y_2, \ldots, y_n\}$ as a set of ANNs, and A as the object under consideration for recognition, $h_k{:}Y \rightarrow [0, 1]$ as the partial evaluation of the object A for class ω_k, that is, $h_k(y_i)$ is an indication of how certain we are in the classification of object A (to be in class ω_k using the NW y_j). Here, 1 indicates absolute certainty that the object A is really in class ω_k and 0 implies absolute certainty that the object A is not in ω_k. Then, corresponding to each y_i the degree of importance, g_i, of how important y_i is in the recognition of the class ω_k should be given. These quantities can be subjectively assigned by an expert (or may be induced from the data set). Since, g_is define the fuzzy density mapping, λ is computed using Equation 10.128. This way the g_λ-FM g is computed. Now, using Equations 10.126 through 10.127 the FI is computed. Then, the class ω_k with the largest FI value is declared as the output class.

10.10.3 GA-Based Approach

We can use GA to search for ensemble members of CNWs, that is, we can optimise the weights of the ensemble NWs with GA. A string in GA must encode $n \times c$ real-valued parameters, R_k^i, in Equation 10.119, that is, each coefficient is encoded by 8 bits and scaled between [0 ~ 1]. The GA then manipulates the most promising strings for improved solution. The GA then operates through a cycle of stages: (a) creation of a sample population of real-valued strings, (b) evaluation of each string with recognition rate on training data (using some fitness function), (c) selection of good strings and (d) genetic manipulation to create the new strings: (i) one-point crossover with probability, say 0.6, and (ii) standard mutation with probability, say 0.01. The GA can be stopped when the recognition rate no longer improves. We can replace all the members of old population with the new ones, and preserve only the best possible solution obtained so far by elitist strategy. The GA takes pieces of weighting coefficients to combine CNWs as such strings.

10.10.4 GA–FL Hybrid Approach

FL and GA are most probably, and perhaps more certainly, very useful techniques that can be combined to produce more powerful intelligent systems, in particular DF systems. One can then devise a hybrid method of FL and GA to give an optimal solution to combine CNWs. The ANNs are used as a baseline system, since they are already well recognised as a powerful I/O mapping operator. FL gives a possibility of incorporating the heuristic knowledge from the expert designer. Thus, human operators can enhance the ANNs by incorporating their own knowledge about the problem with MFs. These MFs (Chapter 3) are then modified via learning process as fine tuning (the defining parameters of these MFS are determined), this is carried out using by ANNs themselves. The FIS is optimised by using the procedure like ANFIS (Chapter 3). The GA is a powerful method for structure optimisation of FL and ANNs which can provide evaluation functions for the GA. A hybrid technique utilises the FI to combine the outputs of separate CNWs with importance of each NW that is assigned by the GA [18].

Let us denote $\hat{g}_\lambda(A)$ be the values provided by the expert human, then $g_\lambda(A)$ and g_i are the identified values. In this hybrid method, the chromosomes encode the fuzzy density values g_i^j by a vector $C_j = \{g_1^j, g_2^j, \ldots, g_k^j; \lambda_j\}$, and the fitness function $f(C_j)$ for chromosome C_j is the sum of the differences between designer-provided FM value $\hat{g}_\lambda(A)$ and FM value obtained by g_i^j and λ_j. This fitness function is given by the following expression:

$$f(C_j) = \sum_{A \in B(X)} \left| \hat{g}_\lambda(A) - \frac{1}{\lambda_j} \{ \prod_{x_i \in A} (1 + \lambda_j g_i^j) - 1 \} \right| \qquad (10.129)$$

In this hybrid method, the genetic operators yield an optimal set of parameters to combine ANNs.

Reference 18 presents the results of the experiments carried out to study the performance of the methods presented in this section. They used the handwritten digit database of Concordia University of Canada that consists of 6000 unconstrained digits (collected from dead letter envelopes by the U.S. Postal Services at different locations in the United States). Their results are very encouraging.

10.11 Image Analysis Using ANFIS: Illustrative Example

In this section, we consider an example of image fusion using the ANFIS system described in Section 3.3. The two input images given to the ANFIS system are shown in Figure 10.10. We see that the top part of input1 pair

Input1   Input2

FIGURE 10.10
Input images of the SARAS aircraft for the ANFIS system.

and the bottom part of the input2 pair are blurred images of an aircraft. The ANFIS requires the output for training for which we use the true image of the SARAS aircraft as shown in Figure 10.11. This aspect can be seen as the situation as the desired image from the ANFIS system. The MATLAB code **fusionanfissaras.m** is used for training and obtaining the fused results. We obtained the following ANFIS information: (i) number of nodes: 35, (ii) number of linear parameters: 27, (iii) number of non-linear parameters: 18, (iv) total number of parameters: 45, (v) number of training data pairs: 262,144, (vi) number of checking data pairs: 0 and (vii) number of fuzzy rules: 9.

The image (intensities) matrices were converted to column (or row) vectors by concatenating the successive columns and the 2D images were stored as one-dimensional (1D) images. These input 1D data were given to the ANFIS system, likewise the training output image was also converted to 1D string of the intensity values. For training, we used three Gaussian/ Bell membership functions and two epoch only. After the ANFIS training the blurred (1D) images were again presented to the ANFIS and the fused (1D) image was obtained. The vector string of the output of the ANFIS was converted back to the image intensity matrix. Figure 10.12 shows the fused image obtained from the ANFIS and the difference image between the true image and the fused image. The percentage fit error (PFE) between the true and the fused image defined as PFE = norm(image difference matrix)*100/ norm(true image matrix) was computed and was found to be 0.623.

This exercise of ANFIS for image fusion is not in the conventional sense, the example of the fusion process because it requires the training using the

True image

FIGURE 10.11
The true image of the SARAS aircraft as the output image used for ANFIS training.

Fused image   Difference image

FIGURE 10.12
The fused image and the difference image.

true image, that is, the 'expected fused' image. However, it can be used to figure out how the so-called fused image is mathematically related to the two input images: fused image (from ANFIS) = $a*input\ image1 + b*input\ image2 + c$. The constant c can be used as the bias parameter. We can attempt to derive the fusion rule (formula) to figure out how the input images might have contributed to the (ANFIS) output image. This process of determining the image fusion rule was illustrated in Chapter 3.

Acknowledgement

Some part of the work presented in Section 10.6 was carried out by Ms Vedshruti as a part of her MTech project (Department of E & C Engineering, MSRIT) under the guidance of Dr. V. Parthasarathy Naidu, and Dr. J. R. Raol.

EXERCISES

10.1 In FFNN what is the role of non-linear function f, that is, the sigmoid function?

10.2 Compare and contrast ANN and FL as modelling processes for dynamic systems.

10.3 Compare ANN with KF and comment on how ANN can be used as a data fuser.

10.4 How does the momentum factor conceptually work in the FFNN learning rule?

10.5 Give at least three approaches of using FL in a KF for the purpose of either tuning and/or improving the performance of the KF.

10.6 Give at least one scheme of parallel processing (two processors: one main and the other as co-processor) of GA.

10.7 Can we use KF for training of FFNN? If yes, how?

10.8 Give one scheme of DF using ANN and GA.

10.9 Give one scheme of DF using ANN and FL.

10.10 Give one scheme of DF using FL and GA.

10.11 If during the computation of the weights (training of ANNs), the value of z in Equation 10.21 becomes 1, what would you do?

References

1. Zadeh, L.A. A definition of soft computing. http://www.soft-computing.de/def.html, accessed April 2013.
2. Eberhart, R.C. and Dobbins, R.W. *Neural Network PC Tools—A Practical Guide.* Academic Press Inc., New York, 1993.
3. Irwin, G.W., Warwick, K. and Hunt, K.J. (Eds.). *Neural Network Applications in Control,* IEE/IET Control Engineering Series, 53. IEE/IET, London, UK, 1995.
4. Raol, J.R., Girija, G. and Singh, J. *Modelling and Parameter Estimation of Dynamic Systems,* IEE/IET Control Engineering Series, 65. IEE/IET, London, UK, 2004.
5. Raol, J.R. and J. Singh. *Flight Mechanics Modeling and Analysis.* CRC Press, FL, 2009.
6. Raol, J.R. *Multisensor Data Fusion with MATLAB.* CRC Press, FL, 2010.
7. Raol, J.R. and Ajith, K.G. (Eds.). *Mobile Intelligent Autonomous Systems.* CRC Press, FL, 2012.
8. Goldberg, D.E. *Genetic Algorithms in Search, Optimization and Machine Learning.* Addison-Wesley Publishing Company Inc., Reading, MA, 1989.
9. Bors, A.G. *Introduction to the Radial Basis Function Networks.* Department of Computer Science, University of York, York, UK. http://www-users.cs.york.ac.uk/adrian/Papers/Others/ OSEE01.pdf, accessed April 2013.
10. Burkey, M. (and Paul), Pan, W., Kou, X., Marler, K.M. and Tsaptsinos, D. *Soft Computing-Fuzzy Logic is a Part of Soft Computing* (ppts). Department of Mathematics, Kingston University. http://www.kingston.ac.uk/~ ma_s435@kingston.ac.uk, accessed April 2013.
11. Ahmad, Z. and Zhang, J. Combination of multiple neural networks using data fusion techniques for enhanced nonlinear process modelling. *Elsevier Journal of Computer and Chemical Engineering,* 30, 295–308, 2005. http://www.elsevier.com/locate/compehemeng.
12. Yadiah, N., Singh, L., Bapi, R.S., Rao, S., Deekshatulu, B.L. and Negi, A. Multisensor data fusion using neural networks. *International Joint Conference on Neural Networks,* Sheraton Vancouver Wall Centre Hotel, Vancouver, BC, Canada, July 16–21, 2006.

13. Mitchell, T.M. *The Discipline of Machine Learning*. CMU-ML-06-108, School of Computer Science, Carnegie Mellon University, Pittsburgh, PA, July 2006.

14. Sewell, M. Machine learning, 2009. http://machine-learning.martinsewell.com/machine-learning.pdf, accessed April 2013.

15. Leonard, D., Lillis, D. and Zhang, L. Applying machine learning diversity to data fusion in information retrieval. http://lill.is/pubs/Leonard2011.pdf, accessed April 2013.

16. Lillis, D., Toolan, F., Collier, R. and Dunnion, J. Extending probabilistic data fusion using sliding windows. *The Proceedings of the 30th European Conference on Information Retrieval (ECIR'08)*, 4956, 358–369, Lecture Notes in Computer Science, Springer, Berlin, 2008.

17. Vyas, N., Farringdon, J., Andre, D. and Stivoric, J.I. Machine learning and sensor fusion for estimating continuous energy expenditure. *Proceedings of the Twenty Third Innovative Applications of Artificial Intelligence Conference*, San Francisco, California, USA. The AAAI Press, Menlo Park, California, 9–11 August 2011. http://www.aaai.org/ocs/index.php/IAAI/IAAI-11/paper/download/.../4018, accessed May 2013.

18. Cho, S.B. Fusion of neural networks with fuzzy logic and genetic algorithm. *Integrated Computer-Aided Engineering*, IOS Press, Amsterdam, The Netherlands, 9(4), 363–372, December 2002. http://sclab.yonsei.ac.kr/publications/Papers/ica00128.pdf, accessed April 2013.

Appendix A: Some Algorithms and/or Their Derivations

A.1 Optimal Filter in Two Parts for State Vector Fusion

The algorithms presented are practical ones. However, there are still possibilities for further refinement and sophistication, which are not dealt with here. We present here the derivation of the filter (of Section 4.8.3) for taking into account the randomly missing data. We consider that the data is missing from 'z' and we denote that variable as 'z_m'. Then we propose the following estimator:

$$\hat{x}(k) = K_1\tilde{x}(k) + K_2 z_m(k) \tag{A.1}$$

Then following the theoretical development of the KF as given in Section 4.2.2.2, we get the following expression:

$$
\begin{aligned}
\underline{x}(k) &= K_1[\underline{x}^*(k) + x(k)] + K_2 bHx(k) + K_2 v(k) - x(k) \\
&= [K_1 + K_2 bH - I]x(k) + K_2 v(k) + K_1 \underline{x}^*(k) \tag{A.2}
\end{aligned}
$$

As $E\{v(k)\} = 0$ and if $E\{\underline{x}^*(k)\} = 0$, for the unbiased a priori estimate, then we have the following condition:

$$E\{\underline{x}(k)\} = E\{(K_1 + K_2 bH - I)\, x(k)\} \tag{A.3}$$

For the unbiased estimate after the measurement is incorporated, we should have $E\{\underline{x}(k)\} = 0$, hence, we get the relation between two weighing factors as

$$K_1 = I - K_2 bH \tag{A.4}$$

By substituting the gain of Equation A.4 into Equation A.1 and simplifying we obtain

$$
\begin{aligned}
\hat{x}(k) &= (I - K_2 bH)\tilde{x}(k) + K_2 z_m(k) \\
&= \tilde{x}(k) + K_2[z_m(k) - bH\tilde{x}(k)] \tag{A.5}
\end{aligned}
$$

Now, we consider the formulation of P (posterior covariance matrix) to determine the covariance of the state error after the measurement is incorporated as

$$\hat{P} = E\{\underline{x}(k)\underline{x}^T(k)\} = E\{(\hat{x}(k) - x(k))(\hat{x}(k) - x(k))^T\}$$
$$= E\{(\tilde{x}(k) - x(k) + K[bHx(k) + v(k) - bH\tilde{x}(k)])(.)^T\} \qquad (A.6)$$

After simplification, following the derivation of the KF, we obtain the following expression for the state-error covariance matrix:

$$\hat{P} = (I - KbH)\tilde{P}(I - KbH)^T + KRK^T \qquad (A.7)$$

In order to get optimal value of gain K, the covariance matrix, the trace of $\hat{P}$ is minimised with respect to K. The cost function is chosen as

$$J = E\{\underline{x}^T(k)\underline{x}(k)\} \qquad (A.8)$$

Then we get the following development:

$$J = trace\{\hat{P}\}$$
$$= trace\{(I - KbH)\tilde{P}(I - KbH)^T + KRK^T\} \qquad (A.9)$$

$$\frac{\partial J}{\partial K} = -2(I - KbH)b\tilde{P}H^T + 2KR = 0 \qquad (A.10)$$

$$KR = b\tilde{P}H^T - b^2KH\tilde{P}H^T$$
$$KR + b^2KH\tilde{P}H^T = b\tilde{P}H^T \qquad (A.11)$$

$$K = b\tilde{P}H^T(b^2H\tilde{P}H^T + R)^{-1} \qquad (A.12)$$

Subsequently we can use the long form of Equation A.7 after computing the value of the optimal gain from Equation A.12. Then the filtering algorithm for optimal filter in two parts for state vector fusion (OFSVF2) is given by Equations 4.144 through 4.148.

A.2 Optimal Filter in Two Parts for Measurement Level Fusion

We present here the derivation of the filter (of Section 4.8.4) for taking into account the randomly missing data. We consider that some data are missing randomly from sensor 2 only, whereas the data from the other sensor are not missing. We consider the measurement data missing as described in Section 4.8.1. As the data are missing at the measurement (data-) level the time-propagation part of the filer remains the same as in the KF, and the filter equations are similar to Equations 4.144 and 4.145. Here, the sensor 1 data are represented by z, and the data from the second sensor are represented by 'z_m'. Then we have the following estimator:

$$\hat{x}(k) = K_1 \tilde{x}(k) + K_2 z_1 + K_3 z_m(k) \tag{A.13}$$

Then following the development of the KF as given in Sections 4.2.2.2 and A.1, we get the following expression:

$$\underline{x}(k) = K_1[\underline{x}^*(k) + x(k)] + K_2(H_1 x(k) + v_1) + K_3(bH_2 x(k) + v_2) - x(k)$$
$$= [K_1 + K_2 H_1 + bK_3 H_2 - I]x(k) + K_2 v_1 + K_3 v_2 + K_1 \underline{x}^*(k) \tag{A.14}$$

As $E\{v(k)\} = 0$ and if $E\{\underline{x}^*(k)\} = 0$, for the unbiased a priori estimate, we have

$$E\{\underline{x}(k)\} = E\{(K_1 + K_2 H_1 + K_3 bH_2 - I) \, x(k)\} \tag{A.15}$$

To obtain the unbiased estimate after the measurement is incorporated, we should have $E\{\underline{x}(k)\} = 0$. We have then the following relation between two weighing factors, for satisfying the condition of unbiasedness for the filter structure:

$$K_1 = I - K_2 H_1 - K_3 bH_2 \tag{A.16}$$

By substituting the gain of Equation A.16 into Equation A.13 and simplifying we get

$$\hat{x}(k) = \tilde{x}(k) + K_2(z_1 - H_1 \tilde{x}) + K_3[z_m(k) - bH_2 \tilde{x}(k)] \tag{A.17}$$

Next, we consider the covariance of the state errors after the measurement is incorporated as

$$\hat{P} = E\{\underline{x}(k)\underline{x}^T(k)\} = E\{(\hat{x}(k) - x(k)) \, (\hat{x}(k) - x(k))^T\}$$
$$= E\{[(\tilde{x}(k) - x(k)) - K_2 H_1(\tilde{x}(k) - x(k)) - bK_3 H_2(\tilde{x}(k) - x(k))$$
$$+ K_2 v_1(k) + K_3 v_2][.]^T\} \tag{A.18}$$

After simplification we get the following expression for the covariance matrix:

$$\hat{P} = (I - K_2 H_1 - bK_3 H_2)\tilde{P}(I - K_2 H_1 - K_3 bH_2)^T + K_2 R_1 K_2^T + K_3 R_2 K_3^t \tag{A.19}$$

Then following the procedure of the minimisation of the trace of the covariance matrix P in Equation A.19, we finally get the following expressions:

$$K_2(H_1\tilde{P}H_1^t + R_1) + K_3(bH_2\tilde{P}H_1^t) = \tilde{P}H_1^t \tag{A.20}$$

$$K_2(bH_1\tilde{P}H_2^t) + K_3(b^2H_2\tilde{P}H_2^t + R_2) = b\tilde{P}H_2^t \tag{A.21}$$

Subsequently, we use the long form for P, after computing the values of the gains from Equations A.20 and A.21, we get the following equation:

$$\hat{P}_f = [I - K_2H_1 - bK_3H_2]\tilde{P}[I - K_2H_1 - bK_3H_2]^t + K_2R_2K_2^t + K_3R_2K_3^t \tag{A.22}$$

Then the filtering algorithm for optimal filter in two parts for measurement level fusion (OFMLF22) is given by Equations 4.149 through 4.154.

A.3 Sum-of-Gaussians Method

We consider the following non-linear dynamic system:

$$\dot{x}(t) = f(x(t),t)) + \Gamma(t) \tag{A.23}$$

$$z_k = h(x_k,t_k)) + v_k \tag{A.24}$$

In Equation A.23, $\Gamma(t)$ represents a zero mean white Gaussian noise with correlation function as $Q\delta(t - \tau)$ [1]. The measurement noise is uncorrelated and zero mean sequence of random noise with known correlation R. The non-linear filtering problem is to find a posterior pdf for the state x_k, given the measurements and the pdf $p(x(t), t|Z_k)$ and $p(x_0, t_0)$ as the prior pdf. In the SOG method, the conditional pdf is represented by a finite sum of Gaussians [1]

$$\hat{p}(x(t),t \mid Z_k) = \sum_{i=1}^{N} w_{t|k}^i N(x(t);\mu_{t|k}^i, P_{t|k}^i) \tag{A.25}$$

In Equation A.25, w, μ and P are the conditional weight, mean and covariance of the ith Gaussian kernel. We also have the following constraint on the weights due to the requirement of positivity and normalisation of the pdf of Equation A.25:

$$\sum_{i=1}^{N} w_{t|k}^{i} = 1 \qquad (A.26)$$

In Equation A.26, each weight is equal to zero or +ve for all t. Due to Gaussian assumption, we use the following EKF equations for propagation of mean and the covariance [1]:

$$\hat{\mu}_{t|k}^{i} = f(\mu_{t|k}^{i}, t),$$
$$A_{t|k}^{i} = \frac{\partial f(x(t), t)}{\partial x(t)} \mid \mu_{t|k}^{i}$$
$$\hat{P}_{t|k}^{i} = A_{t|k}^{i} P_{t|k}^{i} + P_{t|k}^{i} A_{t|k}^{i^{T}} \qquad (A.27)$$
$$\hat{w}_{t|k}^{i} = w_{k|k}^{i}$$

The measurement update is given by the following equations [1]:

$$\hat{\mu}_{k+1|k+1}^{i} = \hat{\mu}_{k+1|k}^{i} + K_k^i(z_k - h(\hat{\mu}_{k+1|k}^{i}, t_k))$$
$$H_k^i = \frac{\partial h(x_k, t_k)}{\partial x_k} \mid \mu_{k+1|k}^{i}$$
$$K_k^i = \hat{P}_{k+1|k}^{i} H_k^{i^{T}} (H_k^i \hat{P}_{k+1|k}^{i} H_k^{i^{T}} + R_k)^{-1}$$
$$\hat{P}_{k+1|k+1}^{i} = (I - K_k^i H_k^i) \hat{P}_{k+1|k}^{i} \qquad (A.28)$$
$$\hat{w}_{k+1|k+1}^{i} = \frac{\hat{w}_{k+1|k}^{i} \beta_k^i}{\sum_{i=1}^{N} \hat{w}_{k+1|k}^{i} \beta_k^i}$$
$$\beta_k^i = N(z_k - h(\hat{\mu}_{k+1|k}^{i}, t_k), H_k^i \hat{P}_{k+1|k}^{i} H_k^{i^{T}} + R_k)$$

Then the point estimate for the state is obtained as the mean of the posterior pdf and the following equations are used [1]:

$$\hat{\mu}_{t|k} = \sum_{i=1}^{N} \hat{w}_{t|k}^{i} \hat{\mu}_{t|k}^{i}$$
$$\hat{P}_{t|k} = \sum_{i=1}^{N} \hat{w}_{t|k}^{i} [\hat{P}_{t|k}^{i} + (\hat{\mu}_{t|k}^{i} - \hat{\mu}_{t|k})(\hat{\mu}_{t|k}^{i} - \hat{\mu}_{t|k})^{T}] \qquad (A.29)$$

The performance of SOG ($\hat{p}(x,t|Z_k)$) would approach to the true pdf, if there is a sufficient number of Gaussian kernels, and the linearisations are valid.

A.4 Particle Filter

The particle filter (PF) is a Monte Carlo (MC) method for state estimation of non-linear dynamic systems with non-linear Gaussian noise. The state and measurement equations are given as [2]

$$x_{t+1} = f_t(x_t) + w_t(x_t)$$
$$z_t = h_t(x_t) + v_t(x_t)$$

(A.30)

The noise process could have any probability distribution, uni- or multi-modal. The output of the filter is $p(x_t)$, the solution is approximate and the process of estimation is computationally slow. The pdf is approximated using point weights as follows:

$$p(x_t) \approx \sum_i^N w_t^i \delta(x - x_t^i)$$

(A.31)

In this process each point is called a particle, and has a positive weight. The number of particles could be say, 1000. The basic algorithm has the steps of: (i) initialisation, (ii) time propagation – that moves the particles ahead in time, (iii) measurement update – change of weights, (iv) re-sampling – if required and (v) go to step (ii) again when the new measurement is made available.

   a. Time propagation

     This obtains one step ahead prediction for each particle

$$x_{t+1}^i = f_t(x_t^i) + w_t(x_t^i)$$

(A.32)

     We note here that, a realisation of the process noise is used for every particle.

   b. Measurement update

     The weights are updated using the measurement

$$w_t^i = w_{t-1}^i p(z_t \mid x_t^i)$$

(A.33)

If the initial distribution is not known, then one can use uniform distribution. As was done in the case of the SOG method the weights are normalised as follows:

$$w_t^i = \frac{w_t^i}{\sum_{i=1}^{N} w_t^i} \tag{A.34}$$

The particles that explain the measurements will gain the weight, and the ones that are far-off from the true state tend to lose the weight. Also, the density of the cloud of the particles will change. Often in the process of filtering, the algorithm could degenerate, and the particles (except one) become very light. This problem is solved by re-sampling so that the weights become equal. A collection of particles is updated for removing the particles with small weights and the ones with important weights are duplicated. The new particles will have new weights. This gives a new approximation to the posterior probability density [3]. The computational complexity is proportional to the number of particles used in the MC simulation of the PF. The approximation error decreases as the number of particles increase. The discretisation step could be 0.01 time unit, depending on the problem at hand.

A.5 Pugachev Filter

The alternative solutions to the non-linear state and parameter estimation problems based on the characteristic function approach, in contradistinction to the conventional conditional probability distribution function, have been studied in References 4–7. These filters are obtained by minimising the mean square error (MSE) in a given class of functions. These functions are called structural functions of the estimators. As a result, the Pugachev filters are conditionally optimal estimators or they are pareto-optimal solutions to the non-linear filtering problems.

Let us consider a non-linear dynamic system described by the following difference equations [5]:

$$x(k + 1) = f(x(k), u(k), k) + q(x(k), k)v(k) \tag{A.35}$$

$$z(k) = h(x(k), u(k), k) + d(x(k), k)v(k) + v_0(k) \tag{A.36}$$

The variables have the usual meanings (Chapters 4 and 5). The solution to this problem is based on the MSE:

$$E\{(x(k) - \hat{x}(k)^T (x(k) - \hat{x}(k)\} \tag{A.37}$$

The structure of the estimator is specified as

$$\hat{x}(k+1) = Cy(k+1)$$
$$y(k+1) = K\xi(y(k), u(k), z(k), k) + b(k)$$

(A.38)

In Equation A.38, we have $C(n \times s, n \geq s)$ as constant class-matrix, K, and b are the optimal gains, and ξ is a pre-assignable vector-values non-linear function (of an arbitrary dimension) and determines the structure of the estimator with an appropriate choice of the class matrix C. This class defines the class of the permissible estimates. The idea of class matrix is that any estimate defined by difference equations can be incorporated into the class of the permissible estimators such that the comparison between the conditionally optimal estimator (by Pugachev) and the other estimators determined by similar equations can be made (feasible). The filtering solutions based on Equation A.38 are conditionally optimal since the optimal gains depend on the class matrix and the structure of the estimator. The optimal values of the gains are given by [5]

$$CK(k)M(k) = L(k)$$
$$Cb(k) = CE_0 - CK(k)E_1$$

(A.39)

The L and M matrices are obtained as a result of minimisation of the MSE

$$L(k) = E\{[x(k+1) - E\{x(k+1)\}]\xi(y(k), u(k), z(k), k)^T\}$$
$$M(k) = E\{[\xi(y(k), u(k), z(k), k) - E\{\xi(y(k), u(k), z(k), k)\}]\xi(y(k), u(k), z(k), k)^T\}$$
$$E_0 = E\{x(k+1)\}; \quad E_1 = E\{\xi(y(k), u(k), z(k), k)\}$$

(A.40)

We note that E is the unconditional mathematical expectation, unlike in the case of KF/EKF wherein the conditional expectation of state x, given the measurements is used for the MSE minimisation. For the conditionally optimal filtering (COF) solutions the expectation E is evaluated by using the joint one-dimensional (1D) characteristic function defined as follows:

$$g(\lambda, \mu) = E\{\exp[i\lambda^T(f(x(k-1), u(k-1), k-1) + q(x(k-1), k-1)v(k-1))$$
$$+ i\mu^T(K(k-1)\xi(y(k-1), u(k-1), h(x(k-1), u(k-1), k-1)$$
$$+ d(x(k-1), k-1)v(k-1) + v_0(k-1), k-1) + b(k-1)]\}$$

(A.41)

The major aspects of these COFs is that we need: (i) to specify the class matrix, (ii) the structural function ξ and (iii) computations of the filters' gains. The computation of the gains needs the evaluation of the unconditional expectation E, which in turn requires the use of the joint 1D characteristic function: of (i) the states, (ii) measurements, (iii) the estimates and (iv) various noise processes. The structural functions can be chosen based on (i) application to suboptimal filtering situations, (ii) specialisation of the problem and (iii) similarity to the non-linear systems. Various ramifications of the Pugachev's non-linear estimators have been presented and investigated in References 5–7, including several special cases based on: (i) specific choices of the structural functions, (ii) correlated process noise, (iii) correlated measurement noise and (iv) non-linear function of the residuals. Also, several similarities and contradistinctions with the existing non-linear filters that are based on conditional probability density functions have also been brought out very clearly in References 5–7. The problem of the computation of the gains can be handled by: (i) using the normal characteristic function with unknown expectation and covariance matrix, (ii) using some known characteristic function with unknown parameters and (iii) expressing the logarithm of the characteristic function in terms of a truncated power series. If all the gain computations can be accomplished a priori, which should be possible since, the expectation E to be evaluated is the unconditional one, then the estimator is the most simple recursive form for the state estimation, since the estimator simply involves the matrix/vector operations. It is suggested here that one can use the concept of particle filtering in order to compute the characteristic function that is required in further computation of various expectation operations in order to compute the filter gains. Thus, it is speculated that the particle filtering approach might be a good and practical MC-based procedure to implement Pugachev's non-linear estimators.

A.6 Extended H-Infinity Filter

The original H-Infinity filter has been devised for linear systems. For a non-linear system, a linearisation procedure is required to be performed as in the case of EKF to derive the appropriate filtering equations for the non-linear system. Such, a filter is called an extended H-Infinity filter (EHIF). We consider the following non-linear time varying system:

$$\dot{X}(t) = f(X,t) + Gv(t)$$
$$Y(t) = CX(t) \qquad\qquad (A.42)$$
$$Z(t) = HX(t) + Dv(t)$$

In Equation A.42, we have X, Y and Z as the state, noiseless measurements and noisy measurements of the given non-linear dynamic system, respectively. $v(t)$ is the deterministic disturbance and follows the constraint [8]:

$$|| \bar{v}(t) ||_2 = || W^{-1/2}v ||_2 \le 1 \tag{A.43}$$

We can consider the X as smooth state space manifold, and f as a smooth vector field on real-valued vector space. Now, the non-linear state equation is expanded into Taylor series about the estimated state as follows:

$$f(X,t) = f(\hat{x},t) + A(t)(X(t) - \hat{x}(t)) + \cdots + \text{Higher Order Terms} \tag{A.44}$$

In Equation A.44, we have now x (-estimate) as the state of the linearised system. We also have the linearised system's coefficient matrix as follows:

$$A(t) = \frac{\partial f(x,t)}{\partial x} \Big|_{x=\hat{x}} \tag{A.45}$$

In the measurement Equation A.42, if the non-linear functions are then considered as is done in the case of EKF, we can linearise these non-linear functions at the current estimate of the state. The linearised state-space equations are given as follows:

$$\dot{x}(t) = Ax + Gv + f(\hat{x},t) - A\hat{x}$$
$$y(t) = Cx(t) \tag{A.46}$$
$$z(t) = Hx(t) + Dv$$

Then, the H-Infinity filter is given as follows [8]:

$$\dot{\hat{x}}(t) = f(\hat{x},t) + K(z - \hat{z})$$
$$\hat{y}(t) = C\hat{x}(t) + L(z - \hat{z}) \tag{A.47}$$
$$\hat{z}(t) = H\hat{x}(t)$$

We now have the following equations for computing the *H-I* gain K, and L:

$$K = PH^T R_v^{-1} \gamma^{-2} PCL$$
$$L - > \gamma^2 I - LR_v L^T \ge 0 \tag{A.48}$$
$$\text{and } R_v = DRD^T$$

Finally, we have the matrix differential equation for the Gramian of state errors (somewhat equivalent to covariance matrix in the KF):

$$\dot{\hat{P}}(t) = \hat{P}(t)A^T(t) + A(t)\hat{P}(t) - \hat{P}(t)(H^T R_v^{-1} H - \lambda^{-2} C^T C)\hat{P}(t) + GQG^T \tag{A.49}$$

References

1. Terejani, G., Singla, P., Singh, T. and Scott, P. D. Adaptive Gaussian sum filter for nonlinear Bayesian estimation. *IEEE Transactions on Automatic Control*, 56(9), 2151–2156, September 2011. http://www.academia.edu/ 677248/.
2. Sundvall, P. An introduction to particle filtering. Presentation in course "Optimal filtering," *Signals, Sensors, Systems, KTH*, November, 2004. PPTs 1–13, http://www. s3.kth.se/~pauls.
3. Baili, H. Online particle filtering of stochastic volatility. *World Congress on Engineering and Computer Science*, WCES, San Francisco, USA, 2010. http://www .researchgate.net/publication/47800386_Online_Particle_Filtering_of_ Stochastic_Volatility.
4. Pugachev, V. S. Conditionally optimal estimation in stochastic differential system. *Automatics*, 18, 685, 1982.
5. Raol, J. R. and Sinha, N. K. Conditionally optimal state estimation for systems governed by difference equations. *Canadian Electrical Engineering Journal*, 12(2), 71–77, 1987.
6. Raol, J. R. and Sinha, N. K. Estimation of states and parameters of stochastic non-linear systems with measurements corrupted by correlated noise. *Problems of Control and Information Theory*, 17(3), 145–158, 1988.
7. Raol, J. R. Stochastic state estimation with application to satellite orbit determination. PhD thesis, McMaster University, Hamilton, Ontario, Canada, 1986.
8. Yuen, H. C. A unified game theory approach to H-Infinity control and filtering. PhD thesis, The University of Hong Kong, October 1997.

Appendix B: Other Methods of DF and Fusion Performance Evaluation Metrics

Here, we briefly present several other approaches to MSDF, image fusion performance metrics and related mathematics.

B.1 Multiple-Criteria Decision-Making Method Based on Multiple-Criterion Data Fusion

The multiple-criteria decision-making (MCDM) is concerned with the ranking of several decision alternatives based on preference judgments. These judgments are made on decision alternatives over a number of criteria [1]. The problem of MCDM is handled with: (i) the data fusion (DF) technology being considered advantageously, and (ii) the fuzzy sets theory and DS theory of evidence being used to deal with MCDM to efficiently treat uncertain information in the process of decision making. The triangular MFs are used for deciding the importance-weights. These weights are transformed into discounting coefficients and the ratings are transformed into basic probability assignments. The final results are obtained via the DS rule of combination. Hence, a new fuzzy evidential MCDM method under uncertain environments is proposed in Reference 1 whereby the linguistic variables are transformed into basic probability assignments. The data from different criteria can be associated and combined using the DS rule.

B.2 Self-Adaptive Data Fusion

In this section, the method of computing optimum weights for fusion rule (for sensor DF/self-adaptive data fusion [SADF]) is given [2]. First the mean and variance of the data from each of the n sensors are computed (Sections 2.2.1 and 2.2.2). Assuming we know the weights to be used for each sensor, we see that the sum of these weights is 1, where i is the number of sensors

$$\sum_{i=1}^{n} w_i = 1 \tag{B.1}$$

Then the fused data is given as

$$z_f = \sum_{i=1}^{n} w_i z_i \tag{B.2}$$

Here, w_i are the weights/factors of the ith sensor, z_i is the measured data of the sensor and the fused data are given by Equation B.2. Now, we obtain the expression for the total mean variance after the DF as follows [2]:

$$\sigma^2 = E[(Z - z_f)^2] = E\left[\left(Z - \sum_{i=1}^{n} w_i z_i\right)^2\right] \tag{B.3}$$

Inserting Equation B.1 in Equation B.3, since the sum of the fusion weights/coefficients is equal to 1, and taking the expression for the weights as a common factor, we obtain the following expression:

$$\sigma^2 = E[(Z - z_f)^2] = E\left[\left(Z - \sum_{i=1}^{n} w_i z_i\right)^2\right]$$

$$= E\left[\left(\sum_{i=1}^{n} w_i(Z - z_i)\right)^2\right] = \sum_{i=1}^{n} w_i^2 \sigma_i^2 \tag{B.4}$$

We use the multi-dimension function theory with the Lagrange multiplier λ and have the following function f:

$$f(w_1, w_2, \ldots, w_n, \lambda) = \sum_{i=1}^{n} w_i^2 \sigma_i^2 + \lambda\left(\sum_{i=1}^{n} w_i - 1\right) \tag{B.5}$$

We finally obtain the following equations [2]:

$$\frac{\partial f}{\partial w_i} = 2w_i \sigma_i^2 - \lambda = 0 \quad (\text{for } i = 1, 2, \ldots, n) \tag{B.6}$$

$$\frac{\partial f}{\partial \lambda} = 1 - \sum_{i=1}^{n} w_i = 0$$

Then solving Equations B.5 and B.6, we obtain the following expression for the weights:

$$w_i = \frac{1}{\sigma_i^2 \left(\sum_{k=1}^{n} \sigma_k^{-2} \right)} \quad \text{(for } i = 1, 2, \ldots, n) \tag{B.7}$$

B.3 Fuzzy Set Data Fusion

A fusion method (fuzzy set/fuzzy set data fusion [FSDF]) based on the nearness of the measured data values for a set of sensors is discussed here [2]. We say that the measured values of the data for each sensor at time k are given as follows:

$$z_i(k); \quad i = 1, 2, \ldots, n \tag{B.8}$$

These measured values of each sensor act as a fuzzy set. The closeness measure between two fuzzy sets is given as

$$\delta_{ij}(k) = \min[z_i(k), z_j(k)] / \max[z_i(k), z_j(k)] \tag{B.9}$$

Based on Equation B.2, we have the approach degree matrix between each sensor at time k is given as [2]

$$\sum(k) = \begin{bmatrix} 1 & \delta_{12}(k) & \cdots & \delta_{1n}(k) \\ \delta_{21}(k) & 1 & \cdots & \delta_{2n}(k) \\ \vdots & \vdots & \cdots & \vdots \\ \delta_{n1}(k) & \delta_{n2}(k) & \cdots & 1 \end{bmatrix} \tag{B.10}$$

With the above definition, the consistent measurement of the measured value between the ith sensor and other sensors at time k is given as follows:

$$r_i(k) = \sum_{j=1}^{n} \delta_{ij}(k) / n \tag{B.11}$$

Next, we define the mean and variance of the ith sensor as follows:

$$\bar{r}_i(k) = \sum_{i=1}^{k} r_i(k) / k \tag{B.12}$$

$$\sigma_i^2(k) = \sum_i^k [r_i(k) - \bar{r}_i(k)]^2 / k \qquad (B.13)$$

We then have the expressions for the weights and the normalised weights as follows [2]:

$$w_i(k) = \bar{r}_i(k)/\sigma_i^2(k) \qquad (B.14)$$

$$W_i(k) = w_i(k) \Big/ \sum_{j=1}^n w_j(k) \qquad (B.15)$$

Finally, we obtain the expression for the fused data values as follows:

$$z_f(k) = \sum_{i=1}^n W_i(k) z_i(k) \qquad (B.16)$$

B.4 Coefficient of Variance Data Fusion

In this method (coefficient of variance data fusion [CVDF]), the fusion weights are obtained by using the so-called coefficient of variance (CV). The CV is obtained as [2]

$$CV_i = \sigma_i/\mu_i \qquad (B.17)$$

In Equation B.1, the ratio is between the standard deviation and the mean value of the data. We then compute the CVs and their inverses for the sensor array. Then we obtain the weights for fusion rule as follows:

$$w_i = CV_i^{-1} \Big/ \sum_{j=1}^n CV_j^{-1}; \quad i = 1, 2, \ldots, n \qquad (B.18)$$

Then the fused data are given as follows:

$$z_f = \sum_{i=1}^n w_i \mu_i \qquad (B.19)$$

The computational complexity of the methods presented in Sections B.2 to B.4 was found to be moderate for SADF method, high for FSDF method and moderate for the CVDF method in the case study presented in Reference 2.

B.5 Measurement Level Data Fusion Method 2

We have discussed the measurement level data fusion method (MLF1) in Sections 4.2.7, 4.8.2 and 4.8.4. Here, we give the MLF2 rule [3] as the modified fusion of MLF1

$$z_k = \left[\sum_{i=1}^{n} R_{ik}^{-1} \right]^{-1} \sum_{i=1}^{n} R_{ik}^{-1} z_{ik} \tag{B.20}$$

$$H_k = \left[\sum_{i=1}^{n} R_{ik}^{-1} \right]^{-1} \sum_{i=1}^{n} R_{ik}^{-1} H_{ik} \tag{B.21}$$

$$R_k = \left[\sum_{i=1}^{n} R_{ik}^{-1} \right]^{-1} \tag{B.22}$$

This rule gives the fused information by weighted measurements. Its performance was found to be better than the MLF1 in terms of MSE as studied for case study of rocket tracking using actual data from four radars (azimuth, elevation and range data) in Reference 3.

B.6 Modified State Vector Fusion Method

We have discussed the conventional state vector level fusion (SVF) method in Sections 4.7.5 and 4.8.5. Here, we give the modified state vector fusion (MSVF) rule [3]. This method is also called modified track-to-track fusion method. In this procedure, the prediction of the Kalman filter (KF) is improved with the fused state estimate at the $k-1$ step. We have the following time-propagation formulation:

$$\hat{x}_{k,k-1} = \Phi_{k-1}\hat{x}_{k-1,k-1} \tag{B.23}$$

Here, we have x, say as the fused state. Then the predicted fused state is combined with the measurements z_k^i and z_k^j to update the prediction to obtain the estimates $\hat{x}_{k,k}^i$ and $\hat{x}_{k,k}^j$. This is obtained with the local KF at the next instant, and the updated estimates are given as follows [3]:

$$z_{k,k-1}^m = H_k^m \hat{x}_{k,k-1}; \ m = i, j \tag{B.24}$$

$$\hat{x}_{k,k}^m = \hat{x}_{k,k-1} + P_{k,k-1}^m(x,z)P_{k,k-1}^m(z,z)^{-1}(z_k^m - \bar{z}_{k,k-1}^m) \tag{B.25}$$

In Equation B.3, we have the following expressions for various covariance matrices:

$$P_{k,k-1}^m(x,z) = E[(x_k - \hat{x}_{k,k-1})(z_k^m - \bar{z}_{k,k-1}^m)^T] \tag{B.26}$$

$$P_{k,k-1}^m(z,z) = E[(z_k^m - \bar{z}_{k,k-1}^m)(z_k^m - \bar{z}_{k,k-1}^m)^T] \tag{B.27}$$

The fused state estimate is given by the following equation:

$$\hat{x}_{k,k} = \hat{x}_{k,k}^i - (P_{k,k}^i - P_{k,k}^{ij})(P_{k,k}^i + P_{k,k}^j - P_{k,k}^{ij} - P_{k,k}^{ji})^{-1}(\hat{x}_{k,k}^i - \hat{x}_{k,k}^{ji}) \tag{B.28}$$

Then the covariance matrix of the fused state estimate is given as follows:

$$P_{k,k} = P_{k,k}^i - (P_{k,k}^i - P_{k,k}^{ij})(P_{k,k}^i + P_{k,k}^j - P_{k,k}^{ij} - P_{k,k}^{ji})^{-1}(P_{k,k}^i - P_{k,k}^{ji}) \tag{B.29}$$

It is noted here that the MSVF is similar to the track-to-track fusion but has different gains.

Its performance was found to be better than the SVF1 terms of MSE as studied for case study of rocket tracking using actual data from four radars (azimuth, elevation and range data) in Reference 3.

B.7 Some More Image Fusion Performance Assessment Metrics

Image quality assessment is a vital task in image processing applications. Image quality is a characteristic of an image that measures the image degradation/deviation, compared to an ideal or perfect image [4]. Image quality depends on several factors: (i) how much the image formation process of the camera deviates from the pinhole model, (ii) the quality of the image measurement process, (iii) the coding artefacts that are introduced in the image and (iv) some external environmental conditions like illumination, clouds, day/night conditions and so on. IQMs (image quality measures/metrics) are figures of merit used for the evaluation of images; these quality measures are divided into two classes: (a) subjective evaluation and (b) objective evaluation. Subjective evaluation is oriented on human vision system (HVS), wherein the subjects view a series of reproduced images and rate them based on the visibility of the artefacts. The subjective quality measurement, mean opinion score (MOS) has been used for many years. There are several other metrics that can be measured objectively and automatically evaluated by a computer program. These are classified as full-reference (FR) or bi-variate methods and no-reference (NR) or univariate methods. In FR image quality assessment methods, the quality of a test image is evaluated by comparing it with a reference image that is assumed to have perfect quality. NR metric tries to assess the quality of an image without any reference to the original one. Image quality cannot be quantified completely by a single metric, and hence, some of the basic image characteristics are used to assess image quality as follows: (a) brightness–brightness related to the illumination system tells how light or dark the image is, (b) clarity demonstrates if the image is blurred or well-defined, (c) resolution is related to the numerical aperture of the objective camera lens (the higher the numerical aperture, the better the resolution) and the wavelength of light passing through the lens (the shorter the wavelength, the better the resolution); it indicates how close two points can be in the image before they are no longer seen as two separate points and (d) contrast is related to the illumination system, specifying the difference in lighting between adjacent areas of the image.

B.7.1 Image Histogram

It is a graphical representation of the pixel distribution in a digital image, where the horizontal axis of the graph represents the pixel variations, and the vertical axis represents the number of pixels available for that particular pixel value, and we can see the entire pixel distribution at a glance. In general, the left-hand side of the horizontal axis represents the black and dark areas, the middle part represents medium grey and the right-hand side represents light and pure white areas, thus, the histogram for a very bright image with few dark areas and/or shadows will have most of its data points on the right side and centre of the graph, and conversely, the histogram for a very dark image will have the majority of its data points on the left-hand side and centre of the graph.

B.7.2 Average Entropy

Entropy is a statistical measure of randomness, which can be used to characterise the texture of an image. Information entropy denoted by a grey image is given as

$$H = -\sum_{k=0}^{255} p(r_k)\log_2 p(r_k); \quad p(r_k) = n_k/n; \quad k = 0,1,2,\dots,255 \qquad \text{(B.30)}$$

Here, n_k is the number of times the kth grey level appears in the image and n is the number of pixels in the image. The entropy for a colour image is the average of information entropies (H) of three colour components red, blue and green and is given as follows:

$$H_{avg} = \frac{\sqrt{H_R^2 + H_G^2 + H_B^2}}{\sqrt{3}} \qquad \text{(B.31)}$$

An image with the ideal equalisation histogram possesses the maximal entropy of 8 bit, and the entropy of an image with one grey value equals to zero, therefore, higher average entropy value signifies better image quality.

B.7.3 Average Contrast

Contrast is a visual-appearance characteristic that makes an object/its representation in an image distinguishable from other objects and the background. In visual perception, contrast is determined by the difference in the colour and brightness of the object and other objects within the same field of view, and is expressed as follows:

$$C_{avg} = \frac{1}{(n-1)(m-1)} \sum_{x=1}^{n-1}\sum_{y=1}^{m-1} |C(x,y)| \qquad \text{(B.32)}$$

Here, n and m are the number of rows and columns in the image. For an IR image, $C(x,y)$, the contrast is the gradient calculated for the image as a single component

$$|C(x,y)| = \sqrt{\nabla^2 I(x,y)}; \quad \nabla I(x,y) = \frac{\partial I(x,y)}{\partial x}i + \frac{\partial I(x,y)}{\partial y}j \qquad \text{(B.33)}$$

Here, ∇ is the gradient operator and $I(x,y)$ is the image pixel value at the point (x,y) coordinates. This average gradient signifies the clarity of an image, and measures the spatial resolution in an image; larger the average gradient, higher the resolution, implying better quality of the image. For a colour image, the colour contrast is given by the average of gradients of red, green and blue considered as follows:

$$|C(x,y)| = \frac{\sqrt{\nabla^2 I_R(x,y) + \nabla^2 I_G(x,y) + \nabla^2 I_B(x,y)}}{\sqrt{3}} \qquad \text{(B.34)}$$

B.7.4 Average Luminance

Luminance specifies the amount of light that passes through (or is emitted from) a particular area, and falls within a given solid angle, and indicates how much luminous power will be perceived by an eye looking at the surface from a particular angle of view. Thus, it is an indicator of how bright the surface will appear, and is expressed as

$$L_{avg} = \frac{1}{nm} \sum_{x=1}^{n} \sum_{y=1}^{m} I(x,y) \tag{B.35}$$

Higher luminance value denotes more brightness in the image. For a colour image it is given as

$$L_{avg} = \frac{1}{nm} \sum_{x=1}^{n} \sum_{y=1}^{m} \frac{I_R(x,y) + I_B(x,y) + I_B(x,y)}{3}$$

B.7.5 Signal-to-Noise Ratio

Signal-to-noise ratio (SNR) is a measure used to quantify how much a signal has been corrupted by noise in the sense of signal power or energy. SNR is referred to, as the ratio of signal (some meaningful information in only the signal/true signal) to the background noise (similar information in false or irrelevant data/noise affecting the same signal). SNR is also calculated as the ratio of mean to standard deviation of a signal or measurement, where the SNR of an image is usually calculated as the ratio of the mean pixel value to the standard deviation of the pixel values over a given neighbourhood: $SNR = \mu/\sigma$

Another definition of the SNR is given by Equation 7.101. In fact, this should be called inverse fusion error square (IFES) metric due to its obvious definition of terms in Equation 7.101, hence, see also Equation 6.91, which is called PSNR in the literature.

B.7.6 Spatial Frequency

Spatial frequency is a characteristic of an image that is periodic across position in space, and is a measure of how often sinusoidal components (of the Fourier transform) of the image repeat per unit distance. Higher value of spatial frequency metric indicates good quality; see Equation 6.94 for the formulae.

B.7.7 Spectral Activity Measure

This metric evaluates a picture quality, and it is a function of discrete Fourier transform (DFT) of an image. The function theory (FT) is a representation of an image as a sum of complex exponentials of varying magnitudes, frequencies and phases, and a DFT is a transform whose input and output values are discrete samples, making it convenient for computer manipulation. The spectral activity metric is given as follows:

$$
\text{SAM} = \frac{(1/nm)\sum_{x=1}^{n}\sum_{y=1}^{m}\{F(x,y)\}^{2}}{\left[\prod_{x=1}^{n}\prod_{y=1}^{m}\{F(x,y)\}^{2}\right]^{1/nm}}
\tag{B.36}
$$

Here, F is DFT of an image I. The SAM ranges from 0 to ∞, and higher values imply higher predictability.

B.7.8 Metrics Based on Co-Occurrence Matrix

The co-occurrence characteristics (COCs) are used to calculate the spatial relationship of pixels in an image by creating a GLCM (grey level co-occurrence matrix or grey level spatial dependence matrix). This is determined by calculating how often a pixel with grey-level value (greyscale intensity) i, occurs horizontally adjacent to a pixel with value j. Each element of i,j in GLCM signifies the number of times that the pixel with value i occurred horizontally adjacent to a pixel with value j. MATLAB function *greycomatrix* is used to create a GLCM with an offset value of [0 1] and the number of levels is 8.

An offset value [row offset, column offset] specifies the relationship or offset, of a pair of pixels: (i) the row offset is the number of rows between the pixel of interest and its neighbour, and (ii) the column offset is the number of columns between the pixel of interest and its neighbour. The offset value of [0 1] indicates horizontal relationship among the pixels at a distance of 1 pixel. The number of levels specify the number of grey levels to use when scaling the greyscale values. NumLevels 8 scales the grey image such that, they are integers between 1 and 8; thus, GLCM is an 8×8 matrix. Then, (i) contrast, (ii) correlation, (iii) energy and (iv) homogeneity metrics are calculated from GLCM, using the function *greycoprops* as follows:

i. Image contrast

Image contrast returns a measure of the intensity contrast between a pixel and its neighbour over the whole image; and for 8 levels GLCM, it ranges from 0 to $(8-1)^2$.

$$
C_{img} = \sum_{i=1}^{8}\sum_{j=1}^{8}|i-j|^{2}g(i,j)
\tag{B.37}
$$

Here, *i,j* are the pixel values in GLCM and *g* is a respective element of the matrix GLCM.

ii. Correlation

Correlation is a measure of how correlated a pixel is to its neighbour. The range of correlation is between –1 and 1. Also, see Section 2.2.3, and Equations 2.4 and 2.5 for related definitions. It is 1 or –1 for a perfect positively or negatively correlated image, and is given as follows:

$$C_{cor} = \sum_{i=1}^{8}\sum_{j=1}^{8}(i - \mu i)(j - \mu j)g(i, j)/(\sigma_i \sigma_j) \tag{B.38}$$

iii. Energy

Energy is the sum of squared elements in the GLCM, and is also known as uniformity, uniformity of energy or angular second moment; it lies between zero and one, and is expressed as

$$E = \sum_{i=1}^{8}\sum_{j=1}^{8} g(i, j)^2 \tag{B.39}$$

iv. Homogeneity

It is a condition in which all the constituents are of the same nature. For image processing and fusion, homogeneity is a value that measures the closeness of the distribution of elements in the GLCM to its diagonal. It signifies if all the pixels in a block are within a specific dynamic range, and this range is from zero to one; for a diagonal GLCM it is 1. It is expressed as follows:

$$H_{img} = \sum_{i=1}^{8}\sum_{j=1}^{8} \frac{g(i, j)}{1+ |i - j|} \tag{B.40}$$

References

1. Yong, D., Felix, T. S. C., Ying, W. and Dong, W. A new linguistic MCDM method based on multiple-criterion data fusion. *Expert Systems with Applications*, 38(6), 6985–6993, 2011. http://dx.doi.org/10.1016/j.eswa.2010.12.016.
2. Liao, Y. H. and Chou J. C. Comparison of pH data measured with a pH sensor array using different data fusion methods. *Sensors*, 12, 12098–12109, 2012. http://www.mdpi.com/journal/sensors. www.mdpi.com/1424-8220/12/9/12098/pdf.

3. Yadiah, N., Singh, L., Bapi, R. S., Rao, S., Deekshatulu, B. L. and Negi, A. Multisensor data fusion using neural networks. *International Joint Conference on Neural Networks*, Sheraton Vancouver Wall Centre Hotel, Vancouver, BC, Canada, July 16–21, 2006.

4. Garlin, L. D. and Naidu, V. P. S. Assessment of color and infrared images using no-reference image quality metrics. *Proceedings of NCATC-2011*, Department of Information Technology, Francis Xavier Engineering College, Tirunelveli, India. International Neural Network Society, India Regional Chapter, Paper No. IP05, pp. 29–35, 6–7 April 2011.

Appendix C: Automatic Data Fusion

The fusing of heterogeneous, duplicate and conflicting data involves several aspects: (i) technical challenges of accessing the data from remote sites, (ii) the need to align heterogeneous schemata of different data, (iii) the discovery of multiple/differing representations of identical objects and (iv) the merging of duplicate data to present a clean and consistent output/result to the user [1]. Some solutions exist for many of these challenging aspects [1]: (a) remote sources access is carried out using Java database connectivity (JDBC), wrappers and web services, (b) present schema mapping methods automatically detect the correspondences among elements of the different schema, (c) the duplication detection is carried out as an individual task → cleansing step in an Extract, Transform and Load (ETL) procedure and (d) data fusion (DF) is increasingly being used in commercial products.

We discuss here the tool called Humboldt Merger (HumMer/HM) that allows ad hoc declarative fusion of such data using sequential query language (SQL) type logic, and performs automatic and virtual ETL allowing for maximum flexibility. The tool visualises each step of DF with user interfacing. It has additional features such as: (i) schema matching can be adjusted, (ii) border-line duplicates can be separated, (iii) data conflict can be resolved manually and (iv) the simple formuation of a DF query and obtaining of the results. The automatic/ad hoc DF is useful in many cases [1]: (a) the integration of finding the corresponding metadata, detecting the entries for identical, say Compact Disks (CDs), and fusing the conflicting data, can be carried out incognito; (b) provision of online data cleansing services, one just receives the clean data if heterogeneous (and dirty) data are submitted and (c) the heterogeneous data, duplicates and data with different levels of accuracies can be fused with the help of the graphical user interface to speed up recovery processes and decision making in situations like tsunami. We hasten to add here that the word 'data fusion' or only 'fusion' used in the context of this tool (for automatic DF) should not be taken to mean the sensor DF used in the rest of the text of the present volume. However, a brief description of the tool is provided here in order that it can provide a model to be followed for WSN data aggregation and perhaps, for the 'sensor DF' itself with some appropriate modification by incorporating the technique of estimation/filtering (as background underlying processes) and sensor 'data fusion' rules. In this way, one may be able to really build an automatic (sensor) DF tool. Also, the present tool can be useful as a part of the Joint Directors Laboratories (JDL) DF process model from the point-of-view of the data base management level, the level 4 processing (Figure 1.1).

The HM tool provides a subset of SQL as query language. It consists of select-project-join (SPJ) queries, and allows sorting, grouping and aggregation. The

FUSE BY statement performs grouping and aggregation. The FUSE FROM defines the tables to be fused using the outer union. The attributes in the FUSE BY serve as object identifier. The schema matching is a process of detecting attribute correlations between two heterogeneous schemes. An underlying algorithm detects a few duplicates in two non-aligned databases, and provides the correlations. The duplicates are detected by considering the text nodes as well as the atoms (sub-elements of the text). The attributes related to the currently considered object, useable by similarity measure, and likely to distinguish duplicates from non-duplicates are considered. After the relevant data of an object have been selected, the comparison is made based on similarity measures. These measures take into account the matched vs. unmatched attributes, edit distance/numerical distance, identification of the power of the data item and contradictory vs. non-specified (say missing) data. Several standard SQL functions (min, max, sum,...) are used in addition to [1]: (i) CHOOSE(SOURCE) returns the value provided by the specific source, (ii) COALESCE takes the first non-null value appearing, (iii) FIRST/LAST takes the first/last value of all values, (iv) VOTE returns the value that appears most often, (v) GROUP returns a set of all conflicting values and leaves to the user, (vi) (ANNOTATED) CONTACT returns the concatenated value, (vii) SHORTEST/LONGEST chooses the value of minimum/maximum length according to a measure and (viii) MOST RCENT evaluates the recency by using another attribute/metadata.

The HumMer tool is Java based with the basic supporting process as the XXL framework. The inference engine (IE) performs the table fetches, joins, unions and groupings, with the top level of graphical user interface. The tool works in query mode via: (i) a basic SQL interface that parses the entire FUSE BY queries and returns the result and (ii) a wizard guiding users in a step by step manner. The tool first generates the relational form of each and passes to the schema matching component. The columns with the same semantics are identified. The data transformation adds an extra sourceID. The resulting table is put in the 'duplicate' and if the tables are part of a larger schema, then the metadata are consulted to fetch the 'atoms' to support the duplication detection. The duplicate detection adds another column to the input table – an objectID specifying n-tuples that represent the same object, and the results are visualised [1]: (i) sure duplicates, (ii) sure non-duplicates and (iii) unsure ones. The user can decide upon individually or in summary all of these. The final table is input to the conflict resolution, where n-tuples with same objectID are fused into a single-tuple and conflicts among them are resolved according to the query specification, and at this point, the relational engine also applies other query predicates. Finally the result is passed to the user to browse/use for further processing.

Another area of interest in developing an automatic DF process or tool is the development of 'algorithm fusion' approach [2]. As we know each fusion method has its own merit/demerits, and that a suitable combination of different fusion schemes has been a useful strategy that might achieve better results. Hitherto the selection/arrangement of the candidate fusion methods/schemes was arbitrary. It depends upon the user's experience. So, an optimal combining strategy for different fusion algorithms, that is, 'algorithm fusion' is required. In this context the aspects that are very important are [2]: (i) design of a general framework for combining various fusion approaches, (ii) development of new methods that can combine aspects of pixel, feature and decision level image fusion and (iii) development of automatic quality assessment (AQA) method for evaluation of the fusion results. An automatic quality assessment scheme or procedure is desirable [2]: (i) to evaluate the benefits of fusion, (ii) to determine an optimal choice of parameters for a certain fusion scheme and (iii) to compare results obtained with various algorithms. Analytical studies are required to evaluate the merits of several quality metrics used thus far: statistical indices such as cross entropy, mean square error, signal-to-noise ratio and image phase congruency. Hence, automatic solution procedure and quality assessment methods are needed to achieve consistently high-quality fusion results for different data sets.

Further aspects and features of an automated DF are [3]: (i) it can reduce the burden on the intelligence analyst, (ii) automation can reduce delivery time of intelligent DF products to the user, (iii) the user sees only the finished products, (iv) he/she need not have a great insight into the sources that contributed to DF results, (v) of course to provide the trust the automated DF scheme requires knowledge of the quality/confidence and pedigree/source and heritage of previous processing. The design and prototype of a standard representation of confidence, pedigree and security classification information are described in Reference 3.

References

1. Bilke, A., Bliholder, J., Bohm, C. and Draba, K. Automatic data fusion with HumMer. *Proceedings of the 31st VLDB Conference*, Trondheim, Norway, 2005 (some material adapted by permission of the Very Large Data Base Endowment). http://www.vldb.org/conf/2005/papers/p1251-bilke.pdf, accessed April 2013.
2. Dong, J., Zhuang, D., Huang, Y. and Fu, J. Advances in multisensor data fusion: Algorithms and applications. *Sensors* 9, 7771–7784, 2009. http://www.mdpi.com/journal/sensors, accessed April 2013.
3. Newman, A. R. Confidence, pedigree and security classification for improved data fusion. *Proceedings of the Fifth International Conference on Information Fusion*, 2, 1408–1415, 2002.

Appendix D: Notes and Information on Data Fusion Software Tools

In this appendix, we give a brief description of some (commercially available and other) software tools useful for target tracking and/or data/image fusion. The listings of some simple MATLAB programs for image fusion are also given.

D.1 FUSion, Estimation and Data Association for Tracking-2

This MATLAB-based software (FUSEDAT: FUSion, Estimation and Data Association for Tracking) provides a set of multi-target multi-sensor tracking and fusion algorithms (*FUSEDAT*-1) [1]. The objective of this software package is to provide users a flexible prototype environment for simulating sensor and target scenarios. This package can also be used to simulate and evaluate the tracking and fusion algorithms. Noisy measurement and clutter environment can be generated based on different sensor models. The nearest neighbour KF, probabilistic data association filter (PDAF), and interacting multiple model (IMM) estimator are available for single sensor tracking. Centralised fusion architecture is available for multi-sensor tracking. Performance evaluation measures such as state estimation errors, track score and track life are included in the package.

D.2 Bearing (Only) Data Association and Tracking

This C-language based code is menu-driven user-friendly software for target motion analysis. This software can be used with bearing only measurements or passive sonar measuring bearing and frequency. The maximum likelihood (ML) estimator along with PDAF algorithm is used to estimate the target motion parameters. This software can be used in simulation or real data modes.

D.3 Multiple Target Data Association and Tracking

This software package provides a user-interactive set of target tracking algorithms for manoeuvring or crossing targets or a splitting target in clutter environment. Single target state estimation algorithms such as nearest neighbor Kalman filter (NNKF), PDAF, and interacting multiple model probabilistic data association are provided. Provision for Monte Carlo runs and plotting of normalised state errors, root mean square state error, root mean square position, velocity and acceleration errors, normalised position, velocity and acceleration errors are available for filter consistency analysis.

D.4 Image Data Association for Tracking

This Microsoft FORTRAN source code (IMDAT: Image Data Association for Tracking) provides a set of user-interactive and user-friendly algorithms for tracking the centroid of an object or target based on grey scale image frames from an imaging sensor. The algorithm performs the following functions: generation of synthetic image sequence, image segmentation, clustering, track initialisation, probabilistic data association filter for state estimation and filter consistency statistics.

D.5 Variable Data Association for Tacking

This Microsoft Visual C++ (VARDAT: Variable Data Association for Tacking) software has the following options: A directly usable 2D assignment algorithm, a 3D assignment algorithm via relaxation and as S-D assignment algorithm (S > 3). This software may be applied for space surveillance, direction finding and target localisation.

Note: We have used many of the above software (SW) tools in our DF exercises and further development of DF solutions to many practical situations earlier. For more information on the five packages (D.1–D.5) please contact: Dr. Yaakov Bar-Shalom, Box U-157, Storrs, CT 06269-2157; Ph: 860-486-4823, Fax: 860-486-5585, email: ybs@ee.uconn.edu.

D.6 Fusion Architectures

This MATLAB code (FuseArch: Fusion Architectures) is provided for 3D target tracking using infra-red search and track (IRST) and radar measurements. Seven different architectures are presented to fuse IRST and radar data to track the target in 3D Cartesian coordinates, with the measurements available in polar coordinates. Performance of these architectures is checked with simulated data.

D.7 DCT-Based Image Fusion

This MATLAB code (DCTIMFUSE: DCT-based image fusion) is provided for image fusion using multi-resolution discrete cosine transform. Fusion quality evaluation metrics are also provided. The demo contains the images to be fused along with ground truth image.

D.8 Interacting Multiple Model Kalman Filter

This MATLAB code (IMMKF: Interacting Multiple Model Kalman Filter) is provided for evaluating the IMMKF performance. Different state estimator performance evaluation metrics are provided.

Note: For more information on the three packages (D.6–D.8) please contact: Dr. V. P. S. Naidu, multi sensor data fusion (MSDF) Lab, FMCD, CSIR-NAL, PB No. 1779, Bangalore-560017, India; email: vpsnaidu@gmail.com.

D.9 Fuzzy Kalman Filter

This MATLAB code (FKF: Fuzzy Kalman Filter) provides fuzzy-based KF for non-linear target state estimator. The performance evaluation of this estimator with standard KF can also be carried out.

D.10 Fuzzy Implication Functions Evaluation Toolbox: ver 1.0

This toolbox is developed for evaluation of fuzzy implication functions (FIF) using various intuitive criteria of the most common inference rules, GMP (Generalised Modus Ponens) and GMT (Generalised Modus Tolens). The toolbox is user interactive and is realised using MATLAB and Graphics on Windows platform. The advantage of having this toolbox is mainly the ease it provides for quick evaluation of FIF through graphics visualisation.

The toolbox enables the user to conclude whether a particular FIF satisfies any of the intuitive criteria or not. This check is necessary in the development process of fuzzy logic-based system using FIF which should at least satisfy one of the existing intuitive criteria of GMP and GMT. Also, the toolbox can be used for finding out new fuzzy implication function by visualising the existing superimposed curves of FIF and corresponding intuitive criteria.

D.11 MsmtDat: Integrated MSST and MSMT Software

Software package *'MsmtDat'* is used for tracking multiple airborne targets such as missile, aircraft, helicopter using data from multiple tracking sensors like ground-based radars and electro-optic transducers (EOTs) including inertial navigation system (INS) and global positioning system (GPS). This software has graphical user interface (GUI)-based front-end panel that provides interface between man and machine. Based on the contents (entered by the user) of front panel, either the multi-sensor single target (MSST) (for single target tracking) or the multi sensor multi target (MSMT) (for multi-target tracking) module will be triggered. For the state estimation, the user can either select the upper triangular diagonal (UD) factorisation-based Kalman filter or the IMM filter by pressing the appropriate buttons present in the front panel.

D.12 Intelligent System for Situation Assessment in Air-Combat

Intelligent system for situation assessment in air-combat (ISSAAC) simulator is an integration of platform models, sensor models, pilot mental models (using FUZZY–Bayesian hybrid technique) and data processing algorithms. It is developed in MATLAB/SIMULINK® environment on Windows XP platform. The objective of ISSAAC simulator is to enhance situational awareness of air combat pilot in beyond visual range (BVR) scenario. The capability of ISSAAC is demonstrated by simulating an air-to-air combat scenario consisting of six targets. The ISSAAC decides what action the pilot should take based on its range, situational geometry and level of threat imposed by targets.

Note: For more information on these four packages (D.9–D.12), please contact: Dr. S. K. Kashyap, MSDF Lab, FMCD, CSIR-NAL, PB No. 1779, Bangalore 560017, India; email: sudesh@nal.res.in.

D.13 Image Fusion

Image registration and fusion algorithms in MATLAB are available at the following link: http://www.imagefusion.org/software.html. A MATLAB code of a simple image fusion algorithm using principal component analysis (PCA) (Chapter 6), named *PCAimfuse_dem.m* is listed below. The aircraft images to be fused are saras51.jpg and saras52.jpg. It is a demonstration program, and to run and generate the results of Figures 6.4 and 6.5 just type: *PCAimfuse_dem* in MATLAB Command window.

```
% Image fusion by PCA-demo program
% PCAimfuse_dem.m
% Dr. VPS Naidu (MSDF Lab, CSIR-NAL, Bangalore)
close all;clear all;home;
% insert images
im1 = double(imread('saras51.jpg'));
im2 = double(imread('saras52.jpg'));
figure(1);subplot(121);imshow(im1,[]);
xlabel('Image Pair 1-Fusion Candidate')
title('Aircraft Image-Top one blurred')
subplot(122);imshow(im2,[]);xlabel('Image Pair 2-Fusion
Candidate')
title('Aircraft Image-Bottom one blurred')
% compute PCA
C = cov([im1(:) im2(:)]);
[V, D] = eig(C);
if D(1,1) >= D(2,2)
        pca = V(:,1)./sum(V(:,1));
else
        pca = V(:,2)./sum(V(:,2));
end
% Image fusion using appropriate PCA components as weights in
the fusion rule
imf = pca(1)*im1 + pca(2)*im2;
figure(2); imshow(imf,[]);
xlabel('Fused Image of Aircraft by PCA')
title('Blurirng is almost gone')
% END OF THE PROGAM
```

D.14 Image Fusion (DT CWT)

Image fusion by dual tree complex wavelet transform (DT CWT) in MATLAB is available at the following link: http://www.mathworks.in/matlabcentral/fileexchange/32086-dt-cwt-based-image-fusion. A MATLAB code of a simple image fusion algorithm using wavelet transform (WT) (Chapter 6), named vpnwtfuseL1demo is listed below. The aircraft candidate images to be fused are saras91.jpg and saras92.jpg. The true image is saras9t.jpg. It is a demonstration program, and to run and generate the results of Figures 6.6 through 6.9 just type *'vpnwtfuseL1demo'* in MATLAB Command window.

```
function[] = vpsnfuseL1demo()
% Dr. VPSN Naidu, FMCD, CSIR-NAL, Bangalore
close all;clear all;home;
% images to be fused
imt = double(imread('saras9t.jpg'));
im1 = double(imread('saras91.jpg'));
im2 = double(imread('saras92.jpg'));
figure(1); imshow(im1,[]);
figure(2); imshow(im2,[]);
[cA1,cH1,cV1,cD1] = dwt2(im1,'db1');
[cA2,cH2,cV2,cD2] = dwt2(im2,'db1');
% Fusion start here
%average approximation coefficients
cAf = 0.5*(cA1 + cA2);
% fusion of detail coefficients
D = (abs(cH1)-abs(cH2)) >= 0;
cHf = D.*cH1 + (~D).*cH2;
D = (abs(cV1)-abs(cV2)) >= 0;
cVf = D.*cV1 + (~D).*cV2;
D = (abs(cD1)-abs(cD2)) >= 0;
cDf = D.*cD1 + (~D).*cD2;
%fused image
imf = idwt2(cAf,cHf,cVf,cDf,'db1');
figure(3);imshow(imf,[]);
% generate the error image pair
imd = imt-imf;
figure(4); imshow(imd,[]);
RMSE = sqrt(mean(imd(:)))
% END of The Program
```

D.15 Image Fusion Toolbox (ver 1.0)

Image fusion toolbox (ver 1.0) in MATLAB includes a set of m-file functions for the pixel-level image fusion of spatially registered grey scale images. It is accompanied by an easy-to-use graphical interface which allows an interactive control over all relevant parameters. The code is available at the following link: http://www.metapix.de/toolbox.htm.

D.16 MATLAB Code of MDCT Used for Image Decomposition (One Level) (Chapter 6)

```
function[I] =mrdct(im)
% multi-Resolution Discrete Cosine Transform
% Dr. VPS Naidu, MSDF Lab, CSIR-NAL, Bangalore
% input: im (input image to be decomposed)
% output: I (Decomposed image)
[m,n] =size(im);
mh =m/2; nh =n/2;
for i =1:m
hdct(i,:) =dct(im(i,:));
end
for i =1:m
hL(i,:) =idct(hdct(i,1:nh));
hH(i,:) =idct(hdct(i,nh +1:n));
end
for i =1:nh
vLdct(:,i) =dct(hL(:,i));
vHdct(:,i) =dct(hH(:,i));
end
for i =1:nh
I.LL(:,i) =idct(vLdct(1:mh,i));
I.LH(:,i) =idct(vLdct(mh +1:m,i));
I.HL(:,i) =idct(vHdct(1:mh,i));
I.HH(:,i) =idct(vHdct(mh +1:m,i));
End
%END
```

D.17 MATLAB Code of IMDCT Used for Image Reconstruction (Chapter 6)

```
function[im] =imrdct(I)
% Inverse Multi-Resolution Discrete Cosine Transform
% Dr. VPS Naidu, MSDF Lab, CSIR-NAL, Bangalore
% input: I (decomposed image)
% output: im (reconstructed image)
[m,n] =size(I.LL);
m2 = m*2;
n2 = n*2;
for i =1:n
ivLdct(:,i) = [dct(I.LL(:,i));dct(I.LH(:,i))];
ivHdct(:,i) = [dct(I.HL(:,i));dct(I.HH(:,i))];
end
```

```
for i = 1:n
ihL(:,i) = idct(ivLdct(:,i));
ihH(:,i) = idct(ivHdct(:,i));
end
for i = 1:m2
hdct(i,:) = [dct(ihL(i,:))  dct(ihH(i,:))];
end
for i = 1:m2
im(i,:) = idct(hdct(i,:));
end
%END
```

D.18 MATLAB Code Used for Image Fusion (Chapter 6)

```
function[imf] = mrdctimfus(im1,im2)
% Image fusion using MDCT
% Dr. VPS Naidu, MSDF Lab, CSIR-NAL, Bangalore
% input: im1 & im2 (images to be fused)
% output: imf (fused image)
% multi-resolution image decomposition
X1 = mrdct(im1);
X2 = mrdct(im2);
% Fusion
X.LL = 0.5*(X1.LL + X2.LL);
D = bdm(X1.LH,X2.LH);
X.LH = D.*X1.LH + (~D).*X2.LH;
D = bdm(X1.HL,X2.HL);
X.HL = D.*X1.HL + (~D).*X2.HL;
D = bdm(X1.HH,X2.HH);
X.HH = D.*X1.HH + (~D).*X2.HH;
% fused image
imf = imrdct(X);
%END
```

D.19 DYNAmic State ESTimation

This MATLAB code (DYNAEST: DYNAmic state ESTimation) is an interactive software. It is basically a tool for Kalman filtering design and simulation, and also for design of very interesting ideas like multiple-models estimators. DYNAEST is companion software for the book *Estimation and Tracking: Principles, Techniques and Software* by Y. Bar-Shalom and X. Li.

   http://vasc.ri.cmu.edu/old_help/Software/Dynaest/dynaest.html, 13-03-2013.

D.20 FastICA

The FastICA is a GUI/MATLAB (7.x and 6.x) based version 2.5 (19 October 2005) tool box (developed by © Hugo Gävert, Jarmo Hurri, Jaakko Särelä and Aapo Hyvärinen) to perform independent component analysis of the mixer of signals (see Section 6.5.2). It has 18 sub-modules. The package is freely downloadable from the net. To run the program, in MATLAB Command window type as follows: 'fasticag' to launch graphical user interface; then give the following command [sig,mixedsig] = demosig; Then use GUI to load the data as follows: click on the Load data, then type 'sig' in the white blank window (of the GUI window) and click on the Load and run the ICA by clicking on 'Do ICA'; select other options as seen on the GUI, and 'plot data'; For example, use for approach 'symmetric', for non-linearity (g) 'pow3' and stabilisation 'off'; 'Plot whitened'; 'Plot ICs', and so on (see Example 6.3).

D.21 General SW Design for Multi-Sensor Data Fusion

General SW design for multi-sensor data fusion [2]. The SW for MSDF is divided into modules: (i) data collection, (ii) database management, (iii) GIS, (iv) target display and (v) alarming, and data simulation for which the primary function, the components and some realisation methods of each module are given. The data exchange among each functional module is performed by: (i) inter-process communication (IPC), (ii) including message queue, (iii) semaphore and (iv) shared memory. Each functional module is executed independently reducing the dependence among these modules. It also helps SW programming and testing. This SW is designed in hierarchical structure. Each module is encapsulated through class structure avoiding SW redundancy and enhances readability.

References

1. Kuo-Chu, C. and Yaakov Bar-Shalom. http://proceedings.spiedigitallibrary .org/proceeding.aspx?articleid=966488, accessed February 2013. *Proceedings of the SPIE* 2235, *Signal and Data Processing of Small Targets* 1994, 497 (6 July 1994); doi:10.1117/12.179074.
2. Zhang, Junliang and Zhao, Yuming. http://proceedings.spiedigitallibrary .org/proceeding.aspx?articleid=989063, *Proceedings of the SPIE* 3719, *Sensor Fusion: Architectures, Algorithms, and Applications III*, 230 (12 March 1999); doi:10.1117/12.341345, accessed February 2013.

Appendix E: Definitions of Sensor DF in Literature

Here, we provide several definitions of sensor data fusion (DF) prevalent in the DF literature [1, p. 1]:

1. JDL (1987). DF is 'a process dealing with the association, correlation, and combination of data and information from single and multiple sources to achieve refined position and identity estimates, and complete and timely assessments of situations and threats, and their significance. The process is characterized by continuous refinements of its estimates and assessments, and the evaluation of the need for additional sources, or modification of the process itself, to achieve improved results'.

2. Hugh Durrant-Whyte (1988). 'The basic problem in multi-sensor systems is to integrate a sequence of observations from a number of different sensors into a single best-estimate of the state of the environment'.

3. Llinas (1988). 'Fusion can be defined as a process of integrating information from multiple sources to produce the most specific and comprehensive unified data about an entity, activity or event. This definition has some key operative words: specific, comprehensive and entity. From an information theoretic point of view, fusion, to be effective as an information processing function, must (at least ideally) increase the specificity and comprehensiveness of the understanding we have about a battlefield entity or else there would be no purpose in performing the function'.

4. Richardson and Marsh (1988). 'Data fusion is the process by which data from a multitude of sensors is used to yield an optimal estimate of a specified state vector pertaining to the observed system'.

5. McKendall and Mintz (1988). '...the problem of sensor fusion is the problem of combining multiple measurements from sensors into a single measurement of the sensed object or attribute, called the parameter'.

6. Waltz and Llinas (1990). 'This field of technology has been appropriately termed data fusion because the objective of its processes is to combine elements of raw data from different sources into a single set of meaningful information that is of greater benefit than the sum of the contributing parts. As a technology, data fusion is actually the

integration and application of many traditional disciplines and new areas of engineering to achieve the fusion of data'.

7. Luo and Kay (1992). 'Multi-sensor fusion, ... , refers to any stage in an integration process where there is an actual combination (or fusion) of different sources of sensory information into one representational format'.

8. Abidi and Gonzalez (1992). 'Data fusion deals with the synergistic combination of information made available by various knowledge sources such as sensors, in order to provide a better understanding of a given scene'.

9. Hall (1992). 'Multi-sensor data fusion seeks to combine data from multiple sensors to perform inferences that may not be possible from a single sensor alone'.

10. DSTO (1994). Data fusion is 'a multilevel, multifaceted process dealing with the automatic detection, association, correlation, estimation, and combination of data and information from single and multiple sources'.

11. Malhotra (1995). 'The process of sensor fusion involves gathering sensory data, refining and interpreting it, and making new sensor allocation decisions'.

12. Hall and Llinas (1997). 'Data fusion techniques combine data from multiple sensors, and related information from associated databases, to achieve improved accuracy and more specific inferences than could be achieved by the use of single sensor alone'.

13. Goodman, Mahler and Nguyen (1997). Data fusion is to 'locate and identify many unknown objects of many different types on the basis of different kinds of evidence. This evidence is collected on an ongoing basis by many possibly allocatable sensors having varying capabilities and to analyze the results in such a way as to supply local and over-all assessments of the significance of a scenario and to determine proper responses based on those assessments'.

14. Paradis, Chalmers, Carling and Bergeron (1997). 'Data fusion is fundamentally a process designed to manage (i.e., organize, combine and interpret) data and information, obtained from a variety of sources that may be required at any time by operators or commanders for decision making...Data fusion is an adaptive information process that continuously transforms available data and information into richer information, through continuous refinement of hypotheses or inferences about real-world events, to achieve a refined (potentially optimal) kinematics and identity estimates of individual objects, and complete and timely assessments of current and potential future situations and threats

(i.e., contextual reasoning), and their significance in the context of operational settings'.

15. Starr and Desforges (1998). 'Data fusion is a process that combines data and knowledge from different sources with the aim of maximizing the useful information content, for improved reliability or discriminant capability, while minimizing the quantity of data ultimately retained'.

16. Wald (1998). 'Data fusion is a formal framework in which are expressed means and tools for the alliance of data of the same scene originating from different sources. It aims at obtaining information of greater quality; the exact definition of greater quality will depend upon the application'.

17. Evans (1998). 'The combining of data from different complementary sources (usually geodemographic and lifestyle or market research and lifestyle) to build a picture of someone's life'.

18. Wald (1999). 'Data fusion is a formal framework in which are expressed the means and tools for the alliance of data originating from different sources'.

19. Steinberg, Bowman and White (1999). 'Data fusion is the process of combining data to refine state estimates and predictions'.

20. Gonsalves, Cunningham, Ton and Okon (2000). 'The overall goal of data fusion is to combine data from multiple sources into information that has greater benefit than what would have been derived from each of the contributing parts'.

21. Hannah, Ball and Starr (2000). 'Fusion is defined materially as a process of blending, usually with the application of heat to melt constituents together (OED), but in data processing the more abstract form of union or blending together is meant. The "heat" is applied with a series of algorithms which, depending on the technique used, give a more or less abstract relationship between the constituents and the finished output'.

22. Dasarathy (2001). 'Information fusion encompasses the theory, techniques, and tools conceived and employed for exploiting the synergy in the information acquired from multiple sources (sensor, databases, information gathered by humans etc.) such that the resulting decision or action is in some sense better (qualitatively and quantitatively, in terms of accuracy, robustness and etc.) than would be possible, if these sources were used individually without such synergy exploitation'.

23. Bloch and Hunter et al. (2001). '...fusion consists in conjoining or merging information that stems from several sources and exploiting that conjoined or merged information in various tasks such as answering questions, making decisions, numerical estimation, etc'.

24. McGirr (2001). 'The process of bringing large amounts of dissimilar information together into a more comprehensive and easily manageable form is known as data fusion'.

25. Bell, Santos and Brown (2002). 'Sophisticated information fusion capabilities are required in order to transform what the agents gather from a raw form to an integrated, consistent and complete form. Information fusion can occur at multiple levels of abstraction'.

26. Challa, Gulrez, Chaczko and Paranesha (2005). Multi-sensor data fusion 'is a core component of all networked sensing systems, which is used either to: join/combine complementary information produced by sensor to obtain a more complete picture or – reduce/manage uncertainty by using sensor information from multiple sources'.

27. Jalobeanu and Gutirrez (2006). 'The data fusion problem can be stated as the computation of the posterior PDF of the unknown single object given all observations'.

28. Mastrogiovanni et al. (2007). 'The aim of a data fusion process is to maximize the useful information content acquired by heterogeneous sources in order to infer relevant situations and events related to the observed environment'.

29. Wikipedia (2007). 'Information Integration is a field of study known by various terms: Information Fusion, Duplication, Referential Integrity and so on. It refers to the field of study of techniques attempting to merge information from disparate sources despite differing conceptual, contextual and typographical representations. This is used in data mining and consolidation of data from semi- or unstructured resources'.

30. Wikipedia (2007). 'Sensor fusion is the combining of sensory data or data derived from sensory data from disparate sources such that the resulting information is in some sense better than would be possible when these sources were used individually. The term better in that case can mean more accurate, more complete, or more dependable, or refer to the result of an emerging view, such as stereoscopic vision (calculation of depth information by combining two-dimensional images from two cameras at slightly different viewpoints). The data sources for a fusion process are not specified to originate from identical sensors. One can distinguish direct fusion, indirect fusion and fusion of the outputs of the former two. Direct fusion is the fusion of sensor data from a set of heterogeneous or homogeneous sensors, soft sensors, and history values of sensor data, while indirect fusion uses information sources like a priori knowledge about the environment and human input. Sensor fusion is also known as (multi-sensor) data fusion and is a subset of information fusion'.

31. MSN Encarta (2007). 'Data integration: the integration of data and knowledge collected from disparate sources by different methods into a consistent, accurate, and useful whole'.

32. Raol, J. R. (2014). DF is an act, that could be additive, multiplicative, operative, and/or logical by which (i) the quantitative information in the sense of Fisher's information matrix is enhanced by fusing/combining data from more than one sensor or source, and/or (ii) the prediction accuracy is enhanced, compared to the usage of a single sensor-data or source.

Reference

1. Anon. Definitions of sensor data fusion in the literature. The Institute of Computer Science Communications and Networked Systems, http://net.cs.uni-bonn.de/de/wg/sensor-data-and-information-fusion/what-is-it/sdf-definitions, accessed April 2013.

Appendix F: Some Current Research Topics in DF

We give here, a list of several current research topics for higher level investigations and studies in the area of data fusion (DF).

1. Investigations of several natural biological systems and their associated fusion strategies.
2. Target-image tracking and fusion using square root information filtering.
3. Fuzzy logic augmented H-infinity filtering for target tracking and kinematic DF.
4. Synergy of missing and delayed measurements (in a filter) for target tracking and kinematic DF.
5. Image fusion using interval type 2 fuzzy logic (IT2FL).
6. Decision fusion using IT2FL for situation assessment.
7. Decision fusion using IT2FL and TSK models for situation assessment.
8. H-infinity-based robust estimation of model error in non-linear systems for DF.
9. Investigation of scale of fluctuation for deriving fusion rules and strategies.
10. Application of matrix factorisation approaches for data and image fusion.
11. Application of (signal/image) component analysis-based DF methods for fusion of acoustic and image data.
12. Determination of fusion rules from fused data/images by inverse (-modelling) methods.
13. Investigations and unification of DF rules and strategies in decision fusion.
14. Investigations and unification of DF rules and strategies in multi-modal systems in wireless sensor networks.
15. Reliability models-based sensor and DF.

Some more research ideas can be obtained from the following literature:

1. Hall, D. L. and Robert, J. L. Survey of commercial software for multisensory data fusion. *Proceedings of the SPIE*, 1956, *Sensor Fusion and Aerospace Applications*, 98, 3 September 1993.

2. Luo, Z. Q. and Tsitsiklis, J. N. Data fusion with minimal communication. *IEEE Transactions on Information Theory*, 40(5), 1551–1563, 1994.

3. Chen, B. and Varshney, P. K. A Bayesian sampling approach to decision fusion using hierarchical models. *IEEE Transactions on Signal Processing*, 50(8), 1809–1818, 2002.

4. Laurence, N. P., Roland C., Sebastien B., Sebastien, C. and Frederic, C. A Bayesian multisensor fusion approach integrating correlated data applied to a real-time pedestrian detection system. *IROS 2008, 2nd Workshop: Planning, Perception and Navigation for Intelligent Vehicles.*

5. Twala, B. Modeling out-of-sequence measurements: A copulas-based approach. In *Mobile Intelligent Autonomous Systems* (Eds. Raol, J. R. and Gopal, A. K.). CRC Press, USA, Chapter 35, 765–770, 2012.

6. Nelsen, R. *An Introduction to Copulas*. Springer-Verlag, New York, USA, 1999.

7. Foo, P. H. and Ng, G. W. High-level information fusion: An overview. *Journal of Advances in Information Fusion*, 8(1), 33–72, June 2013. http://www.isif.org/ node/ 170, accessed December 2014.

8. Madhavan, P. G. Instantaneous scale of fluctuation using Kalman-TFD and applications machine tool monitoring. Proc. SPIE 3162, Advanced Signal Processing: Algorithms, Architectures, and Implementations VII, 78, October 1997); Franklin T. Luk, San Diego, CA, USA, Vol. 3162, pp. 78–89, 1997.

Index